Applied Engineering Mechanics

Applied Engineering Mechanics

Edited by **Derek Pearce**

Willford Press

New York

Published by Willford Press,
118-35 Queens Blvd., Suite 400,
Forest Hills, NY 11375, USA
www.willfordpress.com

Applied Engineering Mechanics
Edited by Derek Pearce

International Standard Book Number: 978-1-68285-033-6 (Hardback)

Printed in the United States of America.

Contents

Preface

Engineering mechanics is a discipline which has a wide range of applications, spread across many disciplines of engineering. This book attempts to understand the multiple branches that fall under engineering mechanics and how such concepts have practical applications. The topics included in this book on engineering mechanics focus on custom instrumentation, diagnostics, experimental techniques, analytical methods, etc. It will serve as a valuable source of reference for graduate and post graduate engineering students.

The information shared in this book is based on empirical researches made by veterans in this field of study. The elaborative information provided in this book will help the readers further their scope of knowledge leading to advancements in this field.

Finally, I would like to thank my fellow researchers who gave constructive feedback and my family members who supported me at every step of my research.

Editor

1

Evolution of the DeNOC-based dynamic modelling for multibody systems

S. K. Saha[1], **S. V. Shah**[2], **and P. V. Nandihal**[1]

[1]Department of Mechanical Engineering, Indian Institute of Technology Delhi, Hauz Khas, New Delhi 110 016, India

[2]Department of Mechanical Engineering, McGill University, Montreal, Canada

Correspondence to: S. K. Saha (saha@mech.iitd.ac.in)

Abstract. Dynamic modelling of a multibody system plays very essential role in its analyses. As a result, several methods for dynamic modelling have evolved over the years that allow one to analyse multibody systems in a very efficient manner. One such method of dynamic modelling is based on the concept of the Decoupled Natural Orthogonal Complement (DeNOC) matrices. The DeNOC-based methodology for dynamics modelling, since its introduction in 1995, has been applied to a variety of multibody systems such as serial, parallel, general closed-loop, flexible, legged, cam-follower, and space robots. The methodology has also proven useful for modelling of proteins and hyper-degree-of-freedom systems like ropes, chains, etc. This paper captures the evolution of the DeNOC-based dynamic modelling applied to different type of systems, and its benefits over other existing methodologies. It is shown that the DeNOC-based modelling provides deeper understanding of the dynamics of a multibody system. The power of the DeNOC-based modelling has been illustrated using several numerical examples.

1 Introduction

Over the last two decades, applications of multibody dynamics have expanded over the fields of robotics, automobile, aerospace, bio-mechanics, and many others. With continuous development in the above mentioned fields, many complex multibody systems have evolved whose dynamics play a pivotal role in their behaviour. Hence, computer-aided dynamic analysis of multibody systems has been a prime motive to the engineers, as high speed computing facilities are readily available. In order to perform computer-aided dynamic analysis, the actual system is represented with its dynamic model which has the information of its link parameters, joint variables and constraints. The dynamic model is nothing but the equations of motion of the multibody system at hand derived from the physical laws of motions. For a system with fewer links, it is easier to obtain explicit expressions for the equations of motion. However, finding equations of motion for complex systems with many links is not an easy task. Sometimes even with 4 or 5 links, say, a 4-bar mechanism, it is difficult to find an explicit expression for the system's inertia in terms of its link lengths, masses, and joint angles. Hence, development of the equations of motion is an essential step for the dynamic analysis.

There are several fundamental methods for the formulation of equations of motion (Greenwood, 1988). For example, Newton-Euler (NE) formulation, Euler-Lagrange principle, Gibbs-Appel approach, Kane's method, D'Alembert's principle, and similar others. All the above mentioned approaches when applied to multibody systems have their own advantages and disadvantages. For example, NE approach, which is one of the classical methods for dynamic formulation, uses the concept of "free-body diagrams". For coupled systems, constrained forces (which are meant here to include both forces and moments) along with those applied externally are included in the free-body diagrams. Mathematically, the NE equations of motion lead to three translational equations of motion of the Centre-of-Mass (COM), and three equations determining the rotational motion of the rigid body. The NE equations of any two free bodies are related through the constraint forces acting at their interface. The constraint forces arise due to the presence of a kinematic pair, e.g., a revolute or a prismatic, between the two neighbouring bodies. For an open-loop multibody system, these constraints along with other unknowns, i.e., the actuating forces can be easily solved recursively. However, for a closed-loop system, the NE equations generally need to be solved simultaneously in order to obtain the driving and constraint forces together. Hence, the use of the NE equations of motion for closed-loop systems is not as efficient as those for open-loop systems.

Euler-Lagrange (EL) formulation is another classical approach which is widely used for dynamic modelling. The EL formulation uses the concept of generalized coordinates instead of Cartesian coordinates. It is based on the minimization of a functional called "Lagrangian" which is nothing but the difference between kinetic energy and potential energy of the system at hand. For open-loop multibody systems, where typically the number of generalized coordinates equals the degree-of-freedom of a system, the constraint forces do not appear in the equations of motion. For closed-loop multibody systems, however, the forces of constraints appear as Lagrange's multipliers.

Kane's formulation (Kane and Levinson, 1983), which is same as the Lagrange's form of D'Alembert's principle, has also been used by many researchers for the development of equations of motion. It is found to be more beneficial than other formulations when used for systems with nonholonomic constraints. Several other methods of dynamic formulations were also proposed in the literature. For example, Khatib (1987) presented the operational-space formulation, whereas Angeles and Lee (1988) presented the natural orthogonal complement (NOC) based approach. Blajer et al. (1994) have also presented an orthogonal complement based formulation for the constrained multibody systems. Park et al. (1995) presented robot dynamics using a Lie group formulation, while Stokes and Brockett (1996) derived the equations of the motion of a kinematic chain using concepts associated with the special Euclidean group. McPhee (1996) showed how to use linear graph theory in multibody system dynamics. Cameron and Book (1997) described a technique based on Boltzmann-Hamel equations to derive dynamic equations of motion. Comprehensive discussion on dynamic formalisms can be found in the seminal text by Roberson and Schwertassek (1988), Schiehlen (1990, 1997), Shabana (2001), and Wittenburg (2008). Recent trends in dynamic formalisms can also be found in the work by Eberhard and Schiehlen (2006).

1.1 Natural Orthogonal Complement (NOC)

It is pointed out here that the Newton-Euler (NE) equations of motion are still found to be popular in the literature of dynamic modelling and analyses. However, it requires solution of the constraint forces which do not play any role in the motion of a system. Hence, extra calculations are required in motion studies. To avoid such extra calculations, there are formulations proposed in the literature where the equations of motion in the Euler-Lagrange (EL) form are obtained from the NE equations. Huston and Passerello (1974) were first to introduce a computer oriented method to reduce the dimension of the unconstrained NE equations by eliminating the constraint forces. Later, Kim and Vanderploeg (1986) derived the equations of motion in terms of relative joint coordinates from Cartesian coordinates through the use of velocity transformation matrix. Velocity transformation matrix relates linear and angular velocities of the links with joint velocities. It is worth noting here that the vector of constraint forces is orthogonal to the columns of the velocity transformation matrix. More precisely, the columns of the velocity transformation matrix span the nullspace of the matrix of velocity constraints. Hence, the said velocity transformation matrix is also referred to as an "orthogonal complement matrix". The phrase "orthogonal complement" was first coined by Hemami and Weimer (1981) for the modelling of nonholonomic systems. Orthogonal complements are not unique. In some approaches, it was obtained numerically, e.g., using singular value decomposition or treating it as an eigen value problem (Wehage and Haug, 1982; Kamman and Huston, 1984, Mani et al., 1985), which are computationally inefficient.

Alternatively, Angeles and Lee (1988) presented a methodology where they derived an orthogonal complement naturally from the velocity constraints. Hence, the name Natural Orthogonal Complement (NOC) was attached to their methodology. The NOC matrix, when combined with the NE equations of motion, leads to the minimal-order constrained dynamic equations of motion by eliminating the constraint forces. This facilitates the representation of the equations of motion in Kane's form that is suitable for recursive computation in inverse dynamics or in the EL form that is suitable for forward dynamics and integration. Later, Angeles and Ma (1988), Cyril (1988), Angeles et al. (1989), and Saha and Angeles (1991) showed the effectiveness of the use of the NOC matrix while applied to systems with holonomic and nonholonomic constraints.

1.2 The Decoupled NOC (DeNOC)

Subsequently, Saha (1995, 1997) presented the decoupled form of the NOC for the serial multibody systems. The two resulting block matrices, namely, an upper block triangular and a block diagonal matrices, are referred to as the Decoupled NOC (DeNOC) matrices. In contrast to the NOC, the DeNOC matrices allow one to recursively obtain the analytical expressions of the vectors and matrices appearing in the equations of motion (Saha, 1999a). This in turn helps to analytically decompose the Generalized Inertia Matrix (GIM) arising out of the constrained equations of motion of the system at hand, allowing one to obtain analytical inverse of the

GIM (Saha, 1999b) and a recursive algorithm for forward dynamics (Saha, 2003). Later, Saha and Schiehlen (2001) showed the power of the DeNOC matrices in obtaining recursive algorithms for the dynamics analyses of closed-loop parallel systems. Subsequently, Khan et al. (2005) illustrated the effectiveness of the DeNOC-based methodology in modelling parallel manipulators. Inspired by the concept of the DeNOC matrices, Dimitrov (2005) used a similar method for dynamic analysis, trajectory planning, and control of space robots. Garcia de Jalon et al. (2005) have also derived matrices which they have pointed out to be similar to the DeNOC matrices of Saha (1995, 1997). The DeNOC matrices have also found an application in the architecture design of a manipulator through its dynamic model simplifications (Saha et al., 2006). More recently, Chaudhary and Saha (2007) have applied the concept of the DeNOC matrices for the dynamic analyses of general closed-loop systems. They have also introduced the concepts like "determinate" and "indeterminate" subsystems which helped to achieve subsystem-level recursions for the inverse dynamics of a general closed-loop system. Systems with closed-loops which are used in automobile steering systems were analyzed by Hanzaki et al. (2009), whereas fuel injection pumps of diesel engines with rolling contacts were analyzed by Sundarranan et al. (2012). Extending the concept of the DeNOC matrices to other type of systems, Mohan and Saha (2007) showed how to derive the DeNOC matrices for a rigid-flexible multibody system. The methodology not only provided efficient dynamic algorithms but also produced numerically stable results. Very recently, Shah et al. (2012a) introduced a concept of "kinematic module" to a tree-type multibody system and derived module-level DeNOC matrices, which provided macroscopic purview of the multibody systems. Moreover, intra- and inter-modular recursive algorithms were derived for the analyses and control of legged robots (Shah, 2011; Shah et al., 2013). It was shown that the concept of Euler-angle-joints (EAJs) (Shah et al., 2012b) coupled with the module-level DeNOC matrices provided very efficient dynamic algorithms for the multibody system consisting of multiple branches and multiple-degrees-of-freedom joints. The algorithms have been implemented in a free software called ReDySim (acronym for Recursive Dynamic Simulator), which can be downloaded free from http://www.redysim.co.nr. ReDySim can be easily used by the students and researchers of multibody dynamics. Note here that the DeNOC-based algorithm was also used by the researchers from other domain, e.g., Patriciu et al. (2004) have adopted the concept for the analysis of conformational dependence of mass-metric tensor determinants in serial polymers with constraints.

The main motivation behind this paper is to bring forth the developments of the DeNOC-based dynamic modelling for multibody systems, which have taken place over more than one and half decades. The paper explains the fundamental principles of the DeNOC-based formulation, their benefits and applications. Rest of the paper is organized as follows: Sect. 2 presents the DeNOC-based dynamic modelling for serial-chain systems, which forms the basis for the dynamic modelling of other type of systems, e.g., tree-type systems explained in Sect. 3. Application to closed-loop systems is explained in Sect. 4, whereas two software, namely, RoboAnalyzer and ReDySim, developed for the use by the students and researchers of multibody dynamics are explained in Sect. 5. The computational aspects are provided in Sect. 6. Finally, conclusions are given in Sect. 7.

2 DeNOC-based dynamic modelling for serial-chain systems

The Natural Orthogonal Complement (NOC) matrix proposed by Angeles and Lee (1988) relates the angular and linear velocities of the rigid bodies in a mechanical system to its associated joint-rates. It is used to develop a set of independent equations of motion from the unconstrained or uncoupled Newton-Euler (NE) equations using free-body diagrams. These independent set of equations was referred by the authors as the Euler-Lagrange equations of motion. Unlike the NOC, its decoupled form, i.e., the DeNOC, proposed by Saha (1995, 1997), allows one to write the expressions of each element of the matrices and vectors associated with the dynamic equations of motion in analytical recursive form.

2.1 Preliminaries and notation

An open-loop serial-chain system, e.g., a robotic manipulator shown in Fig. 1, has a fixed-base, denoted by #0, and n moving rigid bodies or links, indicated with #1, ..., #n, coupled by n single degree-of-freedom (DOF) kinematic pairs or joints numbered as 1, ..., n. The joints are generally revolute or prismatic. In presence of higher-DOF joints, they are modelled as combinations of single-DOF joints. For example, a spherical joint can be modelled as three intersecting revolute joints, whereas a cylindrical joint is modelled as a combination of revolute and prismatic joints. Few terms are defined below which will be used throughout the paper for the derivation of the dynamic models.

The 6-dimensional vectors, twist ($\boldsymbol{t}_i$) of the i-th rigid link undergoing motion in the 3-dimensional Cartesian space and wrench ($\boldsymbol{w}_i$), acting on the i-th link are defined by:

$$\boldsymbol{t}_i \equiv \begin{bmatrix} \boldsymbol{\omega}_i \\ \boldsymbol{v}_i \end{bmatrix} \text{ and } \boldsymbol{w}_i \equiv \begin{bmatrix} \boldsymbol{n}_i \\ \boldsymbol{f}_i \end{bmatrix} \tag{1}$$

where $\boldsymbol{\omega}_i$ is the 3-dimensional vector of angular velocity, and $\boldsymbol{v}_i$ is the 3-dimensional vector of linear velocity of the mass center (C_i) of the i-th link, whereas $\boldsymbol{n}_i$ and $\boldsymbol{f}_i$ are the 3-dimensional vectors of the moment and force applied about and at C_i, respectively. The 6×6 matrices of mass $\mathbf{M}_i$, and angular velocity $\mathbf{W}_i$, of the i-th body are represented by:

$$\mathbf{M}_i \equiv \begin{bmatrix} \mathbf{I}_i & \mathbf{O} \\ \mathbf{O} & m_i\mathbf{1} \end{bmatrix} \text{ and } \mathbf{W}_i \equiv \begin{bmatrix} \boldsymbol{\omega}_i \times \mathbf{1} & \mathbf{O} \\ \mathbf{O} & \mathbf{O} \end{bmatrix} \tag{2}$$

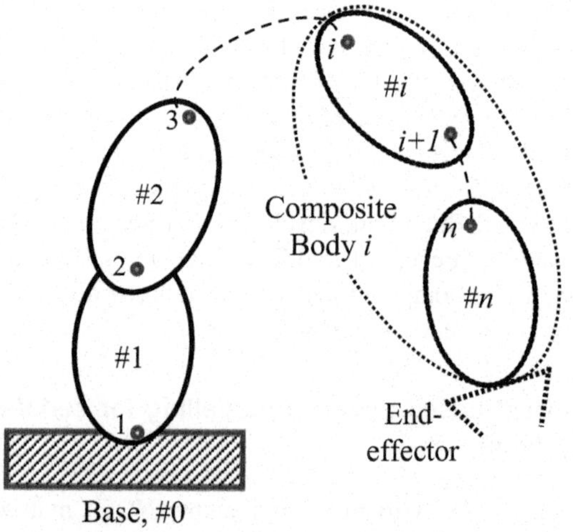

Figure 1. A robot manipulator.

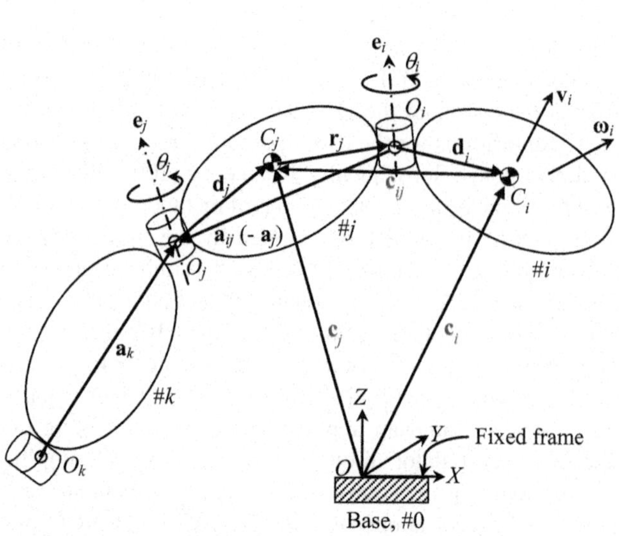

Figure 2. A coupled link system.

where $\omega_i \times \mathbf{1}$ is the 3×3 cross-product tensor associated with the angular velocity vector ω_i which when operates on any 3-dimensional Cartesian vector $\boldsymbol{x}$ leads to the cross-product vector between ω_i and $\boldsymbol{x}$, i.e., $(\omega_i \times \mathbf{1})\boldsymbol{x} \equiv \omega_i \times \boldsymbol{x}$. Also, $\mathbf{1}$ and $\mathbf{O}$ are the 3×3 identity and zero matrices, respectively, whereas $\mathbf{I}_i$ and m_i are the 3×3 inertia tensor about C_i, and the mass of the i-th link, respectively. For the serial-chain mechanical system shown in Fig. 1, the method to obtain the dynamic equations of motion using the DeNOC matrices is as follows:

- Derive the DeNOC matrices.
- Obtain the unconstrained NE equations of motion from the free-body diagrams of each link, and
- Couple the DeNOC matrices with the unconstrained NE equations to obtain a set of constrained independent equations of motion which are same as the system's EL equations of motion.

The above steps are explained next in the following subsections.

2.2 Kinematic constraints

The kinematic constraints in terms of the velocities of two neighbouring links, say, #i and #j, coupled by a revolute joint, as shown in Fig. 2, are given by

$$\omega_i = \omega_j + \dot{\theta}_i \boldsymbol{e}_i \tag{3a}$$

$$\boldsymbol{v}_i = \boldsymbol{v}_j + \omega_j \times \boldsymbol{r}_j + \omega_i \times \boldsymbol{d}_i \tag{3b}$$

where ω_j and $\boldsymbol{v}_j$ are the angular velocity and velocity of the mass of link j, i.e., C_j, respectively. Similarly, ω_i and $\boldsymbol{v}_i$ are defined for the neighbouring link i, whereas $\dot{\theta}_i$ is the joint-rate of the i-th joint. The above six scalar equations can be written in a compact form as

$$\boldsymbol{t}_i = \mathbf{B}_{ij}\boldsymbol{t}_j + \boldsymbol{p}_i\dot{\theta}_i \tag{4}$$

where $\mathbf{B}_{ij}$ is the 6×6 matrix and $\boldsymbol{p}_i$ is the 6-dimensional vector which are given by

$$\mathbf{B}_{ij} \equiv \begin{bmatrix} \mathbf{1} & \mathbf{O} \\ \boldsymbol{c}_{ij} \times \mathbf{1} & \mathbf{1} \end{bmatrix} \text{ and } \boldsymbol{p}_i \equiv \begin{bmatrix} \boldsymbol{e}_i \\ \boldsymbol{e}_i \times \boldsymbol{d}_i \end{bmatrix} \tag{5}$$

Here, $\boldsymbol{c}_{ij}$ is the 3-dimensional position vector from C_i to C_j given by $\boldsymbol{c}_{ij} \equiv -\boldsymbol{d}_i - \boldsymbol{r}_j$, and $\boldsymbol{c}_{ij} \times \mathbf{1}$ is the cross-product tensor associated with vector $\boldsymbol{c}_{ij}$. It is defined similar to $\omega_i \times \mathbf{1}$ of Eq. (2). Moreover, $\boldsymbol{e}_i$ is the unit vector parallel to the axis of rotation of the i-th revolute joint. Interestingly, matrix $\mathbf{B}_{ij}$ and vector $\boldsymbol{p}_i$ have the following interpretations:

- If links #i and #j are rigidly attached, $\mathbf{B}_{ij}$ propagates twist or velocities of #j to #i. Hence, $\mathbf{B}_{ij}$ is termed in Saha (1999a) as the *twist-propagation* matrix, which satisfies

$$\mathbf{B}_{ij}\mathbf{B}_{jk} = \mathbf{B}_{ik} \text{ and } \mathbf{B}_{ii} = \mathbf{1} \tag{6}$$

- On the other hand the vector $\boldsymbol{p}_i$ takes into account the motion of the i-th joint. Hence, vector $\boldsymbol{p}_i$ is termed as the *joint-rate-propagation* vector. The vector $\boldsymbol{p}_i$ in Eq. (5) is defined for a revolute joint. For a prismatic joint, it is given by

$$\boldsymbol{p}_i \equiv \begin{bmatrix} \mathbf{0} \\ \boldsymbol{e}_i \end{bmatrix} \tag{7}$$

Equation (4) can be written for $i = 1, ..., n$, as

$$(\mathbf{1} - \mathbf{B})\,\boldsymbol{t} = \mathbf{N}_d\dot{\boldsymbol{\theta}} \tag{8a}$$

where $\mathbf{1}$ is the $6n \times 6n$ identity matrix, and the $6n \times 6n$ matrix $\mathbf{B}$ has the following representation:

$$\mathbf{B} = \begin{bmatrix} \mathbf{O} & \mathbf{O} & \cdots & \mathbf{O} \\ \mathbf{B}_{21} & \mathbf{O} & \cdots & \mathbf{O} \\ \vdots & \vdots & \ddots & \vdots \\ \mathbf{O} & \cdots & \mathbf{B}_{n,n-1} & \mathbf{O} \end{bmatrix} \tag{8b}$$

It is now simple matter to invert the $6n \times 6n$ matrix, $(\mathbf{1} - \mathbf{B})$, and hence, Eq. (8a) can be rewritten as

$$\boldsymbol{t} = \mathbf{N}\dot{\boldsymbol{\theta}}, \quad \text{where } \mathbf{N} \equiv \mathbf{N}_l\mathbf{N}_d \tag{9a}$$

In Eq. (9a), the matrix $\mathbf{N}$ is the $6n \times n$ Natural Orthogonal Complement (NOC) matrix, as introduced by Angeles and Lee (1988), whereas $\mathbf{N}_l$ and $\mathbf{N}_d$ are the decoupled form of the NOC or the DeNOC matrices proposed first time in Saha (1995). The $6n \times 6n$ matrix $\mathbf{N}_l$ and the $6n \times n$ matrix $\mathbf{N}_d$ are given by

$$\mathbf{N}_l = \begin{bmatrix} \mathbf{1} & \mathbf{O} & \cdots & \mathbf{O} \\ \mathbf{B}_{21} & \mathbf{1} & \cdots & \mathbf{O} \\ \vdots & \vdots & \ddots & \vdots \\ \mathbf{B}_{n1} & \mathbf{B}_{n2} & \cdots & \mathbf{1} \end{bmatrix} \text{ and } \mathbf{N}_d = \begin{bmatrix} \boldsymbol{p}_1 & \mathbf{0} & \cdots & \mathbf{0} \\ \mathbf{0} & \boldsymbol{p}_2 & \cdots & \mathbf{0} \\ \vdots & \vdots & \ddots & \vdots \\ \mathbf{0} & \mathbf{0} & \cdots & \boldsymbol{p}_n \end{bmatrix} \tag{9b}$$

Note that in Eq. (9b), $\mathbf{N}_l$ is a lower block-triangular matrix, whereas $\mathbf{N}_d$ is a block-diagonal matrix, as indicated through their subscripts "l" and "d", respectively. Moreover, $\mathbf{O}$ and $\mathbf{0}$ are the 6×6 matrix of zeros and the 6-dimensional vector of zeros, respectively. The n-dimensional vector $\dot{\boldsymbol{\theta}}$ is defined as

$$\dot{\boldsymbol{\theta}} \equiv \left[\dot{\theta}_1, \cdots, \dot{\theta}_n\right]^T \tag{10}$$

which contains the joint-rates of all the joints in the serial-chain system shown in Fig. 1.

2.3 Unconstrained Newton-Euler (NE) equations

The unconstrained or uncoupled Newton-Euler (NE) equations of motion for the i-th rigid-link (Saha, 1999a) can be written from its free-body diagram, Fig. 3, as

$$\mathbf{I}_i\dot{\omega}_i + \omega_i \times \mathbf{I}_i\omega_i = \boldsymbol{n}_i \tag{11a}$$

$$m_i\dot{\boldsymbol{v}}_i = \boldsymbol{f}_i \tag{11b}$$

where $\dot{\omega}_i$ and $\dot{\boldsymbol{v}}_i$ are the angular acceleration and acceleration of the mass center C_i, respectively. Moreover, $\mathbf{I}_i$ is the 3×3 inertia tensor of i-th link about its mass center C_i, and m_i is its mass. Other variables were defined after Eq. (1). The above six scalar equations can be put in a compact form as

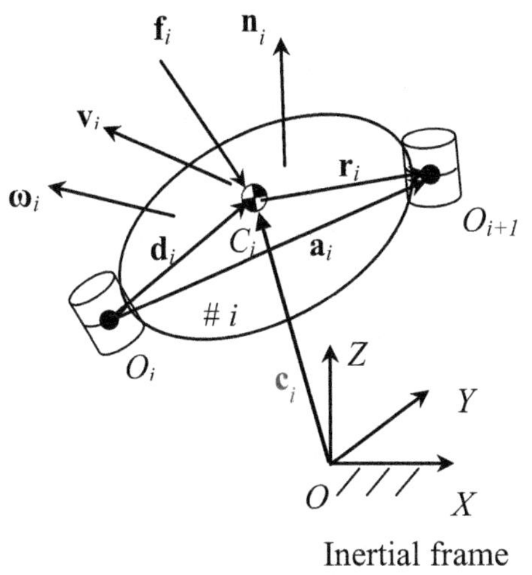

Figure 3. Free-body diagram of the i-th link.

$$\mathbf{M}_i\dot{\boldsymbol{t}}_i + \mathbf{W}_i\mathbf{M}_i\boldsymbol{t}_i = \boldsymbol{w}_i \tag{12}$$

where $\boldsymbol{t}_i$, $\boldsymbol{w}_i$ and $\mathbf{W}_i$, $\mathbf{M}_i$ are defined in Eqs. (1) and (2), respectively. Moreover, $\dot{\boldsymbol{t}}_i$ is the time derivative of the twist $\boldsymbol{t}_i$ of the i-th link. For the whole system of n rigid links, the $6n$ scalar equations (for $i = 1, \ldots, n$, where n is the number of moving rigid links in the serial chain system) can be written as

$$\mathbf{M}\dot{\boldsymbol{t}} + \mathbf{W}\mathbf{M}\boldsymbol{t} = \boldsymbol{w} \tag{13}$$

In Eq. (13), $\dot{\boldsymbol{t}}$ is the time derivative of the generalized twist, $\boldsymbol{t}$. Moreover, $\mathbf{M}$ and $\mathbf{W}$ are the $6n \times 6n$ generalized mass matrix and generalized matrix of angular velocities, respectively, i.e.,

$$\mathbf{M} \equiv \text{diag.}[\mathbf{M}_1, \cdots, \mathbf{M}_n] \text{ and } \mathbf{W} \equiv \text{diag.}[\mathbf{W}_1, \cdots, \mathbf{W}_n] \tag{14}$$

Moreover, $\boldsymbol{w}$ and $\boldsymbol{t}$ are the $6n$-dimensional vectors of generalized wrench and twist, respectively. They are defined as

$$\boldsymbol{w} \equiv \left[\boldsymbol{w}_1^T, \cdots, \boldsymbol{w}_n^T\right]^T \text{ and } \boldsymbol{t} \equiv \left[\boldsymbol{t}_1^T, \cdots, \boldsymbol{t}_n^T\right]^T \tag{15}$$

2.4 Constrained equations using the DeNOC matrices

The kinematic constraints in velocities, i.e., Eq. (9a), then can be incorporated into the unconstrained NE equations of motion, Eq. (13). This is done by pre-multiplying $\mathbf{N}^T$ with the $6n$ unconstrained NE equations of motions of Eq. (13), i.e.,

$$\mathbf{N}^T\left(\mathbf{M}\dot{\boldsymbol{t}} + \mathbf{W}\mathbf{M}\boldsymbol{t}\right) = \mathbf{N}^T\left(\boldsymbol{w}^E + \boldsymbol{w}^C\right) \tag{16}$$

where $\boldsymbol{w}$ is substituted as, $\boldsymbol{w} \equiv \boldsymbol{w}^E + \boldsymbol{w}^C$, in which $\boldsymbol{w}^E$ and $\boldsymbol{w}^C$ are the $6n$-dimensional vectors of external and constraint wrenches, respectively. Since the constraint wrenches do not do any work, $\mathbf{N}^T\boldsymbol{w}^C$ vanishes (Angeles and Lee, 1988). Hence, $\mathbf{N}^T\boldsymbol{w}^C = \mathbf{0}$. Substituting the expression of $\boldsymbol{t}$ from Eq. (9a) and its time derivative, $\dot{\boldsymbol{t}} = \mathbf{N}^T\ddot{\boldsymbol{\theta}} + \dot{\mathbf{N}}\dot{\theta}$ into Eq. (16), one can get the n independent scalar dynamic equations of motion, namely,

$$\mathbf{I}\ddot{\boldsymbol{\theta}} + \mathbf{C}\dot{\boldsymbol{\theta}} = \boldsymbol{\tau} \quad (17)$$

where, $\mathbf{I} \equiv \mathbf{N}^T\mathbf{MN}$: the $n \times n$ generalized inertia matrix (GIM); $\mathbf{C} \equiv \mathbf{N}^T(\mathbf{M\dot{N}} + \mathbf{WMN})$: the $n \times n$ matrix of convective inertia terms (MCI); and $\boldsymbol{\tau} \equiv \mathbf{N}^T\boldsymbol{w}^E$: the n-dimensional vector of generalized forces of driving, and those resulting from gravity, dissipation, and other external forces like foot-ground interaction of a walking robot, etc., if any.

2.5 Analytical expression of the GIM

The analytical expression of the generalized inertia matrix (GIM) appearing in Eq. (17) plays an important role in simplifying, mainly, the forward dynamics agorithm (Saha, 1999a, 2003). In this section, the GIM $\mathbf{I}$ is derived using the expressions of the DeNOC matrices (Saha, 1995, 1997, 1999a, b, 2003). Substituting the expressions of the DeNOC matrices given by Eq. (9b) into the expression of the GIM appearing after Eq. (17), one gets

$$\mathbf{I} = \mathbf{N}_d^T\tilde{\mathbf{M}}\mathbf{N}_d, \quad \text{where } \tilde{\mathbf{M}} \equiv \mathbf{N}_l^T\mathbf{M}\mathbf{N}_l \quad (18)$$

The $6n \times 6n$ symmetric matrix $\tilde{\mathbf{M}}$ can be written as

$$\tilde{\mathbf{M}} \equiv \begin{bmatrix} \tilde{\mathbf{M}}_1 & \mathbf{B}_{21}^T\tilde{\mathbf{M}}_2 & \cdots & \mathbf{B}_{n1}^T\tilde{\mathbf{M}}_n \\ \tilde{\mathbf{M}}_2\mathbf{B}_{21} & \tilde{\mathbf{M}}_2 & \cdots & \mathbf{B}_{n2}^T\tilde{\mathbf{M}}_n \\ \vdots & \vdots & \ddots & \vdots \\ \tilde{\mathbf{M}}_n\mathbf{B}_{n1} & \tilde{\mathbf{M}}_n\mathbf{B}_{n2} & \cdots & \tilde{\mathbf{M}}_n \end{bmatrix} \quad (19)$$

where the 6×6 matrix, $\tilde{\mathbf{M}}_i$, for $i = 1, \cdots, n$, can be obtained recursively, i.e.,

$$\tilde{\mathbf{M}}_i = \mathbf{M}_i + \mathbf{B}_{i+1,i}^T\tilde{\mathbf{M}}_{i+1}\mathbf{B}_{i+1,i} \quad (20)$$

in which $\tilde{\mathbf{M}}_{i+1} \equiv \mathbf{O}$, because there is no $(n+1)$st link in the serial-chain. Hence, $\tilde{\mathbf{M}}_n \equiv \mathbf{M}_n$. The matrix, $\tilde{\mathbf{M}}_i$, is interpreted as the mass matrix of the *Composite Body*, i, that consists of rigidly connected links $\#i, ..., \#n$, as indicated in Fig. 1. Finally, the $n \times n$ GIM $\mathbf{I}$ can be expressed as

$$\mathbf{I} \equiv \begin{bmatrix} i_{11} & & \text{sym} \\ \vdots & \ddots & \\ i_{n1} & \cdots & i_{nn} \end{bmatrix}, \quad \text{where } i_{ij} \equiv \boldsymbol{p}_i^T\tilde{\mathbf{M}}_i\mathbf{B}_{ij}\boldsymbol{p}_j \quad (21)$$

for $i = 1, ..., n$; $j = 1, ..., i$. The term i_{ij} is a scalar and "sym" denotes symmetric elements of the GIM $\mathbf{I}$.

2.6 Recursive inverse dynamics algorithm

The inverse dynamics of a serial-chain system is defined as the process of determining the joint forces/torques when the joint motions of the system are known. The inverse dynamics algorithm calculates the joint torque, τ_i, for $i = 1, ..., n$, in two recursive steps, namely, forward and backward recursions. They are given below.

2.6.1 Step 1: forward recursion

First, the 6-dimensional twist and twist-rate vectors of each link, i.e., $\boldsymbol{t}_i$ and $\dot{\boldsymbol{t}}_i$, respectively, are calculated, for $i = 1, ..., n$, using the following relations:

$$\boldsymbol{t}_i = \mathbf{B}_{i,i-1}\boldsymbol{t}_{i-1} + \boldsymbol{p}_i\dot{\theta}_i \quad (22)$$

$$\dot{\boldsymbol{t}}_i = \mathbf{B}_{i,i-1}\dot{\boldsymbol{t}}_{i-1} + \dot{\mathbf{B}}_{i,i-1}\boldsymbol{t}_{i-1} + \boldsymbol{p}_i\ddot{\theta}_i + \dot{\boldsymbol{p}}_i\dot{\theta}_i \quad (23)$$

$$\boldsymbol{w}_i = \mathbf{M}_i\dot{\boldsymbol{t}}_i + \mathbf{W}_i\mathbf{M}_i\boldsymbol{t}_i \quad (24)$$

In the above equations, $\boldsymbol{t}_0 = \mathbf{0}$ and $\dot{\boldsymbol{t}}_0 = \mathbf{0}$, as link #0 is fixed without any motion.

2.6.2 Step 2: backward recursion

The 6-dimensional vector, $\tilde{\boldsymbol{w}}_i$, and the scalar, τ_i, for $i = n, ..., 1$, are calculated using the following relations:

$$\tilde{\boldsymbol{w}}_i = \mathbf{B}_{i+1,i}^T\tilde{\boldsymbol{w}}_{i+1}, \quad \text{and } \tau_i = \boldsymbol{p}_i^T\tilde{\boldsymbol{w}}_i \quad (25)$$

where for $i = n$, $\tilde{\boldsymbol{w}}_{n+1} = 0$, as there is no $(n+1)$st link in the system. Hence, $\tilde{\boldsymbol{w}}_n = \boldsymbol{w}_n$. The effect of gravity can also be taken into account by providing negative acceleration due to gravity, $\boldsymbol{g}$, to the twist-rate of the first link as an additional term (Kane and Levinson, 1983), i.e.,

$$\dot{\boldsymbol{t}}_1 = \boldsymbol{p}_1\ddot{\theta}_1 + \dot{\boldsymbol{p}}_1\dot{\theta}_1 + \boldsymbol{\rho}, \quad \text{where } \boldsymbol{\rho} \equiv \left[\mathbf{0}^T, -\boldsymbol{g}^T\right] \quad (26)$$

Note that Eqs. (22)–(26) were reported in Saha (1999a) with different notations, which actually have the same interpretations as given above, i.e., twist ($\boldsymbol{t}_i$), twist-rate ($\dot{\boldsymbol{t}}_i$), wrench of composite body ($\tilde{\boldsymbol{w}}_i$), etc. Based on the above mentioned recursive inverse dynamics algorithm, a computer program was developed in C++ which was called RIDIM (Recursive Inverse Dynamic for Industrial Manipulators) (Saha, 1999a). Recently, a similar algorithm has been rewritten in Visual C# and implemented in the "IDyn" module of the newly developed software called RoboAnalyzer (Rajeevlochana and Saha, 2011; Rajeevlochana et al., 2012) which also has 3-dimensional visualisation of the system under study. It is explained in Sect. 6.1, and available free from http://www.roboanalyzer.com for the benefits of students and researchers of multibody dynamics community.

2.7 Recursive forward dynamics algorithm

Forward dynamics of a serial-chain system is defined as the process of determining the joint accelerations when the joint-actuator torques/forces of the system are known. In order to compute the joint accelerations $\ddot{\boldsymbol{\theta}}$ recursively, the GIM, $\mathbf{I}$ of Eq. (17), is decomposed as $\mathbf{I} \equiv \mathbf{UDU}^T$ (Saha, 1995, 1997, 1999b) based on the Reverse Gaussian Elimination (RGE) method, where $\mathbf{U}$ and $\mathbf{D}$ are upper triangular and diagonal matrices, respectively. The $\mathbf{UDU}^T$ decomposition results in an efficient order n, i.e., $O(n)$, computational algorithm in contrast to $O(n^3)$ computations required by the Cholesky decomposition of the GIM (Strang, 1998).

For the development of recursive $O(n)$ forward dynamics algorithm, the constrained dynamics equations of motion, Eq. (17), are rewritten as

$$\mathbf{UDU}^T\ddot{\boldsymbol{\theta}} = \boldsymbol{\varphi} \tag{27}$$

where $\boldsymbol{\varphi} \equiv \boldsymbol{\tau} - \mathbf{C}\dot{\boldsymbol{\theta}}$. Then, three recursive steps are used to calculate the joint accelerations, which are given below.

2.7.1 Step 1

Solution for $\hat{\boldsymbol{\tau}}$, where $\hat{\boldsymbol{\tau}} \equiv \mathbf{DU}^T\ddot{\boldsymbol{\theta}} \equiv \mathbf{U}^{-1}\boldsymbol{\varphi}$. It is found as follows: For $i = n-1, ..., 1$, calculate

$$\hat{\tau}_i = \varphi_i - \boldsymbol{p}_i^T\boldsymbol{\eta}_{i,i+1} \tag{28}$$

where $\boldsymbol{\eta}_{i,i+1}$ is the 6-dimensional vector obtained recursively as

$$\boldsymbol{\eta}_{i,i+1} \equiv \mathbf{B}_{i+1,i}^T\boldsymbol{\eta}_{i+1} \text{ and } \boldsymbol{\eta}_{i+1} \equiv \hat{\tau}_{i+1}\boldsymbol{\psi}_{i+1} + \boldsymbol{\eta}_{i+1,i+2} \tag{29}$$

in which $\boldsymbol{\eta}_{n,n+1} = \mathbf{0}$, and the 6-dimensional vector $\boldsymbol{\psi}_{i+1}$ is evaluated using the following relations:

$$\boldsymbol{\psi}_i = \frac{\hat{\boldsymbol{\psi}}_i}{\hat{m}_i}, \text{ where } \hat{\boldsymbol{\psi}}_i \equiv \hat{\mathbf{M}}_i\boldsymbol{p}_i \text{ and } \hat{m}_i \equiv \boldsymbol{p}_i^T\hat{\boldsymbol{\psi}}_i \tag{30}$$

In Eq. (30), the 6×6 matrix, $\hat{\mathbf{M}}_i$ is obtained recursively as

$$\hat{\mathbf{M}}_i = \mathbf{M}_i + \mathbf{B}_{i+1,i}^T\overline{\mathbf{M}}_{i+1}\mathbf{B}_{i+1,i},$$

$$\text{where } \overline{\mathbf{M}}_{i+1} \equiv \hat{\mathbf{M}}_{i+1} - \hat{\boldsymbol{\psi}}_{i+1}\boldsymbol{\psi}_{i+1}^T \text{and } \hat{\mathbf{M}}_n = \mathbf{M}_n \tag{31}$$

The 6×6 symmetric matrix $\hat{\mathbf{M}}_i$ is the mass matrix of *Articulated Body*, *i*, defined as the links #*i*, ..., #*n*, coupled by the joints *i*+1, ..., *n*. This is in contrast to the definition of the *Composite Body*, *i*, given after Eq. (20), where the links are rigidly connected, i.e., the joints are locked. Note that the mass matrix of the *i*-th *Articulate Body* $\hat{\mathbf{M}}_i$ is nothing but the Articulated-Body-Inertia (ABI) of Featherstone (1987).

2.7.2 Step 2

Solution for $\tilde{\boldsymbol{\tau}}$, where, $\tilde{\boldsymbol{\tau}} \equiv \mathbf{U}^T\ddot{\boldsymbol{\theta}} \equiv \mathbf{D}^{-1}\hat{\boldsymbol{\tau}}$. It is found as follows: for $i = 1, ..., n$,

$$\tilde{\tau}_i = \frac{\hat{\tau}_i}{\hat{m}_i} \tag{32}$$

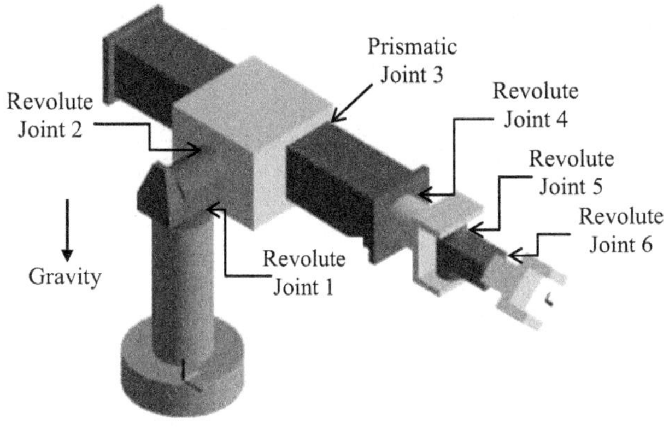

Figure 4. The Stanford arm.

2.7.3 Step 3

Solution for $\ddot{\boldsymbol{\theta}}$, where, $\ddot{\boldsymbol{\theta}} \equiv \mathbf{U}^{-T}\tilde{\boldsymbol{\tau}}$. It is found as follows: For $i = 2, ..., n$,

$$\ddot{\theta}_i = \tilde{\tau}_i - \boldsymbol{\psi}_i^T\boldsymbol{\mu}_{i,i-1} \tag{33}$$

where $\boldsymbol{\mu}_{i,i-1} \equiv \mathbf{B}_{i,i-1}\boldsymbol{\mu}_{i-1}$, $\boldsymbol{\mu}_{i-1} \equiv \boldsymbol{p}_{i-1}\ddot{\theta}_{i-1} + \boldsymbol{\mu}_{i-1,i-2}$, and for $i = 1$, $\boldsymbol{\mu}_{10} \equiv \mathbf{0}$.

Based on the above mentioned forward dynamics algorithm, another C++ program RFDSIM (Recursive Forward Dynamic and Simulation of Industrial Manipulators) was written which was reported in Saha (1999a). A similar algorithm was rewritten in Visual C# and implemented in the "FDyn" module of RoboAnalyzer software (Rajeevlochana et al., 2012; http://www.roboanalyzer.com) with which one can see animation of the systems under study. The numerical integrator used in RoboAnalyzer for the simulation purposes is based on the Runge-Kutta 4th order method (Bathe and Wilson, 1976).

2.8 Numerical example: a 6-DOF Stanford arm

The dynamic analyses of the 6-link 6-DOF serial-chain system with both revolute and prismatic joints, namely, the Stanford arm as shown in Fig. 4, were carried out using RoboAnalyzer. The Denavit and Hartenberg (DH) paramters, which were proposed by Denavit and Hartenberg (1955), and the mass and inertia propoerties are taken from Saha (1999a) as per the notations explained there and in Saha (2008). The numerical values are not reproduced here since the focus of this paper is to review the DeNOC-based formulations and their applicability. However, the joint torques (Joints 1–2, 4–6) and force (Joint 3) obtained from the "IDyn" module of RoboAnalyzer software for the following joint input motions are plotted in Fig. 5:

$$\theta_i = \theta_i(0) + \frac{\theta_i(T) - \theta_i(0)}{T}\left[t - \frac{T}{2\pi}\sin\left(\frac{2\pi}{T}t\right)\right]$$

$$\text{for } i = 1,2,4,5,6 \tag{34}$$

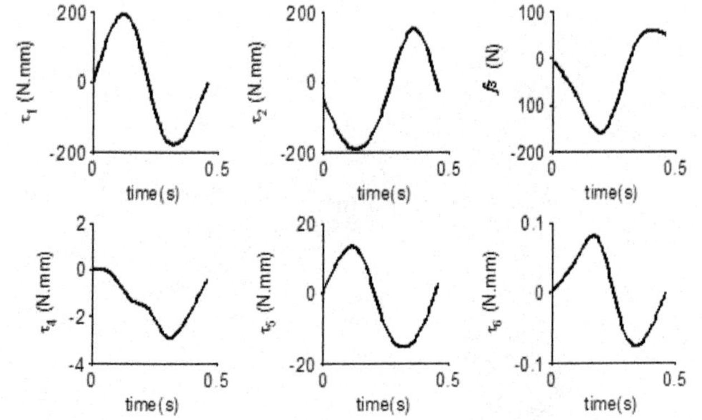

Figure 5. Joint torques (1–2, 4–6) and force (3) for the Stanford arm.

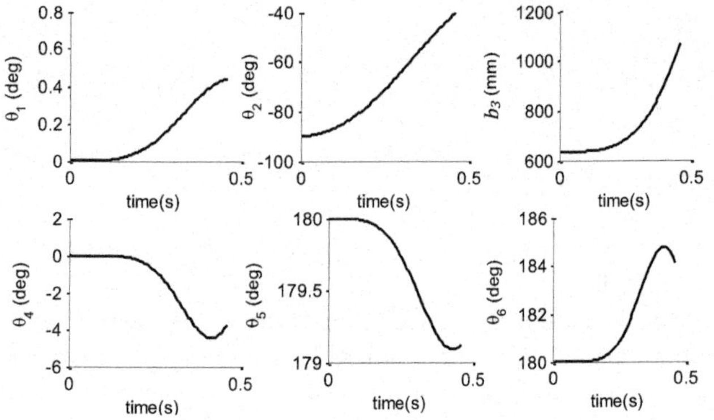

Figure 6. Simulated joint motions for the free-fall of the Stanford arm.

$$b_3 = b_3(0) + \frac{b_3(T) - b_3(0)}{T}\left[t - \frac{T}{2\pi}\sin\left(\frac{2\pi}{T}t\right)\right] \tag{35}$$

where $\theta_i(0) = 0$, for $i = 1$–2, 4–6 and $b_3(0) = 0$ are the variable DH parameters (Saha, 2008) or the joint variables at time $T = 0$, whereas the total time of motion is, $T = 10$ s. Gravity was acting in the negative Z_1-direction. The variable, τ_i, for $i = 1$–2, 4–6, and f_3 in Fig. 5 are the joint torques and force, respectively. The results were verified with those reported in Saha (2008).

The forward dynamics and simulation of the Stanford arm was also performed using "FDyn" module of RoboAnalyzer. The Stanford manipulator was assumed to fall freely under gravity without any external torques and force at the actuating joints. The initial positions were taken same as in the inverse dynamics analysis given after Eq. (35). The results are plotted in Fig. 6, where the variations of the joint motions with respect to time are shown. The results were also verified with those reported in Saha (2008).

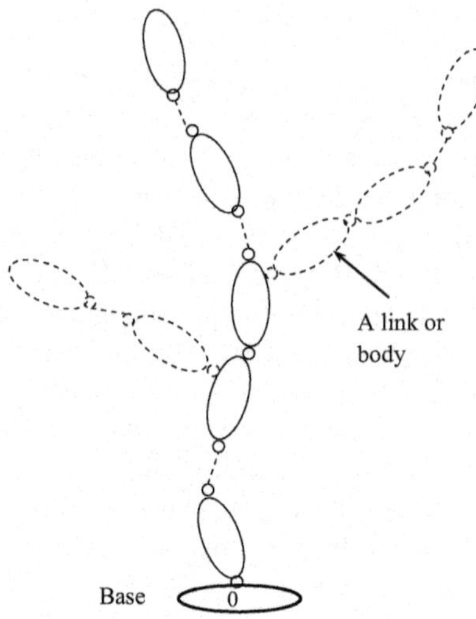

Figure 7. A tree-type system.

3 Tree-type systems

A tree-type system has a set of links connected by kinematic pairs, typically, a revolute or a prismatic joint, as shown in Fig. 7. Other type of joints, say, a universal or spherical, and a cylindrical, can be modelled as a combination of two or three intersecting revolute joints, and a pair of revolute-prismatic joints, respectively, as mentioned in the beginning of Sect. 2.1. Based on the modelling of serial-chain systems, Shah et al. (2011, 2013) extended the methodology to model a tree-type system. For this, the tree-type system was assumed to be a combination of several serial-chain systems called "kinematic modules". Consequently, multi-modular recursive algorithms for the tree-type systems were presented against "full-body-level" recursive dynamics algorithms of Featherstone (1987) and Rodriguez (1992). Each "module" of the tree-type architecture was defined as a set of serially connected links emerges from the last link of its parent module. For example, as indicated in Fig. 8, the parent module of M_i is module M_β.

For the analyses purposes, the tree-type system was first kinematically modularized before its kinematic constraints were derived. The modules are denoted with M_0, M_1, M_2, etc., where a child module bears a number higher than its parent module. Moreover, the links inside any module, say, M_i, are denoted as $\#1^i$, ..., $\#k^i$, ..., $\#\eta^i$, where the superscript i signifies the module number. Considering the tree-type system, there are s number of modules in the system, and there are η^i number of links in the i-th module. The total number of links in the whole system is then obtained by $n = \sum_i^s \eta^i$. The kinematic constraints were next derived at the intra-modular level, i.e., amongst the links inside a module, and inter-modular level, i.e., between the modules. The dynamic analyses were done using intra- (Inside the module)

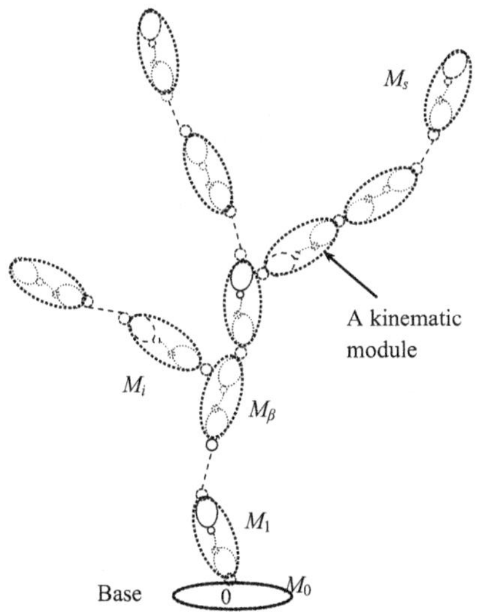

Figure 8. The multi-modular tree-type system.

and inter-modular (between the modules) recursions, as presented in Fig. 9.

3.1 Intra-modular kinematic constraints

Intra-modular kinematic constraints are effectively the velocity constraints between the links of a serial-chain system derived in Eqs. (3–10). Here, however, a little modification is proposed in the definition of each link's linear velocity $\boldsymbol{v}_i$. In contrast to the definition of the velocity of the mass center of the i-th link, C_i, as $\boldsymbol{v}_i$ of Eq. (1), it is defined in this section as the velocity of point O_i where the i-th joint couples the j-th link with the i-th one, as indicated in Fig. 2. Such definition of $\boldsymbol{v}_i$ in twist expression of Eq. (1) was necessitated mainly to take care of the branching issue of the serial-modules in the tree-type system, as shown in Fig. 8. The velocity of the i-th link defined here with respect to O_i (sometimes referred to as the origin of the i-th link). It is actually the velocity of the previous link at its connection point, namely, the last link of the parent module where the first link of the child module is coupled. Hence, where branching occurs no additional computations are required for the calculation of the velocity of the first link belonging to the child module. This was not the case with the definition of the velocity of the i-th link with respect to its mass center C_i in which additional computations would be required to calculate the velocity of the mass center C_i from the origin O_i. Moreover, as the main objective of dynamic analyses is to calculate either joint torques or joint motions, selection of O_i as a reference point, instead of the C_i, can lead to efficient recursive inverse and forward dynamics algorithms, as shown by Shah et al. (2011, 2013). In fact, for the serial-chain systems considered in Sect. 2, the same definition with respect to O_i could have been adopted. This was done with the "IDyn" and "FDyn" modules of the RoboAnalyzer software. In Sect. 2, however, it was shown how the simplest form of the NE equations of motion given by Eq. (1) can be used with the definition of the DeNOC matrices, as demonstrated in the original work of Saha (1995, 1997, 1999a, b, 2003).

Now, with the new definitions of $\boldsymbol{v}_i$ with respect to O_i, Eq. (4) is rewritten as

$$\boldsymbol{t}_i = \mathbf{A}_{ij}\boldsymbol{t}_j + \boldsymbol{p}_i\dot{\theta}_i \tag{36a}$$

where $\boldsymbol{t}_i$ and $\boldsymbol{t}_j$ are the 6-dimensional twist vectors defined in Eq. (1) but with respect to (w.r.t.) the new definition of $\boldsymbol{v}_i$, i.e., w.r.t. point O_i. Accordingly, the 6×6 matrix $\mathbf{A}_{ij}$ is the new twist-propagation matrix. A different notation is used here to distinguish it from $\mathbf{B}_{ij}$ which was defined after Eq. (4) w.r.t. the definition of the velocity of C_i. The 6×6 matrix $\mathbf{A}_{ij}$, and the 6-dimensional joint-rate-propagation vector, $\boldsymbol{p}_i$, are given by

$$\mathbf{A}_{ij} = \begin{bmatrix} \mathbf{1} & \mathbf{O} \\ \boldsymbol{a}_{ij}\times\mathbf{1} & \mathbf{1} \end{bmatrix}, \text{ and}$$

$$\boldsymbol{p}_i = \begin{bmatrix} \boldsymbol{e}_i \\ \mathbf{0} \end{bmatrix} \text{ for revolute; } \boldsymbol{p}_i = \begin{bmatrix} \mathbf{0} \\ \boldsymbol{e}_i \end{bmatrix} \text{ for prismatic} \tag{36b}$$

where the 3-dimensional vector $\boldsymbol{a}_{ij}$ is shown in Fig. 2. Notice the change in the expression of $\boldsymbol{p}_i$ in Eq. (36b) in comparison to the same in Eq. (5) where v_i was defined w.r.t. C_i. For serially connected rigid links in the i-th serial-chain module, one can write the expression for the generalized twist, $\bar{\boldsymbol{t}}_i$, similar to Eq. (9a), as

$$\bar{\boldsymbol{t}}_i = \bar{\mathbf{N}}_i\dot{\bar{\boldsymbol{\theta}}}_i, \text{ where } \bar{\mathbf{N}}_i \equiv [\bar{\mathbf{N}}_l\bar{\mathbf{N}}_d]^i \tag{37}$$

In Eq. (37), the $6\eta^i$-dimensional generalized twist vector $\bar{\boldsymbol{t}}_i$ and the η^i-dimensional generalized joint-rates vector $\dot{\bar{\boldsymbol{\theta}}}_i$ are defined as follows:

$$\bar{\boldsymbol{t}}_i \equiv \begin{bmatrix} \boldsymbol{t}_1 \\ \vdots \\ \boldsymbol{t}_k \\ \vdots \\ \boldsymbol{t}_\eta \end{bmatrix}^i \text{ and } \dot{\bar{\boldsymbol{\theta}}}_i \equiv \begin{bmatrix} \dot{\theta}_1 \\ \vdots \\ \dot{\theta}_k \\ \vdots \\ \dot{\theta}_\eta \end{bmatrix}^i \tag{38}$$

where a bar ("–") over an entity in Eqs. (37) and (38) signifies that the quantity is related to a module and the superscript, i, outside the brackets identifies the module. As a consequence, the generic notation $\boldsymbol{t}_k$ (or $\boldsymbol{t}_{k^i}$) in Eq. (38) is the 6-dimensional twist vector for the kth link in the i-th module. The $6\eta^i\times6\eta^i$ and $6\eta^i\times\eta^i$ DeNOC matrices for the serial-chain module, denoted as $\bar{\mathbf{N}}_{l^i}$ and $\bar{\mathbf{N}}_{d^i}$, respectively, are given by

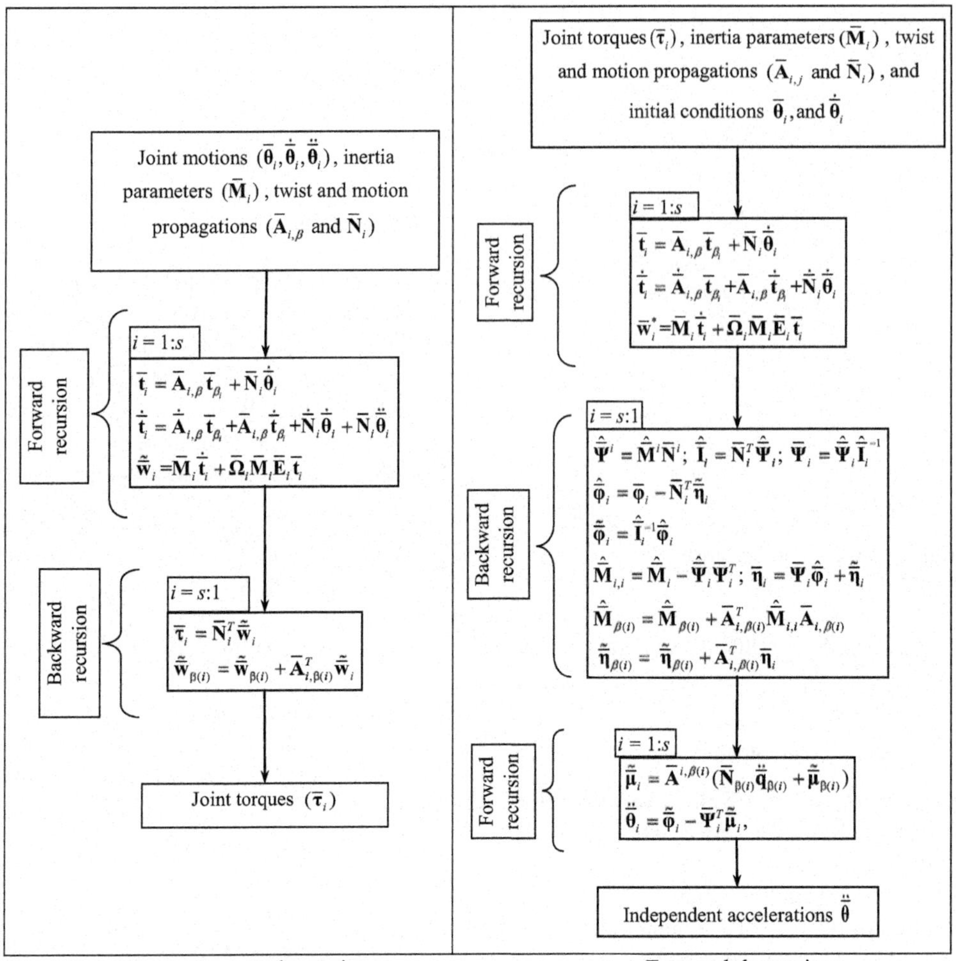

a. Inverse dynamics b. Forward dynamics

Figure 9. Recursive dynamics algorithms (Shah, 2011; Shah et al., 2013).

$$\overline{\mathbf{N}}_{l^i} \equiv \begin{bmatrix} \mathbf{1} & \mathbf{O} & \cdots & \mathbf{O} \\ \mathbf{A}_{21} & \mathbf{1} & \cdots & \mathbf{O} \\ \vdots & \vdots & \ddots & \vdots \\ \mathbf{A}_{\eta 1} & \mathbf{A}_{\eta 2} & \cdots & \mathbf{1} \end{bmatrix}^i \quad \text{and}$$

$$\overline{\mathbf{N}}_{d^i} \equiv \begin{bmatrix} \boldsymbol{p}_1 & \mathbf{0} & \cdots & \mathbf{0} \\ \mathbf{0} & \boldsymbol{p}_2 & \cdots & \mathbf{O} \\ \vdots & \vdots & \ddots & \vdots \\ \mathbf{0} & \mathbf{0} & \cdots & \boldsymbol{p}_\eta \end{bmatrix}^i \tag{39}$$

3.2 Inter-modular kinematic constraints

Having obtained the intra-modular kinematic constraints in the velocity-level, it is now possible to derive the inter-modular kinematic (velocity) constraints, i.e., between two neighbouring serial-chain modules. In a way, each module has been treated similar to a link in a serial-chain module presented in Sect. 2 or Sect. 3.1. For this, module M_β is considered as the parent of module M_i, as shown in Fig. 8. This is similar to link j of Fig. 2 which is the parent of link i. The $6\eta^i$-dimensional generalized twist $\bar{\boldsymbol{t}}_i$ is then obtained from the $6\eta^\beta$-dimensional generalized twist $\bar{\boldsymbol{t}}_\beta$ as

$$\bar{\boldsymbol{t}}_i = \overline{\mathbf{A}}_{i,\beta}\bar{\boldsymbol{t}}_\beta + \overline{\mathbf{N}}_i\dot{\bar{\boldsymbol{\theta}}}_i \tag{40}$$

where $\overline{\mathbf{A}}_{i,\beta}$ is the $6\eta^i \times 6\eta^\beta$ module-twist-propagation matrix which propagates the generalized twist of the parent module (β) to the child module (i) and $\overline{\mathbf{N}}_i$ is the $6\eta^i \times \eta^i$ module-joint-rate propagation matrix, which are given by

$$\overline{\mathbf{A}}_{i,\beta} \equiv \begin{bmatrix} \mathbf{O} & \cdots & \mathbf{O} & \mathbf{A}_{1^i,\eta^\beta} \\ \vdots & \ddots & \vdots & \vdots \\ \mathbf{O} & \cdots & \mathbf{O} & \mathbf{A}_{\eta^i,\eta^\beta} \end{bmatrix} \tag{41a}$$

and

$$\bar{\mathbf{N}}_i = [\bar{\mathbf{N}}_l \bar{\mathbf{N}}_d]^i \equiv \begin{bmatrix} \boldsymbol{p}_1 & 0 & \cdots & 0 \\ \mathbf{A}_{21}\boldsymbol{p}_1 & \boldsymbol{p}_2 & \cdots & 0 \\ \vdots & \vdots & \ddots & \vdots \\ \mathbf{A}_{\eta 1}\boldsymbol{p}_1 & \mathbf{A}_{\eta 2}\boldsymbol{p}_2 & \cdots & \boldsymbol{p}_\eta \end{bmatrix}^i \tag{41b}$$

The vectors $\bar{\boldsymbol{t}}_i$ and $\dot{\bar{\theta}}_i$ are defined in Eq. (38). Next, the $6n$-dimensional generalized twist vector $\boldsymbol{t}$, and the n-dimensional generalized joint-rate vector $\dot{\boldsymbol{\theta}}$, for the whole tree-type system which comprises of s modules and n links are defined as

$$\boldsymbol{t} \equiv \begin{bmatrix} \bar{\boldsymbol{t}}_0^T & \bar{\boldsymbol{t}}_1^T & \cdots & \bar{\boldsymbol{t}}_i^T & \cdots & \bar{\boldsymbol{t}}_s^T \end{bmatrix}^T \text{ and}$$

$$\dot{\boldsymbol{\theta}} \equiv \begin{bmatrix} \dot{\bar{\theta}}_0^T & \dot{\bar{\theta}}_1^T & \cdots & \dot{\bar{\theta}}_i^T & \cdots & \dot{\bar{\theta}}_s^T \end{bmatrix}^T \tag{42}$$

where $\bar{\boldsymbol{t}}_0$ and $\dot{\bar{\theta}}_0$ correspond to the base module M_0 which may not be fixed. For example, in the case of a spacecraft carrying a manipulator, the spacecraft floats with motion of 6-degrees-of-freedom (DOF). For the analysis purposes, its motion need to be specified for further motion analyses of other modules, e.g., the manipulator of the above system.

Upon substitution of the expressions of $\bar{\mathbf{N}}_i$ from Eq. (41b), for $i = 1,..., s$, in Eq. (40), and manipulating the expressions like Eqs. (8)–(9), one obtains the expression of the $6n$-dimensioanl generalized twist $\boldsymbol{t}$ for the whole tree-type system as

$$\boldsymbol{t} = \mathbf{N}_l \mathbf{N}_d \dot{\boldsymbol{\theta}} \tag{43}$$

in which, $\mathbf{N}_l$ and $\mathbf{N}_d$ are the $6(n+1) \times 6(n+1)$ and $6(n+1) \times (n+n_0)$ matrices, respectively, as the tree-type system was assumed to have module M_0 with one-link with n_0 DOF. Matrices $\mathbf{N}_l$ and $\mathbf{N}_d$ for the tree-type system are given by

$$\mathbf{N}_l \equiv \begin{bmatrix} \bar{\mathbf{1}}_0 & & & & \\ \bar{\mathbf{A}}_{10} & \bar{\mathbf{1}}_1 & & \mathbf{O}'s & \\ \bar{\mathbf{A}}_{20} & \bar{\mathbf{A}}_{21} & \bar{\mathbf{1}}_2 & & \\ \vdots & \vdots & \vdots & \ddots & \\ \bar{\mathbf{A}}_{s0} & \bar{\mathbf{A}}_{s1} & \cdots & \cdots & \bar{\mathbf{1}}_s \end{bmatrix},$$

$$\text{where } \bar{\mathrm{A}}_{j,i} \equiv \mathrm{O}, \text{ if } M_j \notin \gamma_i \tag{44}$$

and

$$\mathbf{N}_d \equiv \begin{bmatrix} \bar{\mathbf{N}}_0 & & \mathbf{O}'s & \\ & \bar{\mathbf{N}}_1 & & \\ & & \ddots & \\ & \mathbf{O}'s & & \bar{\mathbf{N}}_s \end{bmatrix} \tag{45}$$

In Eq. (44), $\bar{\mathbf{1}}_i$ is the $6\eta^i \times 6\eta^i$ identity matrix, whereas γ_i stands for the array of all modules including module M_i and outward to it, as shown within dashed line of Fig. 10. The

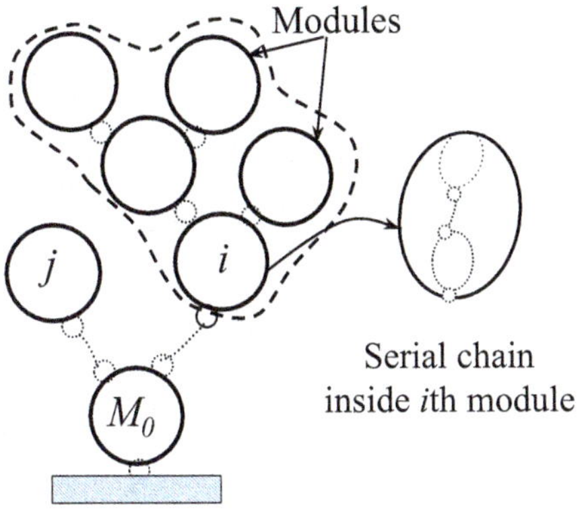

Figure 10. Definition of γ_i.

matrices $\mathbf{N}_l$ and $\mathbf{N}_d$ are the desired Decoupled Natural Orthogonal Compliment (DeNOC) matrices for the whole tree-type system at hand. Note here that the matrices, $\mathbf{N}_l$ and $\mathbf{N}_d$ of Eq. (9b), and $\mathbf{N}_{l^i}$ and $\mathbf{N}_{d^i}$ of Eq. (39), are the special cases of the DeNOC matrices derived in Eqs. (44) and (45), where each module has only one link without any branching.

3.3 Newton-Euler (NE) equations for tree-type systems

In contrast to the expressions for the Newton-Euler (NE) equations of the i-th link given by Eq. (11) or (12), a deviation in their expressions will be observed. This is due to the modified definition of the velocity of the i-th link, i.e., $\boldsymbol{v}_i$, with respect to point O_i. This was mentioned in Sect. 3.1. The NE equations of motion of the k-th link (as the letter "i" will be used to denote module) of the i-th module with respect to point O_k can be expressed as (Shah et al., 2011, 2013)

$$\mathbf{I}_k \dot{\omega}_k + m_k \boldsymbol{d}_k \times \dot{\boldsymbol{v}}_k + \omega_k \times \mathbf{I}_k \omega_k = \boldsymbol{n}_k \tag{46a}$$

$$m_k \boldsymbol{v}_k - m_k \boldsymbol{d}_k \times \dot{\omega}_k - \omega_k \times (m_k \boldsymbol{d}_k \times \omega_k) = \boldsymbol{f}_k \tag{46b}$$

Combining Eqs. (46a)–(46b), one can obtain an expression equivalent to Eq. (12) as

$$\mathbf{M}_k \dot{\boldsymbol{t}}_k + \Omega_k \mathbf{M}_k \mathbf{E}_k \boldsymbol{t}_k = \boldsymbol{w}_k \tag{47a}$$

where the 6×6 matrices $\mathbf{M}_k$, Ω_k, and $\mathbf{E}_k$ are defined as

$$\mathbf{M}_k \equiv \begin{bmatrix} \mathbf{I}_k & m_k \boldsymbol{d}_k \times 1 \\ -m_k \boldsymbol{d}_k \times 1 & m_k 1 \end{bmatrix}, \ \boldsymbol{\Omega}_k \equiv \begin{bmatrix} \omega_k \times 1 & \mathbf{O} \\ \mathbf{O} & \omega_k \times 1 \end{bmatrix},$$

$$\text{and } \mathbf{E}_k \equiv \begin{bmatrix} \mathbf{1} & \mathbf{O} \\ \mathbf{O} & \mathbf{O} \end{bmatrix} \tag{47b}$$

Note in Eq. (47b), that $\mathbf{I}_k$ is the 3×3 mass moment of inertia tensor of the k-th link about O_k. Combining Eq. (47a) for all η^i links of the i-th module and for all s modules, one can write a compact expression equivalent to Eq. (13) as (Shah et al., 2011, 2013)

$$\mathbf{M}\dot{\boldsymbol{t}} + \boldsymbol{\Omega}\mathbf{M}\mathbf{E}\boldsymbol{t} = \boldsymbol{w} \tag{48}$$

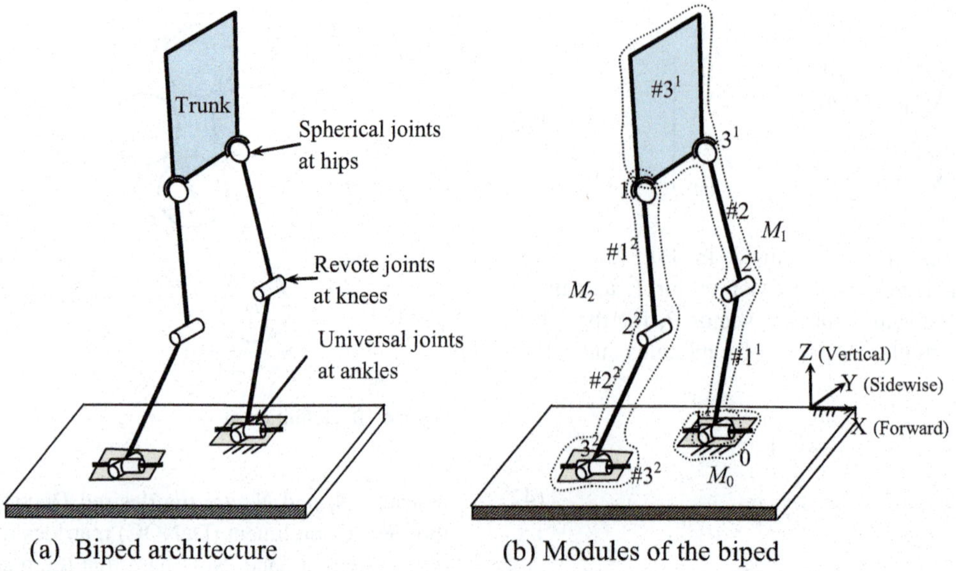

Figure 11. A 7-link spatial biped.

where matrices $\mathbf{M}$, $\mathbf{\Omega}$, and $\mathbf{E}$ are the $6(n+1)\times 6(n+1)$ block-diagonal matrices defined similar to Eq. (14). For details, readers are referred to the Ph.D. thesis of Shah (2011) or the book by Shah et al. (2013).

3.4 Constrained equations for tree-type systems using the DeNOC matrices

The constrained equations of motion for the tree-type systems are derived in this subsection in a similar manner to that of the serial-chain system of Sect. 2, i.e., pre-multiply $\mathbf{N}_d^T \mathbf{N}_l^T$ of Eq. (43) to the unconstrained NE equations given by Eq. (48) to obtain a set of constrained independent equations of motion by eliminating the constraint wrenches. These constrained equations are also referred to as the Euler-Lagrange equations of motion of the tree-type system at hand. They are given by

$$\mathbf{I}\ddot{\boldsymbol{\theta}} + \mathbf{C}\dot{\boldsymbol{\theta}} = \boldsymbol{\tau} \tag{49}$$

where $\mathbf{I}$ is generalized inertia matrix (GIM), $\mathbf{C}$ is the matrix of convective inertia terms (MCI), and $\boldsymbol{\tau}$ is the vector of generalized driving forces, and due to gravity, dissipation, external forces, etc., which have expressions similar to those after Eq. (17).

Note that the expression of Eq. (49) is same as Eq. (17) but the sizes of the corresponding matrices and vectors are different because they represent two different architectures of the multibody systems. Based on Eq. (49), recursive inverse and forward dynamics algorithms for tree-type systems were developed by Shah (2011) and implemented in a software called ReDySim (Recursieve Dynamics Simulator) (Shah et al., 2012c). ReDySim was written in MATLAB environment and available free from http://www.redysim.co.nr.

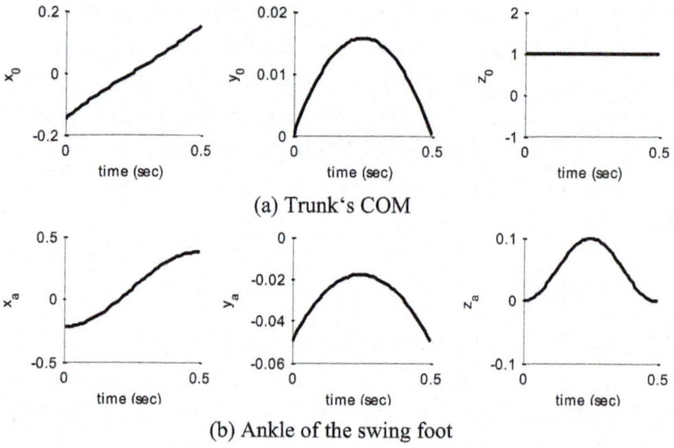

Figure 12. Designed trajectories of the trunk's center of mass (COM) and ankle (Shah, 2011; Shah et al., 2013).

3.5 Numerical example: a spatial biped

In order to illustrate the recursive dynamics algorithms presented in this section, ReDySim was used to analyze a spatial biped shown in Fig. 11. The model parameters were taken from Shah (2011) which will appear in the book by Shah et al. (2013) also. They are not reproduced here due to the reasons cited in Sect. 2.8. However, the designed input motions of the trunk's centre-of-mass (COM) and ankle for stable walking (Shah, 2011) are shown in Figs. 12 and 13, respectively. Based on the inputs of Figs. 12 and 13, the inverse dynamics results were obtained which are shown in Fig. 14.

Forced simulation was performed next, as reported in Shah (2011), where the motion of the biped was studied under the application of joint torques calculated above, i.e.,

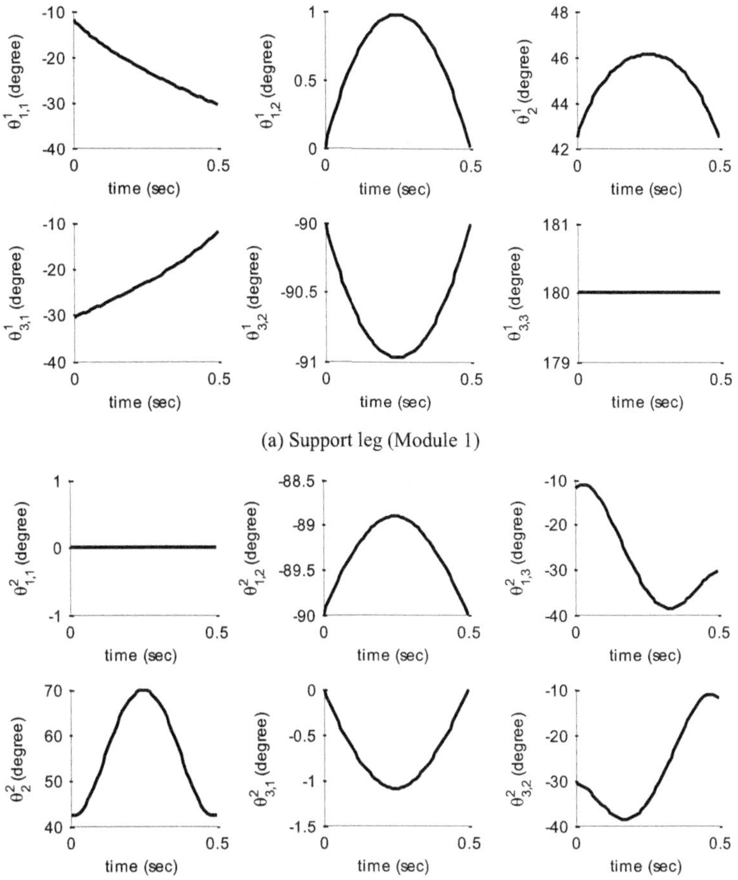

Figure 13. Joint trajectories of the biped obtained from the trajectories of trunk and ankle (Shah, 2011; Shah et al., 2013).

those shown in Fig. 14. The joint motions were calculated using the forward dynamics module of ReDySim. The plots for the simulated joint angles are shown in Fig. 15, along with the desired one. It can be seen that the simulated joint angles match with the desired joint angles up to 0.1 s, i.e., until 0.1 s movement of the biped. After this, the system behaves unexpectedly as evident from the divergent plots of the simulated angles in Fig. 15a. The deviation in the simulated angles is mainly attributed to what is known as zero eigen-value effect (Saha and Schiehlen, 2001). The physical system may also not behave as expected due to disturbances caused by unmodelled parameters like friction, backlash, etc., and non-exact geometrical and inertia parameters. Hence, a control scheme must be considered, as this forms a part and parcel of achieving proper walking. These aspects were explained in detail in Shah (2011) and Shah et al. (2013), and not elaborated further due to space limitation of the paper.

Note several advantages of the concept of the kinematic modules in the dynamics modelling of tree-type systems consisting of serially connected links (Shah, 2011; Shah et al., 2013), which are as follows:

- Extension of the body-to-body velocity transformation relationship to module-to-module velocity transformation relationship.
- Compact representation of the system's kinematic and dynamic models.
- Uniform development of the inverse and forward dynamics algorithms with inter- and intra-modular recursions.
- Module-level analytical expressions of the matrices and vectors appearing in the equations of motion.
- Ease of investigation of any inconsistency in the results of modules without the need to investigate the whole system.
- Possibility of hybrid recursive-parallel algorithms, where each module can be analyzed using recursive relations in parallel.

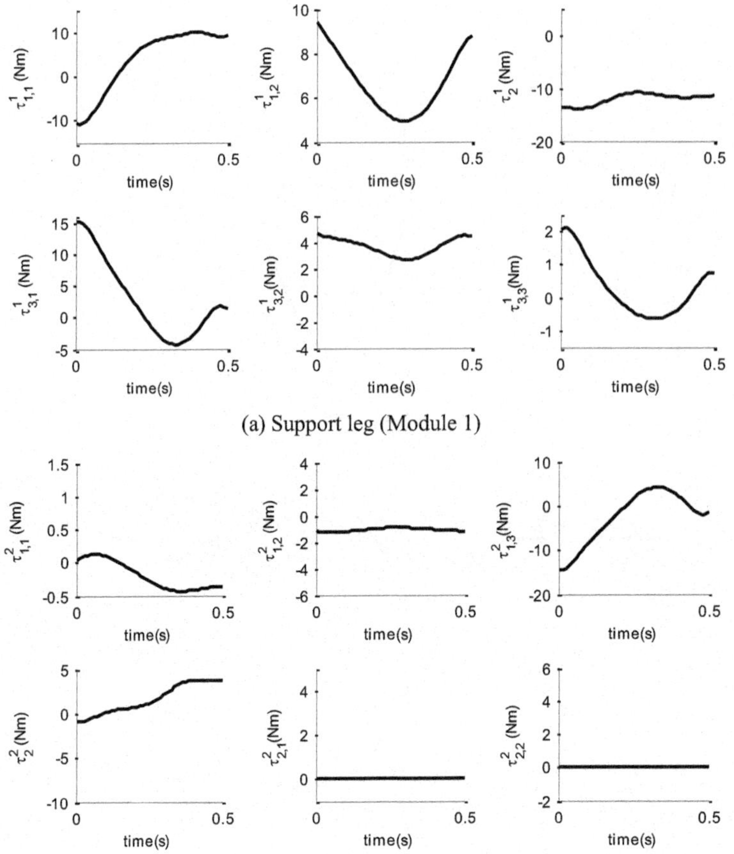

(a) Support leg (Module 1)

(b) Swing leg (Module 2)

Figure 14. Torques at different joints of the biped (Shah, 2011; Shah et al., 2013).

4 Closed-loop systems

The DeNOC-based dynamic modelling of serial-chain and tree-type open-loop systems presented in Sects. 2 and 3, respectively, can be extended to closed-loop systems provided one cuts the closed-loops of a system at suitable locations to make it open. Note that, one needs to use suitable constraint forces at the cut-joints to represent the actual presence of the joints. Such constraint forces are known in the literature as Lagrange multipliers (Chaudhary and Saha, 2007, 2009, and others). The multipliers need to be evaluated from the loop-closure constraints before they can be used as external forces to the resulting open-loop systems. In this section, a planar 4-bar mechanism shown in Fig. 16 is considered to illustrate the concept. However, the methodology is applicable to any general multi closed-loop systems, as shown in Chaudhary and Saha (2007), Shah (2011), and Shah et al. (2013).

Note that to model an open-loop system resulting from a closed-loop system, one needs to re-write Eqs. (13) or (48) as

$$\mathbf{N}^T(\mathbf{M}\dot{\boldsymbol{t}} + \boldsymbol{\Omega}\mathbf{M}\mathbf{E}\boldsymbol{t}) = \mathbf{N}^T(\boldsymbol{w}^{\mathrm{E}} + \boldsymbol{w}^{\lambda} + \boldsymbol{w}^{\mathrm{C}}) \tag{50}$$

where $\boldsymbol{w}^{\lambda}$ is the $6n$-dimensional vector of generalized wrench due to Lagrange multipliers acting at the cut joints. For the 4-bar mechanism shown in Fig. 16a, the two cut-open serial-chain subsystems are shown in Fig. 16b. The resulting open-loop tree-type subsystems have one and two links, respectively, connected by one and two one-DOF revolute joints. Other terms have same meaning as in Sects. 2 and 3. In Eq. (50), $\mathbf{N}^T\boldsymbol{w}^{\mathrm{C}} = 0$ for the reason given after Eq. (16), but $\mathbf{N}^T\boldsymbol{w}^{\lambda} \neq 0$. These terms are now the new unknowns to the inverse and forward dynamics problems that need to be evaluated with the help of loop-closure constraints.

For the closed-loop 1-2-3-4 of the 4-bar mechanism shown in Fig. 16a, one can write

$$\boldsymbol{a}_0 + \boldsymbol{a}_1 = \boldsymbol{a}_2 + \boldsymbol{a}_3 \tag{51}$$

where 2-dimensional vectors of the planar system, $\boldsymbol{a}_i$ for $i = 0, 1, 2, 3$, represent the relative position vectors of the joints in the 4-bar mechanism, Fig. 16a. Differentiating Eq. (51) with respect to time, one obtains

$$\mathbf{J}\dot{\boldsymbol{\theta}} = 0, \text{ where } \dot{\boldsymbol{\theta}} \equiv \begin{bmatrix} \dot{\theta}_1 & \dot{\theta}_2 & \dot{\theta}_3 \end{bmatrix}^T \tag{52a}$$

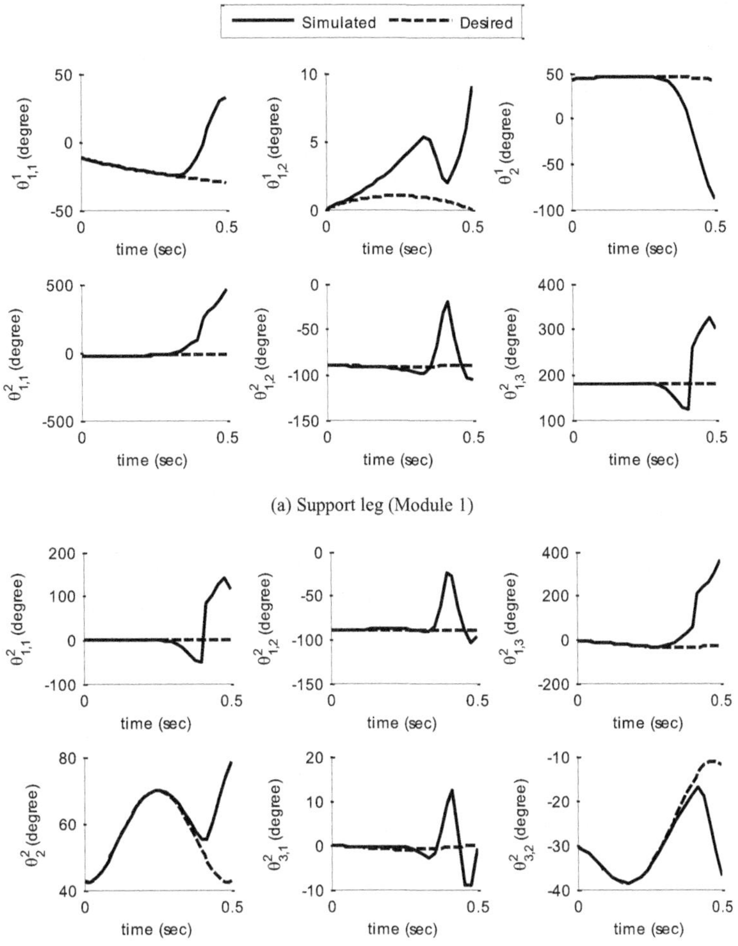

(a) Support leg (Module 1)

(b) Swing leg (Module 2)

Figure 15. Simulated joint angles for the biped (Shah, 2011; Shah et al., 2013).

and the 2×3 Jacobian matrix for the 4-bar mechanism at hand can be given by

$$\mathbf{J} \equiv \begin{bmatrix} -a_1 s_1 - a_2 s_{12} + a_3 s_{123} & -a_2 s_{12} + a_3 s_{123} & -a_3 s_{123} \\ a_1 c_1 + a_2 c_{12} - a_3 c_{123} & a_2 c_{12} + a_3 c_{123} & a_3 c_{123} \end{bmatrix} \tag{52b}$$

where $s_{12} \equiv \sin(\theta_1 + \theta_2)$, $s_{123} \equiv \sin(\theta_1 + \theta_2 + \theta_3)$, and similarly c_{12} and c_{123} etc. Equation (52b) was also derived in Chaudhary et al. (2007, 2009) as

$$\mathbf{J} \equiv \begin{bmatrix} \mathbf{J}_{\mathrm{I}} \\ \mathbf{J}_{\mathrm{II}} \end{bmatrix}, \text{ where } \mathbf{J}_{\mathrm{I}} \equiv \mathbf{A}_{\mathrm{en}}^{\mathrm{I}} \mathbf{N}^{\mathrm{I}} \text{ and } \mathbf{J}_{\mathrm{II}} \equiv \mathbf{A}_{\mathrm{en}}^{\mathrm{II}} \mathbf{N}^{\mathrm{II}} \tag{53}$$

In Eq. (53), $\mathbf{N}^{\mathrm{I}}$ and $\mathbf{N}^{\mathrm{II}}$ are the 6×1 and 12×2 NOC matrices for the two open-chain subsystems I and II, respectively, shown in Fig. 16b, whereas $\mathbf{A}_{\mathrm{en}}^{\mathrm{I}}$ and $\mathbf{A}_{\mathrm{en}}^{\mathrm{II}}$ are the 2×6 and 2×12 twist-propagation matrices for the last link from the point of contact to its previous link to the point where the joint is cut, i.e., Joint 4 of Fig. 16b. Such Jacobians using the DeNOC matrices were also derived in Saha (2008). The resulting constrained dynamic equations of motion for the two subsystems are then written as

$$\mathbf{I}^{\mathrm{I}} \ddot{\theta}^{\mathrm{I}} + \mathbf{C}^{\mathrm{I}} \dot{\theta}^{\mathrm{I}} = \boldsymbol{\tau}^{\mathrm{I}} + (\boldsymbol{\tau}^{\lambda})^{\mathrm{I}} \tag{54a}$$

$$\mathbf{I}^{\mathrm{II}} \ddot{\theta}^{\mathrm{II}} + \mathbf{C}^{\mathrm{II}} \dot{\theta}^{\mathrm{II}} = \boldsymbol{\tau}^{\mathrm{II}} + (\boldsymbol{\tau}^{\lambda})^{\mathrm{II}} \tag{54b}$$

Depending on the type of dynamics problem, i.e., inverse or forward, Eqs. (54a)–(54b) can be solved using recursive "subsystem" or "system" approach. For inverse dynamics, "subsystem recursion" provides a better efficiency, as pointed out by Chaudhary and Saha (2009).

4.1 Numerical example: a 4-bar mechanism

For the numerical results, ReDySim software mentioned in Sect. 3 was used using the lengths of crank (#1), output link (#2), coupler (#3) and fixed-base (#0) as 0.038 m, 0.1152 m, 0.1152 m and 0.0895 m, respectively. The masses

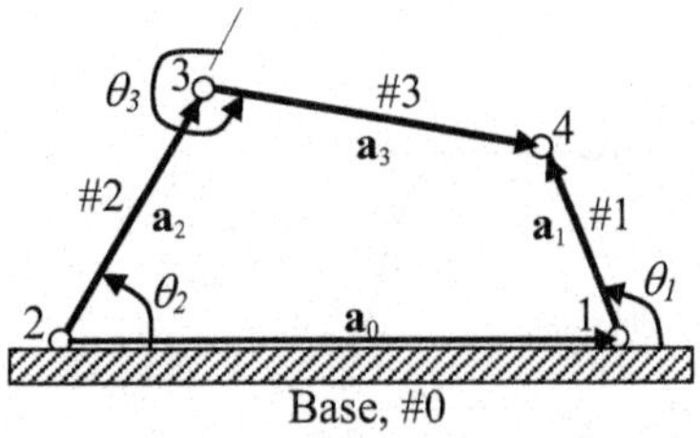

(a) Closed-loop system

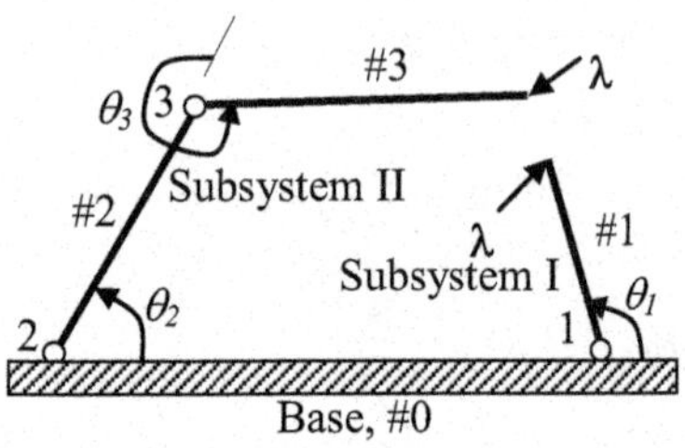

(b) Two equivalent open-loop systems

Figure 16. A 4-bar mechanism.

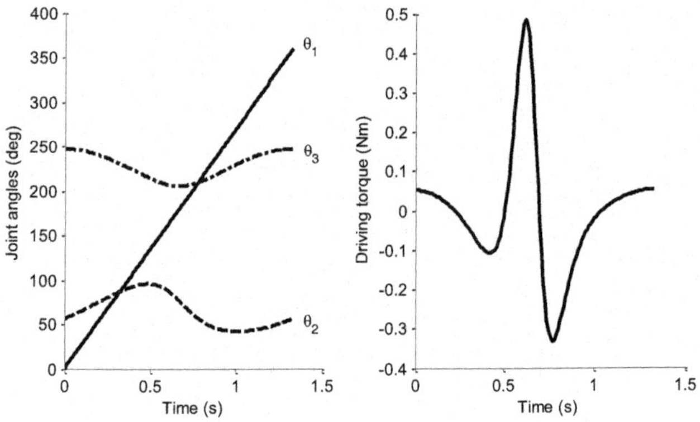

(a) Joint angles (b) Joint torques

Figure 17. Inverse dynamics for a 4-bar mechanism.

of the crank, output link and coupler were taken as 1.5 kgs, 3 kgs and 5 kgs, respectively. The input joint angle and the joint torque at joint 1 are plotted in Fig. 17. The forward dynamics of the 4-bar mechanism was carried out using the same initial configuration, as specified in the inverse dynamics. The simulation was done for the free-fall of the mechanism under gravity without any external torque applied at joint 1. The joint angles and rates are plotted in Fig. 18. The results of inverse dynamics and forward dynamics were validated with MATLAB's SimMechanics model, as reported in Shah et al. (2012c).

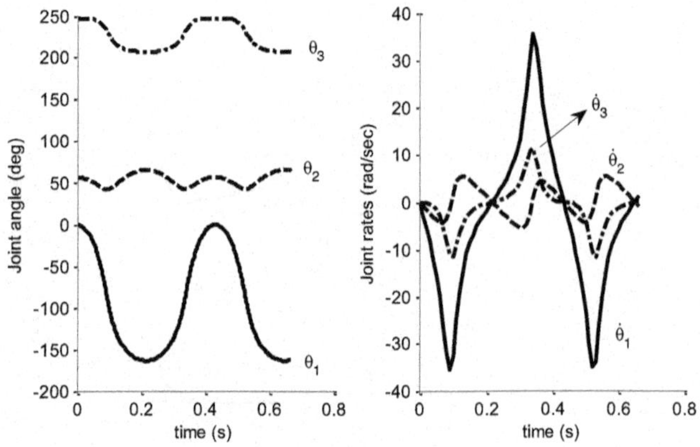

(a) Joint angles (b) Joint rates

Figure 18. Forward dynamics and simulation results for the 4-bar mechanism.

5 Computational efficiency

In this section, computational efficiencies of the DeNOC-based algorithms are investigated. Figure 18 shows comparisons of computational complexities required by several inverse dynamics algorithms, when a system has 1-DOF (Fig. 19a), 2-DOF (Fig. 19b), 3-DOF (Fig. 19c) and equal numbers of 1-2- and 3-DOF joints (Fig. 19d). It may be seen that the recursive inverse dynamics algorithm given in Fig. 9 of Sect. 3 performs as fast as the fastest algorithm available in the literature when the system has only 1-DOF joints, as evident from Fig. 19a. However, when multiple-DOF joints are introduced in the system, the algorithms of Sect. 3 (Shah, 2011), which have been implemented in ReDySim, outperforms the other algorithms available in the literature. This is clear from Fig. 19b–d.

From Fig. 20, it is also clear that the forward dynamics algorithm of Fig. 9 explained in Sect. 3 (Shah, 2011) performs better than any other algorithm available in the literature. More the number of multiple-DOF joints, more the improvement in the computational efficiency, as shown in Fig. 20b–d. This is mainly due the implicit inversion of the GIM based on the Reverse Gaussian Elimination (Saha, 1995, 1997) of the GIM, and simplification of the expressions associated with multiple-DOF joints (Shah et al., 2012b). In the tree-type robotic systems, such as biped, quadruped, etc., where the DOF of the system is more than 30, and the system consists of many multiple-DOF joints, the DeNOC-based algorithms significantly improve the computational efficiency.

6 Software for students and researchers

In order to build up the interest in the areas of multibody dynamics and to provide efficient tools to perform the dynamic analyses, the following multibody simulation tools

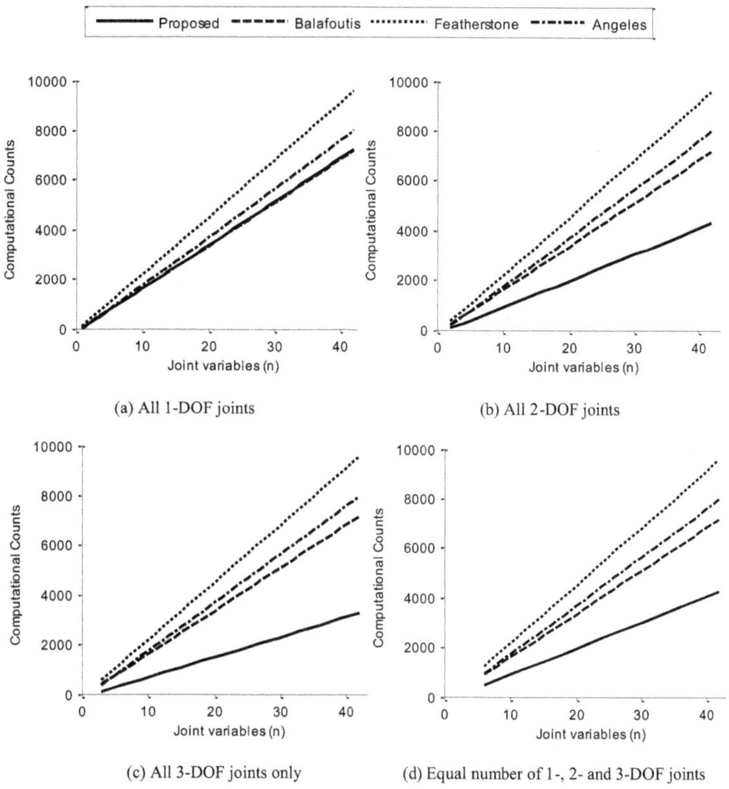

Figure 19. Performance of several inverse dynamics algorithms (Shah, 2011; Shah et al., 2013).

were developed for the students and researchers using the DeNOC-based formulations presented in Sects. 2–4:

- RoboAnalyzer: for serial-chain open-loop systems
- ReDySim (Recursive Dynamic Simulator): for general tree-type systems

These are explained briefly in the following subsections.

6.1 RoboAnalyzer

RoboAnalyzer (Rajeevlochana et al., 2012) is a 3-dimensional model-based software to solve kinematics and dynamics problems of an serial-chain open-loop system. It was developed using Visual C# and OpenGL that take the description of a serial-chain system using the DH parameters (Denavit and Hartenberg, 1955; Saha, 2008), and the mass and inertia properties of each link. In RoboAnalyzer, one can also see the animation of the analyzed systems. For the benefit of the users, CAD models of the standard systems like KUKA, PUMA robots, and others were made available for analysis. The software is freely downloadable from http://www.roboanalyzer.com.

6.2 ReDySim (Recursive Dynamics Simulator)

Recursive Dynamic Simulator (ReDySim) (Shah et al., 2012c) is a multibody dynamics simulation tool which was developed in MATLAB environment. It was developed based on the concept of the DeNOC matrices, and kinematic modules of a tree-type system, as explained in Sect. 3. ReDySim can be used to perform inverse dynamics and simulation of multibody systems. It has two modules, namely, fixed-base and floating-base modules. The latter was not presented in this paper due to limited page restriction of a paper. However, the interested readers can refer to Shah (2011) or Shah et al. (2013) for the dynamics analyses of biped, quadruped and six-legged walking robots using the floating-base concept. Note here that the architectural information of the tree-type systems were provided using the modified-DH parameters (MDH) parameters, as proposed by Khalil and Kleinfinger (1986), instead of the DH parameters defined in Saha (2008). This was done mainly to improve the computational efficiency of the tree-type systems.

ReDySim was also used to solve flexible systems like ropes, etc. which were modelled as hyper-degrees-of-freedom rigid-link systems (Shah, 2011). In fact, the simulation of long chains with the aid of ReDySim showed considerable improvement over commercial software like RecurDyn in terms of the computational time and correctness of

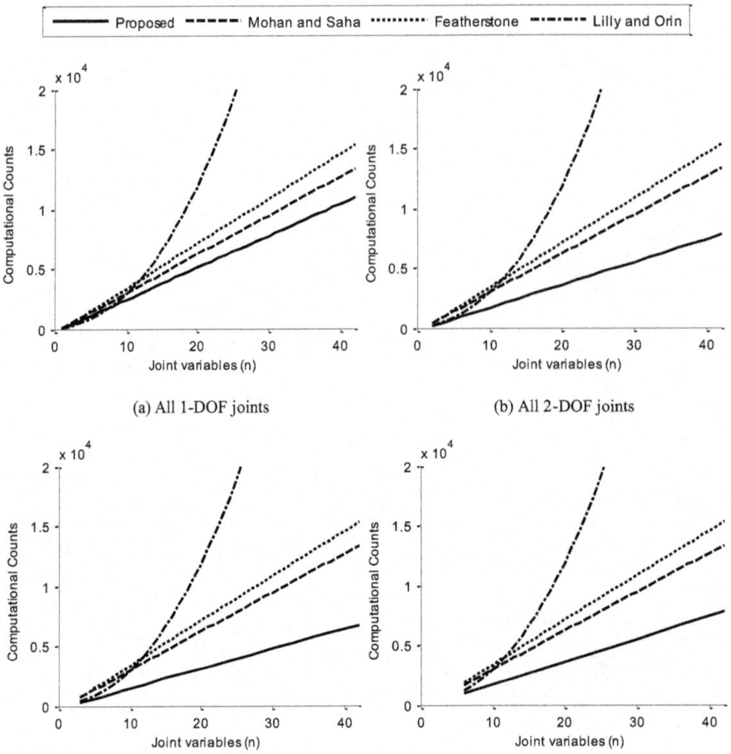

(a) All 1-DOF joints (b) All 2-DOF joints

(c) All 3-DOF joints only (d) Equal number of 1-, 2- and 3-DOF joints

Figure 20. Performance of forward dynamics algorithms (Shah, 2011; Shah et al., 2013).

the results. These results were separately communicated to a journal for publication (Agarwal et al., 2012). ReDySim can be downloaded free from http://www.redysim.co.nr, where the user's manual and some demos are also made available for the benefits of the users.

7 Conclusions

Recursive algorithms are popular due to their efficiency, computational uniformity, and numerical stability. This paper gave insight to the evolution of the Decoupled Natural Orthogonal Complement (DeNOC) matrices for more than one and half decades. It was shown that the use of DeNOC matrices in the dynamic analyses of multibody systems led to the development of several recursive dynamics algorithms for serial-chain, tree-type, and closed-loop systems. Several numerical examples, e.g., the Stanford arm (serial-chain), spatial biped (tree-type), and 4-bar mechanism (closed-loop), were given to explain the concepts presented in this paper. Advantages and computational efficiencies of the DeNOC-based methodologies suggest that it should be used when the DOF of a system is large and the application is real-time. Finally, two types of software meant for serial-chain open-loop systems, and general tree-type systems with abilities to solve closed-loop systems, i.e., RoboAnalyzer and ReDySim, respectively, were explained. It is expected that the algorithms, and more importantly, the software will benefit immensely the students and researchers of multibody dynamics community.

Acknowledgements. The authors would like to thank several ex-students of the Mechatronics Laboratory in the Dept. of Mech. Eng. at IIT Delhi for their contributions in developing several DeNOC-based algorithms. Special thanks are due to two graduate students of the first author, namely, Himanshu Chaudhary, and Amit Jain for the development of dynamics algorithms for closed-loop systems, and implementation of the serial-chain algorithm in RoboAnlyzer software, respectively.

Edited by: A. Müller

References

Agarwal, A., Shah, S. V., Bandyopadhyay, S., and Saha, S. K.: Dynamics of serial chains with large degrees-of-freedom, J. Multibody System Dynamics, under review, 2012.

Angeles, J. and Lee, S.: The formulation of dynamical equations of holonomic mechanical systems using a natural orthogonal complement, ASME J. Appl. Mech., 55, 243–244, 1988.

Angeles, J. and Ma, O.: Dynamic simulation of n-axis serial robotic manipulators using a natural orthogonal complement, Int. J. Robot. Res., 7, 32–47, 1988.

Angeles, J., Ma, O., and Rojas, A.: An algorithm for the inverse dynamics of n-axis general manipulator using Kane's formulation of dynamical equations, Computers and Mathematics with Applications, 17, 1545–1561, 1989.

Bathe, K. J. and Wilson, E. L.: Numerical Methods in Finite Element Analysis, 1st Edn., Prentice-Hall, New Jersy, USA, 1976.

Blajer, W., Bestle, D., and Schiehlen, W.: An orthogonal complement matrix formulation for constrained multibody systems, ASME J. Mech. Design, 116, 423–428, 1994.

Cameron, J. M. and Book, W. J.: Modeling mechanisms with nonholonomic joints using the Boltzmann-Hammel equations, Int. J. Robot. Res., 16, 47–59, 1997.

Chaudhary, H. and Saha, S. K.: Constraint wrench formulation for closed-loop systems using two-level recursions, ASME J. Mech. Design, 129, 1234–1242, 2007.

Chaudhary, H. and Saha, S. K.: Dynamics and Balancing of Multibody Systems, Springer, 2009.

Cyril, X.: Dynamics of Flexible Link Manipulators, Ph.D. thesis, Dept. of Mech. Eng., McGill University, Canada, 1988.

Denavit, J. and Hartenberg, R. S.: A kinematic notation for lower-pair mechanisms based on matrices, ASME J. Appl. Mech., 77, 215–221, 1955.

Dimitrov, D.: Dynamics and Control of Space Manipulators During a Satellite Capturing Operation, Ph.D. thesis, Graduate School of Engineering, Tohoku University, Japan, 2005.

Eberhard, P. and Schiehlen, W.: Computational dynamics of multibody systems: History, formalisms, and applications, ASME J. Comput. Nonlin. Dyn., 1, 3–12, 2006.

Featherstone, R.: Robot Dynamics Algorithms, Kluwer Academic Publishers, 1987.

Garcia de Jalon, J., Alvarez, E., de Ribera, F. A., Rodriguez, I., and Funes, F. J.: A fast and simple semi-recursive formulation for multi-rigid-body systems, in: Advances in Computational Multibody Systems, edited by: Ambrosio, J. A. C., Springer, 1–23, 2005.

Greenwood, D. T.: Principles of Dynamics, Prentice-Hall of India, New Delhi, 1988.

Hanzaki, A. R., Saha, S. K., and Rao, P. V. M.: An improved dynamic modeling of a multibody system with spherical joints, Multibody Syst. Dyn., 21, 325–345, 2009.

Hemami, H. and Weimer, F. C.: Modeling of nonholonomic dynamic systems with applications, ASME J. Appl. Mech., 48, 177–182, 1981.

Huston, R. L. and Passerello, C. E.: On constraint equations – A new approach, ASME J. Appl. Mech., 41, 1130–1131, 1974.

Kamman, J. W. and Huston, R. L.: Constrained multibody system dynamics: An automated approach, Comput. Struct., 18, 999–1003, 1984.

Kane, T. R. and Levinson, D. A.: The use of Kane's dynamical equations for robotics. Int. J. Robot. Res., 2, 3–21, 1983.

Khan, W. A., Krovi, V. N., Saha, S. K., and Angeles, J.: Recursive kinematics and inverse dynamics for a planar 3R parallel manipulator, J. Dyn. Syst.-T. ASME, 127, 529–536, 2005.

Khalil, W. and Kleinfinger, J.: A new geometric notation for open and closed-loop robots, Proc. of the IEEE Int. Conf. on Robotics and Automation, 3, 1174–1179, 1986.

Khatib, O.: Unified approach for motion and force control of robot manipulators: The operational space formulation, IEEE J. Robotics and Automation, RA-3, 43–53, 1987.

Kim, S. S. and Vanderploeg, M. J.: A general and efficient method for dynamic analysis of mechanical systems using velocity transformations, J. Mech. Transm.-T. ASME, 108, 176–182, 1986.

Mani, N. K., Haug, E. J., and Atkinson, K. E.: Application of singular value decomposition for analysis of mechanical system dynamics, J. Mech. Transm.-T. ASME, 107, 82–87, 1985.

McPhee, J. J.: On the use of linear graph theory in multibody system dynamics, Nonlinear Dynam., 9, 73–90, 1996.

Mohan, A. and Saha, S. K.: A recursive, numerically stable, and efficient algorithm for serial robots, Multibody Syst. Dyn., 17, 291–319, 2007.

Park, F. C., Bobrow, J. E., and Ploen, S. R.: A Lie group formulation of robot dynamics, Int. J. Robot. Res., 14, 606–618, 1995.

Patriciu, A., Chirikjian, S. G., and Pappub, R. V.: Analysis of the conformational dependence of mass-metric tensor determinants in serial polymers with constraints, J. Chem. Phys., 121, 12708, doi:10.1063/1.1821492, 2004.

Rajeevlochana, C. G. and Saha, S. K.: RoboAnalyzer: 3D model based robotic learning software, Proc. of the Int. Conf. on Multibody Dynamics, Vijayawada, India, 24–26 February, 3–13, 2011.

Rajeevlochana, C. G., Jain, A., Shah, S. V., and Saha, S. K.: Recursive robot dynamics in RoboAnalyzer, in: Machines and Mechanisms (Proc. of the 15th Nat. Conf. on Machines and Mechanisms), edited by: Bandopadhyay, S., Gurunathan, S. K., and Ramu, P., Narosa Publishing House, New Delhi, 482–490, ISBN: 978-81-8487-192-0, 2012.

Roberson, R. E. and Schwertassek R.: Dynamics of Multibody Systems, Springer, Berlin, 1988.

Rodriguez, G., Jain, A., and Kreutz-Delgado, K.: Spatial operator algebra for multibody system dynamics, J. Astronaut. Sci., 40, 27–50, 1992.

Saha, S. K.: The $\mathbf{UDU}^T$ decomposition of manipulator inertia matrix, Proc. of the IEEE Int. Conf. on Robotics and Automation, Nagoya, Japan, 21–27 May, 3, 2829–2834, 1995.

Saha, S. K.: A decomposition of the manipulator inertia matrix, IEEE Trans. on Robotics and Automation, 13, 301–304, 1997.

Saha, S. K.: Dynamics of serial multibody systems using the decoupled natural orthogonal complement matrices, ASME J. Appl. Mech., 66, 986–996, 1999a.

Saha, S. K.: Analytical expression for the inverted inertia matrix of serial robots, Int. J. Robot. Res., 18, 116–124, 1999b.

Saha, S. K.: Simulation of industrial manipulators based on the $\mathbf{UDU}^T$ decomposition of inertia matrix, Multibody Syst. Dyn., 9, 63–85, 2003.

Saha, S. K.: Introduction to Robotics, Tata McGraw-Hill, New Delhi, 2008.

Saha, S. K. and Angeles, J.: Dynamics of nonholonomic mechanical systems using a natural orthogonal complement, ASME J. Appl. Mech., 58, 238–243, 1991.

Saha, S. K. and Schiehlen, W. O.: Recursive kinematics and dynamics for closed loop multibody systems, Int. J. Mech. Structures Machines, 29, 143–175, 2001.

Saha, S. K., Shirinzadeh, B., and Alici, G.: Dynamic model simplification of serial manipulators, Proc. of the Int. Symp. on Robotics and Automation, San Miguel Regla Hotel, Hgo, Mexico, 25–28 August, 14–19, 2006.

Schiehlen, W.: Multibody Systems Handbook, Springer-Verlag, Berlin, 1990.
Schiehlen, W.: Multibody system dynamics: Roots and perspectives, Multibody Syst. Dyn., 1, 49–188, 1997.
Shabana, A. A.: Computational Dynamics, Wiley, New York, 2001.
Shah, S. V.: Modular Framework for Dynamic Modeling and Analyses of Tree-type Robotics Systems, Ph.D. thesis, Dept. of Mech. Eng., IIT Delhi, 2011.
Shah, S. V., Saha, S. K., and Dutt, J. K.: Modular framework for dynamics of tree-type legged robots, Mech. Mach. Theory, 49, 234–255, 2012a.
Shah, S. V., Saha, S. K., and Dutt, J. K.: Denavit-Hartenberg (DH) parametrization of Euler-angles. ASME J. Nonlinear and Computational Dynamics, 7, 021006, doi:10.1115/1.4005467, 2012b.
Shah, S. V., Nandihal, P. V., and Saha, S. K.: Recursive Dynamics Simulator (ReDySim): A multibody dynamics solver, Theor. Appl., 2, 063011, doi:10.1063/2.1206311, 2012c.
Shah, S. V., Saha, S. K., and Dutt, J. K.: Dynamics of Tree-type Robotics Systems, Springer, 2013.
Stokes, A. and Brockett, R.: Dynamics of kinematic chains, Int. J. Robot. Res., 15, 393–405, 1996.
Strang, G.: Linear Algebra and its Applications, Harcourt, Brace, Jovanovich, Publisher, Florida, 1998.
Sundarraman, P., Saha, S. K., Vasa, N. J., Baskaran, R., Sunilkumar, V., and Raghavendra, K.: Modeling and analysis of a fuel-injection pump used in diesel engines, Int. J. Automot. Techn., 13, 193–203, 2012.
Wehage, R. A. and Haug, E. J.: Generalized coordinate partitioning for dimension reduction in analysis of constrained dynamic systems, ASME J. Mech. Design, 104, 247–255, 1982.
Wittenburg J.: Dynamics of Multibody systems, Springer, Berlin, 2008.

2

Flexible joints in structural and multibody dynamics

O. A. Bauchau and S. Han

University of Michigan – Shanghai Jiao Tong University Joint Institute, Shanghai, 200240, China

Correspondence to: O. A. Bauchau (olivier.bauchau@sjtu.edu.cn)

Abstract. Flexible joints, sometimes called bushing elements or force elements, are found in all structural and multibody dynamics codes. In their simplest form, flexible joints simply consist of sets of three linear and three torsional springs placed between two nodes of the model. For infinitesimal deformations, the selection of the lumped spring constants is an easy task, which can be based on a numerical simulation of the joint or on experimental measurements. If the joint undergoes finite deformations, identification of its stiffness characteristics is not so simple, specially if the joint is itself a complex system. When finite deformations occur, the definition of deformation measures becomes a critical issue. This paper proposes a family of tensorial deformation measures suitable for elastic bodies of finite dimension. These families are generated by two parameters that can be used to modify the constitutive behavior of the joint, while maintaining the tensorial nature of the deformation measures. Numerical results demonstrate the objectivity of the deformations measures, a feature that is not shared by the deformations measures presently used in the literature. The impact of the choice of the two parameters on the constitutive behavior of the flexible joint is also investigated.

1 Introduction

Flexible joints, sometimes called bushing elements or force elements, are found in all multibody dynamics codes. In their simplest form, flexible joints simply consist of sets of three linear and three torsional springs placed between two nodes of a multibody system. For infinitesimal deformations, the selection of the lumped spring constants is an easy task, which can be based on a numerical simulation of the joint or on experimental measurements.

If the joint undergoes finite deformations, identification of its stiffness characteristics is not so simple, specially if the joint is itself a complex system. When finite deformations occur, the definition of the deformation measures becomes a critical issue. Indeed, for finite deformation, the observed nonlinear behavior of materials is partly due to material characteristics, and partly due to kinematics.

For instance, Anand (1979, 1986) has shown that the classical strain energy function for infinitesimal isotropic elasticity is in good agreement with experiment for a wide class of materials for moderately large deformations, provided the infinitesimal strain measure used in the strain energy function is replaced by the Hencky or logarithmic measure of finite strain. This means that the behavior of materials for moderate deformations can be captured accurately using linear constitutive laws, provided that the infinitesimal strain measures are replaced by finite deformation measures that are nonlinear functions of displacement.

These nonlinear deformation measures capture the observed nonlinear behavior associated with the nonlinear kinematics of the problem. Degener et al. (1988) also reported similar findings for the torsional behavior of beams subjected to large axial elongation.

Much attention has been devoted to the problem of synthesizing accurate constitutive properties for the modeling of flexible bushings presenting complex, time-dependent rheological behavior, Ledesma et al. (1996); Kadlowec et al. (2003). It is worth stressing, however, that the literature seldom addresses three-dimensional joint deformations.

Much like multibody codes, most FE codes also support the modeling of lumped structural elements. While linear analysis is easily implemented, problems are encountered when dealing with finite displacements and rotations, as pointed out by Masarati and Morandini (2010). Structural analysis codes, either specifically intended for multibody dynamics analysis, like MSC/ADAMS, or for nonlinear FEA with multibody capabilities, like Abaqus/Standard, allow arbitrarily large absolute displacements and rotations of the nodes and correctly describe their rigid-body motion. When lumped deformable joints are used, relative displacements and rotations are often required to remain moderate, although not necessarily infinitesimal.

Such restrictions occur when using the FIELD element of MSC/ADAMS, a linear element that implements an orthotropic torsional spring based on a constant, orthotropic constitutive matrix. Similarly, the JOINTC element implemented in Abaqus/Standard describes the interaction between two nodes when the second node can "displace and rotate slightly with respect to the first node", because its formulation is based on an approximate relative rotation measure.

The formulations and implementations of flexible joints available in research and commercial codes do not appear to allow arbitrarily large relative displacements and rotations. Moreover, in many cases, the ordering sequence of the nodes connected to the joint matters, because the behavior of the flexible joint is biased towards one of the nodes. This problem is known to experienced analysts using these codes. To the authors' knowledge, these facts are rarely acknowledged in the literature. It appears that little effort has been devoted to the elimination of these shortcomings from the formulations found in research and commercially available codes, although the predictions of these codes might be unexpected.

This paper presents families of finite deformation measures that can be used to characterize the deformation of flexible joints. These deformation measures are closely related to the tensorial parameterization motion developed by Bauchau and Li (2011); Bauchau (2011). Because they are of a tensorial nature, these deformation measures are intrinsic and invariant. Numerical examples demonstrate the invariance of the formulation and the ability to tailor the joint's constitutive behavior in the nonlinear range. Section 2 describes the configuration of the flexible joint. Section 3 presents the proposed deformation measures, which are derived from invariance considerations.

2 Flexible joint configuration

2.1 Kinematics of the flexible joint

Figure 1 shows inertial frame $\mathcal{F}^I = [\mathbf{O}, \mathcal{I} = (\bar{\imath}_1, \bar{\imath}_2, \bar{\imath}_3)]$ and a flexible joint in its reference and deformed configurations. It consists of a three-dimensional elastic body of finite dimension and of two rigid bodies, called handle k and handle ℓ, that are rigidly connected to the elastic body. In the reference configuration, the configuration of the handles is defined by frame $\mathcal{F}_0 = \left[\mathbf{B}, \mathcal{B}_0 = (\bar{b}_{01}, \bar{b}_{02}, \bar{b}_{03})\right]$, where $\mathcal{B}_0$ forms an orthonormal basis. The geometric location of points **K** and **L**, which are material points of handles k and ℓ, respectively, coincides with that of point **B**. The motion tensor that brings frame $\mathcal{F}^I$ to frame $\mathcal{F}_0$ is denoted $\underline{\underline{C}}_0(\underline{u}_0, \underline{\underline{R}}_0)$, where $\underline{u}_0$ is the

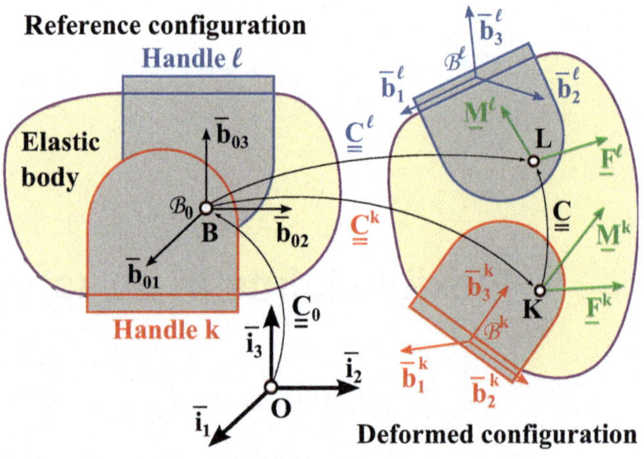

Figure 1. Configuration of the flexible joint.

position vector of point **B** with respect to point **O** and $\underline{\underline{R}}_0$ is the rotation tensor that brings basis $\mathcal{I}$ to basis $\mathcal{B}_0$.

In the deformed configuration, the two handles move to new positions and the elastic body deforms. Materials points **K** and **L** are now at distinct locations. The configurations of the two handles are now distinct and are represented by two distinct frames, $\mathcal{F}^k = \left[\mathbf{K}, \mathcal{B}^k = (\bar{b}_1^k, \bar{b}_2^k, \bar{b}_3^k)\right]$ and $\mathcal{F}^\ell = \left[\mathbf{L}, \mathcal{B}^\ell = (\bar{b}_1^\ell, \bar{b}_2^\ell, \bar{b}_3^\ell)\right]$, respectively. Motion tensors $\underline{\underline{C}}^k$ and $\underline{\underline{C}}^\ell$ bring frame $\mathcal{F}_0$ to frames $\mathcal{F}^k$ and $\mathcal{F}^\ell$, respectively. The displacement vectors from point **B** to points **K** and **L**, are denoted $\underline{u}^k$ and $\underline{u}^\ell$, respectively, and rotation tensors $\underline{\underline{R}}^k$ and $\underline{\underline{R}}^\ell$ bring basis $\mathcal{B}_0$ to bases $\mathcal{B}^k$ and $\mathcal{B}^\ell$, respectively.

Relative motion tensor $\underline{\underline{C}}$ brings frame $\mathcal{F}^k$ to frame $\mathcal{F}^\ell$ and provides an intrinsic representation of the motion of handle ℓ with respect to handle k. The relative displacement vector of point **L** with respect to point **K** is denoted $\underline{u} = \underline{u}^\ell - \underline{u}^k$ and $\underline{\underline{R}} = (\underline{\underline{R}}^\ell \underline{\underline{R}}_0)(\underline{\underline{R}}^k \underline{\underline{R}}_0)^T$ is the relative rotation tensor of basis $\mathcal{B}^\ell$ with respect to basis $\mathcal{B}^k$.

The motion tensors that bring frame $\mathcal{F}^I$ to frames $\mathcal{F}^k$ and $\mathcal{F}^\ell$, denoted $\underline{\underline{\mathbb{C}}}^k$ and $\underline{\underline{\mathbb{C}}}^\ell$, respectively can be expressed as

$$\underline{\underline{\mathbb{C}}}^k = \begin{bmatrix} (\underline{\underline{R}}^k \underline{\underline{R}}_0) & (\widetilde{u}_0 + \widetilde{u}^k)(\underline{\underline{R}}^k \underline{\underline{R}}_0) \\ \underline{\underline{0}} & (\underline{\underline{R}}^k \underline{\underline{R}}_0) \end{bmatrix}, \tag{1a}$$

$$\underline{\underline{\mathbb{C}}}^\ell = \begin{bmatrix} (\underline{\underline{R}}^\ell \underline{\underline{R}}_0) & (\widetilde{u}_0 + \widetilde{u}^\ell)(\underline{\underline{R}}^\ell \underline{\underline{R}}_0) \\ \underline{\underline{0}} & (\underline{\underline{R}}^\ell \underline{\underline{R}}_0) \end{bmatrix}, \tag{1b}$$

respectively. The relative motion tensor then becomes

$$\underline{\underline{C}} = \underline{\underline{\mathbb{C}}}^\ell \underline{\underline{\mathbb{C}}}^{k-1}. \tag{2}$$

The components of relative motion tensor resolved in frames $\mathcal{F}^k$ and $\mathcal{F}^\ell$ are identical,

$$\underline{\underline{C}}^* = \underline{\underline{\mathbb{C}}}^{k-1} \underline{\underline{C}}\, \underline{\underline{\mathbb{C}}}^k = \underline{\underline{\mathbb{C}}}^{\ell-1} \underline{\underline{C}}\, \underline{\underline{\mathbb{C}}}^\ell = \underline{\underline{\mathbb{C}}}^{k-1} \underline{\underline{\mathbb{C}}}^\ell = \begin{bmatrix} \underline{\underline{R}}^* & \widetilde{u}^* \underline{\underline{R}}^* \\ \underline{\underline{0}} & \underline{\underline{R}}^* \end{bmatrix}, \tag{3}$$

where $\underline{\underline{R}}^* = (\underline{\underline{R}}^k \underline{\underline{R}}_0)^T \underline{\underline{R}} (\underline{\underline{R}}^k \underline{\underline{R}}_0) = (\underline{\underline{R}}^\ell \underline{\underline{R}}_0)^T \underline{\underline{R}} (\underline{\underline{R}}^\ell \underline{\underline{R}}_0) = (\underline{\underline{R}}^k \underline{\underline{R}}_0)^T (\underline{\underline{R}}^\ell \underline{\underline{R}}_0)$ are the components of the relative rotation tensor resolved in bases $\mathcal{B}^k$ or $\mathcal{B}^\ell$ and $\underline{u}^* = (\underline{\underline{R}}^k \underline{\underline{R}}_0)^T (\underline{u}^\ell - \underline{u}^k)$.

2.2 Applied loading

The deformation of the flexible joint stems from the applied forces and moments depicted in Fig. 1. At point **K**, the applied force and moment vectors are denoted $\underline{F}^k$ and $\underline{M}^k$, respectively; the corresponding quantities applied at point **L** are denoted $\underline{F}^\ell$ and $\underline{M}^\ell$, respectively. The loading applied to the flexible joint is characterized by arrays $\underline{\mathcal{A}}^k$ and $\underline{\mathcal{A}}^\ell$ that correspond to a translation of these loads to point **O**,

$$\underline{\mathcal{A}}^k = \underline{\underline{\mathcal{T}}}^{k-T} \begin{Bmatrix} \underline{F}^k \\ \underline{M}^k \end{Bmatrix} = \begin{Bmatrix} \underline{F}^k \\ \underline{M}^k + (\widetilde{u}_0 + \widetilde{u}^k)\underline{F}^k \end{Bmatrix}, \tag{4a}$$

$$\underline{\mathcal{A}}^\ell = \underline{\underline{\mathcal{T}}}^{\ell-T} \begin{Bmatrix} \underline{F}^\ell \\ \underline{M}^\ell \end{Bmatrix} = \begin{Bmatrix} \underline{F}^\ell \\ \underline{M}^\ell + (\widetilde{u}_0 + \widetilde{u}^\ell)\underline{F}^\ell \end{Bmatrix}, \tag{4b}$$

where $\underline{\underline{\mathcal{T}}}^k$ and $\underline{\underline{\mathcal{T}}}^\ell$ are the translation tensors from point **O** to points **K** and **L**, respectively, defined as

$$\underline{\underline{\mathcal{T}}}^k = \begin{bmatrix} \underline{\underline{I}} & (\widetilde{u}_0 + \widetilde{u}^k) \\ \underline{\underline{0}} & \underline{\underline{I}} \end{bmatrix}, \tag{5a}$$

$$\underline{\underline{\mathcal{T}}}^\ell = \begin{bmatrix} \underline{\underline{I}} & (\widetilde{u}_0 + \widetilde{u}^\ell) \\ \underline{\underline{0}} & \underline{\underline{I}} \end{bmatrix}, \tag{5b}$$

respectively. Because these loads are both applied at point **O**, the equilibrium condition resulting from Newton's third law simply states

$$\underline{\mathcal{A}}^k + \underline{\mathcal{A}}^\ell = \underline{0}. \tag{6}$$

Note the parallel between vectors $\underline{\mathcal{A}}^k$ and $\underline{\mathcal{A}}^\ell$ and the second Piola-Kirchhoff stress tensor (Malvern, 1969). Indeed, they represent the true loads applied to handle k and ℓ, respectively, in their deformed configurations, but translated to reference point **O**. Although expressing equilibrium of the system in its deformed configuration, Eq. (6) is a linear function of the loads. The joint is assumed to be massless, i.e., inertial forces associated with its motion are neglected.

3 Deformation measures

The differential work done by the applied load will be evaluated in Sect. 3.1, and the nature of deformation measures is further discussed in Sect. 3.2. Section 3.3 then provides invariant deformation measures based on the tensorial parameterization of motion.

3.1 Differential work

The differential work, $\mathrm{d}W$, done by the forces applied to the joint is $\mathrm{d}W = \underline{F}^{kT} \mathrm{d}\underline{u}^k + \underline{M}^{kT} \underline{\mathrm{d}\psi}^k + \underline{F}^{\ell T} \mathrm{d}\underline{u}^\ell + \underline{M}^{\ell T} \underline{\mathrm{d}\psi}^\ell$, where $\mathrm{d}\underline{u}^k$ and $\mathrm{d}\underline{u}^\ell$ are the differential displacements of point **K** and **L**, respectively, and $\underline{\mathrm{d}\psi}^k = \mathrm{axial}(\mathrm{d}\underline{\underline{R}}^k \underline{\underline{R}}^{kT})$ and $\underline{\mathrm{d}\psi}^\ell = \mathrm{axial}(\mathrm{d}\underline{\underline{R}}^\ell \underline{\underline{R}}^{\ell T})$ the differential rotations of handles k and ℓ, respectively. Recasting this expression in a matrix form leads to

$$\begin{aligned} \mathrm{d}W &= \left\{\mathrm{d}\underline{u}^{kT}, \underline{\mathrm{d}\psi}^{kT}\right\} \underline{\underline{\mathcal{T}}}^{kT} \underline{\underline{\mathcal{T}}}^{k-T} \begin{Bmatrix} \underline{F}^k \\ \underline{M}^k \end{Bmatrix} \\ &\quad + \left\{\mathrm{d}\underline{u}^{\ell T}, \underline{\mathrm{d}\psi}^{\ell T}\right\} \underline{\underline{\mathcal{T}}}^{\ell T} \underline{\underline{\mathcal{T}}}^{\ell-T} \begin{Bmatrix} \underline{F}^\ell \\ \underline{M}^\ell \end{Bmatrix} \\ &= \underline{\mathrm{d}\mathcal{U}}^{kT} \underline{\mathcal{A}}^k + \underline{\mathrm{d}\mathcal{U}}^{\ell T} \underline{\mathcal{A}}^\ell, \end{aligned}$$

where the last equality follows from Eq. (4). The differential motions of handles k and ℓ are defined as $\underline{\mathrm{d}\mathcal{U}}^k = \mathcal{A}\mathrm{xial}(\mathrm{d}\underline{\underline{\mathbb{C}}}^k \underline{\underline{\mathbb{C}}}^{k-1})$ and $\underline{\mathrm{d}\mathcal{U}}^\ell = \mathcal{A}\mathrm{xial}(\mathrm{d}\underline{\underline{\mathbb{C}}}^\ell \underline{\underline{\mathbb{C}}}^{\ell-1})$, respectively, leading to

$$\underline{\mathrm{d}\mathcal{U}}^k = \underline{\underline{\mathcal{T}}}^k \begin{Bmatrix} \mathrm{d}\underline{u}^k \\ \underline{\mathrm{d}\psi}^k \end{Bmatrix}, \tag{7a}$$

$$\underline{\mathrm{d}\mathcal{U}}^\ell = \underline{\underline{\mathcal{T}}}^\ell \begin{Bmatrix} \mathrm{d}\underline{u}^\ell \\ \underline{\mathrm{d}\psi}^\ell \end{Bmatrix}, \tag{7b}$$

respectively, where operator $\mathcal{A}\mathrm{xial}(\cdot)$ is defined by Eq. (A3). Finally, introducing Eq. (6), the differential work becomes

$$\mathrm{d}W = \underline{\mathcal{A}}^{\ell T} \left(\underline{\mathrm{d}\mathcal{U}}^\ell - \underline{\mathrm{d}\mathcal{U}}^k\right).$$

The term in parenthesis represents the differential relative motion, $\underline{\mathrm{d}\mathcal{U}} = \underline{\mathrm{d}\mathcal{U}}^\ell - \underline{\mathrm{d}\mathcal{U}}^k$, where $\underline{\mathrm{d}\mathcal{U}} = \mathcal{A}\mathrm{xial}(\mathrm{d}\underline{\underline{C}}\,\underline{\underline{C}}^{-1})$. As expected, the differential work depends on the differential relative motion of the two handles only. The differential work can now be written as $\mathrm{d}W = \underline{\mathcal{A}}^{\ell T} \underline{\mathrm{d}\mathcal{U}} = \underline{\mathcal{A}}^{kT} (-\underline{\mathrm{d}\mathcal{U}})$, where the second equality follows from the equilibrium equation, Eq. (6). This statement simply implies that the differential relative motion of handle ℓ with respect to handle k, denoted $\underline{\mathrm{d}\mathcal{U}}$ is of opposite sign of that of handle k with respect to handle ℓ, as expected. In summary, the differential work is written

$$\mathrm{d}W = \underline{\mathcal{A}}^T \underline{\mathrm{d}\mathcal{U}}, \tag{8}$$

where by convention, $\underline{\mathcal{A}} = \underline{\mathcal{A}}^\ell$ and $\underline{\mathrm{d}\mathcal{U}} = \underline{\mathrm{d}\mathcal{U}}^\ell - \underline{\mathrm{d}\mathcal{U}}^k$ the differential relative motion of handle ℓ with respect to handle k.

Let $\underline{\mathcal{E}}$ be a set of six generalized coordinates that define the relative motion tensor uniquely, i.e., a one-to-one mapping is assumed to exist between these generalized coordinates and the relative motion tensor. It follows that a one-to-one mapping must exist between the relative motion tensor and increments of the generalized coordinates

$$\underline{\mathrm{d}\mathcal{U}} = \underline{\underline{\mathcal{H}}}(\underline{\mathcal{E}})\mathrm{d}\underline{\mathcal{E}}, \tag{9}$$

where tensor $\underline{\underline{\mathcal{H}}}(\underline{\mathcal{E}})$ is the Jacobian or tangent tensor of the coordinate transformation. The differential work done by the forces applied to the joint, Eq. (8), now becomes

$dW = \underline{\mathcal{A}}^T \underline{\underline{\mathcal{H}}}(\underline{\mathcal{E}}) d\underline{\mathcal{E}} = \underline{\mathcal{L}}^T d\underline{\mathcal{E}}$, where the generalized forces associated with the generalized coordinates are defined as

$$\underline{\mathcal{L}} = \underline{\underline{\mathcal{H}}}^T(\underline{\mathcal{E}}) \underline{\mathcal{A}}. \tag{10}$$

It is assumed that the flexible joint is made of an elastic material (Bauchau and Craig, 2009), which implies that the generalized forces can be derived from a potential, the strain energy of the joint, denoted A,

$$\underline{\mathcal{L}} = \frac{\partial A(\underline{\mathcal{E}})}{\partial \underline{\mathcal{E}}}. \tag{11}$$

The differential work now becomes

$$dW = d\underline{\mathcal{E}}^T \frac{\partial A(\underline{\mathcal{E}})}{\partial \underline{\mathcal{E}}} = d(A), \tag{12}$$

and can be expressed as the differential of a scalar function, the strain energy.

3.2 The deformation measures

In the previous section, quantities $\underline{\mathcal{E}}$ were defined as "a set of generalized coordinates that uniquely define the relative motion tensor," but were otherwise left undefined. This implies that these generalized coordinates form a parameterization of the relative motion tensor, i.e., $\underline{\underline{C}} = \underline{\underline{C}}(\underline{\mathcal{E}})$. Because the strain energy of the flexible joint can be expressed in terms of these generalized coordinates they are, in fact, deformation measures for the flexible joint. Consequently, any parameterization of the relative motion tensor provides deformation measures for the flexible joint.

The following notation is introduced

$$\underline{\mathcal{E}} = \begin{Bmatrix} \underline{\varepsilon} \\ \underline{\kappa} \end{Bmatrix}. \tag{13}$$

The first three components of this array form the *stretch vector*, denoted $\underline{\varepsilon}$, and the last three the *wryness vector*, denoted $\underline{\kappa}$. Although any parameterization of the relative motion tensor provides adequate deformation measures for the flexible joint, these deformation measures should be invariant with respect to the choice of coordinate system. This implies that the stretch and wryness vectors should be first-order tensors, and hence, the deformation measure should itself be a first-order tensor.

While the parameterization of rotation has received wide attention (Kane, 1968; Argyris, 1982; Shuster, 1993; Ibrahimbegović, 1997; Bauchau and Trainelli, 2003), much less emphasis has been placed on that of motion (Angeles, 1993; Borri et al., 2000; Merlini and Morandini, 2004; Pennestrì and Stefanelli, 2007). Bauchau and Choi (2003) and Bauchau and Li (2011) have proposed the vectorial parameterization of motion, which consists of a motion parameter vector, $\underline{\mathcal{E}}$, that parameterizes the motion tensor, $\underline{\underline{C}} = \underline{\underline{C}}(\underline{\mathcal{E}})$. Bauchau and Li (2011) and Bauchau (2011) have studied the parameterization of motion, with special emphasis on its tensorial nature. They presented a formal proof that motion parameters vectors are first-order tensors if and only if they are parallel to the eigenvectors of the motion tensor associated with its unit eigenvalues. Furthermore, the tangent tensor is also a second-order tensor for motion parameters vectors.

Deformation measures should be "objective" or "frame-indifferent", and Malvern (1969) gives a precise definition of this concept, which implies that deformation measures should be expressed in terms of tensorial quantities and be invariant under the superposition of a rigid body motion. The relative motion tensor remains invariant under the superposition of a rigid body motion and hence, deformation measures that are functions of the relative motion tensor only will share this property. The previous paragraphs have underlined the fact that the deformation measures should be of a tensorial nature. If this latter property is achieved, the resulting deformation measures will be objective.

In summary, the desired invariance and objectivity of the deformation measure are achieved if and only if this deformation measure is selected to be the vectorial parameterization of motion, which further implies that the deformation measure is parallel to the eigenvector of the motion tensor associated with its unit eigenvalue.

3.3 Explicit expression of the deformation measures

The proposed deformation measures are parallel to the eigenvector of the relative motion tensor associated with its unit eigenvalue. Because this eigenvalue has a multiplicity of two (Bauchau, 2011), two linearly independent eigenvectors exist, which will be selected as

$$\underline{\bar{\mathcal{N}}}_1^\dagger = \begin{Bmatrix} \underline{p} \\ \underline{0} \end{Bmatrix}, \quad \text{and} \quad \underline{\bar{\mathcal{N}}}_2^\dagger = \begin{Bmatrix} \underline{\underline{H}}^{-1}\underline{u} \\ \underline{p} \end{Bmatrix}, \tag{14}$$

where $\underline{p}$ is the Wiener-Milenković rotation parameter vector (Wiener, 1962; Milenković, 1982; Bauchau, 2011) and $\underline{\underline{H}}$ the tangent tensor for this parameterization. The second eigenvector is easily recognized as the Wiener-Milenković motion parameter vector (Bauchau, 2011), i.e.,

$$\underline{\bar{\mathcal{N}}}_2^\dagger = \underline{\mathcal{P}} = \begin{Bmatrix} \underline{q} \\ \underline{p} \end{Bmatrix}. \tag{15}$$

The following notation is introduced: $p = \|\underline{p}\| = 4\tan\phi/4$, $\varrho = \underline{p}^T\underline{q} = pp'd$, where $d = \bar{n}^T\underline{u}$ is the intrinsic displacement, and $\overline{(\cdot)}'$ denoted a derivative with respect to angle ϕ. Note the following limit behaviors: (1) $\lim_{\phi\to 0} p = \phi$, (2) $\lim_{\phi\to 0} \underline{\underline{H}}^{-1} = \underline{\underline{I}}$ and hence, $\lim_{\phi\to 0} \underline{q} = \underline{u}$, and (3) $\lim_{d\to 0} \varrho = 0$;

The most general measure of deformation is a linear combination of vectors $\underline{\bar{\mathcal{N}}}_1^\dagger$ and $\underline{\bar{\mathcal{N}}}_2^\dagger$ defined by Eq. (14), which will be written as

$$\underline{\mathcal{E}} = \begin{bmatrix} \mu \underline{\underline{I}} & \lambda \underline{\underline{I}} \\ \underline{\underline{0}} & \mu \underline{\underline{I}} \end{bmatrix} \underline{\mathcal{P}} = \underline{\underline{\mathcal{Z}}}(\lambda, \mu) \underline{\mathcal{P}}, \tag{16}$$

where matrix $\underline{\underline{\mathcal{Z}}}$ is defined in Eq. (A1), and λ and μ are arbitrary scalars. The stretch and wryness vectors now become $\underline{\varepsilon} = \mu \underline{q} + \lambda \underline{p}$ and $\underline{\kappa} = \mu \underline{p}$, respectively. Mozzi-Chasles' theorem implies that a general motion can be defined in terms of the direction of axis of the motion and in terms of two scalar parameters, ϕ, the magnitude of the relative rotation, and d, the intrinsic relative displacement. Equivalently, scalar functions λ and μ can be expressed in terms of $\varrho = \underline{p}^T \underline{q}$ and $p = \sqrt{\underline{p}^T \underline{p}}$, i.e.,

$$\lambda = \lambda(\varrho, p), \quad \mu = \mu(\varrho, p). \tag{17}$$

Although functions λ and μ are arbitrary, their limit behavior for ϱ and $p \rightarrow 0$ can be obtained based on physical arguments. The wryness vector is written as $\underline{\kappa} = \mu p\ \bar{n}$; since μp must be an odd function of ϕ (or p), it implies that μ is an even function of the same variable. For small values of ϕ (or p), the wryness vector should be equal to the infinitesimal rotation vector, i.e., $\lim_{p \to 0} \underline{\kappa} = \phi \bar{n}$. Introducing the expression for the wryness vector yields $\lim_{\phi \to 0} 4\mu \tan(\phi/4)\ \bar{n} = \lim_{\phi \to 0} \mu\phi\ \bar{n} = \phi \bar{n}$, which implies

$$\lim_{\phi \to 0} \mu = 1. \tag{18}$$

Similarly, for small rotation angles, the stretch vector should equal the displacement vector, i.e., $\lim_{\phi \to 0} \underline{\varepsilon} = \underline{u}$. Introducing the expression for the stretch vector yields $\lim_{\phi \to 0} (\mu \underline{\underline{H}}^{-1} \underline{u} + 4\lambda \tan(\phi/4)\ \bar{n}) = \underline{u}$. Given the limit behaviors of μ and $\underline{\underline{H}}^{-1}$, the stretch vector does indeed converge to the displacement vector. Finally, if the relative motion is planar, the intrinsic displacement vanishes and the stretch vector should be in the plane normal to unit vector $\bar{n}$, i.e., $\lim_{d \to 0} \bar{n}^T \underline{\varepsilon} = 0$. Using the expression for the stretch vector then leads to $\lim_{d \to 0} (\mu p' d + 4\lambda \tan\phi/4) = 0$, which implies

$$\lim_{d \to 0} \lambda = 0. \tag{19}$$

In summary, Eq. (16) defines the proposed deformation measures for flexible joints. These equations are, in fact, the nonlinear deformation-displacement relationships of the problem. They are not fully determined because two scalar functions, λ and μ, appear in their definition. These two functions must satisfy the limit behavior discussed earlier but are otherwise arbitrary. Specific choices of these two scalar functions result in different families of deformation measures.

4 Formulation of flexible joints

The strain energy of the flexible joint is assumed to be a quadratic function of the deformation measures $\underline{\mathcal{E}}^*$,

$$A(\underline{\mathcal{E}}^*) = \frac{1}{2} \underline{\mathcal{E}}^{*T} \underline{\underline{\mathcal{D}}}^* \underline{\mathcal{E}}^*, \tag{20}$$

where $\underline{\underline{\mathcal{D}}}^*$ are the components of the stiffness matrix of the flexible joint resolved in the material frame, which are assumed to be given constants in this frame.

4.1 Elastic forces in the flexible joint

The components of relative motion tensor resolved in the material frame are given by Eq. (3) and its variation becomes

$$\widetilde{\delta \mathcal{U}}^* = \delta \underline{\underline{C}}^* \underline{\underline{C}}^{*-1} = \left(\delta \underline{\underline{\mathbb{C}}}^{k-1} \underline{\underline{\mathbb{C}}}^{\ell} + \underline{\underline{\mathbb{C}}}^{k-1} \delta \underline{\underline{\mathbb{C}}}^{\ell} \right) \underline{\underline{\mathbb{C}}}^{\ell-1} \underline{\underline{\mathbb{C}}}^{k} = \underline{\underline{\mathbb{C}}}^{k-1} \left(\widetilde{\delta \mathcal{U}}^{\ell} - \widetilde{\delta \mathcal{U}}^{k} \right) \underline{\underline{\mathbb{C}}}^{k},$$

where the virtual motion vectors, $\underline{\delta \mathcal{U}}^k$ and $\underline{\delta \mathcal{U}}^\ell$, are defined in Eqs. (7a) and (7b), respectively. It now follows that

$$\underline{\delta \mathcal{U}}^* = \underline{\underline{\mathbb{C}}}^{k-1} \left(\underline{\delta \mathcal{U}}^\ell - \underline{\delta \mathcal{U}}^k \right). \tag{21}$$

The virtual motion vector is related to the virtual changes in the motion parameter vector (Bauchau, 2011) by means of the tangent tensor, $\underline{\underline{\mathcal{H}}}$, as $\underline{\delta \mathcal{U}}^* = \underline{\underline{\mathcal{H}}}(\underline{\mathcal{P}}^*) \delta \underline{\mathcal{P}}^*$. It then follows that $\delta \underline{\mathcal{P}}^* = \underline{\underline{\mathcal{H}}}^{-1}(\underline{\mathcal{P}}^*) \underline{\delta \mathcal{U}}^* = \underline{\underline{\mathcal{H}}}^{-1}(\underline{\mathcal{P}}^k) \underline{\underline{\mathbb{C}}}^{k-1} (\underline{\delta \mathcal{U}}^\ell - \underline{\delta \mathcal{U}}^k)$, where the second equality was obtained with the help of Eq. (21). The motion tensor affords the following multiplicative decomposition (Bauchau, 2011), $\underline{\underline{C}}^*(\underline{\mathcal{P}}^*) = \underline{\underline{\mathcal{H}}}(\underline{\mathcal{P}}^*) \underline{\underline{\mathcal{H}}}^{*-1}(\underline{\mathcal{P}}^*)$. Equation (3) then implies $\underline{\underline{\mathbb{C}}}^\ell \underline{\underline{\mathcal{H}}}^* = \underline{\underline{\mathbb{C}}}^k \underline{\underline{\mathcal{H}}}$, and tensor $\underline{\underline{\mathcal{W}}}$ is defined as

$$\underline{\underline{\mathcal{W}}} = (\underline{\underline{\mathbb{C}}}^k \underline{\underline{\mathcal{H}}})^{-1} = (\underline{\underline{\mathbb{C}}}^\ell \underline{\underline{\mathcal{H}}}^*)^{-1}. \tag{22}$$

Virtual changes in the motion parameter vector now become

$$\delta \underline{\mathcal{P}}^* = \underline{\underline{\mathcal{W}}} (\underline{\delta \mathcal{U}}^\ell - \underline{\delta \mathcal{U}}^k). \tag{23}$$

Variation in strain energy is expressed as $\delta A = \underline{\mathcal{E}}^{*T} \underline{\underline{\mathcal{D}}}^* \delta \underline{\mathcal{E}}^*$ and requires evaluation of the virtual strains, expressed as

$$\delta \underline{\mathcal{E}}^* = \underline{\underline{\mathcal{B}}}^* \delta \underline{\mathcal{P}}^*, \tag{24}$$

where matrix $\underline{\underline{\mathcal{B}}}^*$ is defined by Eq. (C1).

Using Eqs. (24) and (23), variation of the strain energy in the flexible joint becomes $\delta A = (\underline{\delta \mathcal{U}}^{\ell T} - \underline{\delta \mathcal{U}}^{kT}) \underline{\underline{\mathcal{W}}}^T \underline{\underline{\mathcal{B}}}^T \underline{\mathcal{L}}^*$, where the generalized force vector, $\underline{\mathcal{L}}^*$, was defined as

$$\underline{\mathcal{L}}^* = \underline{\underline{\mathcal{D}}}^* \underline{\mathcal{E}}^*. \tag{25}$$

The elastic forces in the joint now become

$$\underline{\mathcal{F}} = \begin{Bmatrix} -\underline{\mathcal{F}}^e \\ \underline{\mathcal{F}}^e \end{Bmatrix}, \tag{26}$$

where

$$\underline{\mathcal{F}}^e = \underline{\underline{\mathcal{W}}}^T \underline{\underline{\mathcal{B}}}^{*T} \underline{\mathcal{L}}^*. \tag{27}$$

4.2 Stiffness matrix of the flexible joint

The stiffness matrix of the flexible joint stems from the linearization of the elastic forces defined by Eq. (26). Linearization of $\underline{\mathcal{F}}^e$ yields

$$\Delta\underline{\mathcal{F}}^e = \Delta\underline{\underline{\mathcal{W}}}^T\underline{\mathcal{N}}^* + \underline{\underline{\mathcal{W}}}^T\Delta\underline{\underline{\mathcal{B}}}^{*T}\underline{\mathcal{L}}^* + \underline{\underline{\mathcal{W}}}^T\underline{\underline{\mathcal{B}}}^{*T}\Delta\underline{\mathcal{L}}^*, \tag{28}$$

where $\underline{\mathcal{N}}^* = \underline{\underline{\mathcal{B}}}^{*T}\underline{\mathcal{L}}^*$. Using Eqs. (24) and (23), the last term of this expression is obtained as $\underline{\underline{\mathcal{W}}}^T\underline{\underline{\mathcal{B}}}^{*T}\Delta\underline{\mathcal{L}}^* = \left(\underline{\underline{\mathcal{B}}}^*\underline{\underline{\mathcal{W}}}\right)^T\underline{\underline{\mathcal{D}}}^*\left(\underline{\underline{\mathcal{B}}}^*\underline{\underline{\mathcal{W}}}\right)(\Delta\underline{\mathcal{U}}^\ell - \Delta\underline{\mathcal{U}}^k)$.

The first term of Eq. (28) is evaluated next. This term is recast as $\Delta\underline{\underline{\mathcal{W}}}^T\underline{\mathcal{N}}^* = \left[\Delta(\underline{\underline{\mathbb{C}}}^k\underline{\underline{\mathcal{H}}})^{-T}\underline{\mathcal{N}}^* + \Delta(\underline{\underline{\mathbb{C}}}^\ell\underline{\underline{\mathcal{H}}}^*)^{-T}\underline{\mathcal{N}}^*\right]/2$ to respect the symmetry of the problem expressed by Eq. (22). Expanding the variations leads to

$$\Delta\underline{\underline{\mathcal{W}}}^T\underline{\mathcal{N}}^* = \left[-\underline{\underline{\mathcal{W}}}^T\underline{\underline{\mathcal{T}}}^*\underline{\underline{\mathcal{W}}} - \frac{\widehat{\mathcal{F}}^e}{2}, \underline{\underline{\mathcal{W}}}^T\underline{\underline{\mathcal{T}}}^*\underline{\underline{\mathcal{W}}} - \frac{\widehat{\mathcal{F}}^e}{2}\right]\begin{Bmatrix}\Delta\underline{\mathcal{U}}^k \\ \Delta\underline{\mathcal{U}}^\ell\end{Bmatrix} \tag{29}$$

where notation $\widehat{(\cdot)}$ is defined by Eq. (A5). Matrix $\underline{\underline{\mathcal{T}}}^* = [\underline{\underline{\mathcal{H}}}^T\underline{\underline{\mathcal{X}}} + \underline{\underline{\mathcal{H}}}^{*T}\underline{\underline{\mathcal{X}}}^*]/2$, where matrices $\underline{\underline{\mathcal{X}}}$ and $\underline{\underline{\mathcal{X}}}^*$ are implicitly defined by the equations $\Delta\underline{\underline{\mathcal{H}}}^{-T}\underline{\mathcal{N}}^* = \underline{\underline{\mathcal{X}}}\Delta\underline{\mathcal{P}}^*$ and $\Delta\underline{\underline{\mathcal{H}}}^{*-T}\underline{\mathcal{N}}^* = \underline{\underline{\mathcal{X}}}^*\Delta\underline{\mathcal{P}}^*$, respectively; explicit expressions of these matrices are given in Eqs. (D4a) and (D4b), respectively.

Finally, the second term of Eq. (28) becomes $\underline{\underline{\mathcal{W}}}^T\Delta\underline{\underline{\mathcal{B}}}^{*T}\underline{\mathcal{L}}^* = \underline{\underline{\mathcal{W}}}^T\underline{\underline{\mathcal{Q}}}^*(\underline{\mathcal{L}}^*, \underline{\mathcal{P}}^*)\Delta\underline{\mathcal{P}}^*$ and Eq. (23) then leads to

$$\underline{\underline{\mathcal{W}}}^T\Delta\underline{\underline{\mathcal{B}}}^{*T}\underline{\mathcal{L}}^* = \underline{\underline{\mathcal{W}}}^T\underline{\underline{\mathcal{Q}}}^*(\underline{\mathcal{L}}^*, \underline{\mathcal{P}}^*)\underline{\underline{\mathcal{W}}}(\delta\underline{\mathcal{U}}^\ell - \delta\underline{\mathcal{U}}^k). \tag{30}$$

where the explicit expression of matrix $\underline{\underline{\mathcal{Q}}}^*$ is given in Eq. (C2).

The linearized elastic forces can now be written explicitly as

$$\Delta\underline{\mathcal{F}} = \underline{\underline{\mathcal{K}}}\begin{Bmatrix}\delta\underline{\mathcal{U}}^k \\ \delta\underline{\mathcal{U}}^\ell\end{Bmatrix}, \tag{31}$$

where the stiffness matrix of the elastic joint is

$$\underline{\underline{\mathcal{K}}} = \begin{bmatrix}\underline{\underline{\mathcal{D}}} + \widehat{\mathcal{F}}^e/2 & -\underline{\underline{\mathcal{D}}} + \widehat{\mathcal{F}}^e/2 \\ -\underline{\underline{\mathcal{D}}} - \widehat{\mathcal{F}}^e/2 & \underline{\underline{\mathcal{D}}} - \widehat{\mathcal{F}}^e/2\end{bmatrix} \tag{32}$$

where $\underline{\underline{\mathcal{D}}} = \underline{\underline{\mathcal{W}}}^T(\underline{\underline{\mathcal{B}}}^{*T}\underline{\underline{\mathcal{D}}}^*\underline{\underline{\mathcal{B}}} + \underline{\underline{\mathcal{T}}}^* + \underline{\underline{\mathcal{Q}}}^*)\underline{\underline{\mathcal{W}}}$.

5 Numerical results

5.1 Change of reference point

To illustrate the concepts developed in the previous sections, consider the system consisting of two beams, each of length $L = 1.2$ m, connected by a flexible joint, as depicted in Fig. 2. The two beams have the same sectional stiffness properties:

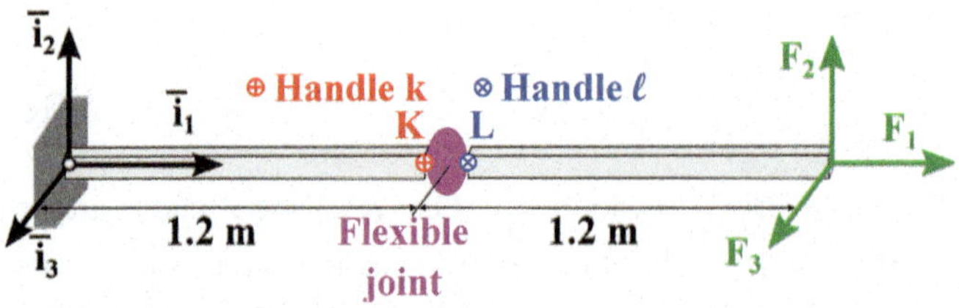

Figure 2. Two beams connected with a flexible joint.

axial stiffness $S = 43.50$ MN; bending stiffness $H_{22} = 23.26$ and $H_{33} = 298.7$ kN m^2 about unit vector $\bar{\imath}_2$ and $\bar{\imath}_3$, respectively; torsional stiffness $H_{11} = 28.05$ kN m^2; and shearing stiffness $K_{22} = 4.0$ and $K_{33} = 2.81$ MN, along unit vectors $\bar{\imath}_2$ and $\bar{\imath}_3$, respectively.

The first beam is clamped at its root and the second is loaded at its tip by concentrated forces tip loads $F_1 = 500\alpha/9$, $F_2 = 250\alpha/9$, and $F_3 = 200\alpha/9$ N, along unit vectors $\bar{\imath}_1$, $\bar{\imath}_2$, and $\bar{\imath}_3$, respectively, as shown in Fig. 2. The load factor $\alpha \in [0, 9]$. The stiffness matrix of flexible joint is given by

$$\underline{\underline{D}}^* = \begin{bmatrix} 2000 & 10 & 10 & & & \\ 10 & 2000 & 10 & & & \\ 10 & 10 & 2000 & & & \\ & & & 1000 & & \\ & & & & 1000 & \\ & & & & & 1000 \end{bmatrix}. \tag{33}$$

Two formulations will be contrasted here. In the first, the deformation measures are selected as

$$\underline{\mathcal{E}}^\dagger = \begin{Bmatrix}\underline{\varepsilon}^\dagger \\ \underline{\kappa}^\dagger\end{Bmatrix} = \begin{Bmatrix}(\underline{\underline{R}}^k\underline{\underline{R}}_0)^T(\underline{u}^\ell - \underline{u}^k) \\ \text{axial}(\underline{\underline{R}}_0^T\underline{\underline{R}}^{kT}\underline{\underline{R}}^\ell\underline{\underline{R}}_0)\end{Bmatrix}. \tag{34}$$

The stretch vector, $\underline{\varepsilon}^\dagger$, corresponds to the components of the relative displacement vector resolved in basis $\mathcal{B}^k$. The wryness vector, $\underline{\kappa}^\dagger$, is the axial part of the relative rotation tensor, resolved on the same basis; note that the components of the wryness vector selected here are identical when resolved in basis $\mathcal{B}^k$ or $\mathcal{B}^\ell$.

Various types of deformation measures have been used to represent the behavior of flexible joints, but those given by Eq. (34) are rather typical; in the following sections, they will be referred to as "typical deformation measures". In contrast, the deformation measures used in this work are defined by Eq. (16) and will be referred as "proposed deformation measures".

The main claim of this paper is that the proposed strain measures are invariant with respect to the choice of reference point, i.e., are objective, a characteristic that is not shared by typical deformation measures. Imagine that points **K** and **L** are interchanged in Fig. 2. Clearly, this does not modify the physical system, and hence, its response under load should be unaffected by this interchange. This basic invariance is satisfied by the proposed deformation measures, but not by their

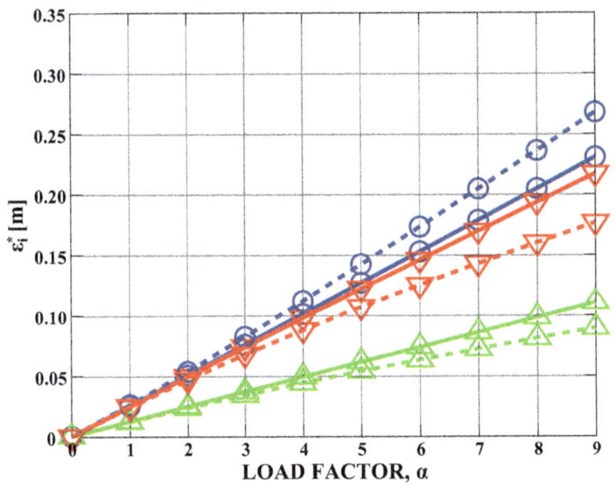

Figure 3. Components of the stretch vector: ε_1^* (○), ε_2^* (△), ε_3^* (▽). Typical deformation measures. Reference point **K**: solid line, point **L**: dashed line.

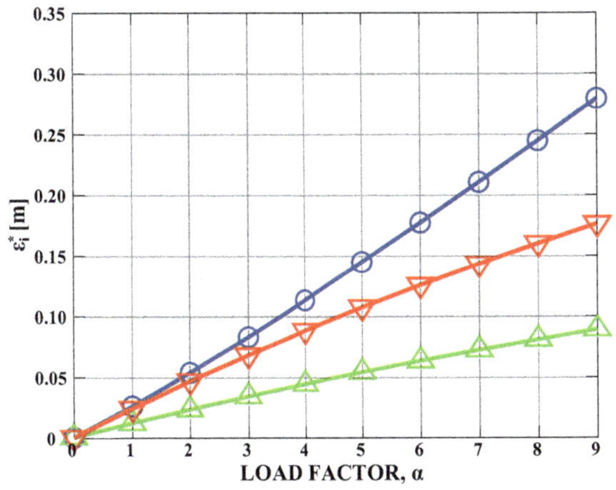

Figure 5. Components of the stretch vector: ε_1^* (○), ε_2^* (△), ε_3^* (▽). Proposed deformation measures.

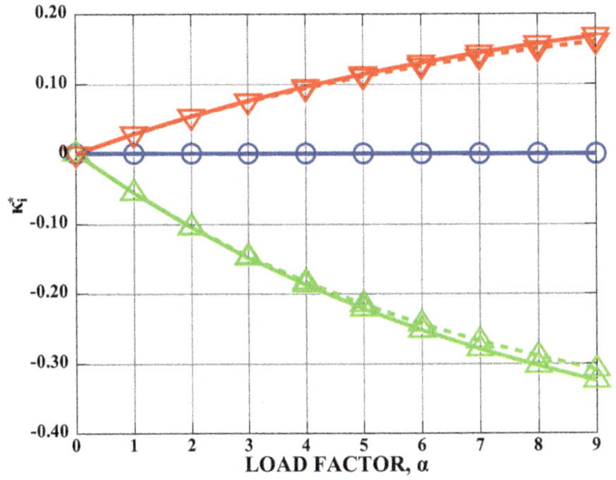

Figure 4. Components of the wryness vector: κ_1^* (○), κ_2^* (△), κ_3^* (▽). Typical deformation measures. Reference point **K**: solid line, point **L**: dashed line.

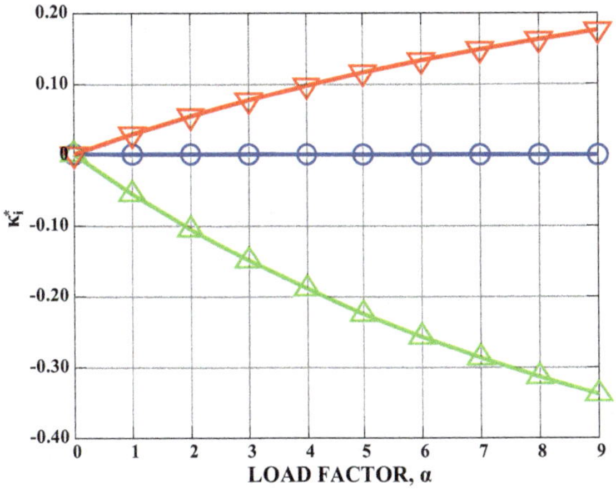

Figure 6. Components of the wryness vector: κ_1^* (○), κ_2^* (△), κ_3^* (▽). Proposed deformation measures

typical counterparts. To illustrate this effect, the following parameters were selected for the proposed deformation measures: $\lambda = 0$ and $\mu = (1 - p^2/16)/(1 + p^2/16)^2$. This choice is not important as the proposed deformation measures are invariant for any choice of these parameters.

For the typical deformation measures, two simulations were performed. In the first run, points **K** and **L** are selected as indicated in Fig. 2 and in the second simulation, these two points were interchanged. Of course, this interchange is a modeling detail, which has no physical meaning. Yet, these two simulations yield different results because the definition of the deformation measures make specific reference to basis $\mathcal{B}^k$, and hence, are inherently "basis sensitive".

Figures 3 and 4 show the components of the stretch and wryness vector components, respectively, versus the load factor, $\alpha \in [0.9]$, for the typical deformation measures defined by Eq. (34). For each deformation component, the response when reference points **K** and **L** are selected as shown in Fig. 2 is shown in solid lines, and the response when points **K** and **L** are interchanged in shown in dashed lines. Clearly, the typical deformation measures do not yield physically meaningful predictions, because different responses are obtained for a given physical system depending on the choice of the labeling convention for points **K** and **L**.

In contrast, the proposed deformation measures yield the same predictions when the roles of points **K** and **L** are interchanged. Figures 5 and 6 show the components of the stretch and wryness vector components, respectively, for the proposed deformation measures defined by Eq. (16).

Of course, for very small deformations, the predictions based on the typical and proposed deformation measures are identical. This result is expected because both typical and proposed deformation measures converge to the same infinitesimal deformation measures and because identical stiffness matrices were used.

Table 1. Choices of parameters λ and μ for the sixteen cases.

Case		a	b	c	d
1	$\lambda = 0$	$\mu = 1$	$1 + p^2$	$1 + p^4$	$1 + p^6$
2	$\lambda = 0$	$\mu = 1$	$1 + \rho$	$1 + \rho^2$	$1 + \rho^3$
3	$\mu = 1$	$\lambda = 0$	p^2	p^4	p^6
4	$\mu = 1$	$\lambda = 0$	ρ	ρ^2	ρ^3

5.2 Choice of λ and μ

The proposed deformation measures are not unique. Rather, they form families dependent on two arbitrary parameters, $\lambda(\rho, p)$ and $\mu(\rho, p)$: each choice of these parameters yields a different deformation measure but for all choices, the objectivity and tensorial nature of the deformation measure is preserved. Assuming that the simple strain energy expression defined by Eq. (20) is used with a given stiffness matrix, the choice of parameters λ and μ will alter the response of the flexible joint under load in the nonlinear regime.

To study the influence of the choice of parameters λ and μ on joint behavior, a very simple example was treated. Handle k of the joint was clamped and forces and moments were applied to handle ℓ; the applied force vector is $\underline{F}^T = \left\{125 \quad 250 \quad 375\right\}\alpha/9$ and $\underline{M}^T = \left\{125 \quad 250 \quad 375\right\}\alpha/9$, where α is the load factor.

A total of sixteen combinations of parameters λ and μ were selected for the study. First, $\lambda = 0$ is selected and cases 1a, 1b, 1c, and 1d correspond to $\mu = 1, \mu = 1 + p^2, \mu = 1 + p^4$, and $\mu = 1 + p^6$, respectively, as listed in the first row of Table 1. For case 1, parameter μ is function of p only. Next, $\lambda = 0$ is selected and cases 2a, 2b, 2c, and 2d correspond to $\mu = 1$, $\mu = 1 + \rho, \mu = 1 + \rho^2$, and $\mu = 1 + \rho^3$, respectively, as listed in the second row of Table 1. For case 2, parameter μ is function of ρ only. Cases 3 and 4 are defined similarly, as listed in the third and fourth row of Table 1, respectively.

In all cases, the stiffness matrix of flexible joint was selected as

$$\underline{\underline{D}}^* = \begin{bmatrix} 1000 & 100 & 100 & & & \\ 100 & 1000 & 100 & & & \\ 100 & 100 & 1000 & & & \\ & & & 1000 & & \\ & & & & 1000 & \\ & & & & & 1000 \end{bmatrix} \tag{35}$$

Figures 7 and 8 show the components of the stretch and wryness vectors, respectively, for cases 1a, 1b, 1c, and 1d. The choice of very simple polynomial expressions for $\mu = \mu(p)$ is arbitrary, but all satisfy the limit behavior expressed by Eq. (18). Similarly, the constant value of $\lambda = 0$ satisfies the limit behavior expressed by Eq. (19).

For small values of the load factor, the flexible joint deformation remains small and identical response is observed for all choices of parameter μ. As larger loads are applied, the deformation is no longer infinitesimal and the joint's nonlinear response is affected by the choice of parameter μ. This effect is particularly pronounced in the wryness response, as shown in Fig. 8.

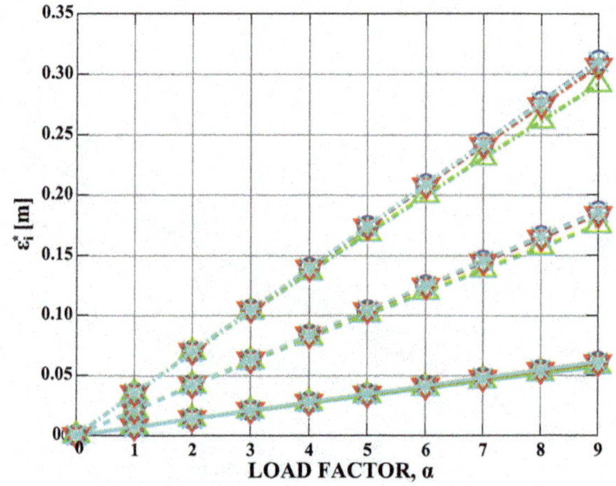

Figure 7. Components of the stretch vector: ε_1^* (solid line), ε_2^* (dashed line), ε_3^* (dash-dotted line). Case 1a (○), 1b (△), 1c (▽), 1d (∗).

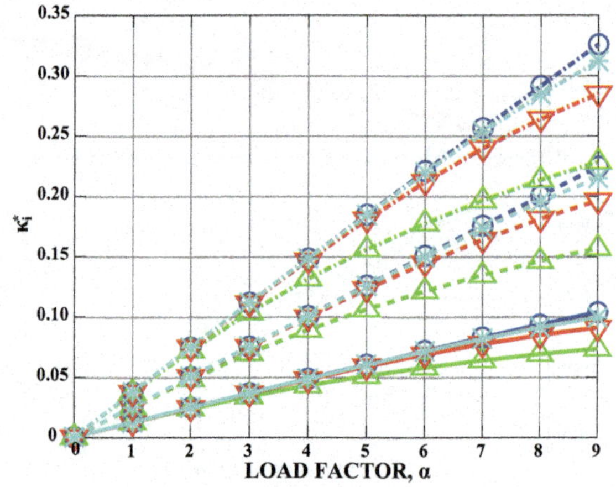

Figure 8. Components of the wryness vector: κ_1^* (solid line), κ_2^* (dashed line), κ_3^* (dash-dotted line). Case 1a (○), 1b (△), 1c (▽), 1d (∗).

Of course, parameters λ and μ are arbitrary functions of both variables ρ and p. Case 1, in which parameter μ is a function of variable p only, is a special case. In case 2, parameter μ is selected to be a simple polynomial function of variable ρ only, as shown in the second row of Table 1. Figures 9 and 10 show the components of the stretch and wryness vectors, respectively, for cases 2a, 2b, 2c, and 2d. The nonlinear response of both stretch and wryness vector components is significantly affected by the choice of parameter μ.

Finally, two additional cases, cases 3 and 4, were treated where parameter λ is selected to be a function of variables p and ρ, respectively, while keeping a constant value of parameter $\mu = 1$, as listed in the last two rows of Table 1,

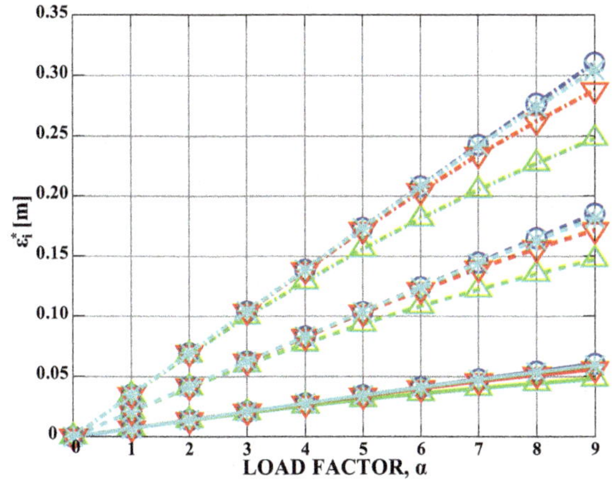

Figure 9. Components of the stretch vector: ε_1^* (solid line), ε_2^* (dashed line), ε_3^* (dash-dotted line). Case 2a (○), 2b (△), 2c (▽), 2d (∗).

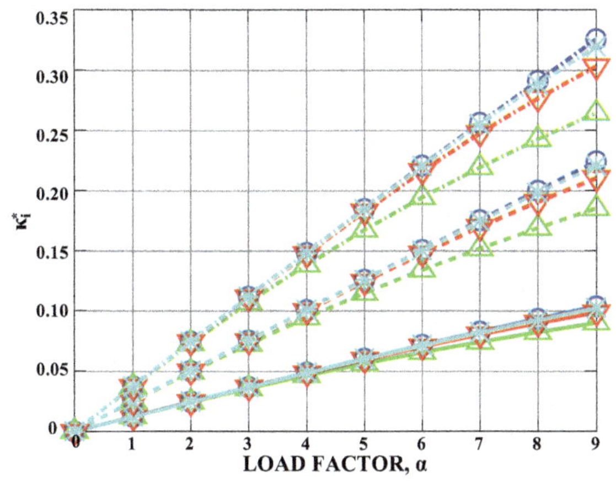

Figure 10. Components of the wryness vector: κ_1^* (solid line), κ_2^* (dashed line), κ_3^* (dash-dotted line). Case 2a (○), 2b (△), 2c (▽), 2d (∗).

respectively. Here again, simple polynomial expressions were selected and the limit behaviors expressed by Eqs. (18) and (19) were satisfied. Figures 11 and 12 show the components of the stretch and wryness vectors, respectively, for cases 3a, 3b, 3c, and 3d. Finally, Figs. 13 and 14 show the components of the stretch and wryness vectors, respectively, for cases 4a, 4b, 4c, and 4d.

The goal of the simple examples presented in the previous paragraphs is to show that the nonlinear behavior of the flexible joint is strongly affected by the functional dependency of parameters λ and μ on variables ρ and p. Of course, in general, the two parameters can be selected to be functions of both variables, *i.e.*, $\lambda = \lambda(\rho, p)$ and $\mu = \mu(\rho, p)$. Softening or stiffening behavior can be obtained by tailoring the functional dependency of parameters λ and μ on variables ρ and p.

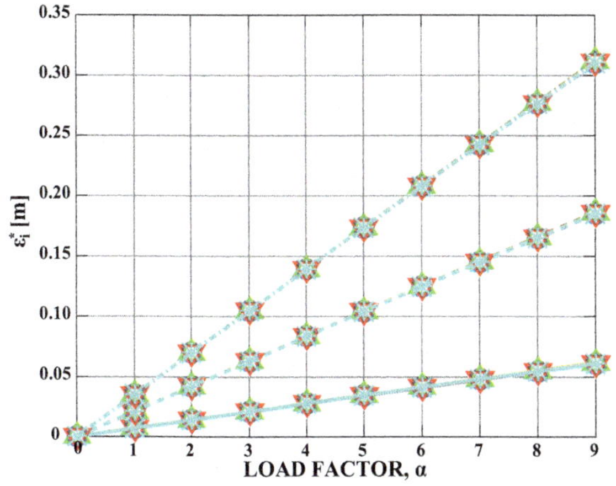

Figure 11. Components of the stretch vector: ε_1^* (solid line), ε_2^* (dashed line), ε_3^* (dash-dotted line). Case 3a (○), 3b (△), 3c (▽), 3d (∗).

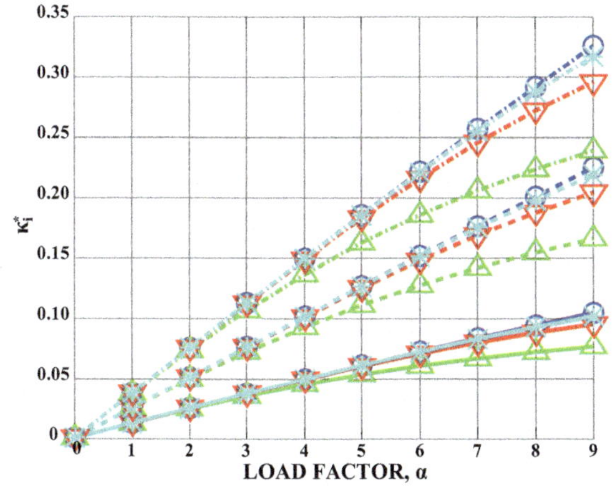

Figure 12. Components of the wryness vector: κ_1^* (solid line), κ_2^* (dashed line), κ_3^* (dash-dotted line). Case 3a (○), 3b (△), 3c (▽), 3d (∗).

The attention has focused thus far on the strain energy given by Eq. (20), which is a quadratic expression of the deformation measures, $A = \underline{\mathcal{E}}^{*T} \underline{\underline{\mathcal{D}}}^* \underline{\mathcal{E}}^*/2$. This leads to the linear relationship between the generalized forces and proposed deformation measures, $\underline{\mathcal{L}}^* = \underline{\underline{\mathcal{D}}}^* \underline{\mathcal{E}}^*$, see Eq. (25). This linear relationship, however, is deceptively simple.

Indeed, Eq. (27) shows that the relationship between the elastic forces in the joint and the proposed deformation measures is $\underline{\mathcal{F}}^e = \underline{\underline{\mathcal{W}}}^T \underline{\underline{\mathcal{B}}}^{*T} \underline{\mathcal{L}}^*$. Next, the relationship between the externally applied loads and proposed deformation measures is obtained with the help of Eq. (4) as

$$\begin{Bmatrix} \underline{F}^k \\ \underline{M}^k \end{Bmatrix} = \underline{\underline{\mathcal{T}}}^{kT} \underline{\underline{\mathcal{W}}}^T \underline{\underline{\mathcal{B}}}^{*T} \underline{\underline{\mathcal{D}}}^* \underline{\underline{\mathcal{Z}}}(\lambda, \mu) \underline{\mathcal{P}}^*, \tag{36a}$$

$$\begin{Bmatrix} \underline{F}^\ell \\ \underline{M}^\ell \end{Bmatrix} = \underline{\underline{\mathcal{T}}}^{\ell T} \underline{\underline{\mathcal{W}}}^T \underline{\underline{\mathcal{B}}}^{*T} \underline{\underline{\mathcal{D}}}^* \underline{\underline{\mathcal{Z}}}(\lambda, \mu) \underline{\mathcal{P}}^*, \tag{36b}$$

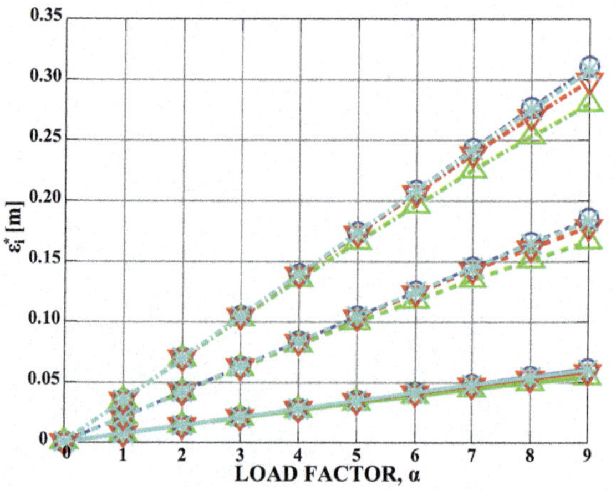

Figure 13. Components of the stretch vector: ε_1^* (solid line), ε_2^* (dashed line), ε_3^* (dash-dotted line). Case 4a (◯), 4b (△), 4c (▽), 4d (∗).

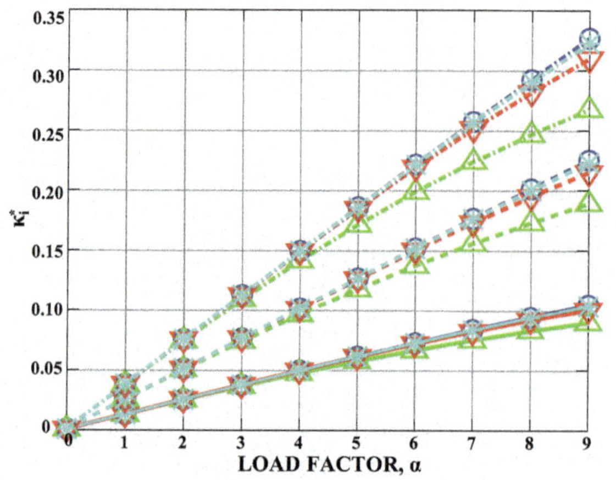

Figure 14. Components of the wryness vector: κ_1^* (solid line), κ_2^* (dashed line), κ_3^* (dash-dotted line). Case 4a (◯), 4b (△), 4c (▽), 4d (∗).

where Eq. (16) was used to express the proposed deformation measures in terms of $\underline{\mathcal{P}}^*$, the relative motion of the joint's two handles. Although the stiffness matrix, $\underline{\underline{\mathcal{D}}}^*$, is constant, the relationship between the externally applied loads and the joint's deformation measures is nonlinear because matrices $\underline{\underline{\mathcal{W}}}$ and $\underline{\underline{\mathcal{B}}}^*$ are nonlinear functions of the deformations measures, see Eqs. (22) and (C1), respectively, and the relative motion of the two handles, $\underline{\mathcal{P}}^*$, is also a nonlinear function of the handle's relative motion. These observations explain the nonlinear load-deformation behavior exhibited in Figs. 4 to 14.

In practice, the joint's constitutive laws can be obtained from a two step procedure. First, experimental measurements must be obtained for infinitesimal deformations of the joint and lead to the identification of the entries of the constant stiffness matrix, $\underline{\underline{D}}^*$. Next, experimental measurements characterizing the joint's behavior in the nonlinear range must be obtained. The optimal functional dependencies of parameters λ and μ on variables ρ and p can then be determined using suitable parameter identification methods.

Of course, it is not guaranteed that suitable functions, $\lambda = \lambda(\rho, p)$ and $\mu = \mu(\rho, p)$, can be found, which will model the observed behavior of the joint accurately. In such case, more complex strain energy expressions could be selected in an attempt to better capture the observed constitutive behavior of the joint; in general, $A = A(\underline{\mathcal{E}}^*)$. Of course, the existence of a strain energy function implies that material behavior is conservative, which will not always be true.

6 Conclusions

This paper has focused on the constitutive behavior of elastic bodies of finite dimension, typically called flexible joints in structural and multibody dynamics. Physically meaningful deformation measures were proposed that are objective and of a tensorial nature; an explicit expression of these measures was derived. Equipped with these deformation measures, constitutive laws for the flexible joint were derived by assuming the existence of a strain energy function that is a quadratic form of these deformation measures. Because all the quantities involved in the formulation are objective and tensorial, the predicted joint behavior presents the required invariance with respect to changes of basis or reference point.

Numerical examples were presented that demonstrate the invariance of the predicted behavior with respect to the choice of reference point, even in the nonlinear range; in contrast, typical formulations found in the literature up to date do not appear to present these desirable characteristics. The proposed deformation measures are not unique: their definition depends on the choice of two parameters, which are functions of the relative rotation and the intrinsic relative displacement at the joint. Numerical examples presented in the paper show that the choice of these two parameters affects the response of the joint in the nonlinear regime significantly. Consequently, the proposed deformation measures form families, and the choice of the functional dependency of the parameter can be selected to tailor the nonlinear response of the joint.

This paper has focused on an expression of the strain energy that depends on the proposed deformation measures in a quadratic manner, leading to a linear relationship between the generalized forces and deformations measure. Despite this linearity, the relationship between externally applied forces and deformations is nonlinear. More general joint constitutive behavior could be obtained by considering more general strain energy expressions. Investigating energy dissipation in flexible joint based on the time rate of change of the proposed deformation measure is another possible extension of the this work.

Appendix A

Notational conventions

To simplify the writing of this seemingly complicated expression, the following notation is introduced. First, tensor $\underline{\underline{\mathcal{Z}}}$, a function of two scalars, α and β, is introduced

$$\underline{\underline{\mathcal{Z}}}(\alpha,\beta) = \begin{bmatrix} \beta \underline{\underline{I}} & \alpha \underline{\underline{I}} \\ \underline{\underline{0}} & \beta \underline{\underline{I}} \end{bmatrix}. \tag{A1}$$

Second, the *generalized vector product tensor* is defined

$$\widetilde{\mathcal{N}} = \begin{bmatrix} \widetilde{n} & \widetilde{m} \\ \underline{\underline{0}} & \widetilde{n} \end{bmatrix}. \tag{A2}$$

Notation $\widetilde{\mathcal{N}}$ does not indicate a 6×6 skew-symmetric tensor, but rather the above 6×6 tensor formed by three skew-symmetric sub-tensors. By analogy to notation $\underline{a} = \text{axial}(\widetilde{a})$, the following operator is introduced

$$\underline{\mathcal{N}} = \mathcal{A}\text{xial}(\widetilde{\mathcal{N}}). \tag{A3}$$

Consider two vectors defined as

$$\underline{\mathcal{V}} = \begin{Bmatrix} \underline{v} \\ \underline{\omega} \end{Bmatrix}, \quad \underline{\mathcal{P}} = \begin{Bmatrix} \underline{p} \\ \underline{q} \end{Bmatrix}.$$

The well-known property of the vector product, $\widetilde{a}\underline{b} = -\widetilde{b}\underline{a}$, then generalizes to

$$\widetilde{\mathcal{V}}\underline{\mathcal{P}} = -\widetilde{\mathcal{P}}\underline{\mathcal{V}}, \tag{A4a}$$

$$\widetilde{\mathcal{V}}^T\underline{\mathcal{P}} = \widehat{\mathcal{P}}\underline{\mathcal{V}}, \tag{A4b}$$

where the following notation was introduced

$$\widehat{\mathcal{P}} = \begin{bmatrix} \underline{\underline{0}} & \widetilde{p} \\ \widetilde{p} & \widetilde{q} \end{bmatrix}. \tag{A5}$$

Finally, the following identity results

$$\widehat{\widehat{\mathcal{P}}\underline{\mathcal{V}}} = \widehat{\mathcal{P}}\widetilde{\mathcal{V}} + \widetilde{\mathcal{V}}^T\widehat{\mathcal{P}}. \tag{A6}$$

Appendix B

Linearization of functions λ and μ

Scalar λ and μ were introduced in Eq. (17) as functions of ϱ and p. Using the chain rule for derivatives, the variation of μ is $\delta\mu = \mu_\varrho \delta\varrho + \mu_p \delta p$, where notations $(\cdot)_\varrho$ and $(\cdot)_p$ indicate derivatives with respect to variable ϱ and angle p, respectively. Because $\varrho = \underline{p}^{*T}\underline{q}^*$ and $p^2 = \underline{p}^{*T}\underline{p}^*$, it follows that $\delta\varrho = \underline{p}^{*T}\delta\underline{q}^* + \underline{q}^{*T}\delta\underline{p}^*$ and $p\delta p = \underline{p}^{*T}\delta\underline{p}^*$, and hence,

$$\begin{Bmatrix} \delta\varrho \\ p\delta p \end{Bmatrix} = \begin{bmatrix} \underline{p}^{*T} & \underline{q}^{*T} \\ \underline{0}^T & \underline{p}^{*T} \end{bmatrix} \delta\underline{\mathcal{P}}^*. \tag{B1}$$

The variations of parameters λ and μ now become

$$\begin{Bmatrix} \delta\lambda \\ \delta\mu \end{Bmatrix} = \underline{\underline{\Lambda}}_1(\lambda,\mu) \begin{bmatrix} \underline{p}^{*T} & \underline{q}^{*T} \\ \underline{0}^T & \underline{p}^{*T} \end{bmatrix} \delta\underline{\mathcal{P}}^*. \tag{B2}$$

where matrix $\underline{\underline{\Lambda}}_1$ is defined as

$$\underline{\underline{\Lambda}}_1(\lambda,\mu) = \begin{bmatrix} \lambda_\varrho & \lambda_p/p \\ \mu_\varrho & \mu_p/p \end{bmatrix}. \tag{B3}$$

Because parameters λ and μ are even functions of variable p, their derivatives with respect to the same variable are odd functions of p and hence, λ_p/p and μ_p/p present no singularity when $p \to 0$.

Consider two arbitrary arrays, $\underline{\mathcal{L}}^T = \left\{\underline{\ell}^T \quad \underline{m}^T\right\}$ and $\underline{\mathcal{P}}^T = \left\{\underline{q}^T \quad \underline{p}^T\right\}$, and two arbitrary scalars, $\alpha = \alpha(\varrho, p)$ and $\beta = \beta(\varrho, p)$. Matrices $\hat{\underline{\underline{\mathcal{Z}}}}$ and $\check{\underline{\underline{\mathcal{Z}}}}$ are defined implicitly by the following two identities, $\delta\underline{\underline{\mathcal{Z}}}(\alpha,\beta)\underline{\mathcal{L}} = \hat{\underline{\underline{\mathcal{Z}}}}(\underline{\mathcal{L}},\alpha,\beta,\underline{\mathcal{P}})\delta\underline{\mathcal{P}}$ and $\delta\underline{\underline{\mathcal{Z}}}^T(\alpha,\beta)\underline{\mathcal{L}} = \check{\underline{\underline{\mathcal{Z}}}}(\underline{\mathcal{L}},\alpha,\beta,\underline{\mathcal{P}})\delta\underline{\mathcal{P}}$, respectively. It is shown easily shown that

$$\delta\underline{\underline{\mathcal{Z}}}(\alpha,\beta)\underline{\mathcal{L}} = \begin{bmatrix} \underline{m} & \underline{\ell} \\ \underline{0} & \underline{m} \end{bmatrix} \begin{Bmatrix} \delta\alpha \\ \delta\beta \end{Bmatrix}, \tag{B4a}$$

$$\delta\underline{\underline{\mathcal{Z}}}^T(\alpha,\beta)\underline{\mathcal{L}} = \begin{bmatrix} \underline{0} & \underline{\ell} \\ \underline{\ell} & \underline{m} \end{bmatrix} \begin{Bmatrix} \delta\alpha \\ \delta\beta \end{Bmatrix}. \tag{B4b}$$

Introducing Eq. (B2) now yields the following explicit expressions for matrices $\hat{\underline{\underline{\mathcal{Z}}}}$ and $\check{\underline{\underline{\mathcal{Z}}}}$,

$$\hat{\underline{\underline{\mathcal{Z}}}}(\underline{\mathcal{L}},\alpha,\beta,\underline{\mathcal{P}}) = \begin{bmatrix} \underline{m} & \underline{\ell} \\ \underline{0} & \underline{m} \end{bmatrix} \underline{\underline{\Lambda}}_1(\alpha,\beta) \begin{bmatrix} \underline{p}^T & \underline{q}^T \\ \underline{0}^T & \underline{p}^T \end{bmatrix}, \tag{B5a}$$

$$\check{\underline{\underline{\mathcal{Z}}}}(\underline{\mathcal{L}},\alpha,\beta,\underline{\mathcal{P}}) = \begin{bmatrix} \underline{0} & \underline{\ell} \\ \underline{\ell} & \underline{m} \end{bmatrix} \underline{\underline{\Lambda}}_1(\alpha,\beta) \begin{bmatrix} \underline{p}^T & \underline{q}^T \\ \underline{0}^T & \underline{p}^T \end{bmatrix}. \tag{B5b}$$

Appendix C

Variations of the strain measures

The proposed strain measures are defined by Eq. (16) and variations of these quantities are expressed as $\delta\underline{\mathcal{E}}^* = \left[\underline{\underline{\mathcal{Z}}}(\lambda,\mu) + \hat{\underline{\underline{\mathcal{Z}}}}(\underline{\mathcal{P}}^*,\lambda,\mu,\underline{\mathcal{P}}^*)\right]\delta\underline{\mathcal{P}}^*$, where matrix $\hat{\underline{\underline{\mathcal{Z}}}}$ is defined by Eq. (B5a). This implies that $\delta\underline{\mathcal{E}}^* = \underline{\underline{\mathcal{B}}}^*\delta\underline{\mathcal{P}}^*$, where matrix $\underline{\underline{\mathcal{B}}}^*$ is defined as

$$\underline{\underline{\mathcal{B}}}^* = \underline{\underline{\mathcal{Z}}}(\lambda,\mu) + \hat{\underline{\underline{\mathcal{Z}}}}(\underline{\mathcal{P}}^*,\lambda,\mu,\underline{\mathcal{P}}^*). \tag{C1}$$

The linearization of the elastic forces requires the evaluation of matrix $\underline{\underline{Q}}^*$, implicitly defined by $\delta \underline{\underline{\mathcal{B}}}^* \underline{\mathcal{L}}^* = \underline{\underline{Q}}^* \delta \underline{\mathcal{P}}^*$. The previous results yield

$$\begin{aligned}\underline{\underline{Q}}^* =& \underline{\underline{Z}}(\epsilon,\gamma) + \hat{\underline{\underline{Z}}}(\underline{\mathcal{P}}^*,\lambda,\mu,\underline{\mathcal{L}}^*) + \hat{\underline{\underline{Z}}}(\underline{\mathcal{L}}^*,\lambda,\mu,\underline{\mathcal{P}}^*) \\ &+ \begin{bmatrix} \underline{p}^* & \underline{q}^* \\ \underline{0} & \underline{p}^* \end{bmatrix} \underline{\underline{S}} \begin{bmatrix} \underline{p}^{*T} & \underline{q}^{*T} \\ \underline{0}^T & \underline{p}^{*T} \end{bmatrix},\end{aligned} \tag{C2}$$

where the following scalar functions were defined

$$\begin{Bmatrix} \alpha \\ \beta \end{Bmatrix} = \begin{bmatrix} \underline{p}^{*T} & \underline{q}^{*T} \\ \underline{0}^T & \underline{p}^{*T} \end{bmatrix} \underline{\mathcal{L}}^*, \quad \begin{Bmatrix} \epsilon \\ \gamma \end{Bmatrix} = \begin{bmatrix} \lambda_\varrho & \lambda_p/p \\ \mu_\varrho & \mu_p/p \end{bmatrix} \begin{Bmatrix} \alpha \\ \beta \end{Bmatrix}, \tag{C3}$$

and

$$\underline{\underline{S}} = \begin{bmatrix} \alpha\lambda_{\varrho\varrho} + \beta\lambda_{\varrho p}/p & \alpha\lambda_{\varrho p}/p + \beta(\lambda_{pp} - \lambda_p/p)/p^2 \\ \alpha\mu_{\varrho\varrho} + \beta\mu_{\varrho p}/p & \alpha\mu_{\varrho p}/p + \beta(\mu_{pp} - \mu_p/p)/p^2 \end{bmatrix}. \tag{C4}$$

Appendix D

Linearization of the tangent tensor

The linearization of the elastic forces also requires linearization of the tangent tensor. When using the Wiener-Milenković motion parameterization, the tangent tensor is expressed as

$$\underline{\underline{\mathcal{H}}}^{-1}(\underline{\mathcal{P}}) = \underline{\underline{Z}}(\bar{\chi}_0,\chi_0) - \widetilde{\mathcal{P}}/2 + \underline{\underline{Z}}(\bar{\chi}_2,\chi_2)\widetilde{\mathcal{P}}\widetilde{\mathcal{P}}, \tag{D1a}$$

$$\underline{\underline{\mathcal{H}}}^{*-1}(\underline{\mathcal{P}}) = \underline{\underline{Z}}(\bar{\chi}_0,\chi_0) + \widetilde{\mathcal{P}}/2 + \underline{\underline{Z}}(\bar{\chi}_2,\chi_2)\widetilde{\mathcal{P}}\widetilde{\mathcal{P}}. \tag{D1b}$$

The linearization of the tangent tensor is achieved by defining matrices $\underline{\underline{X}}$ and $\underline{\underline{X}}^*$ implicitly defined by the following expressions, $\Delta \underline{\underline{\mathcal{H}}}^{-T} \underline{\mathcal{L}} = \underline{\underline{X}}(\underline{\mathcal{L}})\Delta\underline{\mathcal{P}}$ and $\underline{\underline{\mathcal{H}}}^{*-T} \underline{\mathcal{L}} = \underline{\underline{X}}^*(\underline{\mathcal{L}})\Delta\underline{\mathcal{P}}$. Tedious algebra yields explicit equations of these matrices as

$$\begin{aligned}\underline{\underline{X}}(\underline{\mathcal{L}}) =& \check{\underline{\underline{Z}}}(\underline{\mathcal{L}},\bar{\chi}_0,\chi_0,\underline{\mathcal{P}}) + \check{\underline{\underline{Z}}}(\widehat{\mathcal{N}}\underline{\mathcal{P}},\bar{\chi}_2,\chi_2,\underline{\mathcal{P}}) \\ &- \widehat{\mathcal{L}}/2 + \underline{\underline{Z}}^T(\bar{\chi}_2,\chi_2)(\widehat{\mathcal{N}} + \widetilde{\mathcal{P}}^T\widehat{\mathcal{L}}),\end{aligned} \tag{D2a}$$

$$\begin{aligned}\underline{\underline{X}}^*(\underline{\mathcal{L}}) =& \check{\underline{\underline{Z}}}(\underline{\mathcal{L}},\bar{\chi}_0,\chi_0,\underline{\mathcal{P}}) + \check{\underline{\underline{Z}}}(\widehat{\mathcal{N}}\underline{\mathcal{P}},\bar{\chi}_2,\chi_2,\underline{\mathcal{P}}) \\ &+ \widehat{\mathcal{L}}/2 + \underline{\underline{Z}}^T(\bar{\chi}_2,\chi_2)(\widehat{\mathcal{N}} + \widetilde{\mathcal{P}}^T\widehat{\mathcal{L}}),\end{aligned} \tag{D2b}$$

where $\underline{\mathcal{N}} = \widetilde{\mathcal{P}}^T\underline{\mathcal{L}} = \widehat{\mathcal{L}}\underline{\mathcal{P}}$ and matrix $\check{\underline{\underline{Z}}}$ is defined by Eq. (B5b).

The Wiener-Milenković rotation parameterization is defined as $\underline{p} = p(\phi)\bar{n}$, where $p(\phi) = 4\tan\phi/4$. Table D1 lists the expressions for all the scalar functions appearing in the above expression when the Wiener-Milenković motion parameterization is used.

Matrices $\underline{\underline{X}}$ and $\underline{\underline{X}}^*$ defined in Eqs. (D2a) and (D2b), respectively, now simplify considerably. Indeed, Eqs. (A1) and (B5b) imply $\underline{\underline{Z}}^T(\bar{\chi}_2,\chi_2) = \underline{\underline{I}}/8$, $\check{\underline{\underline{Z}}}(\widehat{\mathcal{N}}\underline{\mathcal{P}},\bar{\chi}_2,\chi_2,\underline{\mathcal{P}}) = \underline{\underline{0}}$, and

$$\check{\underline{\underline{Z}}}(\underline{\mathcal{L}},\bar{\chi}_0,\chi_0,\underline{\mathcal{P}}) = \frac{1}{8}\begin{bmatrix} \underline{0} & \underline{\ell} \\ \underline{\ell} & \underline{m} \end{bmatrix} \begin{bmatrix} \underline{p}^T & \underline{q}^T \\ \underline{0}^T & \underline{p}^T \end{bmatrix}, \tag{D3}$$

Table D1. Scalars functions associated with the Wiener-Milenković motion parameterization.

Quantity	Value	Quantity	Value
ν	$1/(1+p^2/16)$	ε	$1/(1-p^2/16)$
p'	$1+p^2/16$	$\bar{\varepsilon}$	$\varrho\varepsilon^2/8$
ζ_1	ν^2/ε	ζ_2	$\nu^2/2$
$\bar{\zeta}_1$	$\varrho\nu^2(1-4\nu)/8$	$\bar{\zeta}_2$	$-\varrho\nu^3/8$
σ_0	ν	σ_2	$\nu^2/8$
$\bar{\sigma}_0$	$-\varrho\nu^2/8$	$\bar{\sigma}_2$	$-\varrho\nu^3/32$
χ_0	$1/\nu$	χ_2	$1/8$
$\bar{\chi}_0$	$\varrho/8$	$\bar{\chi}_2$	0

and finally,

$$\underline{\underline{X}}(\underline{\mathcal{L}}) = \check{\underline{\underline{Z}}}(\underline{\mathcal{L}},\bar{\chi}_0,\chi_0,\underline{\mathcal{P}}) - \widehat{\mathcal{L}}/2 + (\widehat{\mathcal{N}} + \widetilde{\mathcal{P}}^T\widehat{\mathcal{L}})/8, \tag{D4a}$$

$$\underline{\underline{X}}^*(\underline{\mathcal{L}}) = \check{\underline{\underline{Z}}}(\underline{\mathcal{L}},\bar{\chi}_0,\chi_0,\underline{\mathcal{P}}) + \widehat{\mathcal{L}}/2 + (\widehat{\mathcal{N}} + \widetilde{\mathcal{P}}^T\widehat{\mathcal{L}})/8. \tag{D4b}$$

Acknowledgements. This work was not sponsored by any agency as I was transferring from the USA to China and was between the two systems.

Edited by: A. Tasora

References

Anand, L.: On H. Hencky's approximate strain-energy function for moderate deformations, J. Appl. Mech., 46, 78–82, 1979.

Anand, L.: Moderate deformations in extension-torsion of incompressible isotropic elastic materials, J. Mech. Phys. Solids, 34, 293–304, 1986.

Angeles, J.: On twist and wrench generators and annihilators, in: Computer-Aided Analysis Of Rigid And Flexible Mechanical Systems, edited by: Pereira, M. S. and Ambrosio, J., Dordrecht, NATO ASI Series, Kluwer Academic Publishers, 379–411, 1993.

Argyris, J.: An excursion into large rotations, Comput. Method. Appl. M., 32, 85–155, 1982.

Borri, M., Trainelli, L., and Bottasso, C.: On representations and parameterizations of motion, Multibody Syst. Dyn., 4, 129–193, 2000.

Bauchau, O.: Flexible Multibody Dynamics, Dordrecht, Heidelberg, London, New-York, Springer, 2011.

Bauchau, O. and Choi, J.: The vector parameterization of motion, Nonlinear Dynam., 33, 165–188, 2003.

Bauchau, O. and Craig, J.: Structural Analysis with Application to Aerospace Structures, Springer, Dordrecht, Heidelberg, London, New-York, 2009.

Bauchau, O. and Li, L.: Tensorial parameterization of rotation and motion, J. Comput. Nonlin. Dyn., 6, 1–8, 2011.

Bauchau, O. and Trainelli, L.: The vectorial parameterization of rotation, Nonlinear Dynam., 32, 71–92, 2003.

Degener, M., Hodges, D. H., and Petersen, D.: Analytical and experimental study of beam torsional stiffness with large axial elongation, J. Appl. Mech., 55, 171–178, 1988.

Ibrahimbegović, A.: On the choice of finite rotation parameters, Comput. Method. Appl. M., 149, 49–71, 1997.

Kadlowec, J., Wineman, A., and Hulbert. G.: Elastomer bushing response: Experiments and finite element modeling, Acta Mech., 163, 25–38, 2003.

Kane, T.: Dynamics, New York, Holt, Rinehart and Winston, Inc., 1968.

Ledesma, R., Ma, Z.-D., Hulbert, G., and Wineman, A.: A nonlinear viscoelastic bushing element in multibody dynamics, Comput. Mech., 17, 287–296, 1996.

Malvern, L. E.: Introduction to the Mechanics of a Continuous Medium, Prentice Hall, Inc., Englewood Cliffs, New Jersey.

Masarati, P. and Morandini, M.: Intrinsic deformable joints, Multibody Syst. Dyn., 23, 361–386, 2010.

Merlini, T. and Morandini, M.: The helicoidal modeling in computational finite elasticity. Part I: Variational formulation, Int. J. Solids Struct., 41, 5351–5381, 2004.

Milenković, V.: Coordinates suitable for angular motion synthesis in robots, in: Proceedings of the Robot VI Conference, Detroit MI, 2–4 March, Paper MS82-217, 1982.

Pennestrì, E. and Stefanelli, R.: Linear algebra and numerical algorithms using dual numbers, Multibody Syst. Dyn., 18, 323–344, 2007.

Shuster, M. A survey of attitude representations, J. Astronaut. Sci., 41, 439–517, 1993.

Wiener, T.: Theoretical Analysis of Gimballess Inertial Reference Equipment Using Delta-Modulated Instruments, Ph.D. thesis, Massachusetts Institute of Technology, Department of Aeronautical and Astronautical Engineering, Cambridge, Massachusetts, 1962.

3

Earthquake dynamic response of large flexible multibody systems

E. V. Zahariev

Institute of Mechanics, Bulgarian Academy of Sciences, Sofia, Bulgaria

Correspondence to: E. V. Zahariev (evtimvz@bas.bg)

Abstract. In the paper dynamics of large flexible structures imposed on earthquakes and high amplitude vibrations is regarded. Precise dynamic equations of flexible systems are the basis for reliable motion simulation and analysis of loading of the design scheme elements. Generalized Newton–Euler dynamic equations for rigid and flexible bodies are applied. The basement compulsory motion realized because of earthquake or wave propagation is presented in the dynamic equations as reonomic constraints. The dynamic equations, algebraic equations and reonomic constraints compile a system of differential algebraic equations which are transformed to a system of ordinary differential equations with respect to the generalized coordinates and the reactions due to the reonomic constraints. Examples of large flexible structures and wind power generator dynamic analysis are presented.

1 Introduction

Earthquake shaking affects forced motion of the structure basement. The ground motion is registered by so called strong-motion accelerographs and normally consists in three orthogonal components of the ground acceleration. The velocity and displacement of the ground are obtained integrating the data of the accelerogram. They also can be analyzed to obtain direct estimation of peak ground motion, duration shaking and frequency. To specific seismic zones of the earth the so called seismic Zone factor Z is assigned which correspond to a fraction of the earth acceleration. For example for seismic zone 1 $Z = 0.075G$; for zone 4 $Z = 0.4G$ ($G = 9.81$ is the earth acceleration). In Fig. 1 the displacements (all measures are in SI UNITS) of North-South component of El Centro earthquake, California, 1940, as well as, the polynomial approximation (in red) of the numerical data is depicted. To provide guidelines and formulas that constitute minimum legal requirements for design of structures building codes are developed and in use within particular regions. These requirements are intended to achieve satisfactory performance of the structures subject of seismic excitation. Several organizations involved in earthquake-resistant design have published recommendations based on combination of theory, experiments and practical observations, and form the foundation of the official codes (Chopra, 2005).

Two methods for resistant design of buildings and structures are used in practice (Chopra, 2005): the lateral force method, and the dynamic model method. The first method is applicable for seismic zones low seismic risk and low-rise buildings. The buildings are modeled as single degree of freedom systems subject to specific excitation. The dynamic method could be applied to any structure but is normally used for high risk zones and structures over five stories. The structure is modeled as a many degree of freedom discrete system with concentrated masses at each level of the building. In earthquake resistant design of buildings, the maximal response include, displacements, accelerations, shear forces, overturning moments, and torques. Because of the complexity of the dynamics based earthquake resistant design of buildings Chopra and Goel (2002) developed a modal pushover analysis procedure for linearly elastic buildings. The procedure is extended to inelastic buildings and the errors in the procedure relative to a nonlinear response history analysis are documented.

In recent years innovative means of enhancing structure functionality and safety against natural hazards have been in various stages of research and development (Nishitani and

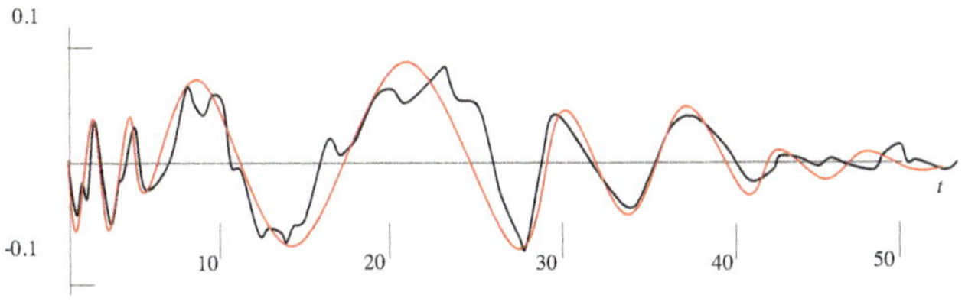

Figure 1. North-South displacement component of El Centro earthquake, California, 1940.

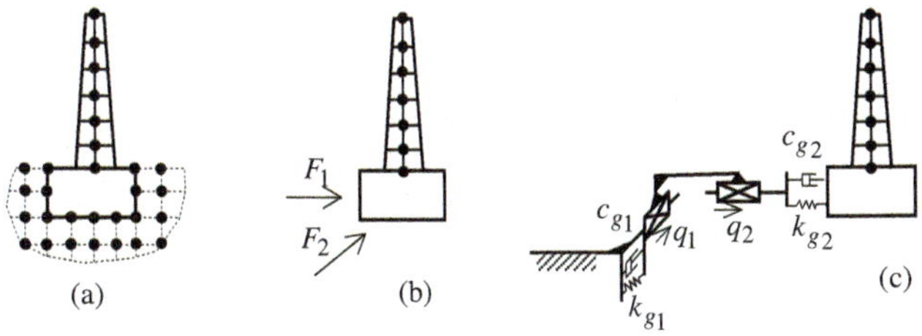

Figure 2. Modeling the ground-basement motion and force interaction.

Inoue, 2001). They can be grouped into three broad areas: base isolation; passive energy dissipation; applications as active control. Base isolation can now be considered a more mature technology with wider compared with the other two. Passive energy dissipation systems encompass a range of materials and devices for enhancing damping, stiffness and strength. In comparison with passive control, active control of structural response could be characterized by the following two features: a certain amount of external power or energy added as a result of a decision making process based on real time measured data (Soong and Dargush, 1997).

The inventive methods for earthquake structure response analysis and design are based on precise dynamic modeling and simulation. The up-to-date methods for multibody system dynamics simulation taking into account nonlinear behavior of the large flexible structures could be classify in three basic groups: floating frame of references (Shabana, 2005), finite elements in relative coordinates (Zahariev, 2006); and absolute nodal coordinate formulation (Shabana, 1996; Shabana and Mikkola, 2003). The large structures and mechanical devices like skyscrapers, antenna, wind power generators (WPG), etc., are typical multibody systems compiled of rigid and flexible parts with nonlinear behavior. Shaking of their foundations because of ground wave propagation, as well as, the nonlinear interaction ground – structure basement could be effectively analyzed using the recent achievements in the multibody system dynamics simulation methodology. The forward dynamic analysis provides the designers with preliminary information for structure behavior and the effect of the design parameters.

In the paper finite elements in relative coordinates and generalized Newton–Euler dynamic equations (Zahariev, 2006) are applied for dynamic simulation of large flexible structures. Simulation of the ground oscillation is implemented using statistical data for earthquake ground acceleration during the disasters. The structure basement oscillations are regarded as reonomic constraints in the system of Differential Algebraic Equations (DAE). The structure response is analyzed integrating the dynamic equations so derived. Several virtual examples of motion simulation and flexible deflections of structures and WPG imposed on earthquakes are presented in the paper.

2 Ground-structure basement interaction

The earthquake wave propagation is well studied and sufficient data collection is at the disposal of the scientists and engineers. But, according to the author opinion, information about the structure basement response because of ground motion and wave propagation is still not well studied and additional investigations are needed. Laboratory experiments together with the simulation results could help for revealing the whole picture of the ground-basement interaction during an earthquake. In some cases the ground, imposed on wave propagation, could behave as a viscous material and the interaction with the basement are to be analyzed by means of hydrodynamics methodology. Obviously, further investigations in this direction are to be conducted.

In the paper, for simulation purposes only, the ground is considered as environment with specific elastic and damping properties, which in certain conditions could experience plastic deflections. In Fig. 2 the two dimensional (2-D) kinematic model of ground-basement interaction of a structure imposed of earthquake is shown. The ground is presented by finite element (FE) discretization (Fig. 2a). The structure is compiled of rigid body basement and a flexible tower of FEs. The precise dynamic model should consider the wave propagation of the ground and the interaction with the basement. Using the methodology of the hydrodynamics the reduced forces (F_1 and F_2, Fig. 2b) loading the basement could be calculated and included in the dynamic equations of the structure. Implementation of this method requires enormous statistical data for the ground properties and local earthquake wave propagation. The simulation of the basement motion is much simpler when it is built on a rigid ground, for example rocks. Then the basement motion could be considered coincident with the ground motion read from the statistical data.

A simplified model for ground-basement interaction is presented in Fig. 2c. Virtual springs and dampers represent the elasticity and the damping properties of the ground, respectively. The spring properties are presented as a resistant force that depends on the deflections. Since the ground could exhibit residual deflections (virtual plastic deflections) the structure could change its location. The linear ground motion is 3-D and additional 2-D rotation (variation of the basement orientation) is possible.

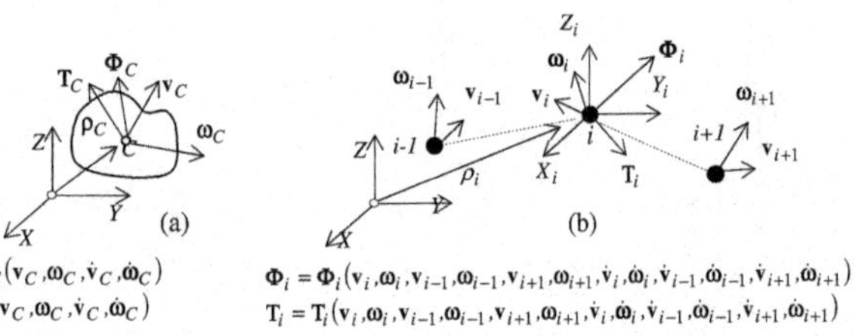

Figure 3. Inertia forces in a rigid body and nodes of finite elements discretization.

3 Generalized quasi-static dynamics of rigid and flexible systems

The quasi-static dynamics assumes the inertia forces as external loading. The inertia forces in the dynamic equations of finite element discretization are not precisely defined in the finite element theory that is discussed by Zahariev (2006). In Fig. 3 a free rigid body, as well as, a node i of a free flexible element and its neighbor nodes $i-1$ and $i+1$ are presented. The linear and angular velocities of the rigid body mass center ($\mathbf{v}_C$, ω_C) and of the nodes ($\mathbf{v}_i$, ω_i for a node i) are also shown. The body inertia forces Φ_C and torques $\mathbf{T}_C$ depend on its own velocities and accelerations, while the inertia forces $\mathbf{\Phi}_i$ and torques $\mathbf{T}_i$ loading the node i depend not only on its own velocities and accelerations but on these of the neighbor nodes $i-1$ and $i+1$ too. A node of a flexible element is a coordinate system which motion does not depend on the coordinates of the other nodes. The only thing that keeps the flexible element as a common set of nodes it is the stiffness of the material and the stiffness forces between the neighbor nodes.

In the literature of the finite element theory the inertia forces loading the nodes of a flexible element are defined by the expression (Doyle, 2001) $\mathbf{F} = \mathbf{M} \cdot \ddot{\mathbf{A}}$, where $\mathbf{F} = \begin{bmatrix} \mathbf{\Phi}_1^{\backslash} & \mathbf{T}_1^{\backslash} & \ldots & \mathbf{\Phi}_i^{\backslash} & \mathbf{T}_i^{\backslash} & \ldots & \mathbf{\Phi}_n^{\backslash} & \mathbf{T}_n^{\backslash} \end{bmatrix}^{\backslash}$ is the matrix vector of the inertia forces and torques in all nodes, $\mathbf{M}$ is the global mass matrix of the element, $\ddot{\mathbf{A}} = \begin{bmatrix} \dot{\mathbf{v}}_1^{\backslash} & \dot{\omega}_1^{\backslash} & \ldots & \dot{\mathbf{v}}_i^{\backslash} & \dot{\omega}_i^{\backslash} & \ldots & \dot{\mathbf{v}}_n^{\backslash} & \dot{\omega}_n^{\backslash} \end{bmatrix}^{\backslash}$. The inertia forces so computed do not include the velocity dependent terms as it is for the rigid body. The generalized Newton–Euler dynamic equations derived by Zahariev (2006) define the inertia forces in a flexible element and the velocity dependent terms are taken into account, i.e.:

$$\mathbf{F} = \mathbf{M} \cdot \ddot{\mathbf{A}} + \dot{\mathbf{A}}^{\otimes} \cdot \mathbf{M} \cdot \dot{\mathbf{A}} - \mathbf{M} \cdot \dot{\mathbf{\Theta}} = \begin{bmatrix} \cdots \\ \mathbf{\Phi}_{i-1} \\ \mathbf{T}_{i-1} \\ \mathbf{\Phi}_i \\ \mathbf{T}_i \\ \mathbf{\Phi}_{i+1} \\ \mathbf{T}_{i+1} \\ \cdots \end{bmatrix} = \mathbf{M} \cdot \begin{bmatrix} \vdots \\ \dot{\mathbf{v}}_{i-1} \\ \dot{\omega}_{i-1} \\ \dot{\mathbf{v}}_i \\ \dot{\omega}_i \\ \dot{\mathbf{v}}_{i+1} \\ \dot{\omega}_{i+1} \\ \vdots \end{bmatrix}$$

$$+ \begin{bmatrix} \cdots & \cdots & \cdots & \cdots & \cdots & \cdots & \cdots & \cdots \\ \cdots & \omega_{i-1}^{\times} & \mathbf{0} & \mathbf{0} & \mathbf{0} & \mathbf{0} & \mathbf{0} & \cdots \\ \cdots & \mathbf{v}_{i-1}^{\times} & \omega_{i-1}^{\times} & \mathbf{0} & \mathbf{0} & \mathbf{0} & \mathbf{0} & \cdots \\ \cdots & \mathbf{0} & \mathbf{0} & \omega_i^{\times} & \mathbf{0} & \mathbf{0} & \mathbf{0} & \cdots \\ \cdots & \mathbf{0} & \mathbf{0} & \mathbf{v}_i^{\times} & \omega_i^{\times} & \mathbf{0} & \mathbf{0} & \cdots \\ \cdots & \mathbf{0} & \mathbf{0} & \mathbf{0} & \mathbf{0} & \omega_{i+1}^{\times} & \mathbf{0} & \cdots \\ \cdots & \mathbf{0} & \mathbf{0} & \mathbf{0} & \mathbf{0} & \mathbf{v}_{i-1+}^{\times} & \omega_{i+1}^{\times} & \cdots \\ \cdots & \cdots & \cdots & \cdots & \cdots & \cdots & \cdots & \cdots \end{bmatrix} \cdot \mathbf{M}$$

$$\cdot \begin{bmatrix} \vdots \\ \mathbf{v}_{i-1} \\ \omega_{i-1} \\ \mathbf{v}_i \\ \omega_i \\ \mathbf{v}_{i+1} \\ \omega_{i+1} \\ \vdots \end{bmatrix} - \mathbf{M} \cdot \begin{bmatrix} \vdots \\ \omega_{i-1}^{\times} \cdot \mathbf{v}_{i-1} \\ \mathbf{0} \\ \omega_i^{\times} \cdot \mathbf{v}_i \\ \mathbf{0} \\ \omega_{i+1}^{\times} \cdot \mathbf{v}_{i+1} \\ \mathbf{0} \\ \vdots \end{bmatrix} \tag{1}$$

where $\dot{\mathbf{A}}^{\otimes} = \mathrm{diag}\left(\dot{\mathbf{A}}_1^{\otimes}, \dot{\mathbf{A}}_2^{\otimes}, \cdots, \dot{\mathbf{A}}_n^{\otimes}\right)$; $\dot{\mathbf{A}}_i^{\otimes} = \begin{bmatrix} \omega_i^{\times} & \mathbf{0} \\ \mathbf{v}_i^{\times} & \omega_i^{\times} \end{bmatrix}$, similarly to definition of a skew-symmetric matrix of a vector angular velocity $\omega_i^{\times}$, is a generalized skew-symmetric matrix of the linear and angular velocities of node i; $\dot{\mathbf{\Theta}} = \begin{bmatrix} \dot{\mathbf{\Theta}}_1^{\backslash} & \dot{\mathbf{\Theta}}_2^{\backslash} & \cdots & \dot{\mathbf{\Theta}}_n^{\backslash} \end{bmatrix}^{\backslash}$; $\dot{\Theta}_i = \begin{bmatrix} \omega_i^{\times} \cdot \mathbf{v}_i \\ \mathbf{0} \end{bmatrix}$; $\mathbf{0}$ is zero-matrix. As it could be seen from Eq. (1) the inertia forces are expressed with respect to the quasi-velocities and accelerations and do not depend on the kind of coordinates. That is significant advantage for application of the numerical algorithm for computer code generation of the dynamic equations and expansion of the numerical method of the Newton–Euler equations in flexible systems. Absolute velocity values are invariant relative to the kind of coordinate this approach, actually, could be applied to every kind of the coordinates ($\omega_i = \omega_i(\mathbf{q})$; $\mathbf{v}_i = \mathbf{v}_i(\mathbf{q})$, Where $\mathbf{q}$ is $n \times 1$ matrix-vector of the coordinates. Using the principle of the virtual work all external forces, including the inertia, stiffness and damping forces are reduced to the system coordinates to derive the dynamic equation. Applying the general partitioning methods or the augmented Lagrangian approach the system of differential equations subject to algebraic constraints could be reduced to a system of $n \times n$ ODE with respect to the generalized coordinates $\mathbf{q}$, i.e.:

$$\mathbf{M} \cdot \ddot{\mathbf{q}} + \mathbf{B}(\mathbf{q}, \dot{\mathbf{q}}) = \mathbf{S}. \tag{2}$$

$\mathbf{M}$ is the mass matrix, $\mathbf{q}$ are the generalized coordinates, $\mathbf{S}$ are the generalized forces, $\mathbf{B}(\mathbf{q},\dot{\mathbf{q}})$ is the velocity depend term.

4 Dynamics of multibody systems subject to reonomic constraints

Reonomic are the kinematic constraints that depend on time. For a multibody system with degree of freedom n which motion is described by ODE, Eq. (2), subject to m reonomic constraints

$$\ddot{q}_i = \ddot{q}_i(t), i = k+1, k+2, ..., k+m, \tag{3}$$

the dynamic equations are presented as follows:

$$\begin{aligned}&\mathbf{M}\cdot\left[\begin{array}{ccccccc} \ddot{q}_1 & \cdots & \ddot{q}_{k+1} & \cdots & \ddot{q}_{k+m} & \cdots & \ddot{q}_n \end{array}\right]^{\backslash}\\ &+\mathbf{B}(\mathbf{q},\dot{\mathbf{q}}) = \left[\begin{array}{ccccccc} S_1 & \cdots & S_{k+1} & \cdots & S_{k+m} & \cdots & S_n \end{array}\right]^{\backslash}\end{aligned} \tag{4}$$

In Eq. (4) the coordinates $\ddot{q}_i = \ddot{q}_i(t), i = k+1, k+2, .., k+m$ are known, while the generalized forces $S_i = S_i(t), i = k+1, k+2..., k+m$ are unknown. In other words, the solution of equations system, Eq. (4), results in solution of mixed direct and inverse dynamic problem to find the values of the coordinates $\ddot{q}_i = \ddot{q}_i(t), i = 1, ..., k, k+m+1, ..., n$ and the generalized forces $S_i = S_i(t), i = k+1, k+2..., k+m$. The dynamic equations, subject to reonomic constraints, Eq. (4), are transformed with respect to the unknown parameters as follows:

$$\begin{aligned}&\underline{\mathbf{M}}\cdot\left[\begin{array}{ccccccc} \ddot{q}_1 & \cdots & S_{k+1} & \cdots & S_{k+m} & \cdots & \ddot{q}_n \end{array}\right]^{\backslash}\\ &+\mathbf{B}(\mathbf{q},\dot{\mathbf{q}}) = -\underline{\underline{\mathbf{M}}}\cdot\left[\begin{array}{ccc} \ddot{q}_{k+1} & \cdots & \ddot{q}_{k+m} \end{array}\right]^{\backslash} + \underline{\mathbf{S}}\end{aligned} \tag{5}$$

where the matrices $\underline{\mathbf{S}}, \underline{\underline{\mathbf{M}}}, \underline{\mathbf{M}}$ and are as follows:

$$\underline{\mathbf{S}} = \left[\begin{array}{ccccccccc} S_1 & \cdots & S_k & 0 & \cdots & 0 & S_{k+m+1} & \cdots & S_n \end{array}\right]^{\backslash};$$

$$\underline{\underline{\mathbf{M}}} = \left[\begin{array}{cccc} m_{1,k+1} & m_{1,k+2} & \cdots & m_{1,k+m} \\ m_{2,k+1} & m_{2,k+2} & \cdots & m_{2,k+m} \\ \cdots & \cdots & \cdots & \cdots \\ m_{n,k+1} & m_{n,k+2} & \cdots & m_{n,k+m} \end{array}\right];$$

$$\underline{\mathbf{M}} = \left[\begin{array}{ccccccc} m_{1,1} & \cdots & 0 & \cdots & 0 & \cdots & m_{1,n} \\ \cdots & \cdots & \cdots & \cdots & \cdots & \cdots & \cdots \\ m_{k+1,1} & \cdots & -1 & \cdots & 0 & \cdots & m_{k+1,n} \\ \cdots & \cdots & \cdots & \cdots & \cdots & \cdots & \cdots \\ m_{k+m,1} & \cdots & 0 & \cdots & -1 & \cdots & m_{k+m,n} \\ \cdots & \cdots & \cdots & \cdots & \cdots & \cdots & \cdots \\ m_{n,1} & \cdots & 0 & \cdots & 0 & \cdots & m_{n,n} \end{array}\right].$$

5 Earthquake dynamic response of a WPG

The WPG is compiled of large flexible bodies (the blades) connected to a shaft (rigid or flexible) that transmits the torque to a gear with elastic clutch and electric generator machine. This structure together with the large flexible tower over which it is placed implements complex motion in space. It is imposed of wind load with transient behavior and is very sensitive of the changing working conditions of the environment.

Mainly two wind turbine concepts with their control strategies have been applied in practice, namely: active stall control wind turbine with induction generator; variable speed, variable pitch wind turbine with doubly-fed induction generator. The first group operates with almost constant speed. In this case, the generator directly couples the grid to drive train. The second one operates with variable speed. In this case, the generator does not directly couple the grid to drive train. Thereby, the rotor is permitted to rotate at any speed by introducing power electronic converters between the generator and the grid. The constant speed configuration is characterized by stiff power train dynamics due to the fact that electrical generator is locked to the grid; as a result, just a small variation of the rotor shaft speed is allowed. The construction and performance of this system are very much dependent on the mechanical characteristic of the mechanical subsystems. In addition, the turbulence and tower shadow induces rapidly fluctuation loads that appear as variations in the power. These variations are undesired for grid-connected wind turbine, since they result in mechanical stresses that decrease the lifetime of wind turbine and decrease the power quality.

Alternatively, variable speed configurations provide the ability to control the rotor speed. This allows the wind turbine system to operate constantly near to its optimum tip-speed ratio. Although the main disadvantage of the variable-speed configuration are the additional cost and the complexity of power converters required to interface the generator and the grid, its use has been increased due the above mentioned advantages.

There are not enough studies on the influence of the seismic loading to the WPG and in many cases the influence of such loading is not taken into account in the design process. The main purpose of the structural model of a wind turbine is to be able to determine temporal variation of the loads in various components in order to estimate fatigue damage. To calculate the deflections and velocities of various components in the wind turbine in the time domain, a precise dynamic model including the inertia terms is needed. The wind turbines are large flexible structures (more than 150 m in diameter) which blades implement complex motion in space with high velocities and rapidly changing accelerations of the blade deflections. The vibrations cause unstable working conditions, noise and random loads of the units. Because of the changing external loading over the blades the turbine shaft is imposed on bending and torsion including also impacts in the bearings.

Structures of WPGs imposed on earthquakes are investigated here in order to prove the reliability of the numerical algorithms for dynamic analysis subject to reonomic constraints. An example of the dynamic model of a WPG and its response because of prescribed ground motion is presented.

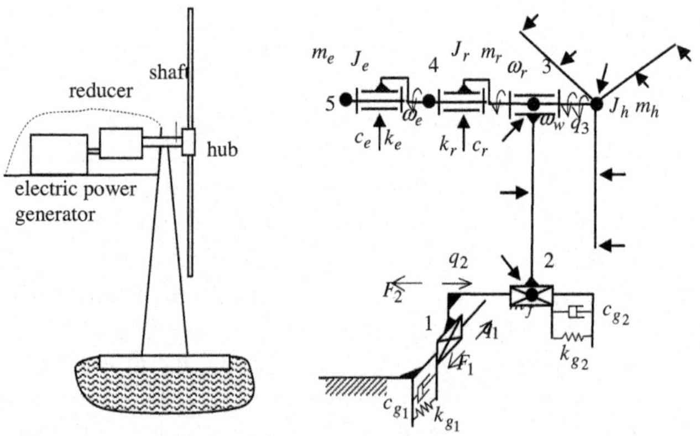

Figure 4. A design scheme and the discretized structure of a wind power generator.

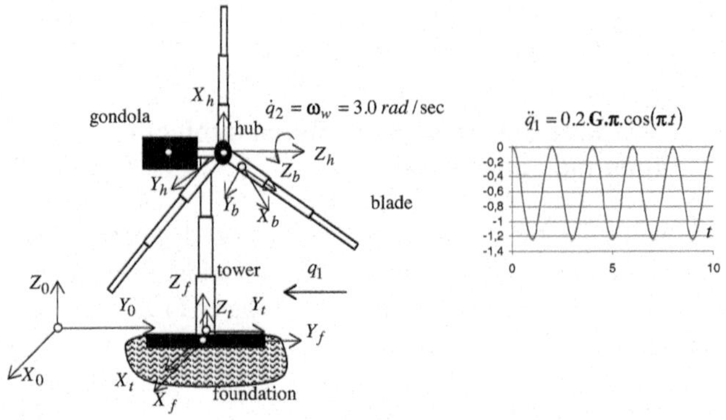

Figure 5. Example of motion simulation of constant speed WPG subject to ground shaking.

In Fig. 4 a simplified presentation of a WPG design scheme, as well as, its finite element discretization are shown. The gear increases the blade angular velocity ω_w to an optimal value ω_r for the electric power generator, ω_e. The elasticity of the gear shaft and the clutch causes different values of ω_r and ω_e. The reducer is assumed a rigid block with stiffness k_r and damping c_r, which values are provided by the manufacturer. The power unit is adjusted to an elastic tower and rigid body foundation. 2-D ground motion, respectively, two coordinates q_1 and q_2 are regarded. The coordinate q_3 is the shaft rotation of the blades. Friction forces F_1, F_2 appear between the ground and the foundation. The elastic and damping coefficients of the ground are $k_g\, c_g$. The flexible nodes are pointed out by arrows. The dense circles present the rigid bodies. Each flexible node adds six degree of freedom (dof). Some of them could be neglected (for example the longitudinal deflections). The translational joints numbered with 1 and 2 present the displacements of the ground. The mass of the foundation is m_f. The flexible blades are connected to a common hub with mass m_h and inertia J_h. Joints 3 represents the flexible shaft gear and the hub. The mass properties of the gear are m_r, J_r. Joint 4 represents the flexibility of the gear. Jount 5 represnts the flexibility of the clutch between the electric generator and the gear (stiffness k_e and damping c_e). The inertia proprties of the generator are m_e, J_e. Details for discretization of a rigid and flexible multibody system, as well as, for deriving the dynamic equations could be found in the paper of Zahariev (2006). The dynamic equations of the WPG (Fig. 4) with n dof, for which the kinematic constraints are transformed or the generalized partitioning approach is applied, are presented by second order ODE, Eq. (2). The reonomic constraints presenting 2-D ground motion are as follows:

$$\ddot{q}_1 = \ddot{q}_1(t); \quad \ddot{q}_2 = \ddot{q}_2(t). \tag{6}$$

Since the values of q_1 and q_2 are defined at every instant the Eqs. (2) and (6) are transformed to linear equation system with respect to $\ddot{q}_i, i = 3, 4, ..., n$ and the generalized forces in joints 1 and 2 (S_1, S_2), i.e.:

$$\begin{bmatrix} -\mathbf{E} & \underline{\underline{\mathbf{M}}} \\ \mathbf{0} & \end{bmatrix} \cdot \begin{bmatrix} S_1 \\ S_2 \\ \underline{\ddot{q}} \end{bmatrix} + \mathbf{B}(\mathbf{q}, \dot{\mathbf{q}}) = -\underline{\underline{\mathbf{M}}} \cdot \begin{bmatrix} \ddot{q}_1 \\ \ddot{q}_2 \end{bmatrix} + \underline{\mathbf{S}} \tag{7}$$

$\mathbf{E}$ and $\mathbf{0}$ are 2×2 and $n-2 \times n-2$ unity and zero matrices, respectively, $\underline{\underline{\mathbf{M}}}$ is the matrix compiled of the first two colunms of $\mathbf{M}$, $\underline{\mathbf{S}} = \begin{bmatrix} 0 & 0 & S_3 & \dots & S_n \end{bmatrix}^{\backslash}$. Two types of WPG are used in practice, i.e.: constant speed and variable speed (constant torque) WPG. For the first case there is an additional reonomic constraint of the type $\omega_w = \dot{q}_3 = \text{const.}$ For the second type WPG the torque of the blade shaft is approximately constant that is force constraint imposed on the dynamic equations. Integrating numerically Eq. (7) one obtains the values of the generalized coordinates, the forces between the ground and the foundation and, as a result, the global motion of the structure with respect to time.

In Fig. 5 and in Table 1 the discretization and mass distribution scheme of a virtual WPG are presented, as well as, its design and mass parameters. The structure is compiled of three rigid bodies, foundation, hub and gondola. The tower and the three blades are assumed flexible. Each of them is divided into three substructures with decreasing size from the basement and the hub to the tips. It is assumed two reonomic constraints. The first one $q_1 = q_1(t)$ is a ground motion perpendicular to the plain of the blades and the second one $q_2 = q_2(t)$ is prescribed constant rotation of the blade shaft. It is shown the orientation of the coordinate systems of the corresponding substructures and bodies.

Two numerical experiments are conducted. The first one (Fig. 6) is simulation of the WPG earthquake response and the resulting tower and blade tips displacements in case of no rotation of the blades (still position of the turbine). The second one (Fig. 7) is in case of turbine rotation with the prescribed angular velocity $\omega_w = 3\,\text{rad}\,\text{s}^{-1}$. As it could be seen from the examples the deflections of the tower and the blades are several times higher when the turbine rotates. This is as a result of the inertia forces caused by the velocity dependent

Table 1. Design and mass parameters of the WPG virtual structure (all measures are in SI).

RIGID BODIES							
	mass		inertia tensor		relative mass center		
foundation	5000		diag(100, 100, 100)		$[x\,y\,z]^{\backslash} = [0\;0\;0]^{\backslash}$		
hub	80		diag(1, 1, 1)		$[x\,y\,z]^{\backslash} = [0\;0\;0.3]^{\backslash}$		
gondola	300		diag(5, 5, 5)		$[x\,y\,z]^{\backslash} = [0\;0\;-1.8]^{\backslash}$		
FLEXIBLE ELEMENTS							
TOWER							
	length	cross section area	module of elasticity E	angular elasticity G	I_x	I_y	I_c
substructure 1	10.0	0.4	2.1×10^{12}	0.7×10^{11}	0.1	0.1	0.2
substructure 2	10.0	0.3	2.1×10^{12}	0.7×10^{11}	0.09	0.09	0.18
substructure 3	10.0	0.2	2.1×10^{12}	0.7×10^{11}	0.08	0.08	0.16
BLADES							
substructure 1	10.0	0.1	2.1×10^{12}	0.7×10^{11}	0.04	0.04	0.08
substructure 2	10.0	0.09	2.1×10^{12}	0.7×10^{11}	0.035	0.035	0.07
substructure 2	10.0	0.08	2.1×10^{12}	0.7×10^{11}	0.03	0.03	0.06

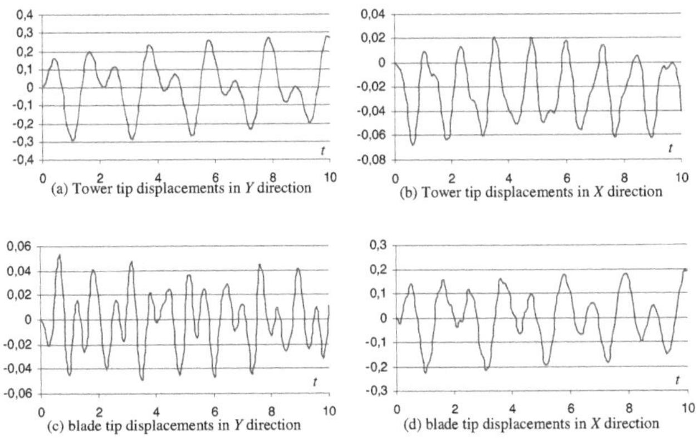

Figure 6. Displacements of the WPG imposed on earthquake without rotation of the turbine.

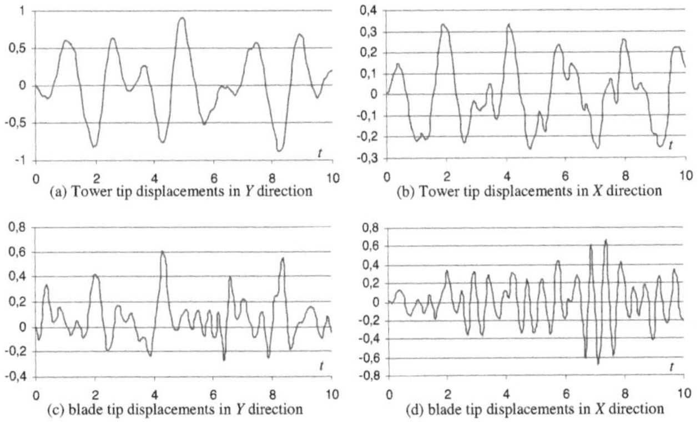

Figure 7. Displacements of the rotating WPG imposed on earthquake.

terms and the mutual influence of the ground motion and the rotation of the turbine. The examples demonstrate the effectiveness of the numerical algorithm, as well as, the necessity of the preliminary analysis of large flexible structures in various working conditions and scenarios.

6 Conclusions

Numerical algorithm for dynamic simulation of large flexible structures imposed on earthquakes is developed. Precise dynamic model of the flexible system is derived using generalized Newton–Euler dynamic equations for rigid and flexible bodies. A procedure for transformation of DAE subject to reonomic constraints to a system of ordinary ODE is proposed. Examples of motion simulation of WPG imposed on earthquake are depicted.

Acknowledgements. The author acknowledges the financial support of Ministry of Education, Youth and Science, contract No. DUNC-01/3, 12 March 2009.

Edited by: A. Müller

References

Chopra, A.: Dynamics of structures. A primer. Earthquake. Engineering Research Institute, 2005.

Chopra, A. and Goel, A.: Modal pushover analysis procedure for estimating seismic demands for buildings, Earthquake Engineering and Structural Dynamics, 31, 561–582, 2002.

Doyle, F.: Nonlinear Analysis of Thin-Walled Structures, Springer-Verlag, 2001.

Nishitani, A. and Inoue, Y.: Overview of the application of active/semiactive control to building structures in Japan, Earthquake Engineering and Structural Dynamics, 30, 1565–1574, 2001.

Shabana, A.: An absolute nodal coordinate formulation for the large rotation and deformation analysis of flexible bodies, Technical Report No. MBS96 1-UIC, Department of Mechanical Engineering, University of Illinois, Chicago, 1996.

Shabana, A.: Dynamics of multibody systems, 3rd Edn., Cambridge University Press, 2005.

Shabana, A. and Mikkola, A.: Use of the finite element absolute nodal coordinate formulation in modeling slope discontinuity, ASME J. Mech. Design, 125, 342–350, 2003.

Soong, T. and Dargush, G.: Passive energy dissipation systems in structural engineering, London, Wiley, 1997.

Zahariev, E.: Generalized finite element approach to dynamics modeling of rigid and flexible systems, Mech. Based Des. Struc., 34, 81–109, 2006.

4

Chrono: a parallel multi-physics library for rigid-body, flexible-body, and fluid dynamics

H. Mazhar[1], **T. Heyn**[1], **A. Pazouki**[1], **D. Melanz**[1], **A. Seidl**[1], **A. Bartholomew**[1], **A. Tasora**[2], **and D. Negrut**[1]

[1]Simulation Based Engineering Lab, Department of Mechanical Engineering, University of Wisconsin, Madison, WI, 53706, USA

[2]Department of Industrial Engineering, University of Parma, V.G.Usberti 181/A, 43100, Parma, Italy

Correspondence to: D. Negrut (negrut@engr.wisc.edu)

Abstract. The last decade witnessed a manifest shift in the microprocessor industry towards chip designs that promote parallel computing. Until recently the privilege of a select group of large research centers, Teraflop computing is becoming a commodity owing to inexpensive GPU cards and multi to many-core x86 processors. This paradigm shift towards large scale parallel computing has been leveraged in Chrono, a freely available C++ multi-physics simulation package. Chrono is made up of a collection of loosely coupled components that facilitate different aspects of multi-physics modeling, simulation, and visualization. This contribution provides an overview of Chrono::Engine, Chrono::Flex, Chrono::Fluid, and Chrono::Render, which are modules that can capitalize on the processing power of hundreds of parallel processors. Problems that can be tackled in Chrono include but are not limited to granular material dynamics, tangled large flexible structures with self contact, particulate flows, and tracked vehicle mobility. The paper presents an overview of each of these modules and illustrates through several examples the potential of this multi-physics library.

1 Introduction

Over the last decade there has been a manifest trend in the hardware industry to increase flop rates by increasing the number of cores available on a processor. To a very large extent, the tide that propelled sequential computing for several decades is subsiding. The frequency at which cores are operated today has at best plateaued; in many cases, it went down in an attempt to tame power dissipation and overheating. Instruction level parallelism advances that ensured respectable gains through pipelining and out of order execution have largely fulfilled their potential. The bright spot in this evolving hardware landscape has been the growing impetus behind parallel computing hardware. If anything has held steady over the last four decades, it has been the pace at which transistors are packed per unit area in computer chips. This trend allows today chip designs that draw on 22 nm feature length. Intel's road map calls for 14 nm technology in 2014, 10 nm in 2016, 7 nm in 2018, and 5 nm in 2020. In other words, the number of transistors per unit area will continue to double every two years for the current decade. This will translate into immediate access to commodity chips that host multiple compute cores. Given the stagnation in processor operating frequency, an ever growing gap between CPU speed and memory speed, and the waning of instruction level parallelism gains, it becomes apparent that the only way we can continue to enjoy reduced simulation times or ability to rely on refined models is to fall back on parallel computing. There are two major directions in which parallel computing has evolved. The x86 architecture has defined a solution that evolved as a steady and predictable process in which the number of cores on a chip increased over time: AMD produces today 16 core chips, while Intel has 12 core processors. Leveraging these chips requires a low entry point that calls for programming against relatively mature libraries such as OpenMP, MPI, pthreads, cilk, TBB, etc. At memory bandwidths of 75 GB s^{-1} and flop rates of 0.3 TFlop s^{-1}, this has traditionally represented the conservative choice for entering the parallel computing arena. With the release of CUDA 1.0 in 2006, NVIDIA offered a second alternative to leveraging parallel computing by programming the ubiquitous video cards available on millions of desktops worldwide. This path to parallel computing is less conventional as it requires one to get familiar with the hardware layout and memory hierarchy associated with GPUs. Today, an Nvidia GPU has close to seven billion transistors. Priced at about $6000, an Nvidia Kepler K20x delivers a memory bandwidth of 250 GB s^{-1} and 1.3 TFlop s^{-1} by virtue of using more than 2800 Scalar Processors. It is used side by side with a regular CPU processor, which means that heterogeneous computing, on the CPU and GPU, can lead to substantial speed gains. In this framework, the GPU plays the role of an accelerator by boosting the floating point performance of the CPU. A similar setup is offered now by Intel; i.e., CPU plus accelerator, owing to its recent release of the Knights Corner architecture. A Knights Corner chip has about 60 cores, can deliver up to 320 GB s^{-1} and 1 TFlop s^{-1}, and uses the x86 instruction set architecture, which translates into an easier adoption path provided one is familiar with OpenMP or MPI.

It becomes apparent that in the immediate future, any increase in simulation speed or model complexity in Computational Science will be fueled by parallel computing. This paper outlines an ongoing effort in the area of computational mutlibody dynamics that is motivated by this belief. It starts with a description of a core simulation engine that aims at simulation of many-body dynamics problems with friction and contact. Chrono::Engine handles both rigid and flexible bodies and draws on MPI and/or GPU computing. It then discusses Chrono::Fluid, a GPU parallel simulation tool that aims at fluid-solid interaction problems, which is singled out as an application area that has been largely ignored until recently due to an excessive computational burden incurred by the simulation of systems of practical relevance. Finally, the papers outlines a rendering pipeline that is used for postprocessing of big data. Chrono::Render is capable of using 320 cores and is built around Pixar's RenderMan. All these components combine to produce Chrono, a multi-physics simulation environment that is designed to take advantage of commodity parallel computing made available by many-core and GPU architectures.

2 Chrono::Engine

The Chrono::Engine software is a general-purpose simulator for three dimensional multi-body problems (Tasora and Anitescu, 2011). Specifically, the code is designed to support the simulation of very large systems such as those encountered in granular dynamics, where the number of interacting elements can be in the millions. Target applications include tracked vehicles operating on granular terrain (Heyn, 2009) or the Mars Rover operating on discrete granular soil. In these applications, it is desirable to model the granular terrain as a collection of many thousands or millions of discrete bodies interacting through contact, impact, and friction. Note that such systems also include mechanisms composed of rigid bodies and mechanical joints. These challenges require an efficient and robust simulation tool, which has been developed in the Chrono simulation package. Chrono::Engine was initially developed leveraging the Differential Variational Inequality (DVI) formulation as an efficient method to deal with problems that encompass many frictional contacts – a typical bottleneck for other types of formulations (Anitescu and Tasora, 2010; Tasora and Anitescu, 2010). This approach enforces non-penetration between rigid bodies through constraints, leading to a cone-constrained quadratic optimization problem which must be solved at each time step (Negrut et al., 2012). Chrono::Engine has since been extended to support the Discrete Element Method (DEM) formulation for handling the frictional contacts present in granular dynamics problems (Cundall, 1971; Cundall and Strack, 1979). This formulation computes contact forces by penalizing small interpenetrations of colliding rigid bodies. Various contact force models can be used depending on the application (Mindlin and Deresiewicz, 1953; Kruggel-Emden et al., 2007).

The remainder of this section describes the features of Chrono::Engine, starting with the structure of the code. Next, several sub-sections describe the use of GPU computing in the collision detection task, the use of MPI for distributed solution of large systems, and validation work which has been done to assess the accuracy of the simulation tool.

2.1 Code structure of Chrono::Engine

The core of Chrono::Engine is built around the concept of middleware, namely a layer of classes and functions that can be used by third-party developers to create complex mechanical simulation software with little effort (Tasora et al., 2007). Because of this, graphical user interfaces and end-user tools are not the main focus of the Chrono::Engine core project; it is assumed that programs with graphical interfaces are built on top of such middleware, or should be considered as additional, or optional, modules.

Given the complexity of the project, approaching half a million lines of code, the software is organized in classes and namespaces as recommended by the Object Oriented Programming paradigm, targeting modularity, encapsulation, reusability and polymorphism. The libraries of Chrono::Engine are thread safe, fully re-entrant, and include more than six hundred C++ classes. Objects from these classes can be instantiated and used to define models and simulations that run in third party software, for instance vehicle simulators, CAD tools, virtual reality applications, or robot simulators.

Chrono::Engine is completely platform-independent, hence libraries are available for Windows, Linux and Mac OSx, for both 32 bit and 64 bit versions. Moreover, we followed a modular approach, splitting the libraries in modules that can be dynamically loaded only if necessary, thus minimizing issues of dependency from other libraries and reducing memory footprint. For instance, we developed libraries for MATLAB interoperability, for real-time visualization through OpenGL, for interfacing with post-processing tools, etc. (see Fig. 1).

Classes and objects have been tested and profiled for fast execution, in order to achieve real-time performance when possible. Modern programming techniques have been adopted, like metaprogramming, class templating, class factories, memory leak trackers and persistent-transient data mapping. C++ operator overloading has been used to provide a compact algebra to manage quaternions, static and moving coordinate systems, and OS-agnostic classes are used for logging, streaming/checkpointing and exception handling.

We embraced an intense object-oriented approach, therefore most C++ objects that define parts of the multibody model are inherited from a base class called `ChPhysicsItem`, which defines the essential interfaces for all items that have some degrees of freedom. For example, specialized classes that inherit the `ChPhysicsItem` are the `ChBody` class, which is used for 3-D rigid bodies as shown in Fig. 2, `ChShaft`, which is used for 1-D concentrated parameter models of power trains, `ChLinkLockRevolute` that is a joint between rigid bodies, and so on. A set of more than thirty mechanical constraints are part of this class hierarchy. Furthermore, the architecture is open to further definition of new specialized classes for user-customized parts and joints. An object of `ChSystem` class stores a list of all moving parts and performs the simulation.

Each `ChPhysicsItem`-inheriting class can encapsulate a variable number of `ChLcpVariable` objects and/or a variable number of `ChLcpConstraint` objects, that are fed to the solver for Cone Complementarity Problems (CCP) at each time step of the DVI integration; this helps the development of black-box CCP solvers that are independent from the data structures of the physical layer. Also, these data structures represent the sparse data for the model description, which is completely matrix- and vector-free for the sake of a small memory footprint and fast linear algebra. Specifically, tthe system matrices for mass, Jacobians, etc. are never explicitly assembled. The objects of most of the above mentioned classes are managed by smart (shared) pointers with automatic deletion.

This relieves the programmer from the burden of taking care of object's lifetime, given that the relationships between objects can be quite complex as illustrated in Fig. 3. A large portion of the C++ classes are available also as Python modules; this enables the use of most simulation features in a scripted environment. Since novice users are more comfortable with Python than with C++, the Python interface proved to be optimal for teaching purposes. The Python interface was produced using the SWIG utility, a process that automatically generates the code for the Python wrapper.

The software architecture has been designed to accommodate an expandable system for handling assets (meshes, textures, CAD models), with multiple paths from pre-processing to post-processing. To this end, we also provide a C# add-in for a parametric 3-D CAD package (SolidWorks) that can be used to export models into Chrono::Engine without programming efforts (see Fig. 4).

2.2 Collision detection in Chrono::Engine

This section describes the collision detection algorithm designed and implemented for the Chrono::Engine package. Recall that problems of interest are focused on granular dynamics, such as sand flowing inside an hourglass, a rover running over sandy terrain, an excavator/frontloader digging/loading granular material, etc. In this context, the collision detection task is performed on a rather small collection of rigid and/or deformable bodies of complex geometry (hourglass wall, wheel, track shoe, excavator blade, dipper), and a very large number of bodies (millions to billions) that make up the granular material. On this scale, the collision detection task, particularly when dealing with the granular material, fits perfectly the Single Instruction Multiple Data (SIMD) computation paradigm. Specifically, the same sequence of instructions needs to be applied to every individual body and/or contact in the granular material. Therefore, a collision detection algorithm capable of leveraging the SIMD computational power of commodity Graphics Processing Units (GPUs) was developed and implemented to remove collision detection as the bottleneck in large granular dynamics simulations.

The parallel collision detection algorithm is separated into two phases, broadphase, and narrowphase. The broadphase algorithm quickly determines a list of potential contact pairs while the narrowphase algorithm determines actual contact information. A brief outline of the parallel collision detection algorithm is presented below, for more details see (Mazhar et al., 2011; Pazouki et al., 2012, 2010).

2.2.1 Broad-Phase algorithm

The Broad-Phase algorithm is used to compute whether two bodies might be in contact at a given time. The purpose of the broad-phase algorithm is not to find actual contact information, but rather to determine if a contact could potentially occur based on the Axis Aligned Bounding Boxes of the bodies involved.

An Axis Aligned Bounding Box (AABB) is a special case of a bounding box that is always aligned to the global reference frame, simplifying collision detection as the bounding box cannot rotate. Because of this, the volume enclosed by the bounding box will always be equal to or greater than the volume of the shape it encloses. AABB generation is simple and can be easily paralellized on a per object basis. See Fig. 5 for an example of AABB computation for a cylinder in 3-D space.

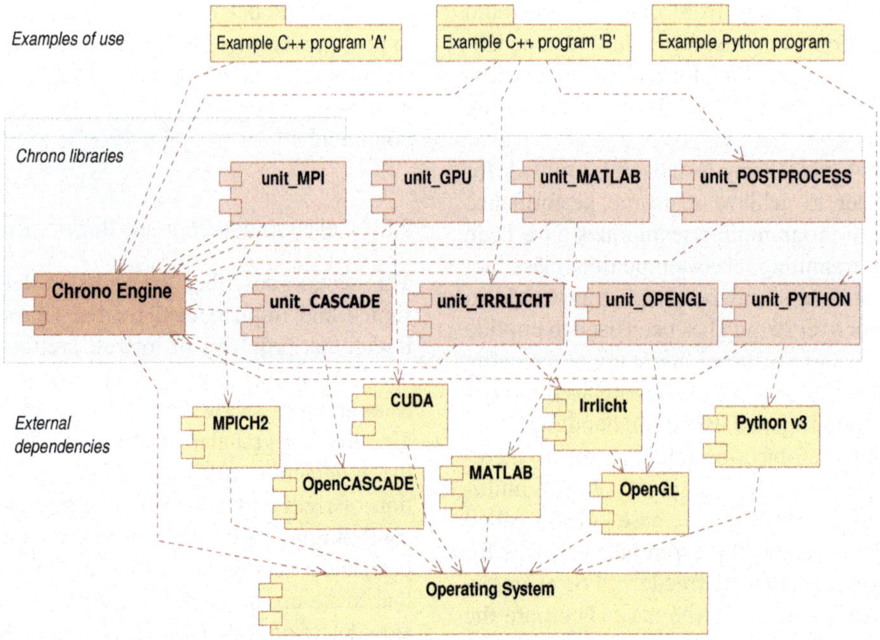

Figure 1. UML graph of dependencies between module libraries.

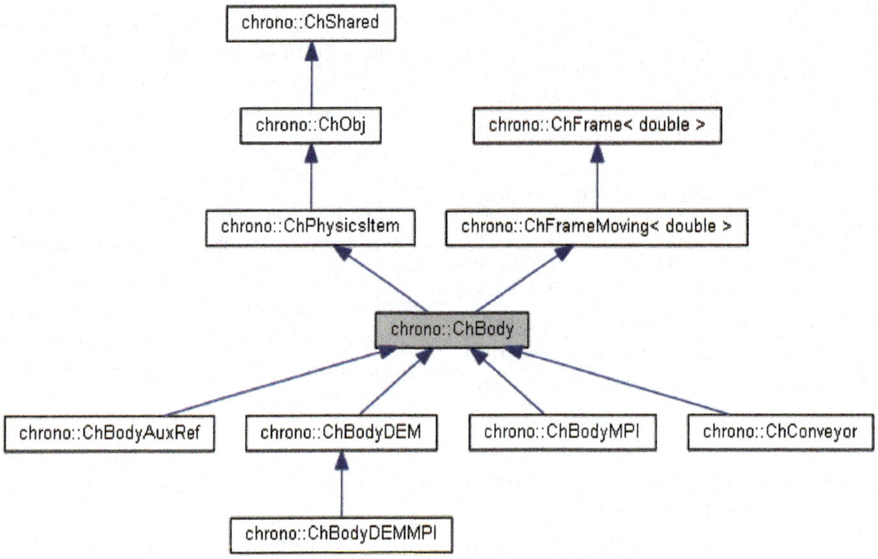

Figure 2. Class ineritance diagram for objects of ChBody type.

2.2.2 Spatial Subdivision algorithm

A high-level overview of the GPU-based collision detection is as follows. The collision detection process starts by identifying the intersections between AABBs and bins (see Fig. 6 for a visual representation of a bin). The AABB-bin pairs are subsequently sorted by bin id. Next, each bin's starting index is determined so that the bins' AABBs can be traversed sequentially. All AABBs touching a bin are subsequently checked against each other for collisions.

2.2.3 Narrow-Phase algorithm

Once potential contacts have been determined from the broad-phase collision detection stage, the Narrow-Phase algorithm needs to process each possible contact and determine if it actually occurs. To this end an algorithm capable of determining contacts between convex geometries was implemented on the GPU. This algorithm, called "XenoCollide" (Snethen, 2007), is based upon Minkowski Portal Refinement (MPR) (Snethen, 2008).

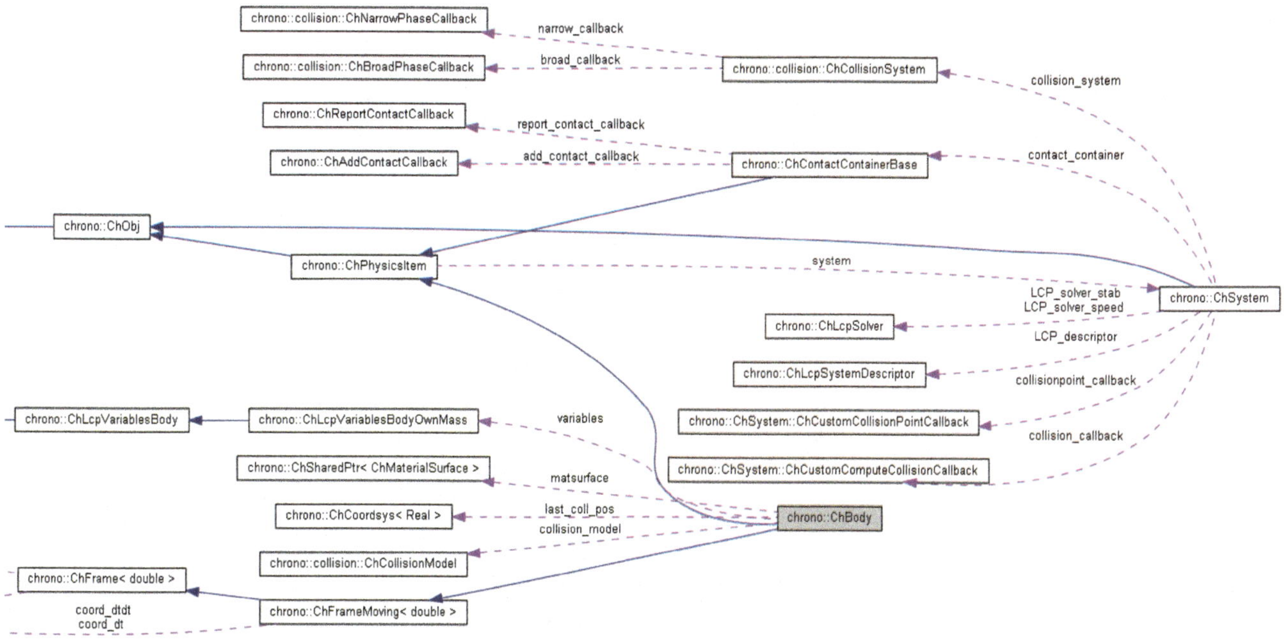

Figure 3. Collaboration graph between classes: example for ChBody and ChSystem.

2.3 Using MPI for distributed Chrono

Chrono has been further extended to allow the use of CPU parallelism for certain problems. To efficiently simulate large systems, a domain decomposition approach has been developed to allow the use of many-core compute clusters. In this approach, we divide the simulation domain into a number of sub-domains in a lattice structure. Each sub-domain manages the simulation of all bodies contained therein. Note that bodies may span the boundary between adjacent sub-domains. In this case, the body is considered shared and its dynamics may be influenced by the participating sub-domains. The implementation leverages the MPI standard (Gropp et al., 1999) to implement the necessary communication and synchronization between sub-domains.

This approach enables the simulation of large systems in two ways. First, it relies on the power of parallel computing since one computer core can be assigned to each MPI process (and therefore to each sub-domain). These processes can execute in parallel, constrained only by the required communication and synchronization. Second, it allows access to the larger memory pool available on distributed memory architectures. Whereas a single node or GPU card may have about 6 GB of memory, a distributed memory cluster may have on the order of 1 TB of memory, enabling the simulation of vastly larger problems.

Note that the domain decomposition approach currently uses the discrete element method to resolve friction and contact forces between elements in the system. The approach also supports constraints between bodies in the simulation by considering an assembly of constrained rigid bodies as a unit which must always be kept together. Therefore, if any body in a chain of constrained bodies is contained in a given sub-domain, all bodies in the chain are considered by that sub-domain and used to correctly solve the constraint equations.

2.3.1 Sub-division and set-up

A pre-processing step is used to discretize the simulation domain into a specified number of sub-domains, set up the communication conduits between processes, and initialize the sub-domains as appropriate. The sub-division is based on a cubic lattice with support for arbitrary sized divisions. The sub-domain boundaries are aligned with the global cartesian coordinate system, and their locations are user-specified. Separate MPI processes are mapped to each sub-domain. Note that at this time, the sub-division is static and does not change during the simulation. Therefore, the user should be careful to set up the discretization to maintain the best possible load balancing.

In terms of communication, each sub-domain in the grid can communicate with all other sub-domains. These communication pathways are set up and initialized during the preprocessing step and persist throughout the simulation.

Note that this implementation relies heavily on inheritance and the class-based structure of Chrono. For example, ChSystem is extended to ChSystemMPI by including the code to perform communication and synchronize the subdomains.

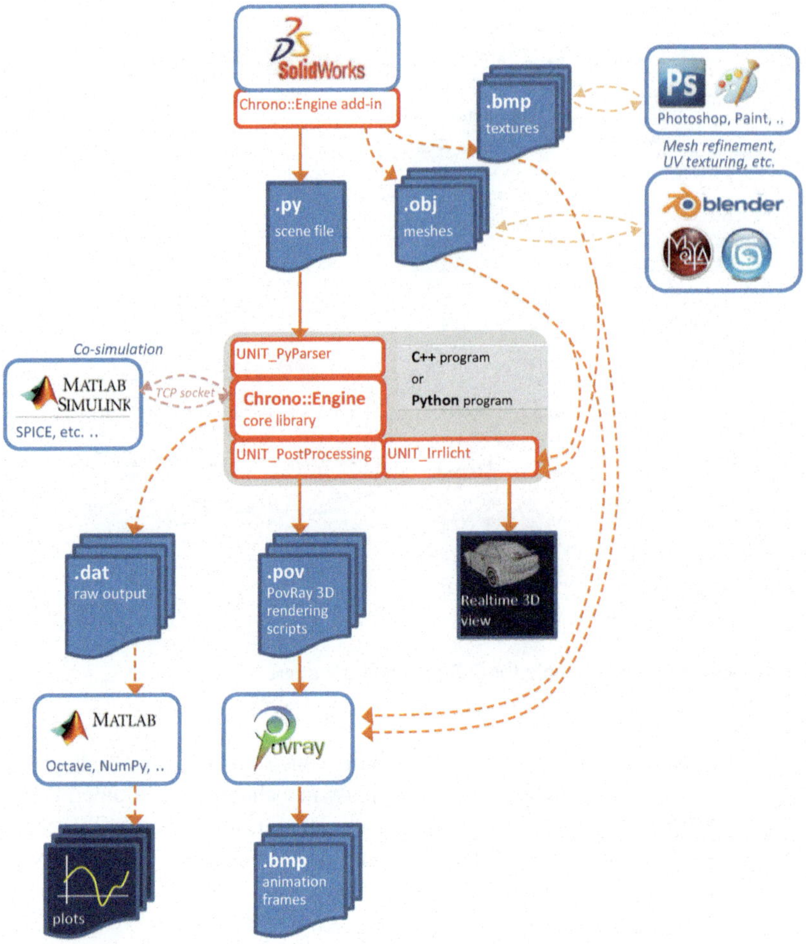

Figure 4. Network of asset workflows.

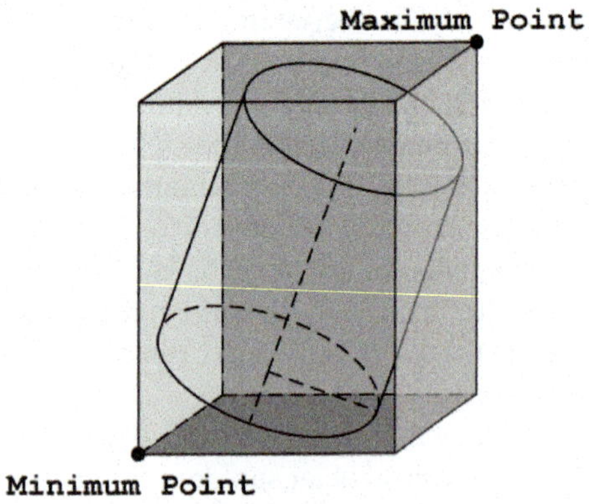

Figure 5. Example of AABB generation for 3-D cylinder.

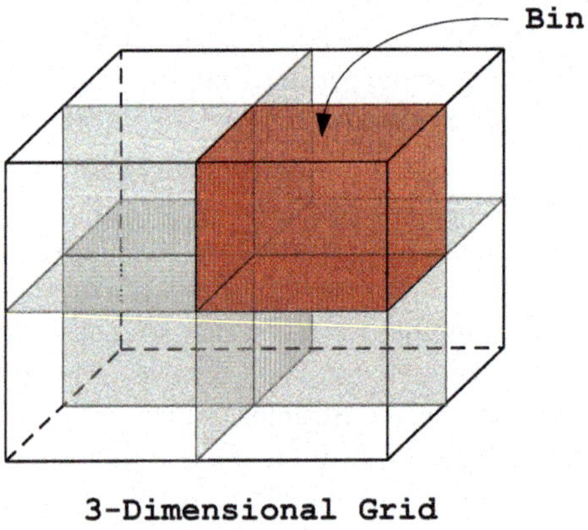

Figure 6. Example of 3-D space divided into bins.

2.3.2 Simulation and communication

Each sub-domain is now represented by a `ChSystemMPI` object and an associated MPI process. For example, assume a simulation is discretized into a set S of m sub-domains. In this case, let $S = \{A, B, C, D\}$ and $m = 4$, and map an MPI rank to each sub-domain so that A is mapped to MPI rank 0, $B \rightarrow 1$, $C \rightarrow 2$, and $D \rightarrow 3$. Each sub-domain maintains at all times $m+1$ lists of objects. The first list contains all objects which are even partially contained in the associated sub-domain. These are the objects which must be considered

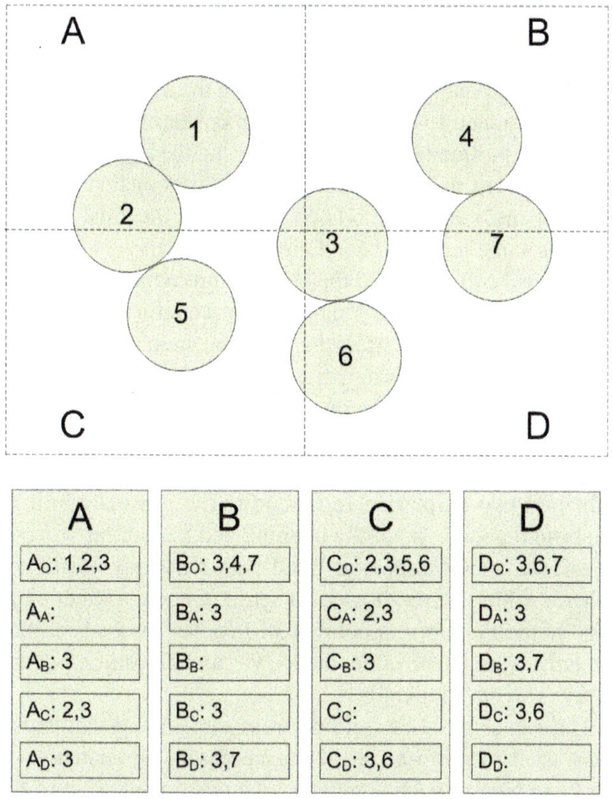

Figure 7. (Top) Sample 2-D simulation domain with four sub-domains and seven objects. (Bottom) Corresponding object lists for each sub-domain.

when computing contact forces, for example. The next m lists contain bodies which are shared with other sub-domains.

In our example, sub-domain B maintains the lists B_O, B_A, B_B, B_C, and B_D. B_O is the list of all objects that intersect (touch) sub-domain B, while B_A is the list of objects which are in sub-domain A and B. Note that sub-domain A has a list A_B which should contain the same objects as B_A. Further, list B_B is not used but is created for the sake of generality. All lists are maintained in order sorted by object ID number (see Fig. 7).

The sub-domains are now ready for time-stepping. Each sub-domain X performs collision detection among all objects in list X_O and computes the associated collision forces based on the DEM force model. Then, sub-domain X computes the net force on each object in list X_O, taking into account the contact forces, gravitational forces, and applied forces.

Next, mid-step communication occurs. Sub-domain X should send to each sub-domain Y the net force on each body in list X_Y. Similarly, X should receive from each Y the net force on each body in list X_Y. Finally, X should compute the total force on each body in list X_O. Note that X may receive force contributions for a given body from any or all of the other sub-domains in the system.

At this point, each sub-domain X has the true net force on each body in its list X_O. Each sub-domain can advance the state of its bodies in time by one time step by computing the new accelerations, velocities, and positions of all objects in the sub-domain given their mass/inertia properties and the set of applied forces. We perform an end-of-step communication to synchronize object states among sub-domains. All sub-domains which share a given body should compute its new state identically, but due to the potential for round-off error we synchronize the state from the master sub-domain (where the center-of-mass is located) to all others. The final stage is to process the $m+1$ lists in each sub-domain, as objects may enter or leave a given sub-domain or be shared between a different set of sub-domains, necessitating updates of the contents of the lists.

2.3.3 Example simulation

In this example we simulate a Mars Rover type wheeled vehicle operating on granular terrain. The vehicle is composed of a chassis and six wheels connected via revolute joints. The wheels are driven with a constant angular velocity of π rad s^{-1}. The granular terrain is composed of 2 016 000 spherical particles. The simulation is divided into 64 sub-domains and uses a time step of 10^{-5} s. This small time step is necessary due to the use of the DEM approach to compute contact forces – a stiff force model is used to achieve small normal interpenetration, requiring a small step size to maintain stability. A snapshot from the simulation can be seen in Fig. 8. In the figure, note that the wheels of the rover are checkered blue and white. This signifies that the master copy of the rover assembly is in the blue sub-domain and the rover spans into adjacent sub-domains. In Fig. 8, the rover has settled into the granular terrain and is starting to move forward. The rear wheels displace more granular material than the front wheels because the center of mass of the rover is closer to the rear of the vehicle.

2.4 Validation and demonstration of technology

This section describes a validation effort in which experimental results were compared to simulation results obtained from Chrono::Engine. To this end, a test rig was designed and fabricated to measure the rate at which granular material flowed out of a slit due to gravity. Chrono::Engine was used to set up a corresponding simulation to match the experimental results. For more detail, see Melanz et al. (2010).

2.4.1 Experimental model

The experimental set-up consisted of a fixed base, a movable wall (angled at 45°), a translational stage, a linear actuator, and a scale (see schematic in Fig. 9). The linear actuator was capable of quickly opening a precise gap, out of which the granular material would flow due to gravity. The

Figure 8. Snapshot of Mars Rover simulation with 2 016 000 terrain particles using 64 sub-domains. Bodies are colored by sub-domain, with shared bodies (those which span sub-domain boundaries) colored white.

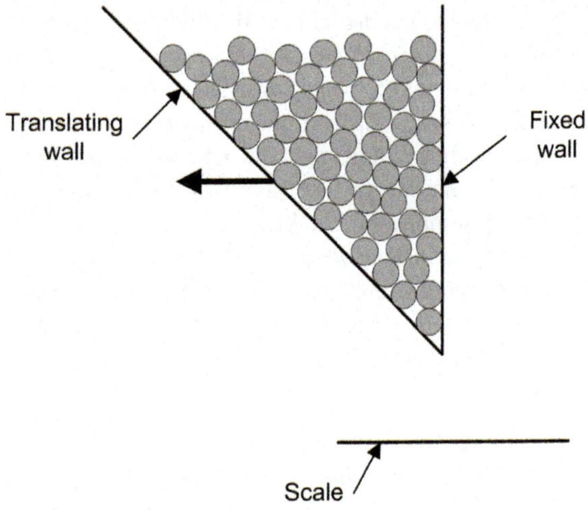

Figure 9. Schematic of validation experiment. A linear actuator and translational stage moved the left angled side a fixed amount, opening a precise gap from which the particles flowed. The mass flow rate was measured by the scale. Schematic not to scale.

scale recorded the mass of collected granular material as a function of time. The granular material consisted of approximately 40 000 uniform glass disruptor beads with diameter of 500 microns. Experiments were performed for gap sizes of 1.5 mm, 2 mm, 2.5 mm, and 3 mm. At least 5 experiments were performed for each gap size.

2.4.2 Simulation model

Chrono::Engine was used to build a model representing the experimental set up described above. In the model, the trough was represented by four rectangular boxes of finite dimensions. The motion of the box representing the angled side was captured from the data sheet of the translational stage. The granular material was modeled as perfect, identical spheres with the same mass and coefficient of friction.

The load cell measured the outflow through the gap. In the simulation, the scale was modeled by counting the number of spheres below a certain height. The number of spheres multiplied by the mass and gravity yielded the weight which was compared with experimental results. A plane was used to contain the spheres after they had been counted.

In order to save computational time, the simulation was split into two parts: one representing the process of filling the trough and the other the opening and measuring process. In this way, the trough was filled with randomly positioned spheres which were allowed to settle. Once the kinetic energy of the system was below 0.001 Joules and had reached a relatively constant value, the x-, y- , and z-position of each sphere was saved to a file.

The same initial conditions from the settling simulation were used to perform all of the necessary simulations. At the beginning of each simulation the position data set of the spheres was loaded into the model and the spheres were created at the same positions they appeared in the filling process. The motion was applied to the translating side to achieve the desired gap size, and the material began to flow.

The simulations setup consisted of 39 000 rigid body spheres with a radius of 2.5×10^{-4} m and a mass of 1.631×10^{-7} kg. The following parameters were set for this simulation. A time step of 10^{-4} s with 500 CCP iterations, and a tolerance of 10^{-7} for the maximum velocity correction. Simulations were generally run for 8 s. SI units were used for all parameters.

2.4.3 Procedure used to select the friction coefficient

The friction coefficient of a certain material is not a constant value. It can depend on various environmental influences such as humidity, surface quality, temperature etc. The friction coefficient of glass was an unknown in the validation process and needed to be determined before further observations could be done. To achieve this, one experiment at a gap size of 1.5 mm was performed and multiple simulations with the same setup and different friction coefficients were performed. The simulation results were compared to the experimental test results to determine which friction coefficient resulted in the best match, see Fig. 10. It was determined that $\mu = 0.15$ most closely matched the experimental results. This value was used for all subsequent simulations.

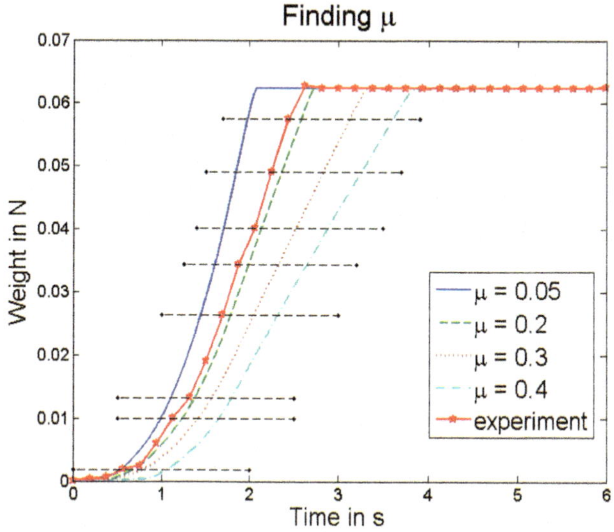

Figure 10. Selection of μ.

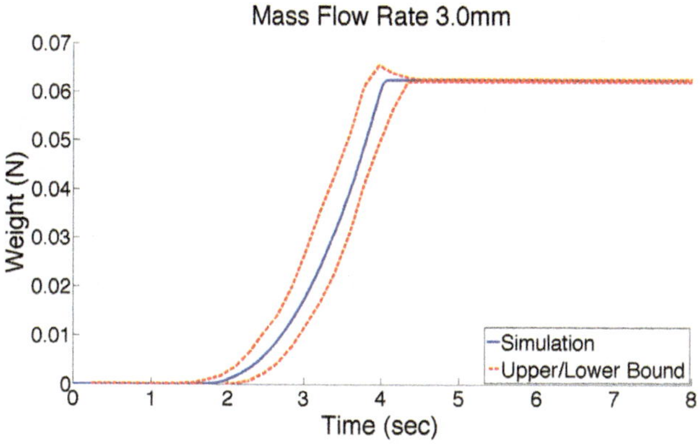

Figure 11. Weight vs. time for a gap size of 3 mm.

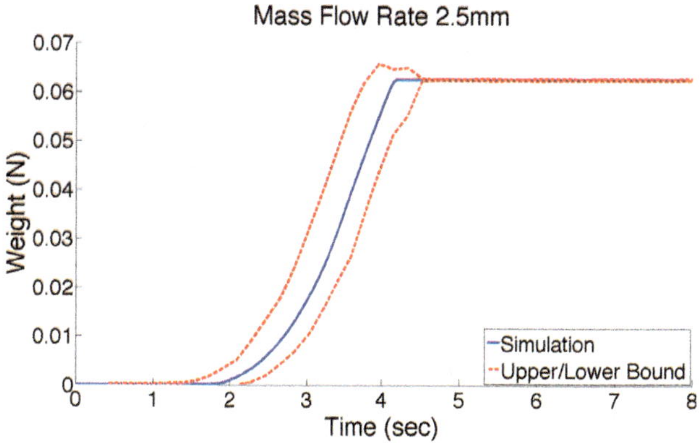

Figure 12. Weight vs. time for a gap size of 2.5 mm.

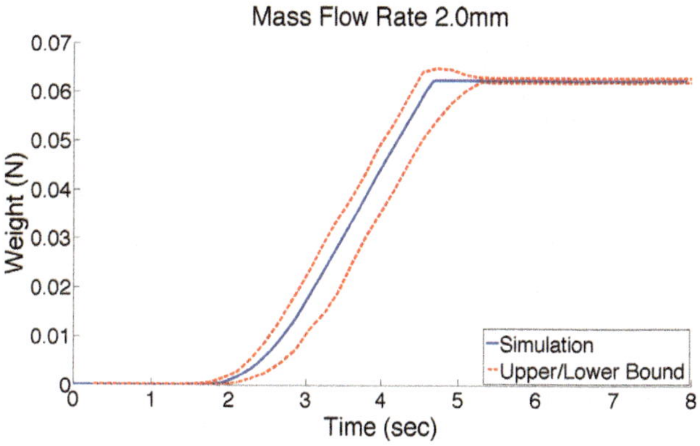

Figure 13. Weight vs. time for a gap size of 2 mm.

2.4.4 Results

The weight of the collected granular material is plotted versus time for various gap sizes in Fig. 11 through Fig. 14 using the friction coefficient determined in Fig. 10. For each experiment, the result from the simulation in Chrono::Engine, shown by the solid line, is overlaid on top of the standard deviation of the experimental runs, shown by the dashed line. Note that the simulated result lies within a single standard deviation of the experimental data.

3 Chrono::Flex

The Chrono::Flex software is a general-purpose simulator for three dimensional flexible multi-body problems and provides a suite of flexible body support. The features included in this module are multiple element types, the ability to connect these elements with a variety of bilateral constraints, multiple solvers, and contact with friction. Additionally, Chrono::Flex leverages the GPU to accelerate solution of large problems.

3.1 Element types

Chrono::Flex includes two element types implemented using the Absolute Nodal Coordinate Formulation (ANCF) (Berzeri et al., 2001; von Dombrowski, 2002). The gradient-deficient beam element and the gradient-deficient plate element are described below.

3.1.1 Gradient-deficient beam elements

This implementation uses gradient deficient ANCF beam elements to model slender beams, examples of which are shown in Fig. 15. These are two node elements with one position vector and only one gradient vector used as nodal coordinates. Each node thus has 6 coordinates: three components of the global position vector of the node and three components of the position vector gradient at the node. This formulation displays no shear locking problems for thin and stiff beams and is computationally more efficient compared to the original ANCF due to the reduced number of nodal coordinates (Gerstmayr and Shabana, 2006). The gradient deficient

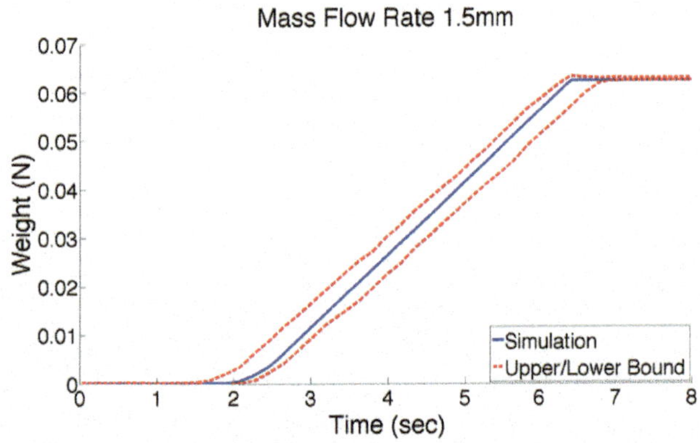

Figure 14. Weight vs. time for a gap size of 1.5 mm. This was the test case that was used for calibration.

Figure 15. Two models with friction and contact using Chrono::Flex beam elements: a ball sitting on grass-like beams and a ball hitting a net.

ANCF beam element does not describe a rotation of the beam about its own axis so the torsional effects cannot be modeled.

3.1.2 Gradient-deficient plate elements

Much like beams, numerical difficulties are encountered in the fully parameterized plate element when the system has very thin and stiff components (Dufva and Shabana, 2005). The high frequencies that are induced along the thin direction of the element require an extremely small time step, resulting in longer simulation times. In the case where the aspect ratio (length divided by thickness) of the element is high, plane stress assumptions can be made that allow a reduced-order element to be accurate. Specifically, Kirchhoff's plate theory, which does not account for shear deformation, is used and results in an element with 36 degrees of freedom, or nodal coordinates, are shown in Fig. 16.

3.2 Kinematic constraints

Several types of mechanical joints are modeled in Chrono::Flex. A spherical joint (Shabana, 2005) between two nodes of any two bodies will require the position vector of each node to be identical. A revolute joint will have two additional constraints to the spherical joint constraints. In this case, the gradient vectors of the two nodes will remain

Figure 16. Two models with friction and contact using Chrono::Flex plate elements: a cloth hanging on a sphere and a closed contour shaped like a tire.

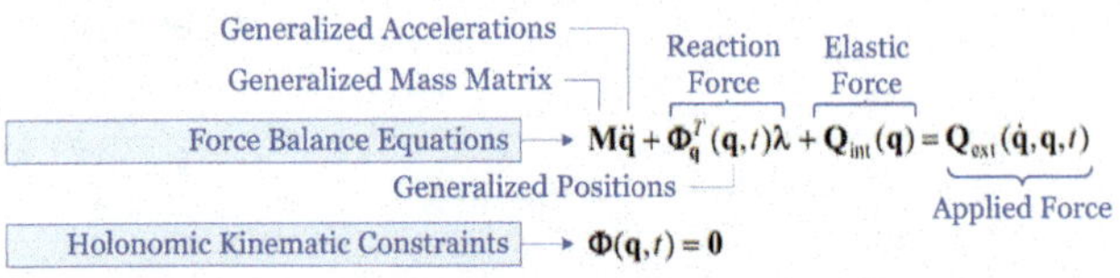

Figure 17. The equations of motion for Chrono::Flex.

in a plane perpendicular to the axis of revolute joint. There are also additional constraints due to the element connectivity in each beam. The element connectivity can be modeled as a fixed joint between the nodes. Here the common node between two elements is treated as two different nodes attached to each other through the fixed joint. This fixed joint requires all the nodal coordinates of the two nodes be identical. The generalized coordinates of the system change in time under the effect of applied forces such that these constraint equations are satisfied at all times. The time evolution of the system is governed by the Lagrange multiplier form of the constrained equations of motion.

3.3 Solvers

The equations shown in Fig. 17 form a system of index-3 Differential Algebraic Equations (DAEs). Although several low order numerical integration schemes have been effectively used to solve index-3 DAEs, Chrono::Flex utilizes the Newmark integration scheme (Hussein et al., 2008). Originally used in the structural dynamics community for the numerical integration of a linear set of second order ODEs, it was adapted for the discretization of DAEs. This implicit solver was proved to have convergence of order 1 or 2, depending on the choice of parameters γ and β.

At each time step, the numerical solution commences by solving the nonlinear set of equations shown in Fig. 18. The numerical solution of the nonlinear algebraic system falls back on a Newton-type iterative algorithm that requires the computation of its sensitivity matrix. Advancing the numerical solution in time draws on three loops: the outer-most loop marches forward in time, while at each time step the second loop solves the algebraic discretization problem in Fig. 18. Each iteration in this second loop launches a third

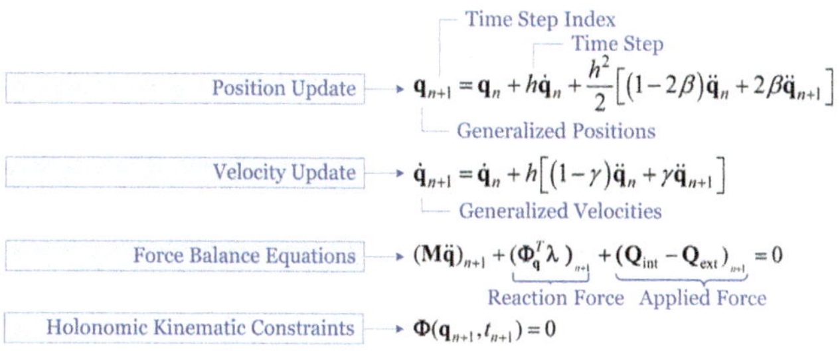

Figure 18. The discretized equations of motion for Chrono::Flex (fully implicit).

loop whose role is that of producing a vector of corrections for the acceleration and Lagrange multipliers. The corrections are computed using the BiCGStab iterative solver (Yang and Brent, 2002), which also provides for a matrix-free solution. A serial solver was implemented using the Armadillo Matrix Algebra Library (Sanderson, 2010) and a GPU parallel solver was implemented using CUSP (Bell and Garland, 2012), a linear algebra library built on top of CUDA. Chrono::Flex was validated in Khude et al. (2011) as well as in Melanz (2012) against the commercial code ADAMS (MSC.Software, 2012), and the nonlinear finite element analysis code ABAQUS (ABAQUS, 2004).

4 Chrono::Fluid

The Chrono::Fluid component aims at leveraging GPU computing to efficiently simulate fluid dynamics and fluid-solid interaction problems. Fluid-Solid Interaction (FSI) covers a wide range of applications, from blood and polymer flow to tanker trucks and ships. Simulation of the FSI problem requires two components: Fluid and Solid simulations. Simulation of the Solid phase, either rigid or flexible, in an HPC fashion, is described in previous sections. To leverage the existing solid phase simulation, the fluid flow simulation should satisfy some conditions, introduced by the aforementioned target problems. First, the fluid flow may experience large domain deformation due to the motion of the solid phase. Second, the two phases should be coupled via an accurate algorithm. Third, target problems may experience free surface as well as internal flow. Finally, the whole simulation should be capable of an HPC implementation to maintain the scalability of the code.

Fluid flow can be simulated in either an Eulerian or a Lagrangian framework. Provided that the interfacial forces are captured thoroughly, the Lagrangian framework is capable of tracking the domain deformation introduced by the motion of the solid phase at almost no extra cost. Smoothed Particle Hydrodynamics (SPH) (Lucy, 1977; Gingold and Monaghan, 1977; Monaghan, 2005), its modifications (Monaghan, 1989; Dilts, 1999), and variations (Koshizuka et al., 1998) have been widely used for the simulation of the fluid domain in a Lagrangian framework. The main evolution equations of the fluid flow using SPH are expressed as

$$\frac{\mathrm{d}\rho_a}{\mathrm{d}t} = \rho_a \sum_b \frac{m_b}{\rho_b} (v_a - v_b) . \nabla_a W_{ab} \quad (1)$$

$$\frac{\mathrm{d}v_a}{\mathrm{d}t} = -\sum_b m_b \left((\frac{p_b}{{\rho_a}^2} + \frac{p_a}{{\rho_b}^2}) \nabla_a W_{ab} - \frac{(\mu_a + \mu_b) r_{ab} . \nabla_a W_{ab}}{{\bar{\rho}_{ab}}^2 (r_{ab}^2 + \varepsilon \bar{h}_{ab}^2)} v_{ab} \right) + f_a \quad (2)$$

which are solved in conjunction with

$$\mathrm{d}x/\mathrm{d}t = v \quad (3)$$

to update the fluid properties. In Eqs. (1) to (3), ρ, v, and p are local fluid density, velocity, and pressure, respectively, m is the representative fluid mass assigned to the SPH marker, W is a kernel function which smooths out the local fluid properties within a resolution length $l = \kappa h$, and r_{ab} is the distance between two fluid markers denoted by a and b. Fluid flow evolution equations, defined by Eqs. (1) to (3), are solved explicitly, where pressure is related to density via an appropriate state equation to maintain the compressibility below 1 %. To increase the accuracy and stability of the simulation, an XSPH modification (Monaghan, 1989) and Shephard filtering (Dalrymple and Rogers, 2006) were applied.

4.1 FSI with Smoothed Particle Hydrodynamics: a quick overview

A proper choice of fluid-solid coupling should satisfy the no-slip and impenetrability conditions on the surface of the solid obstacles. By attaching Boundary Condition Enforcing (BCE) markers on the surface of the solid objects, the local relative velocity, i.e., at the markers location, of the two phases will be zero (see Fig. 19). The position and velocity of the BCE markers are updated according the motion of the solid phase, which results in the propagation of the solid motion to the fluid domain. On the other hand, the interaction forces on the BCE markers are used to calculate the total force and torque exerted by the fluid on the solid object.

4.1.1 FSI with Smoothed Particle Hydrodynamics: proximity computation

The overall simulation of the FSI framework is performed in parallel, where each thread handles the force calculation of a fluid or BCE marker first, and a rigid body later. Next, the parallel threads perform the kinematics update of the fluid markers, rigid bodies, and BCE markers, respectively. An essential part of the force calculation stage is the proximity computation, which will be explained briefly herein.

Proximity computation used in our work leverages the algorithm provided in CUDA SDK (NVIDIA Corporation, 2012), where the computation domain is divided into bins

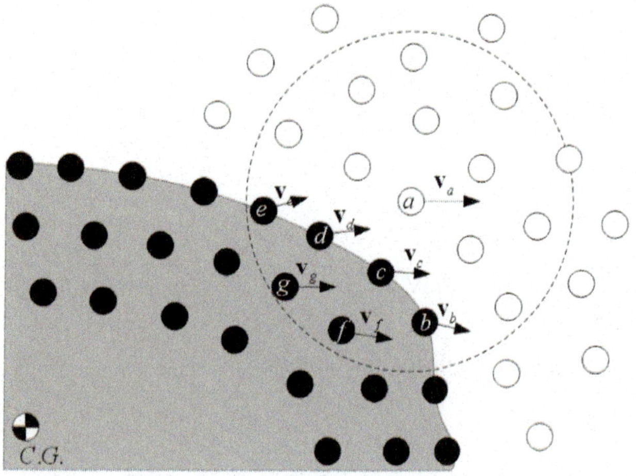

Figure 19. Coupling of the fluid and solid phases. BCE and fluid markers are represented by black and white circles, respectively.

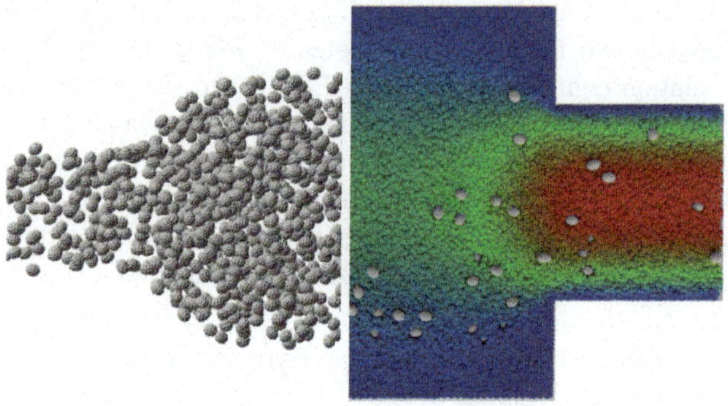

Figure 20. Simulation of rigid bodies inside a fluid flow: rigid ellipsoids with their BCE markers are shown in the left image while the fluid's velocity contours and rigid ellipsoids at the mid-section of the channel are shown in the right image.

whose sizes are the same as the resolution length of the SPH kernel function. A hash value is assigned to each marker based on its location with respect to the bins. Markers are sorted based on their hash value. The sorted properties are stored in independent arrays to improve the memory access and cache coherency. To compute the forces on a marker, the lists of the possible interacting markers inside its bin and all 26 neighbor bins are called. The hash values of the bins are used to access the relevant segments of the sorted data.

4.2 Validation and demonstration of technology

The aforementioned FSI simulation engine was used to validate the lateral migration of cylindrical particles in plane Poiseuille flow, spherical particles in pipe flow, and particle distribution in Poiseuille flow of suspension (Pazouki and Negrut, 2012,?). Due to the scalability of Chrono::Fluid in both fluid and solid phases, increasing the number of rigid bodies, which translates into decreasing the number of fluid

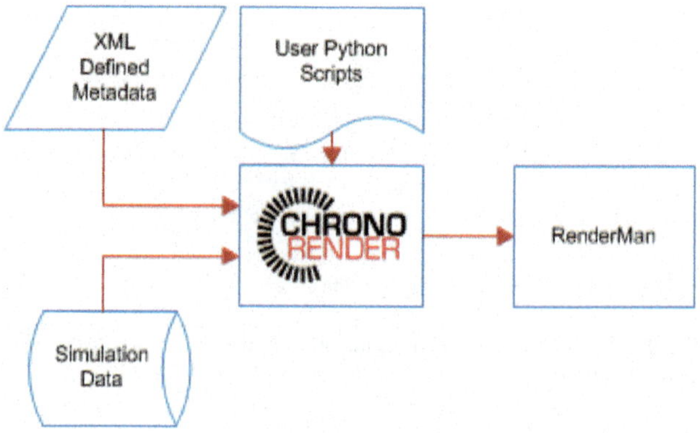

Figure 21. Chrono::Render architecture.

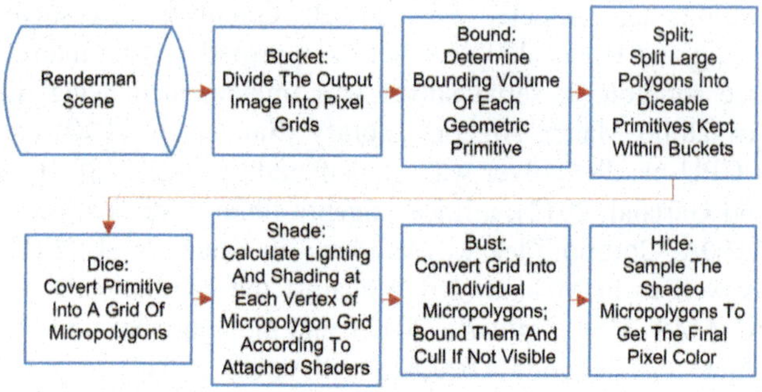

Figure 22. An overview of the REYES Pipeline.

particles, does not affect the simulation time. Therefore, the simulation of a highly dense suspension is possible. Figure 20 shows the result of the simulation of the flow of suspension including 1500 particles through a channel. A similar scenario with 13 000 particles in suspension was simulated in Chrono::Fluid.

5 Chrono::Render

Chrono::Render is a software package that enables simple, streamlined, and fast visualization of arbitrary data using Pixar's RenderMan (Pixar, 1988, 1989, 2000, 2005). Specifically, Chrono::Render contains a hybrid of processing binaries and Python scripting modules that seek to abstract away the complexities of rendering with RenderMan. Additionally, Chrono::Render is targeted for providing rendering as an automated post-processing step in a remote simulation pipeline, hence it is controlled via a succinct XML specification for "gluing" together rendering with arbitrary processes. As seen in Fig. 21, Chrono::Render combines simulation data, XML describing how to use the data, and optional user-defined Python scripts into a complex, visually-rich scene to be rendered by RenderMan.

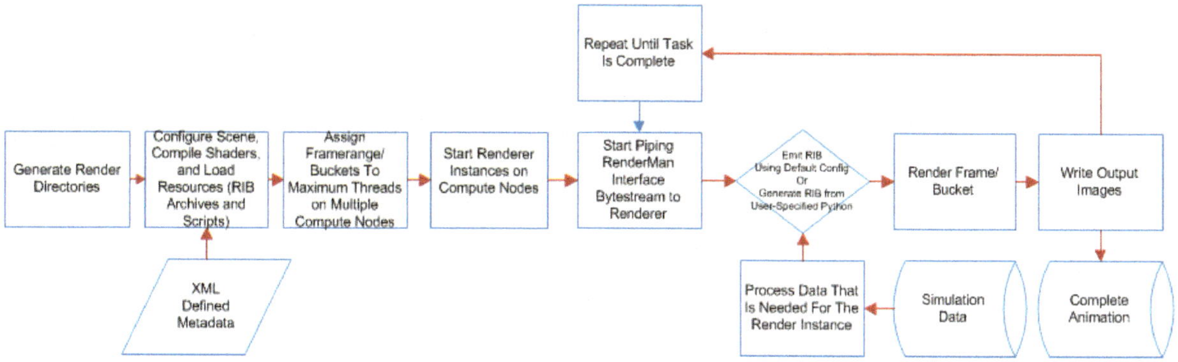

Figure 23. Chrono::Render execution workflow.

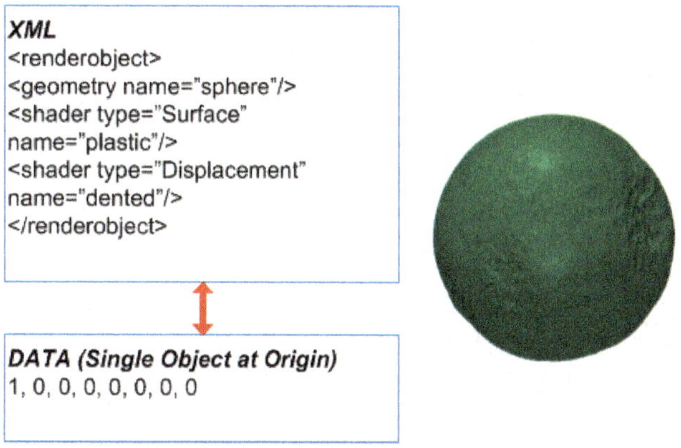

Figure 24. Simple XML for a sphere with a Surface and Displacement shader.

5.1 On the choice of RenderMan

Using RenderMan for rendering is motivated by the scope of arbitrary data sets and the potentially immense scene complexity that results from big data; REYES, the underlying architecture for RenderMan is ideally suited for this task. REYES works by dividing each surface in the scene into a grid of micropolygons and shades at the grid vertices (Cook et al., 1987) (see Fig. 22).

This results in tractable rendering for complex scenes because: (a) only a small portion of the scene needs to be in memory at any given time; (b) grid-based computation leads to optimal memory access patterns; (c) non-visible objects need not be loaded into memory; (d) fully-rendered objects can be removed from memory; and (e) objects are tessellated according to size on the screen; less complex geometry is dynamically loaded whenever possible.

REYES is perfectly suited for parallel processing since it scales linearly with the number of cores. Considering that REYES needs only a handful of relevant scene elements at a time, this data can be parsed into low-memory buckets and distributed amongst cores for parallel rendering; thus REYES' low memory-footprint and efficient concurrent resource usage for the complex scenes makes it a great renderer for a distributed-computing platform.

5.2 Accessibility of high-quality graphics

Although REYES can manage the issue of scene complexity, leveraging this power is difficult without computer graphics expertise. The guiding principle of Chrono::Render is to make high-quality rendering available to researchers, most of whom do not have the background or bandwidth to spend time learning how to use complex graphics applications or make sense of REYES' intricacies. Consequently, Chrono::Render encapsulates into the XML specification the complicated steps needed to make interesting visual effects, such as multipass rendering. The user must only instance the correct XML components to achieve high-quality renders. The program flow of Chrono::Render is shown in Fig. 23.

The XML specification allows for the concise expression of salient features and scene objects. For example, the snippet in Fig. 24 illustrates the XML file that translates a single line from a comma-separated value (CSV) data file into a RenderMan sphere using two shaders.

Although simple, the render is visually rich. This description is often enough to visualize most generic data, but it cannot handle all arbitrary visualizations, so in order to maintain generality we make use of Python scripts and wrappers to enable simplified procedural RenderMan Interface Bytestream generation. Any XML element can be scripted such that at runtime, the script output will be piped into the same rendering context. This makes it possible to perform processing for specialized data as well as modularize the rendering of specific effects. Obviously this adds more complexity for defining the scene, but Chrono::Render provides Python modules with methods and classes intended to ease this programming as much as possible. Additionally, most of the Chrono::Render Python modules wrap C++ functions and classes with the purpose of exploiting speed while still making use of the syntactical/type-free simplicity of Python. Figure 25 gives an example of combining XML with Python scripts to achieve a more complicated render.

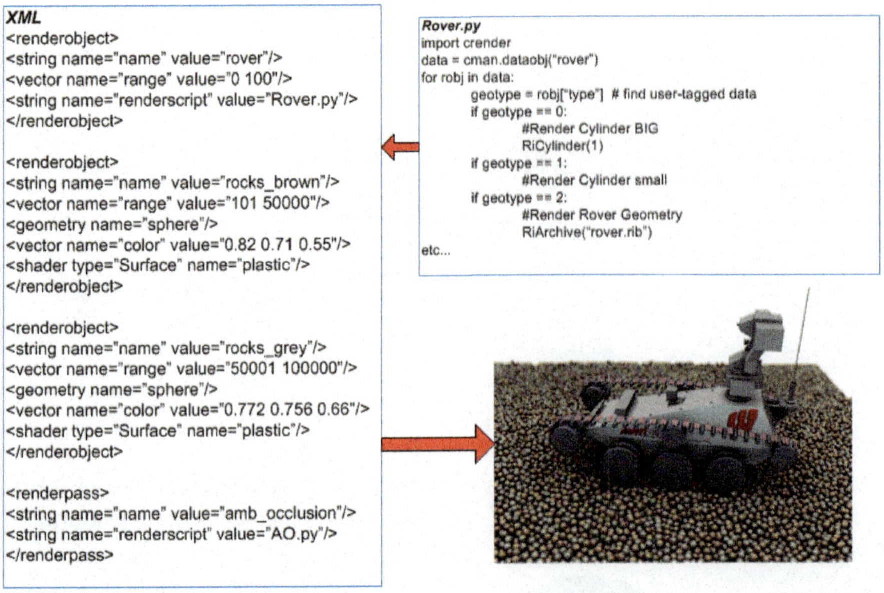

Figure 25. General purpose rendering with Chrono::Render. The Rover body contains multiple shape descriptions of which are generated from a Python script. Data is tagged with a name which can be later be accessed using some of Chrono::Render's Python functionality.

5.3 Other capabilities

Beyond interpreting parameters and data into RenderMan calls, Chrono::Render provides tools for bootstrapping rendering projects. Chrono::Render can: (a) construct directory structures for localizing and managing scene resources; (b) automate distribution of rendering across a multi-node network; (c) convert common graphics file formats into RenderMan file formats such as Wavefront Objs and Mtls to RenderMan RIBs and Shaders; (d) generate XML for automatically adding parameters to the scene description for describing advanced visual effects such as subsurface scattering, ambient occlusion, reflections, etc.; (e) mesh point-clouds, particularly useful for particle-based fluid simulations; and (f) dump the generated RenderMan calls to disk for reuse.

Chrono:Render is currently available for free download as a pre-built binary for Linux. Members of the Wisconsin Applied Computing Center can use this capability remotely as a service by leveraging 320 AMD cores on which Chrono::Render is currently deployed.

6 Conclusions and future work

The Chrono simulation package is composed of a collection of components designed to perform multi-physics simulations leveraging emerging high-performance computing hardware. Chrono::Engine provides support for rigid body dynamics, focusing on large granular dynamics problems, Chrono::Flex enables simulation of flexible beam and plate elements interacting through contact and bilateral constraints, while Chrono:Fluid allows the simulation of fluid flows and fluid-solid interaction problems. Finally, Chrono::Render provides high-quality visualization of arbitrary simulation data from the other Chrono components. These components have been designed to leverage high-performance computing hardware whenever possible. Chrono::Engine supports CPU parallelism through a domain-decomposition approach, while Chrono::Engine, Chrono::Flex, and Chrono::Fluid all support GPU parallelism to further improve simulation performance.

While these components provide useful simulation capabilities on their own, ongoing work seeks to further integrate the various Chrono components.

Chrono availability

Major releases of the Chrono::Engine software are available from the Chrono::Engine website at http://chronoengine.info. Chrono in its entirety can be downloaded from http://sbel.wisc.edu/chrono. The latter site also displays the nightly build status for various platforms and unit testing results.

Acknowledgements. Financial support for the Wisconsin authors was provided in part by the National Science Foundation Award 0840442 and Army Research Office W911NF-12-1-0395. Financial support for A.Tasora was provided in part by the Italian Ministry of Education under the PRIN grant 2007Z7K4ZB. We thank NVIDIA and AMD for sponsoring our research programs in the area of high-performance computing.

Edited by: A. Müller

References

ABAQUS: User Manual – Version 6.5, Hibbitt, Karlsson and Sorensen, Inc., Pawtucket, RI, 2004.

Anitescu, M. and Tasora, A.: An iterative approach for cone complementarity problems for nonsmooth dynamics, Comput. Optim. Appl., 47, 207–235, doi:10.1007/s10589-008-9223-4, 2010.

Bell, N. and Garland, M.: CUSP: Generic Parallel Algorithms for Sparse Matrix and Graph Computations, http://cusp-library.googlecode.com, version 0.3.0, 2012.

Berzeri, M., Campanelli, M., and Shabana, A. A.: Definition of the Elastic Forces in the Finite-Element Absolute Nodal Coordinate Formulation and the Floating Frame of Reference Formulation, Multibody Syst. Dyn., 5, 21–54, 2001.

Cook, R. L., Carpenter, L., and Catmull, E.: The Reyes Image Rendering Architecture, SIGGRAPH 1987 Proceedings, 95–102, 1987.

Cundall, P.: A computer model for simulating progressive large-scale movements in block rock mechanics, in: Proceedings of the International Symposium on Rock Mechanics, Nancy, France, 1971.

Cundall, P. and Strack, O.: A discrete element model for granular assemblies, Geotechnique, 29, 47–65, 1979.

Dalrymple, R. and Rogers, B.: Numerical modeling of water waves with the SPH method, Coast. Eng., 53, 141–147, 2006.

Dilts, G.: Moving-least-squares-particle hydrodynamics I. Consistency and stability, Int. J. Numer. Meth. Eng., 44, 1115–1155, 1999.

Dufva, K. and Shabana, A.: Analysis of thin plate structures using the absolute nodal coordinate formulation, P. I. Mech. Eng. K-J. Mul., 219, 345–355, 2005.

Gerstmayr, J. and Shabana, A.: Analysis of thin beams and cables using the absolute nodal co-ordinate formulation, Nonlinear Dynam., 45, 109–130, 2006.

Gingold, R. and Monaghan, J.: Smoothed particle hydrodynamics-theory and application to non-spherical stars, Mon. Not. R. Astron. Soc., 181, 375–389, 1977.

Gropp, W., Lusk, E., and Skjellum, A.: Using MPI: Portable Parallel Programming with the Message-Passing Interface, 2nd Edn., MIT Press, 1999.

Heyn, T.: Simulation of Tracked Vehicles on Granular Terrain Leveraging GPUComputing, M.S. thesis, Department of Mechanical Engineering, University of Wisconsin-Madison, http://sbel.wisc.edu/documents/TobyHeynThesis_final.pdf, 2009.

Hussein, B., Negrut, D., and Shabana, A.: Implicit and explicit integration in the solution of the absolute nodal coordinate differential/algebraic equations, Nonlinear Dynam., 54, 283–296, 2008.

Khude, N., Melanz, D., Stanciulescu, I., and Negrut, D.: A Parallel GPU Implementation of the Absolute Nodal Coordinate Formulation With a Frictional/Contact Model for the Simulation of Large Flexible Body Systems, ASME Conference on Multibody Systemss and Nonlinear Dynamics, 2011.

Koshizuka, S., Nobe, A., and Oka, Y.: Numerical analysis of breaking waves using the moving particle semi-implicit method, Int. J. Numer. Meth. Fl., 26, 751–769, 1998.

Kruggel-Emden, H., Simsek, E., Rickelt, S., Wirtz, S., and Scherer, V.: Review and extension of normal force models for the discrete element method, Powder Technol., 171, 157–173, 2007.

Lucy, L.: A numerical approach to the testing of the fission hypothesis, Astron. J., 82, 1013–1024, 1977.

Mazhar, H., Heyn, T., and Negrut, D.: A scalable parallel method for large collision detection problems, Multibody Syst. Dyn., 26, 37–55, doi:10.1007/s11044-011-9246-y, 2011.

Melanz, D.: On the Validation and Applications of a Parallel Flexible Multi-body Dynamics Implementation, M.S. thesis, University of Wisconsin-Madison, 2012.

Melanz, D., Tupy, M., Smith, B., Turner, K., and Negrut, D.: On the Validation of a Differential Variational Inequality Approach for the Dynamics of Granular Material-DETC2010-28804, in: Proceedings to the 30th Computers and Information in Engineering Conference, edited by: Fukuda, S. and Michopoulos, J. G., ASME International Design Engineering Technical Conferences (IDETC) and Computers and Information in Engineering Conference (CIE), 2010.

Mindlin, R. and Deresiewicz, H.: Elastic spheres in contact under varying oblique forces, J. Appl. Mech., 20, 327–344, 1953.

Monaghan, J.: On the problem of penetration in particle methods, J. Comput. Phys., 82, 1–15, 1989.

Monaghan, J.: Smoothed particle hydrodynamics, Rep. Prog. Phys., 68, 1703–1759, 2005.

MSC.Software: ADAMS: Automatic Dynamic Analysis of Mechanical Systems, Ann Arbor, Michigan, 2012.

Negrut, D., Tasora, A., Mazhar, H., Heyn, T., and Hahn, P.: Leveraging parallel computing in multibody dynamics, Multibody Syst. Dyn., 27, 95–117, doi:10.1007/s11044-011-9262-y, 2012.

NVIDIA Corporation: NVIDIA CUDA Developer Zone, available at: https://developer.nvidia.com/cuda-downloads, 2012.

Pazouki, A. and Negrut, D.: Direct simulation of lateral migration of bouyant particles in channel flow using GPU computing, in: Computers and Information in Engineering, CIE32, ASME, Chicago, IL, USA, 2012a.

Pazouki, A. and Negrut, D.: A numerical study of the effect of rigid body rotation, size, skewness, mutual distance, and collision on the radial distribution of suspensions in pipe flow, in review, 2013.

Pazouki, A., Mazhar, H., and Negrut, D.: Parallel Ellipsoid Collision Detection with Application in Contact Dynamics-DETC2010-29073, in: Proceedings to the 30th Computers and Information in Engineering Conference, edited by: Fukuda, S. and Michopoulos, J. G., ASME International Design Engineering Technical Conferences (IDETC) and Computers and Information in Engineering Conference (CIE), 2010.

Pazouki, A., Mazhar, H., and Negrut, D.: Parallel collision detection of ellipsoids with applications in large scale multibody dynamics, Math. Comput. Simulat., 82, 879–894, doi:10.1016/j.matcom.2011.11.005, 2012.

Pixar: The RenderMan Interface, Technical specification, Pixar, 1988, 1989, 2000, 2005.

Sanderson, C.: Armadillo: An open source C++ linear algebra library for fast prototyping and computationally intensive experiments, Tech. rep., Technical report, NICTA, 2010.

Shabana, A. A.: Dynamics of Multibody Systems, Cambridge University Press, 3rd Edn., 2005.

Snethen, G.: XenoCollide Website, http://www.xenocollide.com, 2007.

Snethen, G.: XenoCollide: Complex Collision Made Simple, in: Game Programming Gems 7, edited by: Jacobs, S., Charles River

Media, 165–178, 2008.
Tasora, A. and Anitescu, M.: A convex complementarity approach for simulating large granular flows, J. Comput. Nonlin. Dyn., 5, 1–10, doi:10.1115/1.4001371, 2010.
Tasora, A. and Anitescu, M.: A matrix-free cone complementarity approach for solving large-scale, nonsmooth, rigid body dynamics, Comput. Method. Appl. M., 200, 439–453, doi:10.1016/j.cma.2010.06.030, 2011.
Tasora, A., Righettini, P., and Silvestri, M.: Architecture of the Chrono::Engine physics simulation middleware, in: Proceedings of ECCOMAS 2007 Multibody Conference, 2007.
von Dombrowski, S.: Analysis of Large Flexible Body Deformation in Multibody Systems Using Absolute Coordinates, Multibody Syst. Dyn., 8, 409–432, doi:10.1023/A:1021158911536, 2002.
Yang, L. and Brent, R.: The improved BiCGStab method for large and sparse unsymmetric linear systems on parallel distributed memory architectures, in: Algorithms and Architectures for Parallel Processing, 2002. Proceedings. Fifth International Conference on, IEEE, 324–328, 2002.

5

Internal redundancy: an approach to improve the dynamic parameters around sharp corners

S. S. Parsa[1]**, J. A. Carretero**[1]**, and R. Boudreau**[2]

[1]Department of Mechanical Engineering, University of New Brunswick, Fredericton, NB, Canada
[2]Département de génie mécanique, Université de Moncton, Moncton, NB, Canada

Correspondence to: J. A. Carretero (juan.carretero@unb.ca)

Abstract. In recent years, redundancy in parallel manipulators has been studied broadly due to its capability of overcoming some of the drawbacks of parallel manipulators including small workspaces and singular configurations. Internal redundancy, first introduced for serial manipulators, refers to the concept of adding movable masses to some links so as to allow to control the location of the centre of mass and other dynamic properties of some links. This concept has also been referred to as variable geometry. This paper investigates the effects of internal redundancy on the dynamic properties of a planar parallel manipulator while performing a family of trajectories. More specifically, the 3-RRR planar manipulator, where a movable mass has been added to the distal link, is allowed to trace trajectories with rounded corners and different radii. The proposed method uses the manipulator's dynamic model to actively optimise the location of the redundant masses at every point along the trajectory to improve the dynamic performance of the manipulator. Numerical examples are shown to support the idea.

1 Introduction

Redundancy in parallel manipulators is normally divided into kinematic redundancy, actuation redundancy and branch redundancy (Lee and Kim, 1993; Zanganeh and Angeles, 1994; Merlet, 1996; Ruggiu and Carretero, 2009; Boudreau and Nokleby, 2012). Actuation redundancy consists of replacing passive joints with active ones (Zanganeh and Angeles, 1994; Cheng et al., 2003, 2011; Nokleby et al., 2005) where the number of degrees-of-freedom or mobility of the manipulator does not change. Although actuation redundancy can help either eliminate or reduce singular configurations, issues such as force interference make the manipulators more complex to analyze, design and control (Firmani and Podhorodeski, 2004; Garg et al., 2009). The second type of redundancy is called branch redundancy where an extra actuated branch is added to the manipulator (Firmani et al., 2007). Branch redundancy can improve the force capabilities of the manipulator and reduce the number of singular configurations. The third type of redundancy is called kinematic redundancy where active joints and links are added to one or more branches of the manipulator (Merlet, 1996; Wang and Gosselin, 2004). This type of redundancy can enhance the dexterity of the manipulator as well as enlarge the workspace. Additionally, kinematic redundancy allows to plan trajectories far from certain singular configurations as the inverse displacement problem has an infinite number of solutions (thus often allowing for the manipulator to remain as far as possible from singular configurations) (Ebrahimi et al., 2008).

Redundant parallel manipulators have been widely used to improve the trajectories of parallel robots. For instance, Cha et al. (2007) showed that kinematically redundant manipulators can effectively avoid singular configurations thus increasing the singularity-free workspace of the parallel manipulator.

Ruggiu and Carretero (2010) applied an optimisation procedure on a kinematically redundant parallel manipulator to minimise the acceleration of the actuators while following certain trajectories. The method was applied on a kinematically redundant parallel manipulator following square paths with rounded corners. They showed that the accelerations of

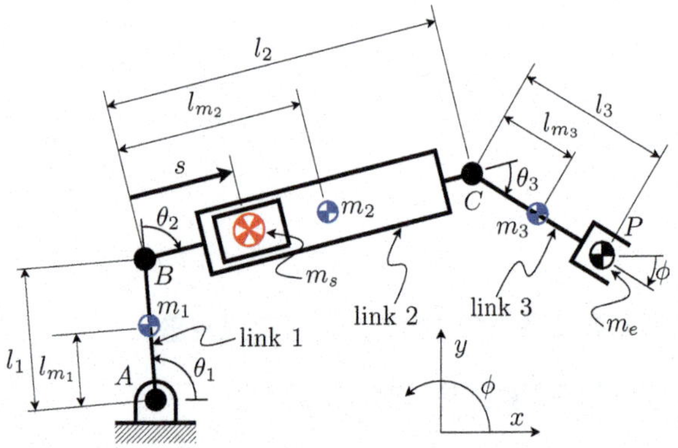

Figure 1. Serial manipulator with internal redundancy in link 2. The location of mass m_s can be changed without altering the position of the end effector (point p).

the actuated joints on the kinematically redundant manipulator are significantly less than the ones needed for a non-redundant manipulator.

Recently, a new type of redundancy called internal redundancy has been the focus of some attention in the context of serial manipulators (Vukobratović et al., 2000). Similar to the types of redundancy described earlier, a new set of degrees of freedom (DOF) is added to the serial manipulator. However, in contrast with the redundant actuators and/or links described earlier, the new DOF is used to change the internal geometry of a link resulting in the change of the location of the link's centre of mass and its inertial mass distribution parameters (i.e., its mass moment of inertia). Since the changes are made within the internal members of the link, the redundant DOF does not have a direct effect on the end effector pose (i.e., position and orientation). More specifically, in reference to Fig. 1, the position of the mass m_s in link 2 can be changed without altering the pose of the end effector. This allows for different internal motions for a given trajectory of the end effector thus adding control to some dynamic parameters of the manipulator to attempt to improve its dynamic performance for specific tasks.

In this paper, the concept of internal redundancy is applied to a planar parallel manipulator. First, a 3-RRR manipulator with internal redundancy in all three branches is described and its kinematic and dynamic equations (Sects. 2 and 3) are derived. Then, an optimisation problem is formulated where the displacement of each of the portable masses at every point throughout a trajectory is sought to minimise the torques at the base actuators (Sect. 4). The architectural parameters and trajectory planning algorithm are explained through a numerical example and are presented in Sect. 5 and then discussed in more detail in Sect. 6. Finally, Sect. 7 presents the conclusions and briefly discusses potential future work.

2 The 3-RRR manipulator with internal redundancy

A 3-DOF planar parallel manipulator shown in Fig. 2a is chosen to investigate the effect of internal redundancy in parallel manipulators. The manipulator is a symmetrical 3-RRR manipulator with base ($G_1G_2G_3$) and end effector ($A_1A_2A_3$) as equilateral triangles. The three revolute actuators to move the manipulator's end effector are located at G_i, the base joint of each branch. The length of the proximal links, i.e., links G_iB_i ($i = 1,2,3$), has been denoted by l_1 while the length of the distal link, i.e., B_iA_i ($i = 1,2,3$) has been denoted by l_2.

In order to study the concept of internal redundancy, a portion of the distal link (the portion from B_i to A_i') protrudes on the opposite side of the revolute joint at B_i and creates a linear track from A_i to A_i' where the redundant mass m_s can slide on (see Fig. 2b). The position of the mass relative to the elbow joint B_i is given by s_i and is measured in the direction of A_i. Since the masses m_s are mounted on tracks or prismatic joints, their position along A_iA_i' can be actively controlled. More specifically, the distance s_i from elbow joint B_i to the centre of mass m_s can be actively controlled thus changing the overall dynamic properties and effects of links $A_i'A_i$.

To help complete the dynamic model, each element has been given a mass while symmetry has been assumed to simplify the analysis. Moreover, the links have been modelled as slender rods. The proximal links have all been assigned a mass m_1 with their centre of mass located halfway between G_i and B_i while all three distal links have been assigned a mass m_2 with their centre of mass located halfway between A_i' and A_i. The moving platform has been assigned a mass m_e with its barycentre located at the centroid of the moving platform.

3 Dynamic model of the redundant 3-RRR manipulator

The inverse dynamic problem of a 3-RRR planar parallel manipulator with 3-DOF of internal redundancy is developed in this section. The dynamic model is obtained using the Principle of Virtual Work as well as d'Alembert's principle. The derivation is similar to that presented in Wu et al. (2011). For this purpose, a complete kinematic model of the manipulator needs to be developed to derive the velocity and acceleration equations. In addition to that, inertial forces and moments of all the links need to be calculated.

Note that in what follows, the equations are derived for each of the three legs. Therefore, index i in the equations that follow is assumed to respectively take the values 1, 2 and 3 when deriving the equations for legs 1, 2 and 3.

3.1 Kinematics

The base coordinate frame O-xy (denoted by $\{O\}$) shown in Fig. 2 is fixed on point G_1. Also, a moving coordinate frame P-x_Ny_N (denoted by $\{N\}$) is attached to the barycentre of the

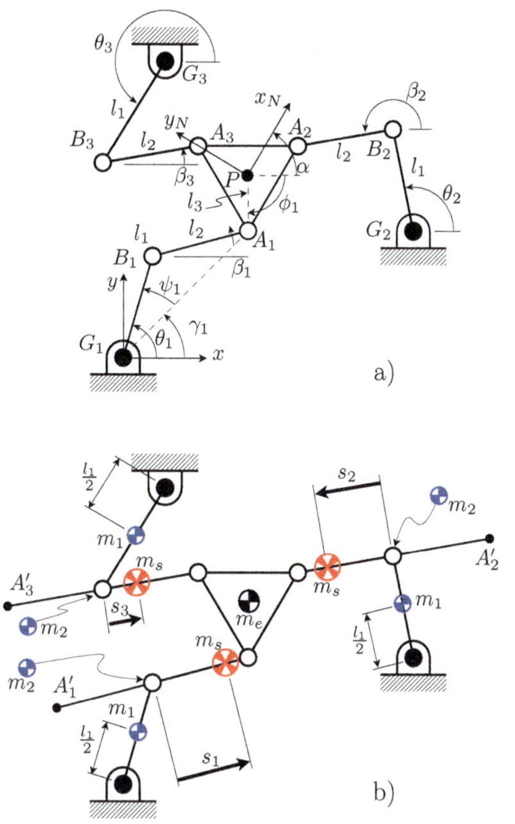

Figure 2. 3-RRR planar manipulator: **(a)** basic kinematic parameters and **(b)** location of the centre of mass of each component (l_i are fixed values while s_i are variable).

moving platform. The position vector of B_i in the base coordinate frame is defined as follows:

$$\mathbf{r}_{B_i} = \mathbf{r}_{G_i} + l_1 \begin{bmatrix} \cos\theta_i \\ \sin\theta_i \end{bmatrix} \tag{1}$$

where $\mathbf{r}_{B_i}$ describes the position vector of point B_i, $\mathbf{r}_{G_i}$ is the position vector of point G_i and θ_i is the angle link G_iB_i makes with the x-axis (i.e., the actuation variable for the motor located at G_i). The position vector of A_i is expressed as follows:

$$\mathbf{r}_{A_i} = \mathbf{r}_{B_i} + l_2 \begin{bmatrix} \cos\beta_i \\ \sin\beta_i \end{bmatrix} = \mathbf{r}_p + \mathbf{R}\mathbf{r}^N_{A_i} \tag{2}$$

where $\mathbf{r}_{A_i}$ is the position vector of point A_i, β_i describes the angle of link B_iA_i with respect to the horizontal x direction, $\mathbf{r}_p$ is the position vector of point p and $\mathbf{r}^N_{A_i}$ is the position vector of A_i expressed in frame N. The rotation matrix $\mathbf{R}$ describing frame $\{N\}$ relative to frame $\{O\}$ is defined as follows:

$$\mathbf{R} = \begin{bmatrix} \cos\alpha & -\sin\alpha \\ \sin\alpha & \cos\alpha \end{bmatrix}. \tag{3}$$

The constraint equation of motion is written as follows:

$$\|\mathbf{r}_{A_i} - \mathbf{r}_{B_i}\| = l_2 \tag{4}$$

where l_2 is the portion of the distal link from B_i to A_i.

3.2 Inverse displacement problem

The inverse displacement problem of the 3-RRR planar manipulator is written as follows:

$$\theta_i = \gamma_i \pm \psi_i \tag{5}$$

where γ_i is defined as follows:

$$\gamma_i = \text{atan2}(x_{A_i}, y_{A_i}) \tag{6}$$

where atan2 is the quadrant corrected inverse tangent function, while x_{A_i} and y_{A_i} are the Cartesian components of the position of A_i relative to G_i and are written as follows:

$$x_{A_i} = x - l_3\cos\phi_i - x_{G_i} \tag{7}$$

$$y_{A_i} = y - l_3\sin\phi_i - y_{G_i} \tag{8}$$

where x and y are Cartesian positions of point P, l_3 is the radius of the moving platform (i.e., the distance between P and A_i) and ϕ_i is given by

$$\phi_1 = \alpha + \frac{\pi}{6} \quad \phi_2 = \alpha + \frac{5\pi}{6} \quad \phi_3 = \alpha - \frac{\pi}{2}. \tag{9}$$

The equation for ψ_i is written as follows:

$$\psi_i = \cos^{-1}\left(\frac{l_1^2 - l_2^2 + x_{A_i}^2 + y_{A_i}^2}{2l_1(x_{A_i}^2 + y_{A_i}^2)}\right). \tag{10}$$

3.3 Velocity and acceleration

Taking the time derivative of Eq. (1) leads to

$$\dot{\mathbf{r}}_{B_i} = l_1\dot{\theta}_i \begin{bmatrix} -\sin\theta_i \\ \cos\theta_i \end{bmatrix} \tag{11}$$

where $\dot{\theta}_i$ is the angular velocity of actuator i. The linear velocity of point A_i is written as follows:

$$\dot{\mathbf{r}}_{A_i} = \mathbf{v}_P + \dot{\alpha}\mathbf{N}\mathbf{R}\mathbf{r}^N_{A_i} = \dot{\mathbf{r}}_{B_i} + l_2\dot{\beta}_i \begin{bmatrix} -\sin\beta_i \\ \cos\beta_i \end{bmatrix} \tag{12}$$

where $\mathbf{v}_P$ is the vector describing the linear velocity of point P. Matrix $\mathbf{N}$ is defined as follows:

$$\mathbf{N} = \begin{bmatrix} 0 & -1 \\ 1 & 0 \end{bmatrix}. \tag{13}$$

The angular velocity of link B_iA_i is derived from Eq. (12) and is written as follows:

$$\dot{\beta}_i = \begin{bmatrix} \frac{-\sin\beta_i}{l_2} & \frac{\cos\beta_i}{l_2} \end{bmatrix} (\dot{\mathbf{r}}_{A_i} - \dot{\mathbf{r}}_{B_i}) \tag{14}$$

The linear acceleration of points B_i and A_i is obtained as the time derivative of Eqs. (11) and (12):

$$\mathbf{a}_{B_i} = l_1\ddot{\theta}_i \begin{bmatrix} -\sin\theta_i \\ \cos\theta_i \end{bmatrix} - l_1\dot{\theta}_i^2 \begin{bmatrix} \cos\theta_i \\ \sin\theta_i \end{bmatrix} \tag{15}$$

$$\mathbf{a}_{A_i} = \mathbf{a}_P + \ddot{\alpha}\mathbf{N}\mathbf{R}\mathbf{r}^N_{A_i} - \dot{\alpha}^2\mathbf{R}\mathbf{r}^N_{A_i} \tag{16}$$

where $\mathbf{a}_P$ is the linear acceleration of point P. The time derivative of Eq. (14) leads to

$$\ddot{\beta}_i = \begin{bmatrix} \frac{-\sin\beta_i}{l_2} & \frac{\cos\beta_i}{l_2} \end{bmatrix} (\mathbf{a}_{A_i} - \mathbf{a}_{B_i}) - \dot{\beta} \begin{bmatrix} \frac{\cos\beta_i}{l_2} & \frac{\sin\beta_i}{l_2} \end{bmatrix} (\dot{\mathbf{r}}_{A_i} - \dot{\mathbf{r}}_{B_i}). \quad (17)$$

In order to generate the Jacobian matrix, the time derivative of Eq. (4) yields

$$\mathbf{J} = \begin{bmatrix} \frac{-a_1}{c_1} & \frac{-b_1}{c_1} & \frac{-d_1}{c_1} \\ \frac{-a_2}{c_2} & \frac{-b_2}{c_2} & \frac{-d_2}{c_2} \\ \frac{-a_3}{c_3} & \frac{-b_3}{c_3} & \frac{-d_3}{c_3} \end{bmatrix} \quad (18)$$

where the elements of this Jacobian matrix are as follows:

$$a_i = -h_1(x - x_{G_i}) + h_3\cos\theta_i + h_1\cos\phi_i \quad (19)$$

$$b_i = -h_1(y - y_{G_i}) + h_1\sin\theta_i + h_2\sin\phi_i \quad (20)$$

$$c_i = h_3[(y - y_{G_i})\cos\theta_i - (x - x_{G_i})\sin\theta_i] + \sin(\theta_i - \phi_i) \quad (21)$$

$$d_i = h_2[(y - y_{G_i})\cos\phi_i - (x - x_{G_i})\sin\phi_i] - \sin(\theta_i - \phi_i) \quad (22)$$

where x_{G_i} and y_{G_i} are the Cartesian components of the position of point G_i (Gosselin and Angeles, 1988) while $h_1 = \frac{1}{l_1 l_3}$, $h_2 = \frac{1}{l_1}$ and $h_3 = \frac{1}{l_3}$.

3.4 Link Jacobian matrices

Since the Principle of Virtual Work is applied to develop the dynamic model of the 3-RRR manipulator, link Jacobian matrices have to be derived. When the end effector is subjected to a virtual displacement, the link Jacobian sub-matrix related to the linear velocity provides the virtual displacement of a point on a link, while the link Jacobian sub-matrix related to angular velocity produces the virtual angular displacement of a link (also referred to as partial velocity and partial angular velocity matrices by some authors, Wu et al., 2009, 2011). Points G_i, B_i and P are considered as the pivotal points of links G_iB_i, B_iA_i and the moving platform, respectively. The link Jacobian sub-matrix related to the angular velocity of link G_iB_i is written as follows:

$$\mathbf{G}_{i1} = \begin{bmatrix} \frac{-a_i}{c_i} & \frac{-b_i}{c_i} & \frac{-d_i}{c_i} \end{bmatrix}. \quad (23)$$

The link Jacobian sub-matrix related to the linear velocity of point G_i is zero since the velocity of that point is zero. The link Jacobian sub-matrix related to the linear velocity of point G_i is thus written as:

$$\mathbf{H}_{i1} = \mathbf{0} \quad (24)$$

The link Jacobian sub-matrix related to the linear velocity of point B_i and the link Jacobian sub-matrix related to the angular velocity of link B_iA_i are obtained from $\mathbf{J}$ (Eq. 18) and from Eq. (14), respectively and are written as follows:

$$\mathbf{H}_{i2} = \frac{l_1}{c_i} \begin{bmatrix} a_i\sin\theta_i & b_i\sin\theta_i & d_i\sin\theta_i \\ -a_i\cos\theta_i & -b_i\cos\theta_i & -d_i\cos\theta_i \end{bmatrix} \quad (25)$$

$$\mathbf{G}_{i2} = \begin{bmatrix} \frac{-\sin\beta_i}{l_2} & \frac{\cos\beta_i}{l_2} \end{bmatrix} \left(\begin{bmatrix} \mathbf{e}_1 & \mathbf{e}_2 \end{bmatrix} + \mathbf{NRr}_{A_i}\mathbf{e}_3^T - \begin{bmatrix} -l_1\sin\theta_i \\ l_1\cos\theta_i \end{bmatrix} \mathbf{G}_{i1} \right) \quad (26)$$

where $\mathbf{e}_1 = [1\ 0\ 0]^T$, $\mathbf{e}_2 = [0\ 1\ 0]^T$ and $\mathbf{e}_3 = [0\ 0\ 1]^T$.

The link Jacobian sub-matrix related to the angular velocity of the moving platform and the link Jacobian sub-matrix related to the linear velocity of point P are written as follows:

$$\mathbf{G}_N = \mathbf{e}_3^T \quad (27)$$

$$\mathbf{H}_N = \begin{bmatrix} 1 & 0 & 0 \\ 0 & 1 & 0 \end{bmatrix}. \quad (28)$$

3.5 Inertial force and inertial moment

Here, the Newton-Euler formulation is applied to develop the inertial forces and the inertial moments of each moving body about its centre of mass. Then, these inertial forces and moments are calculated about pivotal points (i.e., points A_i, B_i and G_i). The inertial force and moment of link G_iB_i about pivotal point G_i are written as follows:

$$\mathbf{F}_{i1} = -m_1\left(\frac{l_1}{2}\ddot{\theta}_i[-\sin\theta_i\ \cos\theta_i]^T - \frac{l_1}{2}\dot{\theta}_i^2[\cos\theta_i\ \sin\theta_i]^T\right) \quad (29)$$

$$M_{i1} = -\ddot{\theta}_i I_{i1} \quad (30)$$

where θ_i, $\dot{\theta}_i$ and $\ddot{\theta}_i$ are the angular displacement, angular velocity and angular acceleration of actuator i, and I_{i1} is the moment of inertia of link G_iB_i about point G_i.

The influence of internal redundancy appears in the inertial force and moment of the distal links where the moment of inertia and mass centre of the links vary with respect to the position of m_s. The equations for the inertial force and moment about point B_i of the distal links are written as follows:

$$\mathbf{F}_{i2} = -m_{2\mathrm{tot}}\left(\mathbf{a}_{B_i} + r_{i2}\ddot{\beta}_i[-\sin\beta_i\ \cos\beta_i]^T - r_{i2}\dot{\beta}_i^2[\cos\beta_i\ \sin\beta_i]^T\right) - m_s\ddot{s}_i[\cos\beta_i\ \sin\beta_i]^T - 2m_s\dot{s}_i\dot{\beta}_i[-\sin(\beta_i)\cos(\beta_i)]^T \quad (31)$$

$$M_{i2} = -\ddot{\beta}_i I_{i2} - m_{2\mathrm{tot}} r_{i2}[-\sin\beta_i\quad \cos\beta_i]\mathbf{a}_{B_i} - 2m_s s_i\dot{s}_i\dot{\beta}_i \quad (32)$$

where β_i, $\dot{\beta}_i$ and $\ddot{\beta}_i$ are the displacement, angular velocity and angular acceleration of the passive joints and $m_{2\mathrm{tot}}$ is the total mass of link $A_i'A_i$, i.e., $m_{2\mathrm{tot}} = m_2 + m_s$. Also, $\mathbf{a}_{B_i}$ describes the linear acceleration of point B_i, r_{i2} is the distance between the centre of mass of link $A_i'A_i$ and point B_i while I_{i2} is the moment of inertia of the distal link with respect to B_i. The distance from point B_i to the barycentre of the redundant mass is s_i while $\dot{s}_i$ and $\ddot{s}_i$ describe the velocity and acceleration of m_s. The position of the centre of mass of the distal link and its moment of inertia vary with respect to the position of the portable mass and are written as follows:

$$r_{i2} = \frac{m_s s_i + m_2 r_{G_2}}{m_2 + m_s} \tag{33}$$

$$I_{i2} = I_{A_i'A_i} + m_s (s_i)^2 \tag{34}$$

where r_{G_2} is the position of the centre of mass of the distal link (excluding m_s) and is equal to zero for the case when B_iA_i is equal to B_iA_i', and $I_{A_i'A_i}$ is the moment of inertia of link $A_i'A_i$ about its centre of mass (excluding m_s).

The inertial force and moment of the moving platform about point P is written as follows:

$$\mathbf{F}_N = -m_n \mathbf{a}_P \tag{35}$$

$$M_N = -\ddot{\alpha} I_N \tag{36}$$

where $\mathbf{a}_P$ and $\ddot{\alpha}$ are the linear acceleration of point P and the angular acceleration of the moving platform, respectively while m_n and I_N represents the mass and the moment of inertia of the moving platform.

3.6 Dynamic model

The dynamic equation of the 3-$\underline{R}$RR is written as follows:

$$\mathbf{J}^T \boldsymbol{\tau} + \sum_{i=1}^{3}\sum_{j=1}^{2} \begin{bmatrix}\mathbf{H}_{ij}^T & \mathbf{G}_{ij}^T\end{bmatrix} \begin{bmatrix}\mathbf{F}_{ij} & \mathbf{M}_{ij}\end{bmatrix}^T + \begin{bmatrix}\mathbf{H}_N^T & \mathbf{G}_N^T\end{bmatrix} [\mathbf{F}_N \quad \mathbf{M}_N]^T = 0 \tag{37}$$

where $\mathbf{J}$ is the Jacobian matrix of the manipulator, $\boldsymbol{\tau}$ presents the torque vector, $\mathbf{H}_{ij}$ are the link Jacobian sub-matrices related to velocity and $\mathbf{G}_{ij}$ are the link Jacobian sub-matrices related to the angular velocity of the links, $\mathbf{H}_N$ and $\mathbf{G}_N$ represent the link Jacobian sub-matrix related to velocity and the link Jacobian sub-matrix related to the angular velocity of the moving platform, $\mathbf{F}_{ij}$ and $\mathbf{M}_{ij}$ are inertial forces and moments of the robot links and $\mathbf{F}_N$ and $\mathbf{G}_N$ represent the inertial force and moment of the moving platform.

4 Trajectory optimisation

When planning a trajectory in the Cartesian space, the displacement, velocity and acceleration of the end effector are known. These can be used to calculate the kinematic properties of all active joints for every point in the trajectory while the dynamic equations can be used to compute the actuator torques. Since the necessary torques to move the end effector are a function of the position, velocity and acceleration of the portable masses, moving the redundant masses (i.e., changing s_i, $\dot{s}_i$ and $\ddot{s}_i$ for $i = 1,2,3$) will also have a direct effect on the torques at the base-mounted actuators.

Here, variables s_i are optimised to minimise the manipulator's total torque at a specific time step within the trajectory. The optimisation problem is written as follows:

$$\min_{s_i} \quad \sum_{i=1}^{3} (\tau_i(s_i) - \lambda \bar{\tau}_i)^2 \tag{38}$$

$$\text{subject to} \quad -l_2 \le s_i \le l_2 \tag{39}$$

$$-\dot{s}_{\max} \le \dot{s}_i \le \dot{s}_{\max} \tag{40}$$

$$-\ddot{s}_{\max} \le \ddot{s}_i \le \ddot{s}_{\max} \tag{41}$$

where τ_i refers to the optimised torque of actuator i at every time step, $\bar{\tau}_i$ is the torque value obtained when a similar manipulator without internal redundancy is used and λ is a coefficient between 0 and 1 which makes the objective function flexible on the percentage of the optimised torque value with respect to the torques of the non-redundant manipulator. The optimisation variable (i.e., s_i) is the distance from joint B_i to the barycentre of the redundant mass. In Eq. (39), the value of s_i has been constrained so as to keep it within track $A_i'A_i$. Also, the rate of change of s_i (i.e., $\dot{s}_i$) is bounded in the positive and negative directions to a maximum absolute value $\dot{s}_{\max}$ (with $\dot{s}_{\max} > 0$). In addition to that, the rate of change of $\dot{s}_i$ (i.e., $\ddot{s}_i$) is bounded to a maximum absolute value $\ddot{s}_{\max}$. These limits prevent any sudden changes in the motion of the portable masses. The choice of the objective function will be clearer when the results are presented.

During the optimisation procedure, the position of m_{si}, i.e., variable s_i, changes to minimise the sum of the squared actuator torques within that specific time step. To achieve this, the following steps are followed:

1. **Define the reference trajectory (point-to-point):** the desired trajectory is planned in Cartesian space and the displacement, velocity and acceleration of the actuators are obtained using the corresponding inverse kinematic solutions.

2. **Calculate the torques of the non-redundant manipulator:** the torque values of the manipulator without internal redundancy is calculated for the defined reference trajectory.

3. **Define the search space:** the displacements of the redundant actuators through the trajectory are used as the design variables for the optimisation process.

4. **Define the bounds**: based on the current position of the portable masses and a user-defined maximum velocity and maximum acceleration of the redundant actuators, the upper and lower bounds of the optimisation variables are calculated.

5. **Define the initial condition:** the initial position of the portable masses needs to be adjusted as it affects the optimisation results.

6. **Define the optimisation stopping criteria:** the difference between two consequent optimisation search variables (i.e., displacement of the portable masses) as well

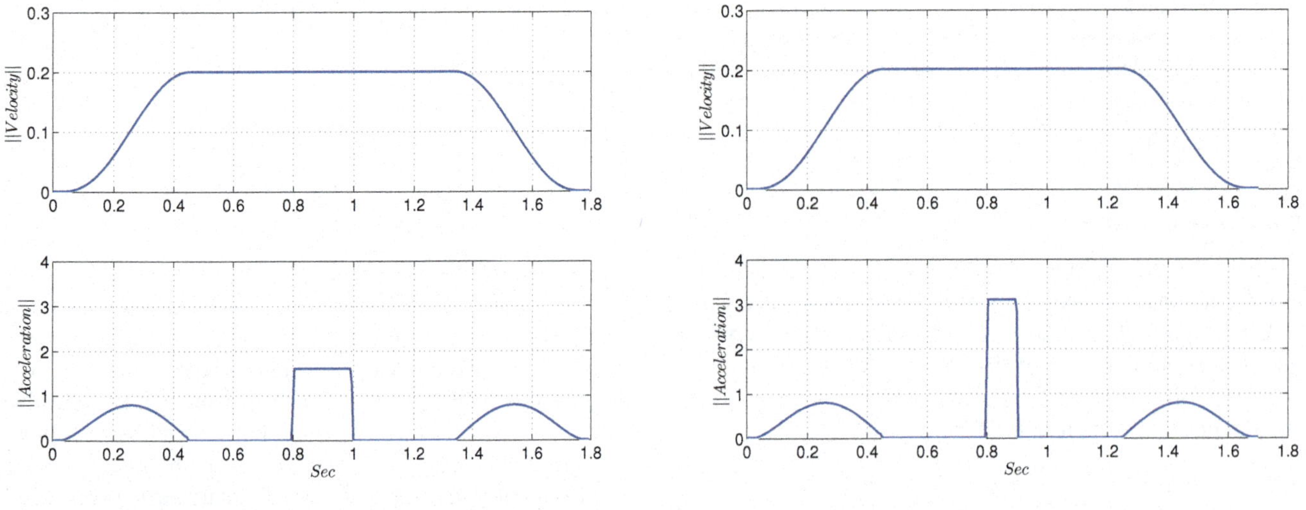

(a) Reference trajectory $r = 0.025$ m (b) Reference trajectory $r = 0.013$ m

Figure 3. The norm of the velocity (in m s^{-1}) and the acceleration (in m s^{-2}) of reference trajectories.

as the difference between their objective function values are monitored at every iteration of optimisation. Once they have met the pre-defined user threshold, the optimisation procedure stops.

7. **Optimise the position of the portable mass:** a non-linear multi-variable constrained optimisation is conducted to minimise the active-joint torques in Eq. (38).

   - The displacement, velocity and acceleration of the base actuators are calculated at every step of the optimisation procedure.
   - The current velocity and acceleration of the redundant actuators are calculated using the time history of the redundant actuators.
   - The objective function value is determined.

5 Numerical example

5.1 Architectural parameters and analysed trajectory

The manipulator's architectural parameters for the current example are as follows: all proximal link lengths are set to 1 m (i.e., $l_1 = 1$ m for all legs). Also, all distal link lengths are set to 1 m (i.e., $l_2 = 1$ m for all legs) where a track has been attached to every distal link to allow the portable mass to move from $s_i = -1$ to 1 m. The base and moving platforms are equilateral triangles inscribed in circles of 1 m and 0.25 m in radius, respectively. The mass m_1 of each of the proximal links is 1 kg while the distal links have a mass $m_2 = 1$ kg (including the mass of the track) and the end effector has mass $m_e = 0.5$ kg and $m_s = 3$ kg.

5.2 Trajectory planning

The procedure has been studied on two trajectories with rounded corners which have been planned in the Cartesian space. For both trajectories, the end effector moves on a straight line with an initial velocity of 0 m s^{-1} while keeping the end effector with constant orientation. As the tracking velocity reaches a user defined velocity in a specified time (0.2 m s^{-1} in 0.4 s), the end effector tracks the trajectory with a constant velocity. The abrupt acceleration change between $t = 0.8$ and $t = 1.0$ s occurs when the end-effector enters the rounded corner segment and normal acceleration occurs. The end effector decelerates in (0.4 s) to come to a stop in the last point of the trajectory. However, the radii of the rounded corners of the trajectories are different.

The trajectory's initial position is $p_1 = [1\ 0.4]^T$. Also, the radii of the round corners are $r = 0.025$ m and $r = 0.013$ m. Each trajectory starts from point p_1 and goes in the positive Y direction. Once the end effector moves 0.07 m in the Y direction, the rounded corner commences (the rounded corner is a quarter of a circle). Thereafter, the end effector travels 0.07 m in the negative x direction. The norm of the Cartesian velocity and the acceleration of the end effector is presented in Fig. 3. Since the radii of the rounded corners are different, the total length of the trajectories are not the same.

The optimisation problem was implemented in Matlab. The function *fmincon* was used to perform the constrained local optimisation in Eqs. (38) to (40). More particularly, the Sequential Quadratic Programming (SQP) with Hessian update option within *fmincon* was used. The SQP method is an alternative approach for handling inequality constraints in non-linear programming where SQP finds the minimum of a sequence of quadratic programming sub-problems. The objective function is estimated with a quadratic function and

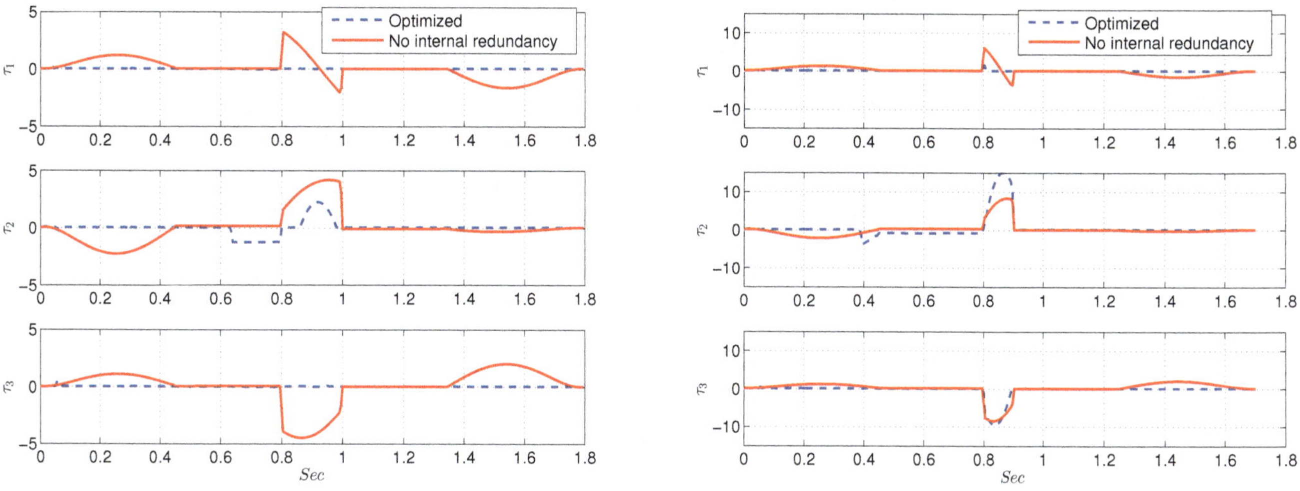

(a) Base joint torque for $r = 0.025$ m

(b) Base joint torque for $r = 0.013$ m

Figure 4. The torques of the ground actuators (in Nm) for $\lambda = 0$.

is minimised subject to the linearised constraints. In this method, the Hessian of the Lagrangian function is estimated at every iteration using a quasi-Newton update method. This approximation is used to create a quadratic programming sub-problem and its solution is applied to generate a search direction for the line search procedure (Fletcher, 1987).

In the current numerical example, the velocity of the portable masses is allowed to vary in the range between $-1\,\mathrm{m\,s^{-1}}$ and $+1\,\mathrm{m\,s^{-1}}$. The maximum absolute value of the acceleration of the portable masses is considered as $7\,\mathrm{m\,s^{-2}}$ and $m_{s_i} = 3$ kg for $i = 1, 2, 3$.

6 Results and discussion

Figure 4a illustrates the comparison between the torque values obtained from the optimisation routine and the manipulator without internal redundancy (i.e., $m_{s_i} = I_{B_i A'_i} = 0$) for the trajectory with $r = 0.025$ m as the radius of the rounded corner and $\lambda = 0$. As shown in Fig. 4a, the manipulator with internal redundancy can follow the reference trajectory with significantly lower torques (approximately 10^{-1} Nm) in both the accelerating and decelerating phases as well as the rounded corner area. However, the optimised torque for joint two is greater than the non-optimised one at $t = 0.65$ s. As can be seen in Fig. 6a, the acceleration of the portable mass is zero at $t = 0.65$ s which means the velocity of the portable mass meets the limits. Consequently, the effect of $\ddot{s}_1$ is eliminated from the dynamic equation at that instant. Also, the optimised torque for joint two at $t = 0.9$ s is slightly greater than the optimised torque of actuators one and three at the same time instant. As it is shown in Fig. 6a, the acceleration of the second portable mass meets the limit at $t = 0.9$ s. Consequently, the inertial force that is produced due to the motion of the portable mass remains constant as well as its improving effect on the torque.

Figure 4b presents the result of the optimised torques against non-optimised ones for the trajectory with a smaller rounded corner radius (i.e., $r = 0.013$ m). As can be seen in the torque plot of joints 2 and 3, the optimised value of the torques are greater than the non-optimised value when the end effector goes through the rounded corner. This is due to the acceleration of the portable mass (i.e., $\ddot{s}_1$) which meets the pre-defined threshold (see Fig. 6b). At this point, the accelerations of the portable masses remain constant as well as their effect on the inertial force of the distal link. Similar to the results for the trajectory with $r = 0.025$ m, the velocity of the portable mass two meets the limit at $t = 0.4$ s and the corresponding acceleration drops to zero. It has been noticed that the optimised value of the torque of joint 2 and 3 will be less than non-optimised one if the limit of the acceleration of the portable mass is increased to $12\,\mathrm{m\,s^{-2}}$. Also, there is a small jump at the optimised torque value of joint 1 at $t = 0.8$ s. This is due to the acceleration of the portable mass that meets the limit.

Figure 5a shows the result of optimisation of the torques for the trajectory with $r = 0.025$ m wile $\lambda = 0.5$. Since the optimised torque values need to be as small as half of the torque values of non-redundant manipulator, the portable masses need to produce smaller inertial forces and moments in comparison with the scenario with $\lambda = 0$ (Fig. 4a). Consequently, the portable masses move with smaller velocity and acceleration (Fig. 6c) which prevents them from meeting the limits. As it is seen in Fig. 6c, all portable masses move with relatively smaller acceleration in comparison with Fig. 6a. Also, the second portable mass does not meet the velocity limit at $t = 0.65$ s. The displacement of the portable masses is shown

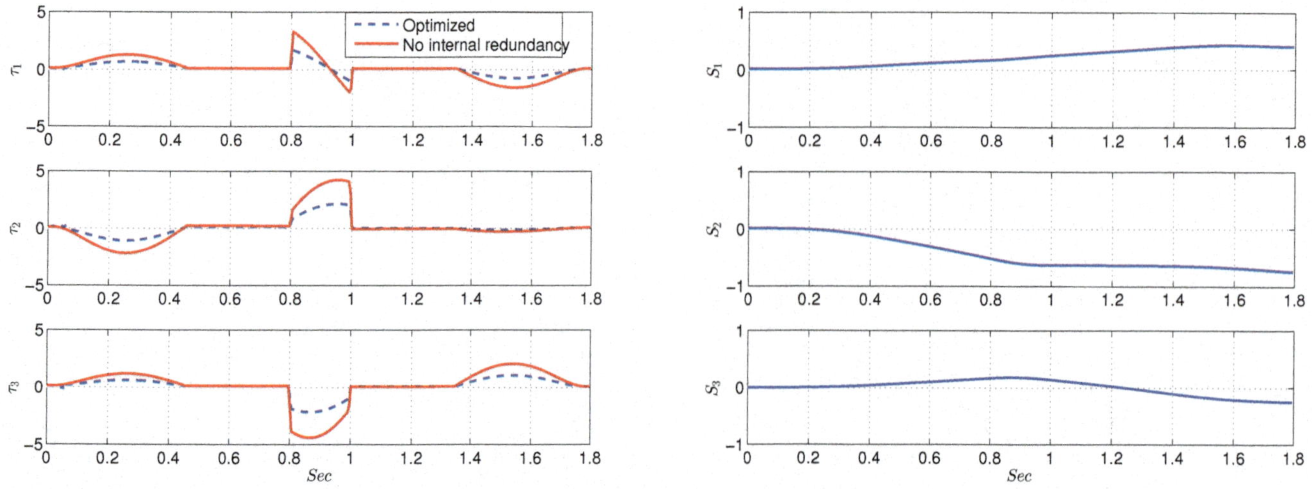

(a) Base joint torque (b) Displacement of portable masses

Figure 5. The torques of the ground actuators (in Nm) and displacement of the portable masses (in m) for $r = 0.025$ m and $\lambda = 0.5$.

in Fig. 5b where they are initial at 0 and are allowed to move between -1 and 1.

The torque values of the joints are relatively small in all cases when the end effector moves with a constant velocity. When the acceleration of the end effector is zero, the inertial forces and moments of the links decrease. In addition to that, the inertial force of the end effector will turn to zero.

7 Conclusions

The dynamic model of a 3-RRR planar parallel manipulator involving a portable mass on the distal links is developed. The total of the squared actuators torques is investigated. An optimisation algorithm is implemented to find the optimal position of the portable masses while the end effector undergoes an arbitrary trajectory with a rounded corner.

The concept was tested on two trajectories with different rounded corners using the same Cartesian velocity. The results of the conducted tests suggest that the motion of the portable masses can improve (i.e., reduce) the ground actuator torques for both accelerating and decelerating sections. Also, the base actuator torques improve when the end effector tracks the rounded corner with $r = 0.025$ m. However, the optimised torques are greater than the the non-optimised ones around the rounded corner for the trajectory with $r = 0.013$ m. Since the trajectory with sharper corner imposes greater torque values on the ground joints, the motion of the portable masses need to generate greater inertial forces and moments on the distal links to improve the torque values at rounded corner. However, the changes in inertial forces and moments of the distal links are limited due the limits that have been defined for the velocity and acceleration of the portable masses.

The objective function is flexible to determine the percentage of improvement of the optimised torques with respect to the torque values of the same manipulator without internal redundancy. As greater improvement of the torques requires higher limits of the velocity and the acceleration for portable masses, the objective function can be adjusted to keep the optimisation variables away from the limits.

The obtained simulation results suggest that if a manipulator can not follow a trajectory with a rounded corner due to the torque limits of the ground joints, it will be feasible through application of internal redundancy (without altering the ground actuators). This is possible as the dynamic forces required to perform the more demanding trajectories are shared by both the base actuators as well as the additional actuators on the distal links.

There are a few parameters that affect the the simulation such as Cartesian velocity of the end effector, the radius of the rounded corner and the allowed limits of the velocity and acceleration of the portable masses. For instance, having a relatively large end effector velocity demands greater torque values at the ground joints. Consequently, the portable masses need to generate greater forces and moments on the distal links which is proportional to the limits of the velocity and acceleration of the portable masses. Moreover, due the aforementioned force sharing effect, the balance between the contribution of the two sets of actuators to the specific task needs to be carefully considered (e.g., using an objective function that considers both sets of actuators).

As future work, it is suggested to look at the trajectory globally rather than point-to-point motion planning. In that case, the position of the portable masses can be adjusted with respect to the any up-coming critical situation (i.e., rounded corner).

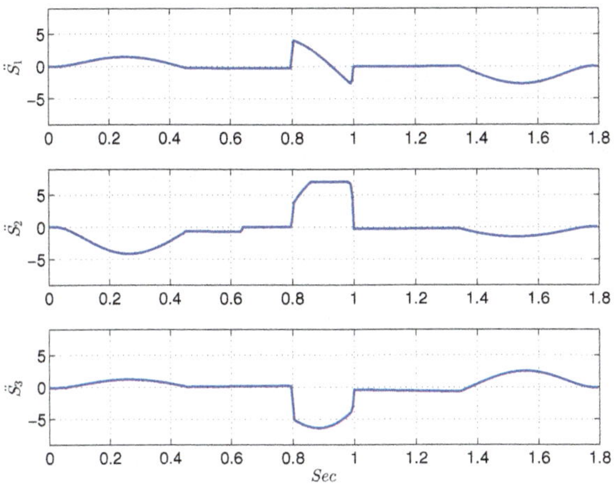

(a) $r = 0.025$ (m) and $\lambda = 0$

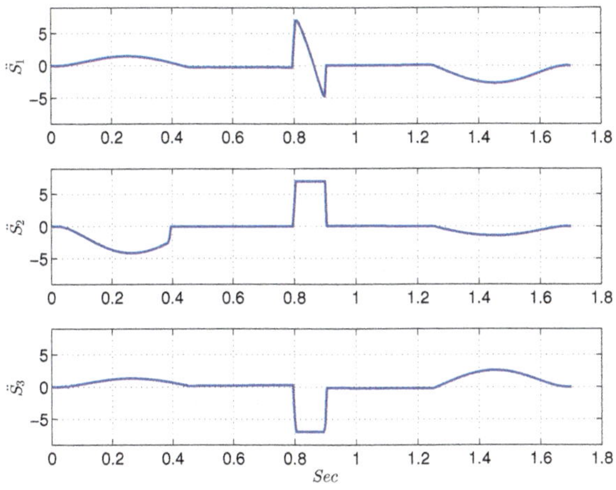

(b) $r = 0.013$ (m)and $\lambda = 0$

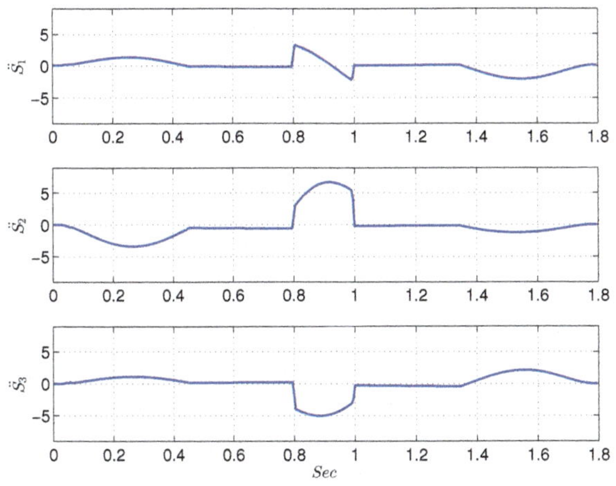

(c) $r = 0.025$ (m) and $\lambda = 0.5$

Figure 6. The acceleration of the portable masses (in m s^{-2}).

Acknowledgement. The authors acknowledge the financial support from the Natural Science and Engineering Research council of Canada through their Discovery Grant program.

Edited by: A. Müller

References

Boudreau, R. and Nokleby, S.: Force Optimisation of kinematically-redundant planar parallel manipulators following a desired trajectory, Mech. Mach. Theory, 56, 138–155, 2012.

Cha, S., Lasky, T. A., and Velinsky, S. A.: Kinematically-Redundant Variations of the 3-RRR Mechanism and Local Optimization-Based Singularity Avoidance, Mechanism Based Design of Structures and Machines, 35, 15–38, 2007.

Cheng, H., Yiu, Y.-K., and Li, Z.: Dynamics and Control of Redundantly Actuated Parallel Manipulators, IEEE/ASME Transactions on Mechatronics, 8, 483–491, doi:10.1109/TMECH.2003.820006, 2003.

Cheng, H., Liu, G. F., Yiu, Y. K., Xiong, Z. H., and Li, Z.: Advantages and dynamics of parallel manipulators with redundant actuation, in: Proceedings of the IEEE/RSJ International Conference on Intelligent Robots and Systems, 171–176, Maui, USA, 2011.

Ebrahimi, I., Carretero, J. A., and Boudreau, R.: Kinematic analysis and path planning of a new kinematically redundant planar parallel manipulator, Robotica, 26, 405–413, 2008.

Firmani, F. and Podhorodeski, R. P.: Force-unconstrained poses for a redundantly actuated planar parallel manipulator, Mech. Mach. Theory, 39, 459–476, 2004.

Firmani, F., Zibil, A., Nokleby, S. B., and Podhorodeski, R. P.: Force-Moment Capabilities of Revolute-Jointed Planar Parallel Manipulators with Additional Actuated Branches, T. Can. Soc. Mech. Eng., 31, 469–481, http://www.tcsme.org/Vol31-No4.html, 2007.

Fletcher, R.: Practical Methods of Optimization, John Wiley and Sons, 1987.

Garg, V., Carretero, J. A., and Nokleby, S. B.: A New Method to Determine the Force and Moment Workspaces of Actuation Redundant Spatial Parallel Manipulators, Journal of Mechanisms and Robotics, 1, 1–8, doi:10.1115/1.3147184, 2009.

Gosselin, C. and Angeles, J.: The optimum kinematic design of a planar three-degree-of-freedom parallel manipulator, Journal of Mechanism, Transmissions, and Automation in Design, 110, 35–41, 1988.

Lee, S. and Kim, S.: Kinematic analysis of generalized parallel manipulator systems, in: Proceedings of the IEEE Conference on Decision and Control, 2, 1097–1102, 1993.

Merlet, J.-P.: Redundant parallel manipulators, Lab. Robotics Automat., 8, 17–24, 1996.

Nokleby, S. B., Fisher, R., Podhorodeski, R. P., and Firmani, F.: Wrench capabilities of redundantly-actuated parallel manipulators, Mech. Mach. Theory, 40, 578–599, doi:10.1016/j.mechmachtheory.2004.10.005, 2005.

Ruggiu, M. and Carretero, J. A.: Kinematic Analysis of the 3-P̲RP̲R Redundant Planar Parallel Manipulator, in: Proceedings of the 2009 CCToMM Symposium on Mechanisms, Machines, and Mechatronics, Québec, Canada, 2009.

Ruggiu, M. and Carretero, J. A.: Actuation strategy based on the acceleration model for the 3-PRPR redundant planar parallel manipulator, in: Advances in Robot Kinematics: Analysis and Design, edited by: Lenarčič, J. and Stanisic, M. M., Springer, Piran-Portorož, Slovenia, 2010.

Vukobratović, M., Potkonjak, V., and Matijević, V.: Internal redundancy – the way to improve robot dynamics and control performances, J. Intell. Robot. Syst., 27, 31–66, 2000.

Wang, J. and Gosselin, C. M.: Kinematic Analysis and Design Of Kinematically Redundant Parallel Mechanisms, J. Mech. Design, 126, 109–118, 2004.

Wu, J., Wang, J., Wang, L., and Tiemin, L.: Dynamics and control of a planar 3-DOF parallel manipulator with actuation redundancy, Mech. Mach. Theory, 44, 835–849, 2009.

Wu, J., Wang, J., and You, Z.: A comparison study on the dynamics of planar 3-DOF 4-RRR, 3-RRR and 2-RRR parallel manipulators, Journal of Robotics and Computer Integrated Manufacturing, 27, 150–156, 2011.

Zanganeh, K. E. and Angeles, J.: Instantaneous kinematics and design of a novel redundant parallel manipulator, in: Proceedings of IEEE Conference on Robotics and Automation, 3043–3048, San Diego, USA, 1994.

6

Dynamic modelling of a 3-CPU parallel robot via screw theory

L. Carbonari, M. Battistelli, M. Callegari, and M.-C. Palpacelli

Università Politecnica delle Marche, Via Brecce Bianche, 60131, Ancona, Italy

Correspondence to: L. Carbonari (l.carbonari@univpm.it)

Abstract. The article describes the dynamic modelling of I.Ca.Ro., a novel Cartesian parallel robot recently designed and prototyped by the robotics research group of the Polytechnic University of Marche. By means of screw theory and virtual work principle, a computationally efficient model has been built, with the final aim of realising advanced model based controllers. Then a dynamic analysis has been performed in order to point out possible model simplifications that could lead to a more efficient run time implementation.

1 Introduction

Many approaches are available for the dynamic modelling of multi-body mechanical systems (Kovecses et al., 2003; Moon, 2008; Papastavridis, 2012) and in the last years, many most of them have been investigated by robotics researchers to achieve efficient models of robots dynamics. Indeed, the efficiency in computation of inverse dynamics of robotic manipulators has a fundamental importance if such tools are involved in the implementation of model based control algorithms whose effectiveness is strongly affected by the computational efficiency of the mathematical model (Lin and Song, 1990; Wang et al., 2007). Thus, it is interesting to investigate the possibility to build simplified dynamics models, especially for parallel kinematic machines that are characterized by an inherent toughness due to the closed kinematic structure. Such peculiarity often complicates the computation of the dynamic model and sometimes prevents the use of model based controls. This inherent complexity is the main reason why only few dynamic models of parallel robots are presently available in scientific literature in symbolic form (Dasgupta and Mruthyunjaya, 1998; Tsai, 2000; Caccavale et al., 2003).

The traditional Newton-Euler formulation, which has been widely used in the past (Do and Yang, 1988; Dasgupta and Mruthyunjaya, 1998) and is still used for specific tasks by some researchers (Kunquan and Rui, 2011; Khalil and Ibrahim, 2007), hardly adapts to the particular case of parallel kinematics machines.

As a matter of fact, all mechanical principles have been used to carry on dynamic analysis of robotic systems, such as the generalized momentum approach (Lopes, 2009), the Hamilton's principle (Miller, 2004), the Lagrange formulation (Wronka and Dunnigan, 2011; Di Gregorio and Parenti-Castelli, 2004) and the virtual work principle (Zhang and Song, 1993).

This last method was proposed in 1993 by Zhang and Song, who used it for the inverse dynamic modelling of open-loop manipulators; later Wang and Gosselin in 1998 expanded the approach to the study of closed kinematics mechanical chains and it is still much used for the modelling of PKMs (Daun et al., 2010). Due to its computational efficiency, such approach is often used when the dynamic modelling aims at the realization of model-based control algorithms. In fact, even if all methods lead to equivalent dynamic equations, these equations present different levels of complexity and associated computational loads; minimizing the number of operations involved in the computation of the manipulator dynamics model has been the main goal of recently proposed techniques (Abdellatif and Heimann, 2009; Yang et al., 2012): since by the use of the virtual work principle constraint forces and moments do not need to be computed, this approach leads to faster computational algorithms, which is a very important advantage for the purpose of robot control. Furthermore, the vector approach specific of the virtual work principle is particularly feasible for computer implementation.

In order to formulate the dynamic model of a mechanical system, the knowledge of its position kinematics is strictly necessary. As a matter of fact, the solution of the forward kinematics problem (FKP) of a parallel platform represents a challenging issue that not necessarily yields to a closed form solution, especially when the robot end-effector is allowed to perform motions of rotation.

As argued by authors in past works (Carbonari, 2012; Carbonari and Callegari, 2012) the 3-CPU parallel architecture can provide the end-effector with different kinds of mobility, depending on the mutual configuration of the joints that compose the leg's kinematic chain. Carbonari et al. (2013) also demonstrated that a reconfiguration of the universal joint allows to modify the kinematic behaviour of a 3-CPU parallel robot, switching from a pure rotational to a pure translational kinematic behaviour.

This paper focuses on the dynamic modelling of a pure translational 3-CPU architecture, called I.Ca.Ro. by Callegari and Palpacelli (2008), aimed at the realization of a nonlinear model based control scheme. The main object of the present work is to produce a numerically efficient dynamic model of the machine, suitable to be used for the realization of a control algorithm. To this aim, the differential kinematics of the manipulator has been tackled taking advantage of a screw based approach (Gallardo et al., 2003). For the seek of completeness, the position kinematics is also presented here in order to improve the reader's understanding of the problem.

2 Robot kinematics

The I.Ca.Ro. parallel robot is a pure translational Cartesian tripod whose limbs are built of a C-P-U (cylindrical-prismatic-universal) joints chain. The first body of each leg is connected to the robot chassis by means of a cylindrical joint, realized through a prismatic actuated pair and a revolute passive joint coaxial to the first one (refer to Fig. 1). The second body is linked to the first one through a prismatic joint, perpendicular to the axis of the cylindrical joint. At last, the second body is connected to the moving platform through a universal joint composed of two revolutes: the first revolute joint is parallel to the second link and the second is perpendicular to the first one. The last revolute of each leg connects the manipulator with the respective limb; the axes of such joints are coplanar.

Due to the robot kinematic architecture, the I.Ca.Ro. parallel manipulator is only able to provide the end-effector with pure translations. In fact, by means of screw theory it can be observed that each leg exerts a constraint wrench made of a pure torque along the direction of its passive prismatic pair. The connection of the three legs to the mobile platform produces a wrench system of three orthogonal torques, whose dual space is spanned by a basis of three linearly independent pure translations. Thus, the orientation of a reference frame that solidly moves with it remains constant notwithstanding the displacement that the actuated joints perform.

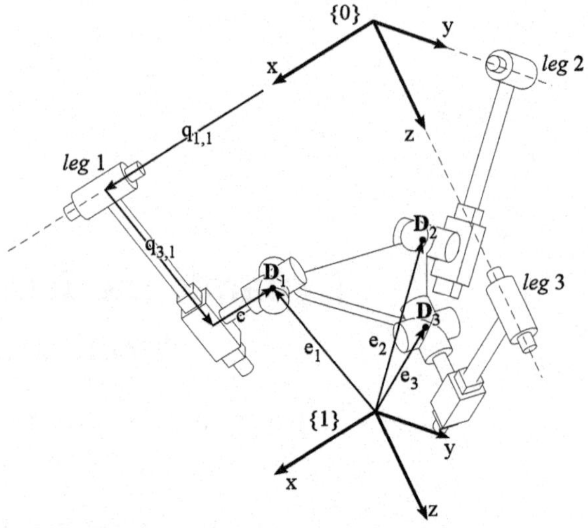

Figure 1. Kinematics of the 3-CPU pure translational parallel robot.

With respect to the notation introduced in Fig. 1, the homogeneous transformation matrix that describes the configuration of reference frame {1} with respect to reference frame {0} can be expressed as:

$$
{}^{0}\mathbf{T}_1 = \begin{bmatrix} 1 & 0 & 0 & p_x \\ 0 & 1 & 0 & p_y \\ 0 & 0 & 1 & p_z \\ 0 & 0 & 0 & 1 \end{bmatrix} \tag{1}
$$

where p_x, p_y and p_z denote the position of the centre of the moving frame {1} and the 3 by 3 identity rotation matrix suggests that the orientation of the manipulator remains constant.

The forward and inverse kinematics problems of the robot can be easily solved taking advantage of three loop closure equations. The positions of the three attachment points $\boldsymbol{D}_i$ between the end-effector and the three limbs can be simply reached through the transformation matrix ${}^{0}\mathbf{T}_1$, being their position fixed with respect to reference frame {1}: $\boldsymbol{D}_i = {}^{0}\mathbf{T}_1 \begin{bmatrix} \boldsymbol{e}_i^T & 1 \end{bmatrix}^T$, where $\boldsymbol{e}_1 = -e\begin{bmatrix} 0 & 1/\sqrt{2} & 1/\sqrt{2} \end{bmatrix}^T$, $\boldsymbol{e}_2 = -e\begin{bmatrix} 1/\sqrt{2} & 0 & 1/\sqrt{2} \end{bmatrix}^T$, $\boldsymbol{e}_3 = -e\begin{bmatrix} 1/\sqrt{2} & 1/\sqrt{2} & 0 \end{bmatrix}^T$.
As it is shown in the following, the coordinates of such points can be also reached through the use of legs' joints displacements. The comparison between the different expressions of these three points provides the solution of the problem. The reference frames shown in Fig. 2 are used to define legs kinematics.

Starting with frame {0}, which is the frame attached to manipulator chassis, a displacement along its x-axis and a rotation about the same axis are needed to describe the position of the second frame $\{A\}$. Transformation matrix will look

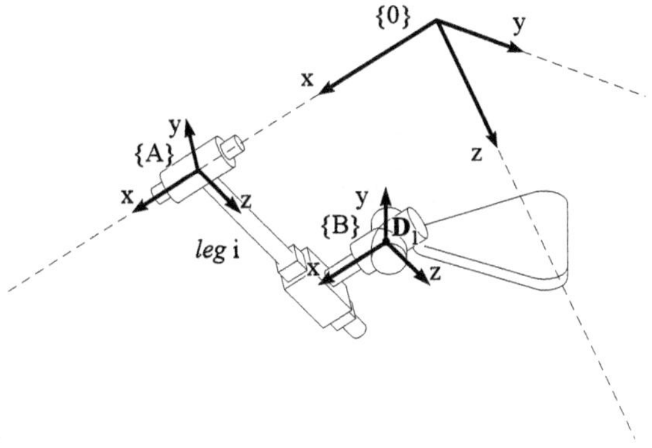

Figure 2. Reference frames along leg structure.

like:

$$^{0}\mathbf{T}_A = \begin{bmatrix} 1 & 0 & 0 & q_{1,1} \\ 0 & cq_{2,1} & -sq_{2,1} & 0 \\ 0 & sq_{2,1} & cq_{2,1} & 0 \\ 0 & 0 & 0 & 1 \end{bmatrix} \tag{2}$$

where $q_{2,1}$ denotes the rotation of the cylindrical pair of the leg 1 and shorthand notation is used for trigonometric functions. To reach the configuration of the third reference frame $\{B\}$, thus the position of attachment point $\boldsymbol{D}_1$, a translational transformation matrix is sufficient:

$$^{A}\mathbf{T}_B = \begin{bmatrix} 1 & 0 & 0 & -c \\ 0 & 1 & 0 & 0 \\ 0 & 0 & 1 & q_{3,1} \\ 0 & 0 & 0 & 1 \end{bmatrix} \tag{3}$$

with $q_{3,1}$ denoting the displacement performed by prismatic pair of the first leg. Coordinates of point $\boldsymbol{D}_1$ can now be achieved as:

$$\boldsymbol{D}_1 = {}^{0}\mathbf{T}_A {}^{A}\mathbf{T}_B \begin{bmatrix} 0 & 0 & 0 & 1 \end{bmatrix}^T \tag{4}$$

Since the kinematic chain is identical for each leg, it is not necessary to explicit their homogeneous transformations as specific cases. Indeed, Eq. (4) can be exploited if pre multiplied by a transformation that rotates the starting reference frame. In particular, it is possible to define the matrices:

$$^{0}\mathbf{T}_{\text{leg2}} = \begin{bmatrix} 0 & 0 & 1 & 0 \\ 1 & 0 & 0 & 0 \\ 0 & 1 & 0 & 0 \\ 0 & 0 & 0 & 1 \end{bmatrix} \quad {}^{0}\mathbf{T}_{\text{leg3}} = \begin{bmatrix} 0 & 1 & 0 & 0 \\ 0 & 0 & 1 & 0 \\ 1 & 0 & 0 & 0 \\ 0 & 0 & 0 & 1 \end{bmatrix} \tag{5}$$

such that the coordinates of the remaining attachment points can be expressed as:

$$\begin{aligned} \boldsymbol{D}_2 &= {}^{0}\mathbf{T}_{\text{leg2}} {}^{\text{leg2}}\mathbf{T}_A {}^{A}\mathbf{T}_B \begin{bmatrix} 0 & 0 & 0 & 1 \end{bmatrix}^T \\ \boldsymbol{D}_3 &= {}^{0}\mathbf{T}_{\text{leg3}} {}^{\text{leg3}}\mathbf{T}_A {}^{A}\mathbf{T}_B \begin{bmatrix} 0 & 0 & 0 & 1 \end{bmatrix}^T \end{aligned} \tag{6}$$

Even if it is not specified, it should be evident that transformations $^{\text{leg2}}\mathbf{T}_A$, $^{\text{leg3}}\mathbf{T}_A$ and $^{A}\mathbf{T}_B$ of the 2nd and 3rd equations in (6) are expressed in terms of the respective joints variables $q_{1,i}$, $q_{2,i}$ and $q_{3,i}$.

In order to give a general expression of legs kinematics, it is also introduced the matrix $^{0}\mathbf{T}_{\text{leg1}}$, an identity 4 by 4 matrix that allows to write the Eq. (4) as:

$$\boldsymbol{D}_1 = {}^{0}\mathbf{T}_{\text{leg1}} {}^{\text{leg1}}\mathbf{T}_A {}^{A}\mathbf{T}_B \begin{bmatrix} 0 & 0 & 0 & 1 \end{bmatrix}^T \tag{7}$$

where $^{\text{leg1}}\mathbf{T}_A = {}^{0}\mathbf{T}_A$.

Expansion of (6) and (7) yields to the expression of such points coordinates in terms of joints variables, that are:

$$\boldsymbol{D}_1 = \begin{bmatrix} q_{1,1} - c \\ q_{3,1}sq_{2,1} \\ -q_{3,1}cq_{2,1} \\ 1 \end{bmatrix} \quad \boldsymbol{D}_2 = \begin{bmatrix} -q_{3,2}cq_{2,2} \\ q_{1,2} - c \\ q_{3,2}sq_{2,2} \\ 1 \end{bmatrix} \quad \boldsymbol{D}_3 = \begin{bmatrix} q_{3,3}sq_{2,3} \\ -q_{3,3}cq_{2,3} \\ q_{1,3} - c \\ 1 \end{bmatrix} \tag{8}$$

Equations (8) can be inverted to achieve the expression of legs joints variables as functions of coordinates of the three attachment points $\mathbf{D}_i$:

$$\begin{aligned} q_{1,1} &= D_{1,x} + c & q_{2,1} &= \tan^{-1}\frac{D_{1,y}}{-D_{1,z}} & q_{3,1} &= \frac{D_{1,y}}{\sin q_{2,1}} \\ q_{1,2} &= D_{2,y} + c & q_{2,2} &= \tan^{-1}\frac{D_{2,z}}{-D_{2,x}} & q_{3,2} &= \frac{D_{2,z}}{\sin q_{2,2}} \\ q_{1,3} &= D_{3,z} + c & q_{2,3} &= \tan^{-1}\frac{D_{3,x}}{-D_{3,y}} & q_{3,3} &= \frac{D_{3,x}}{\sin q_{2,3}} \end{aligned} \tag{9}$$

It is worth to remark that such coordinates are uniquely determined if the pose of the manipulator is known.

3 Velocity kinematics

In order to build the Jacobian matrices of a PKM, both position and orientation of the joints axes are needed. Firstly it is necessary to define an appropriate number of reference frames preferably attached in convenient points of the kinematic chain. It is worth to remember that the screws must be expressed with respect to a reference frame attached to robot manipulator and whose orientation is constant and coincident to that of robot absolute reference frame $\{0\}$. Furthermore, in order to define the dynamical model of the machine, the moving frame should be centred at the c.o.m. of the body object of the velocity analysis.

In the case of the pure translational robot I.Ca.Ro. the moving platform reference frame $\{1\}$ represents a feasible choice for expressing the screw coordinates. In fact, it is solid with the end-effector and it does not rotate; moreover, due to the pure translational mobility of the end-effector, the origin of frame $\{1\}$ moves with the same velocity and acceleration of the end-effector c.o.m. even if it is not centred on it.

The screw coordinates of every kinematic pair involved in leg kinematic chain must be expressed: to this aim, convenient local frames must be arranged along the legs as shown in Fig. 3.

Two new frames $\{C\}$ and $\{D\}$ must be attached to the revolute joints that compose the universal joints. The homogeneous transformations denoting configuration of such frames are defined as:

$$
{}^{B}\mathbf{T}_C = \begin{bmatrix} 1 & 0 & 0 & 0 \\ 0 & cq_{4,1} & -sq_{4,1} & 0 \\ 0 & sq_{4,1} & cq_{4,1} & 0 \\ 0 & 0 & 0 & 1 \end{bmatrix}
\quad
{}^{C}\mathbf{T}_D = \begin{bmatrix} cq_{5,1} & 0 & sq_{5,1} & 0 \\ 0 & 1 & 0 & 0 \\ -sq_{5,1} & 0 & cq_{5,1} & 0 \\ 0 & 0 & 0 & 1 \end{bmatrix} \tag{10}
$$

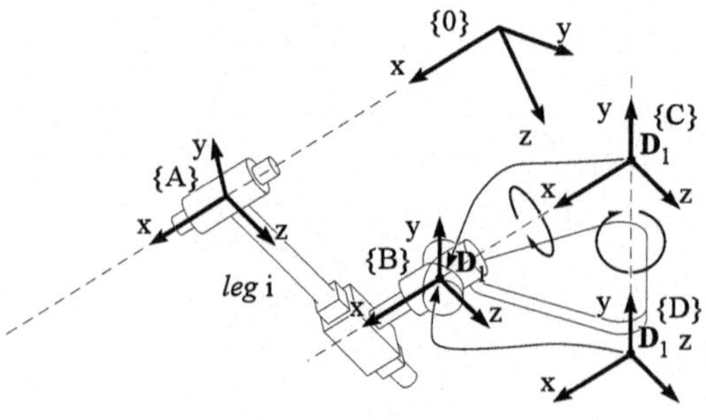

Figure 3. Local frames used for definition of joints unit screws.

The pose of frames $\{A\}$, $\{B\}$, $\{C\}$ and $\{D\}$ is described by the following homogeneous transformations:

$$
\begin{aligned}
{}^{1}\mathbf{T}_{A,1} &= {}^{1}\mathbf{T}_0{}^{0}\mathbf{T}_{\text{leg1}}{}^{\text{leg1}}\mathbf{T}_A \\
{}^{1}\mathbf{T}_{B,1} &= {}^{1}\mathbf{T}_0{}^{0}\mathbf{T}_{\text{leg1}}{}^{\text{leg1}}\mathbf{T}_A{}^{A}\mathbf{T}_B \\
{}^{1}\mathbf{T}_{C,1} &= {}^{1}\mathbf{T}_0{}^{0}\mathbf{T}_{\text{leg1}}{}^{\text{leg1}}\mathbf{T}_A{}^{A}\mathbf{T}_B{}^{B}\mathbf{T}_C \\
{}^{1}\mathbf{T}_{D,1} &= {}^{1}\mathbf{T}_0{}^{0}\mathbf{T}_{\text{leg1}}{}^{\text{leg1}}\mathbf{T}_A{}^{A}\mathbf{T}_B{}^{B}\mathbf{T}_C{}^{C}\mathbf{T}_D
\end{aligned} \tag{11}
$$

The unit vectors of joints' axes can be easily expressed in the global frame by means of the proper mapping between the local frames and the global one:

- **Joint 1**, prismatic:

$$
\begin{bmatrix} \boldsymbol{s}_{1,i} \\ 1 \end{bmatrix} = {}^{1}\mathbf{T}_{A,i} \begin{bmatrix} 1 \\ 0 \\ 0 \\ 1 \end{bmatrix} \quad \rightarrow \quad \boldsymbol{S}_{1,i} = \begin{bmatrix} \mathbf{0} \\ \boldsymbol{s}_{1,i} \end{bmatrix} \tag{12}
$$

- **Joint 2**, revolute:

$$
\begin{bmatrix} \boldsymbol{s}_{2,i} \\ 1 \end{bmatrix} = {}^{1}\mathbf{T}_{A,i} \begin{bmatrix} 1 \\ 0 \\ 0 \\ 1 \end{bmatrix} \quad \begin{bmatrix} \boldsymbol{r}_{2,i} \\ 1 \end{bmatrix} = {}^{1}\mathbf{T}_{A,i} \begin{bmatrix} 0 \\ 0 \\ 0 \\ 1 \end{bmatrix} \rightarrow \boldsymbol{S}_{2,i} = \begin{bmatrix} \boldsymbol{s}_{2,i} \\ \boldsymbol{r}_{2,i} \times \boldsymbol{s}_{2,i} \end{bmatrix} \tag{13}
$$

- **Joint 3**, prismatic:

$$
\begin{bmatrix} \boldsymbol{s}_{3,i} \\ 1 \end{bmatrix} = {}^{1}\mathbf{T}_{A,i} \begin{bmatrix} 0 \\ 0 \\ 1 \\ 1 \end{bmatrix} \quad \rightarrow \quad \boldsymbol{S}_{3,i} \begin{bmatrix} \mathbf{0} \\ \boldsymbol{s}_{3,i} \end{bmatrix} \tag{14}
$$

- **Joint 4**, first revolute of the universal joint:

$$
\begin{bmatrix} \boldsymbol{s}_{4,i} \\ 1 \end{bmatrix} = {}^{1}\mathbf{T}_{B,i} \begin{bmatrix} 1 \\ 0 \\ 0 \\ 1 \end{bmatrix} \quad \begin{bmatrix} \boldsymbol{r}_{4,i} \\ 1 \end{bmatrix} = {}^{1}\mathbf{T}_{B,i} \begin{bmatrix} 0 \\ 0 \\ 0 \\ 1 \end{bmatrix} \rightarrow \boldsymbol{S}_{4,i} = \begin{bmatrix} \boldsymbol{s}_{4,i} \\ \boldsymbol{r}_{4,i} \times \boldsymbol{s}_{4,i} \end{bmatrix} \tag{15}
$$

- **Joint 5**, second revolute of the universal joint:

$$
\begin{bmatrix} \boldsymbol{s}_{5,i} \\ 1 \end{bmatrix} = {}^{1}\mathbf{T}_{C,i} \begin{bmatrix} 0 \\ 1 \\ 0 \\ 1 \end{bmatrix} \quad \begin{bmatrix} \boldsymbol{r}_{5,i} \\ 1 \end{bmatrix} = {}^{1}\mathbf{T}_{C,i} \begin{bmatrix} 0 \\ 0 \\ 0 \\ 1 \end{bmatrix} \rightarrow \boldsymbol{S}_{5,i} = \begin{bmatrix} \boldsymbol{s}_{5,i} \\ \boldsymbol{r}_{5,i} \times \boldsymbol{s}_{5,i} \end{bmatrix} \tag{16}
$$

In order to simplify expressions of manipulator Jacobian matrices it is possible to use three screws which have the main characteristic of being reciprocal to all unit screws of the leg, with the exception of the actuated joints screws. Such screws are here called $\boldsymbol{S}_{r,i}$.

To this aim, it is possible to make use of the unit screw $\boldsymbol{S}_{4,i}$ shown in Fig. 4, which turns reciprocal to each non-actuated screw present in the leg kinematic chain due to the fact that it is coplanar to both $\boldsymbol{S}_{2,i}$ and $\boldsymbol{S}_{5,i}$, it intersects the axis of the prismatic joint described by $\boldsymbol{S}_{3,i}$ and, by definition, it is reciprocal to itself. Furthermore, it is not reciprocal to $\boldsymbol{S}_{1,i}$ being parallel but not aligned to the axis of the actuated prismatic joint.

At this point vector expressions have been given for joints screws and for legs reciprocal screws. The Jacobian matrices can be formulated in order to achieve an expression for the velocity problem which has the well known form $\mathbf{J}_X\dot{\boldsymbol{x}} = \mathbf{J}_Q\dot{\boldsymbol{q}}$, where $\dot{\boldsymbol{x}}$ is the velocity vector of the platform that, in a general way, can be expressed as $\dot{\boldsymbol{x}} = \begin{bmatrix} \omega_x & \omega_y & \omega_z & \dot{q}_x & \dot{q}_y & \dot{q}_z \end{bmatrix}^T$.

Firstly it is introduced the Jacobian matrix $\mathbf{J}_X$, whose expression can be formulated as a function of reciprocal screws:

$$
\mathbf{J}_X = \begin{bmatrix} \boldsymbol{S}_{r,1}^T \\ \boldsymbol{S}_{r,2}^T \\ \boldsymbol{S}_{r,3}^T \end{bmatrix} = \begin{bmatrix} \boldsymbol{S}_{4,1}^T \\ \boldsymbol{S}_{4,2}^T \\ \boldsymbol{S}_{4,3}^T \end{bmatrix}
= \begin{bmatrix} 0 & q_{3,1}cq_{2,1}-p_z & q_{3,1}sq_{2,1}+p_y & 1 & 0 & 0 \\ q_{3,2}sq_{2,2}+p_z & 0 & q_{3,2}cq_{2,2}-p_x & 0 & 1 & 0 \\ q_{3,3}cq_{2,3}-p_y & q_{3,3}sq_{2,3}+p_x & 0 & 0 & 0 & 1 \end{bmatrix} \tag{17}
$$

The moving platform of I.Ca.Ro. PKM is only allowed to perform pure translations; this implies that the first three components of the vector $\dot{\boldsymbol{x}}$, i.e. ω_x, ω_y and ω_z, are identically null. As a consequence, such components and the first three columns of matrix $\mathbf{J}_X$ can be eliminated due to the fact that the do not give any contribution to equation $\mathbf{J}_X\dot{\boldsymbol{x}} = \mathbf{J}_Q\dot{\boldsymbol{q}}$. Thus, the Jacobian $\mathbf{J}_X$ is a three by three identity matrix which multiplies the end-effector velocity vector $\dot{\boldsymbol{x}} = \begin{bmatrix} \dot{q}_x & \dot{q}_y & \dot{q}_z \end{bmatrix}^T$.

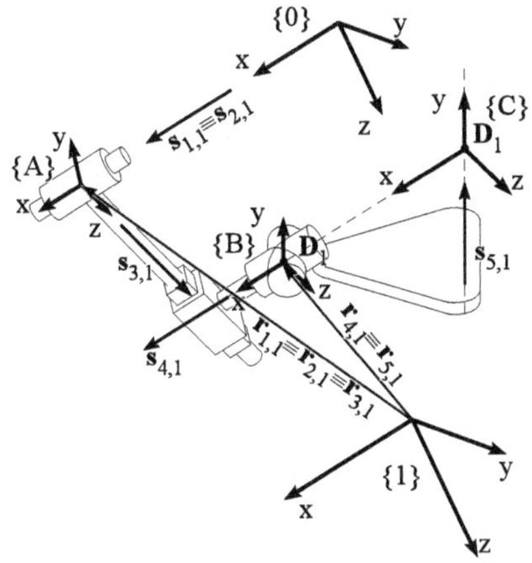

Figure 4. Unit screws of each kinematic joints.

Finally matrix $\mathbf{J}_Q$ is introduced:

$$\mathbf{J}_Q = \begin{bmatrix} \boldsymbol{S}_{r,1}^T \boldsymbol{S}_{1,1} & 0 & 0 \\ 0 & \boldsymbol{S}_{r,2}^T \boldsymbol{S}_{1,2} & 0 \\ 0 & 0 & \boldsymbol{S}_{r,3}^T \boldsymbol{S}_{1,3} \end{bmatrix} = \begin{bmatrix} 1 & 0 & 0 \\ 0 & 1 & 0 \\ 0 & 0 & 1 \end{bmatrix} \tag{18}$$

Due to pure translational robot kinematic behaviour, the end-effector velocity problem can be simply expressed as:

$$\begin{bmatrix} 1 & 0 & 0 \\ 0 & 1 & 0 \\ 0 & 0 & 1 \end{bmatrix} \begin{bmatrix} \dot{p}_x \\ \dot{p}_y \\ \dot{p}_z \end{bmatrix} = \begin{bmatrix} 1 & 0 & 0 \\ 0 & 1 & 0 \\ 0 & 0 & 1 \end{bmatrix} \begin{bmatrix} \dot{d}_1 \\ \dot{d}_2 \\ \dot{d}_3 \end{bmatrix} \tag{19}$$

Thanks to the screw based approach, the velocities of passive joints have been eliminated from the formulation of end-effector velocity kinematics. Nevertheless this information is needed to perform other types of analysis such as the study of robot dynamics.

Thus, the knowledge of the velocity vectors of each member composing the legs is necessary and it can be achieved through the computation of passive joints velocity as functions of active joints rates. To this aim, robot architecture constraint equations are exploited to build a matrix **A** relating prismatic actuated joints velocities to all other rates:

$$\dot{\boldsymbol{q}}_{\mathrm{p}} = \mathbf{A}\dot{\boldsymbol{q}}_{\mathrm{a}} \tag{20}$$

where $\dot{\boldsymbol{q}}_{\mathrm{p}}$ is a vector collecting velocities of passive joints and $\dot{\boldsymbol{q}}_{\mathrm{a}}$ is the vector of actuated joints rates.

The main aim of this computation is to provide the needed tools for the dynamic modelling of the robotic system. Therefore, a simplification based on influence of each body is introduced yielding a relevant computational simplification. The mass of the elements that compose the revolute joints is negligible if compared with masses of legs linkages and translating parts of prismatic actuators. Hence, it is supposed here that they only marginally affect the whole dynamic behaviour of the robot: thus, their contribution is not considered. This simplification is reasonably acceptable and enormously simplifies robot model because of the complexity introduced by the velocity expressions of these elements.

Such simplification allows us to reduce the dimension of matrix **A**. If the actual number of active and passive joints is considered, Eq. (20) can be expanded to:

$$\begin{bmatrix} \dot{q}_{2,1} \\ \dot{q}_{2,2} \\ \dot{q}_{2,3} \\ \dot{q}_{3,1} \\ \dot{q}_{3,2} \\ \dot{q}_{3,3} \end{bmatrix} = \mathbf{A} \begin{bmatrix} \dot{q}_{1,1} \\ \dot{q}_{1,2} \\ \dot{q}_{1,3} \end{bmatrix} \tag{21}$$

where $\dot{q}_{1,i}$ is the translation rate of cylindrical pair of i-th leg, $\dot{q}_{2,i}$ is the rotation rate of the same pair and $\dot{q}_{3,i}$ is the translation rate of the passive prismatic joint.

The constraint matrix **A** can be built considering the mobility of each attachment point between legs and manipulator; indeed, the velocity of such points is known and equal to the velocity of the moving platform due to the fact that they solidly move with the end-effector which only performs pure translational motions. The velocities of passive joints can be related to the components of the velocity vectors as visible in Fig. 5 for a general mobility.

The component of $v_{D,i}$ along the direction perpendicular to leg plane is expressed by:

$$v_{D,i}^{\perp} = \boldsymbol{v}_{D,i}^T(s_{2,i} \times s_{3,i}) \tag{22}$$

The velocity along this direction is fully due to the rotation of the cylindrical joint, so that:

$$\dot{q}_{2,i} = \frac{\boldsymbol{v}_{D,i}^T(s_{2,i} \times s_{3,i})}{q_{3,i}} \tag{23}$$

The component of velocity that lies on the leg plane is due to both the actuated and non actuated prismatic joints:

$$\begin{aligned} \dot{q}_{1,i} &= \boldsymbol{v}_{D,i}^T s_{1,i} \\ \dot{q}_{3,i} &= \boldsymbol{v}_{D,i}^T s_{3,i} \end{aligned} \tag{24}$$

The first equation in (24) simply relates the velocity along the axis of the cylindrical joint to the actuation rate; thus, it is not useful for the construction of the constraint matrix. On the other hand, the second equation can be used for the scope.

Equation (23) and the second equation in (24) can be expanded and written in the matrix form (20). In the case of a pure translational robot the velocities of legs attachment points correspond to the velocity of the origin of end effector reference frame. Exploiting Eqs. (22) and the second of (24), expressions of non actuated joints rates are achievable. In this case, the simplification introduced by end-effector mobility allows to show which is the actual shape of such expressions.

For the revolute joints it is:

$$\begin{aligned} \dot{q}_{2,1} &= \frac{-\dot{q}_{1,2}cq_{2,1} - \dot{q}_{1,3}sq_{2,1}}{q_{3,1}} \\ \dot{q}_{2,2} &= \frac{-\dot{q}_{1,1}sq_{2,2} - \dot{q}_{1,3}cq_{2,2}}{q_{3,2}} \\ \dot{q}_{2,3} &= \frac{-\dot{q}_{1,1}cq_{2,3} - \dot{q}_{1,2}sq_{2,3}}{q_{3,3}} \end{aligned} \tag{25}$$

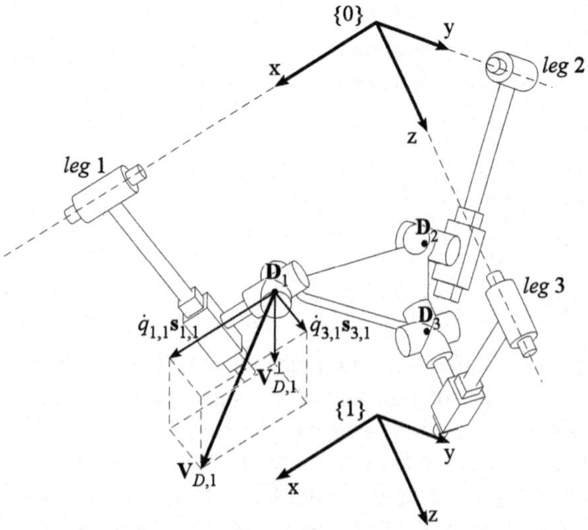

Figure 5. Velocity of attachment points between legs and moving platform.

while prismatic joints rates are:

$$\begin{array}{rcl}\dot{q}_{3,1} & = & -\dot{q}_{1,2}\mathrm{s}q_{2,1}+\dot{q}_{1,3}\mathrm{c}q_{2,1}\\ \dot{q}_{3,2} & = & \dot{q}_{1,1}\mathrm{c}q_{2,2}-\dot{q}_{1,3}\mathrm{s}q_{2,2}\\ \dot{q}_{3,3} & = & -\dot{q}_{1,1}\mathrm{s}q_{2,3}+\dot{q}_{1,2}\mathrm{c}q_{2,3}\end{array} \quad (26)$$

Hence, the matrix formulation of the 6 passive velocities is:

$$\begin{bmatrix}\dot{q}_{2,1}\\ \dot{q}_{2,2}\\ \dot{q}_{2,3}\\ \dot{q}_{3,1}\\ \dot{q}_{3,2}\\ \dot{q}_{3,3}\end{bmatrix}=\begin{bmatrix}0 & \frac{-\mathrm{c}q_{2,1}}{q_{3,1}} & \frac{-\mathrm{s}q_{2,1}}{q_{3,1}}\\ \frac{-\mathrm{s}q_{2,2}}{q_{3,2}} & 0 & \frac{-\mathrm{c}q_{2,2}}{q_{3,2}}\\ \frac{-\mathrm{c}q_{2,3}}{q_{3,3}} & \frac{-\mathrm{s}q_{2,3}}{q_{3,3}} & 0\\ 0 & -\mathrm{s}q_{2,1} & \mathrm{c}q_{2,1}\\ \mathrm{c}q_{2,2} & 0 & -\mathrm{s}q_{2,2}\\ -\mathrm{s}q_{2,3} & \mathrm{c}q_{2,3} & 0\end{bmatrix}\begin{bmatrix}\dot{q}_{1,1}\\ \dot{q}_{1,2}\\ \dot{q}_{1,3}\end{bmatrix} \quad (27)$$

In the remainder of this work, the constraint matrix is used to express the velocity of the reference frames attached to the robot bodies. To do that, further Jacobian matrices are introduced. In particular, the velocity of the c.o.m. of each body is written according to the general formulation:

$$\dot{\boldsymbol{x}}=\mathbf{J}\dot{\boldsymbol{q}}_{\mathrm{a}} \quad (28)$$

where $\dot{\boldsymbol{q}}_{\mathrm{a}}$ is the vector of actuated joints rates.

It is important to remark that the target of this section is the definition of legs bodies velocities, whose serial kinematics chain does not allow the simplification of passive joints rates. Thus, the Jacobian formulation $\dot{\boldsymbol{x}}=\tilde{\mathbf{J}}\begin{bmatrix}\dot{\boldsymbol{q}}_{\mathrm{a}}^T & \dot{\boldsymbol{q}}_{\mathrm{p}}^T\end{bmatrix}^T$ involves also the velocity of non actuated joints. Nevertheless, the influence of such joints can be explicited by means of the constraint matrix (27). Equation (28) becomes:

$$\dot{\boldsymbol{x}}=\tilde{\mathbf{J}}\begin{bmatrix}\mathbf{I}\\ \mathbf{A}\end{bmatrix}\dot{\boldsymbol{q}}_{\mathrm{a}} \quad (29)$$

where $\mathbf{I}$ is a 3×3 identity matrix; the matrices $\tilde{\mathbf{J}}$ can be very quickly expressed taking advance of joints screws; thus the formulation of legs velocities turns out to be an immediate iterative process based on collection of already introduced vectors.

Firstly, the velocities of the three sliders are achieved: for the sake of conciseness, these bodies are denoted as sl_1, sl_2 and sl_3 with reference to the leg which they are part of. The serial chain that allows reaching their screw is composed only by the actuated prismatic pair. In this case the body is not allowed to rotate, so that the position of the screw axis does not influence the screw expression:

$$\dot{\boldsymbol{x}}_{sl,i}=\boldsymbol{S}_{1,i}\dot{q}_{i,1}=\begin{bmatrix}\mathbf{0}\\ \boldsymbol{s}_{1,i}\end{bmatrix}\dot{q}_{i,1} \quad (30)$$

Equations (30) can be written according to the generic formulation (29):

$$\begin{array}{l}\dot{\boldsymbol{x}}_{sl,1}=\begin{bmatrix}\boldsymbol{S}_{1,1} & \mathbf{0} & \mathbf{0} & \mathbf{0} & \mathbf{0} & \mathbf{0} & \mathbf{0} & \mathbf{0} & \mathbf{0}\end{bmatrix}\begin{bmatrix}\mathbf{I}\\ \mathbf{A}\end{bmatrix}\dot{\boldsymbol{q}}_{\mathrm{a}}=\mathbf{J}_{sl1}\dot{\boldsymbol{q}}_{\mathrm{a}}\\ \dot{\boldsymbol{x}}_{sl,2}=\begin{bmatrix}\mathbf{0} & \boldsymbol{S}_{1,2} & \mathbf{0} & \mathbf{0} & \mathbf{0} & \mathbf{0} & \mathbf{0} & \mathbf{0} & \mathbf{0}\end{bmatrix}\begin{bmatrix}\mathbf{I}\\ \mathbf{A}\end{bmatrix}\dot{\boldsymbol{q}}_{\mathrm{a}}=\mathbf{J}_{sl2}\dot{\boldsymbol{q}}_{\mathrm{a}}\\ \dot{\boldsymbol{x}}_{sl,3}=\begin{bmatrix}\mathbf{0} & \mathbf{0} & \boldsymbol{S}_{1,3} & \mathbf{0} & \mathbf{0} & \mathbf{0} & \mathbf{0} & \mathbf{0} & \mathbf{0}\end{bmatrix}\begin{bmatrix}\mathbf{I}\\ \mathbf{A}\end{bmatrix}\dot{\boldsymbol{q}}_{\mathrm{a}}=\mathbf{J}_{sl3}\dot{\boldsymbol{q}}_{\mathrm{a}}\end{array} \quad (31)$$

In a very similar way, velocities of the first links (called here $l1_i$) can be achieved. The serial kinematic chain characteristic of such bodies is composed by the actuated prismatic pair and the non actuated revolute joint:

$$\dot{\boldsymbol{x}}_{l1,i}=\boldsymbol{S}_{1,i}\dot{q}_{i,1}+\boldsymbol{S}_{2,i}\dot{q}_{i,2}=\begin{bmatrix}\mathbf{0}\\ \boldsymbol{s}_{1,i}\end{bmatrix}\dot{q}_{i,1}+\begin{bmatrix}\boldsymbol{s}_{2,i}\\ \boldsymbol{r}_{2,i}\times\boldsymbol{s}_{2,i}\end{bmatrix}\dot{q}_{i,2} \quad (32)$$

The position of the screw axis is relevant for the computation of the revolute joint unit screw. Even though its expression does not coincide with the previously found value, axis position is quickly computable using homogeneous transformation matrix and the position vector $\boldsymbol{p}_{l1,i}$ of the center of mass of body $l1,i$ with respect to reference frame $\{A\}$. The position of the screw axis is computable as the difference between absolute position $\boldsymbol{O}_{A,i}$ of frame $\{A\}$ and absolute position of center of mass:

$$\begin{array}{l}\boldsymbol{P}_{l1,i}={}^{0}\mathbf{T}_{A}\boldsymbol{p}_{l1,i}\\ \boldsymbol{O}_{A,i}={}^{0}\mathbf{T}_{A}\begin{bmatrix}0 & 0 & 0 & 1\end{bmatrix}^T\end{array} \rightarrow \boldsymbol{r}_{2,i}=\boldsymbol{O}_{A,i}-\boldsymbol{P}_{l1,i} \quad (33)$$

Expansion of (32) for all robot legs yields:

$$\begin{array}{l}\dot{\boldsymbol{x}}_{l1,1}=\begin{bmatrix}\boldsymbol{S}_{1,1} & \mathbf{0} & \mathbf{0} & \boldsymbol{S}_{2,1} & \mathbf{0} & \mathbf{0} & \mathbf{0} & \mathbf{0} & \mathbf{0}\end{bmatrix}\begin{bmatrix}\mathbf{I}\\ \mathbf{A}\end{bmatrix}\dot{\boldsymbol{q}}_{\mathrm{a}}=\mathbf{J}_{l1,1}\dot{\boldsymbol{q}}_{\mathrm{a}}\\ \dot{\boldsymbol{x}}_{l1,2}=\begin{bmatrix}\mathbf{0} & \boldsymbol{S}_{1,2} & \mathbf{0} & \mathbf{0} & \boldsymbol{S}_{2,2} & \mathbf{0} & \mathbf{0} & \mathbf{0} & \mathbf{0}\end{bmatrix}\begin{bmatrix}\mathbf{I}\\ \mathbf{A}\end{bmatrix}\dot{\boldsymbol{q}}_{\mathrm{a}}=\mathbf{J}_{l1,2}\dot{\boldsymbol{q}}_{\mathrm{a}}\\ \dot{\boldsymbol{x}}_{l1,3}=\begin{bmatrix}\mathbf{0} & \mathbf{0} & \boldsymbol{S}_{1,3} & \mathbf{0} & \mathbf{0} & \boldsymbol{S}_{2,3} & \mathbf{0} & \mathbf{0} & \mathbf{0}\end{bmatrix}\begin{bmatrix}\mathbf{I}\\ \mathbf{A}\end{bmatrix}\dot{\boldsymbol{q}}_{\mathrm{a}}=\mathbf{J}_{l1,3}\dot{\boldsymbol{q}}_{\mathrm{a}}\end{array} \quad (34)$$

Finally, the velocity of the last link that composes the leg (here called body $l2_i$) is a linear combination of the elementary screws of the actuated prismatic joint, the first passive revolute joint and the non actuated prismatic joint of each leg:

$$\begin{array}{rcl}\dot{\boldsymbol{x}}_{l2,i} & = & \boldsymbol{S}_{1,i}\dot{q}_{i,1}+\boldsymbol{S}_{2,i}\dot{q}_{i,2}+\boldsymbol{S}_{3,i}\dot{q}_{i,3}\\ & = & \begin{bmatrix}\mathbf{0}\\ \boldsymbol{s}_{1,i}\end{bmatrix}\dot{q}_{i,1}+\begin{bmatrix}\boldsymbol{s}_{2,i}\\ \boldsymbol{r}_{2,i}\times\boldsymbol{s}_{2,i}\end{bmatrix}\dot{q}_{i,2}+\begin{bmatrix}\mathbf{0}\\ \boldsymbol{s}_{3,i}\end{bmatrix}\dot{q}_{i,3}\end{array} \quad (35)$$

The position $\boldsymbol{r}_{2,i}$ of the screw axis is once again different from the previously exposed case. Nevertheless, its expression is achievable by the definition of a position vector $\boldsymbol{p}_{l2,i}$ of the center of mass of bodies $l2_i$ with respect to their attached frame $\{B\}$. The distance from this point to the axis of the revolute joint is given by:

$$\begin{matrix} \boldsymbol{p}_{l2,i} = {}^0\mathbf{T}_B \boldsymbol{p}_{l2,i} \\ \boldsymbol{O}_{B,i} = {}^0\mathbf{T}_B \begin{bmatrix} 0 & 0 & 0 & 1 \end{bmatrix}^T \end{matrix} \quad \rightarrow \quad \boldsymbol{r}_{2,i} = \boldsymbol{O}_{B,i} - \boldsymbol{p}_{l2,i} \tag{36}$$

As done in previous case, the Jacobian formulation can be plainly reached also for velocities of bodies $l2_i$:

$$\begin{aligned} \dot{\boldsymbol{x}}_{l2,1} &= \begin{bmatrix} S_{1,1} & \mathbf{0} & \mathbf{0} & S_{2,1} & \mathbf{0} & \mathbf{0} & S_{3,1} & \mathbf{0} & \mathbf{0} \end{bmatrix} \begin{bmatrix} \mathbf{I} \\ \mathbf{A} \end{bmatrix} \dot{\boldsymbol{q}}_{\mathrm{a}} = \mathbf{J}_{l2,1} \dot{\boldsymbol{q}}_{\mathrm{a}} \\ \dot{\boldsymbol{x}}_{l2,2} &= \begin{bmatrix} \mathbf{0} & S_{1,2} & \mathbf{0} & \mathbf{0} & S_{2,2} & \mathbf{0} & \mathbf{0} & S_{3,2} & \mathbf{0} \end{bmatrix} \begin{bmatrix} \mathbf{I} \\ \mathbf{A} \end{bmatrix} \dot{\boldsymbol{q}}_{\mathrm{a}} = \mathbf{J}_{l2,2} \dot{\boldsymbol{q}}_{\mathrm{a}} \\ \dot{\boldsymbol{x}}_{l2,3} &= \begin{bmatrix} \mathbf{0} & \mathbf{0} & S_{1,3} & \mathbf{0} & \mathbf{0} & S_{2,3} & \mathbf{0} & \mathbf{0} & S_{3,3} \end{bmatrix} \begin{bmatrix} \mathbf{I} \\ \mathbf{A} \end{bmatrix} \dot{\boldsymbol{q}}_{\mathrm{a}} = \mathbf{J}_{l2,3} \dot{\boldsymbol{q}}_{\mathrm{a}} \end{aligned} \tag{37}$$

4 Acceleration kinematics

The dynamic modelling of a mechanical system requires a complete knowledge of machine kinematics. Therefore, acceleration of each body must be studied.

Manipulator velocity kinematics has been formulated through the well known Jacobian formulation $\dot{\boldsymbol{q}}_{\mathrm{a}} = \mathbf{J}\dot{\mathbf{x}}$, where $\mathbf{J} = \mathbf{J}_Q^{-1}\mathbf{J}_X$ is in this case a 3 by 3 identity matrix. Direct differentiation of such expression yields:

$$\ddot{\boldsymbol{q}}_a = \dot{\mathbf{J}}\dot{\boldsymbol{x}} + \mathbf{J}\ddot{\boldsymbol{x}} \tag{38}$$

where $\dot{\mathbf{J}}_X$ is the derivative of the jacobian matrix, which is a constant matrix. Thus, in the case of a pure translational machine, the derivation of $\dot{\mathbf{J}}_X$ yields $\dot{\mathbf{J}}_X = \mathbf{0}$.

The acceleration kinematics of other robot members is easily achievable by direct differentiation of the velocity kinematics previously defined:

$$\ddot{\boldsymbol{X}}_{j,i} = \dot{\mathbf{J}}_{j,i}\dot{\boldsymbol{q}}_{\mathrm{a}} + \mathbf{J}_{j,i}\ddot{\boldsymbol{q}}_a \tag{39}$$

were $\dot{\mathbf{J}}_{j,i}$ is the time derivative of the respective Jacobian matrix. Expansion of (39) yields to very a long formulation that, for sake of conciseness, is not shown here.

5 Virtual work principle

The virtual work principle approach for dynamic modelling requires the definition of the 6 dimensional vector $\boldsymbol{F}_{j,i}$, whose components collect resultants of both active and inertial forces and torques acting on the j-th body of i-th leg, computed with respect to the center of mass of the member:

$$\boldsymbol{F}_{j,i} = \begin{bmatrix} \boldsymbol{n}_{j,i} \\ \boldsymbol{F}_{j,i} \end{bmatrix} = \begin{bmatrix} -{}^0\mathbf{I}_{j,i}\dot{\omega}_{j,i} - \omega_{j,i} \times \left({}^0\mathbf{I}_{j,i}\omega_{j,i}\right) \\ m_{j,i}\left(\boldsymbol{g} - \dot{\boldsymbol{v}}_{j,i}\right) \end{bmatrix} \tag{40}$$

In a similar way, it is introduced the vector $\boldsymbol{F}_{\mathrm{EE}}$ that collects forces and torques acting on robot's end-effector and computed with respect to manipulator centre of mass:

$$\boldsymbol{F}_{\mathrm{EE}} = \begin{bmatrix} \boldsymbol{n}_{\mathrm{EE}} \\ \boldsymbol{F}_{\mathrm{EE}} \end{bmatrix} = \begin{bmatrix} -{}^0\mathbf{I}_{\mathrm{EE}}\dot{\omega}_{\mathrm{EE}} - \omega_{\mathrm{EE}} \times \left({}^0\mathbf{I}_{\mathrm{EE}}\omega_{\mathrm{EE}}\right) \\ m_{\mathrm{EE}}\left(\boldsymbol{g} - \dot{\boldsymbol{v}}_{\mathrm{EE}}\right) \end{bmatrix} \tag{41}$$

Virtual works principle allows writing:

$$\boldsymbol{\delta q}_{\mathrm{a}}^T \boldsymbol{\tau} + \delta \boldsymbol{x}^T \boldsymbol{F}_{\mathrm{EE}} + \sum_{i,j} \delta \boldsymbol{x}_{j,i}^T \boldsymbol{F}_{j,i} = 0 \tag{42}$$

where vector $\boldsymbol{\delta q}_{\mathrm{a}}$ represents the virtual displacements of actuated joints, $\boldsymbol{\tau}$ is the vector of actuation torques and $\delta\boldsymbol{x}$ is the virtual displacement of rotation/displacement of respective body.

The differential kinematics of the manipulator, whose formulation has been introduced in previous sections, is usefully exploited to relate actuated joints displacements to other bodies twists. Indeed, end-effector differential kinematics expression allows writing:

$$\boldsymbol{\delta q}_{\mathrm{a}} = \mathbf{J}_X \delta \boldsymbol{x} \tag{43}$$

In a similar way, the twist of other robot members can be expressed through the respective Jacobian matrices:

$$\delta \boldsymbol{x}_{j,i} = \mathbf{J}_{j,i} \boldsymbol{\delta q}_{\mathrm{a}} \tag{44}$$

When Eq. (43) is invertible, i.e. when determinant of the Jacobian matrix is not null, the differential of end effector twist can be expressed in terms of actuated joints translations, being $\delta\boldsymbol{x} = \mathbf{J}_X^{-1}\boldsymbol{\delta q}_{\mathrm{a}}$. Dynamics Eq. (42) becomes:

$$\boldsymbol{\delta q}_{\mathrm{a}}^T \boldsymbol{\tau}_A + \boldsymbol{\delta q}_{\mathrm{a}}^T \mathbf{J}_X^{-T} \boldsymbol{F}_{\mathrm{EE}} + \boldsymbol{\delta q}_{\mathrm{a}}^T \sum_{i,j} \mathbf{J}_{j,i}^T \boldsymbol{F}_{j,i} = 0 \tag{45}$$

For non null virtual displacements $\boldsymbol{\delta q}_{\mathrm{a}}$, such term can be collected and eliminated, yielding:

$$\boldsymbol{\tau} + \mathbf{J}_X^{-T} \boldsymbol{F}_{\mathrm{EE}} + \sum_{i,j} \mathbf{J}_{j,i}^T \boldsymbol{F}_{j,i} = 0 \tag{46}$$

Equation (46) can be collected in the canonical form:

$$\boldsymbol{\tau} + \mathbf{M}(\boldsymbol{q}_{\mathrm{a}})\ddot{\boldsymbol{q}}_{\mathrm{a}} + \boldsymbol{V}(\boldsymbol{q}_{\mathrm{a}}, \dot{\boldsymbol{q}}_{\mathrm{a}}) + \boldsymbol{G}(\boldsymbol{q}_{\mathrm{a}}) = \mathbf{0} \tag{47}$$

As known, each component of this equation includes different contributions to the dynamics of the manipulator: $\mathbf{M}(\boldsymbol{q}_{\mathrm{a}})\ddot{\boldsymbol{q}}_{\mathrm{a}}$ called later $\boldsymbol{\tau}_{\mathbb{M}}$ is a contribution due to inertial effects of bodies masses, $\boldsymbol{V}(\boldsymbol{q}_{\mathrm{a}}, \dot{\boldsymbol{q}}_{\mathrm{a}})$, hereby called $\boldsymbol{\tau}_{\mathbb{V}}$, is due to Coriolis and centripetal accelerations and, at last, $\boldsymbol{G}(\boldsymbol{q}_{\mathrm{a}})$ are the forces deriving from gravitational action on robot members, called here $\boldsymbol{\tau}_{\mathbb{G}}$.

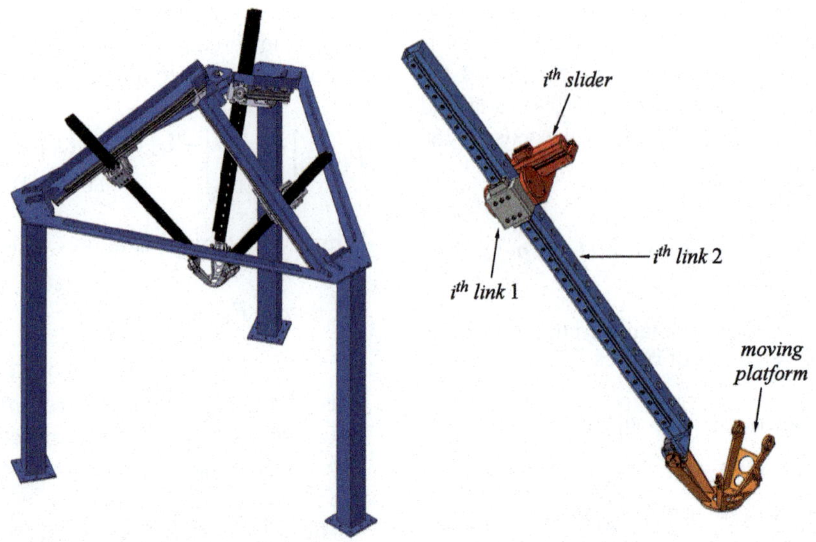

Figure 6. Multibody model of I.Ca.Ro. parallel manipulator.

Table 1. Phisical characteristics of the I.Ca.Ro. robot members.

body	c.o.m. [m]	mass [kg]	inertia matrix $\left[\text{kg m}^2\right]$
slider	not relevant	5.19	not relevant
link 1	$1\times10^{-3}\left[-0.04 \quad -43.69 \quad 0\right]^T$	2.62	$\begin{bmatrix} 0.003 & \sim 0 & \sim 0 \\ \sim 0 & 0.004 & \sim 0 \\ \sim 0 & \sim 0 & 0.003 \end{bmatrix}$
link 2	$1\times10^{-3}\left[32.25 \quad 13.16 \quad -552.57\right]^T$	11.12	$\begin{bmatrix} 1.405 & -4.388\times10^{-4} & -0.018 \\ -4.388\times10^{-4} & 1.405 & -0.008 \\ -0.018 & -0.008 & 0.008 \end{bmatrix}$
moving platform	not relevant	1.60	not relevant

6 Model verification

In this section a verification of the inverse dynamics model is proposed. With this aim, a multibody model of the 3-CPU pure translational parallel platform has been settled up. Under hypothesis of coherence between the two models in terms of geometrical and mass parameters, a perfect correspondence on actuation forces should be noticed when an identical motion law is used.

The multibody model (see Fig. 6) of the parallel platform is based on a graphical CAD representation of robot members. Definition of joints between the bodies allows the software to reproduce machine kinematic and dynamic behaviour. Each member composing the robot has been measured through mass geometry instruments provided by the CAD environment. In order to improve readers' understanding on the correspondence between multibody model and mathematical model of the platform, Fig. 6 also shows the members of each leg with different colors. A characterization of interesting physical properties of each member is given in Table 1; it is worth to remark that the multibody model has been built with the maximum respect of the actual I.Ca.Ro. prototype, in order to give a description as much as possible reliable of the mechanical system. For the sake of conciseness, magnitudes that are not useful for dynamic modelling of the manipulator are not reported.

It should be remarked that the inertia matrices expressed for each body refer to a reference frame centred in the centre of mass of and attached to the respective body. Since the model needs these matrices to be expressed with respect to the fixed reference frame {0}, a coordinates change must be made. Then, for those bodies that are allowed to roatate it is ${}^0\mathbf{I}_i = {}^0\mathbf{R}_i {}^i\mathbf{I}_i {}^0\mathbf{R}_i^T$ where ${}^0\mathbf{R}_i$ denote the orientation of the i-th body with respect to reference frame {0}.

The physical properties described in Table 1 have been used also for the mathematical model in order to perform a direct comparison with results provided by the multibody environment. As an example, results are reported deriving from a particular set of actuation displacements profiles: a harmonic time history has been chosen for each prismatic joint displacement; each slider is moved with a different frequency. Details on the used functions are shown in Fig. 7.

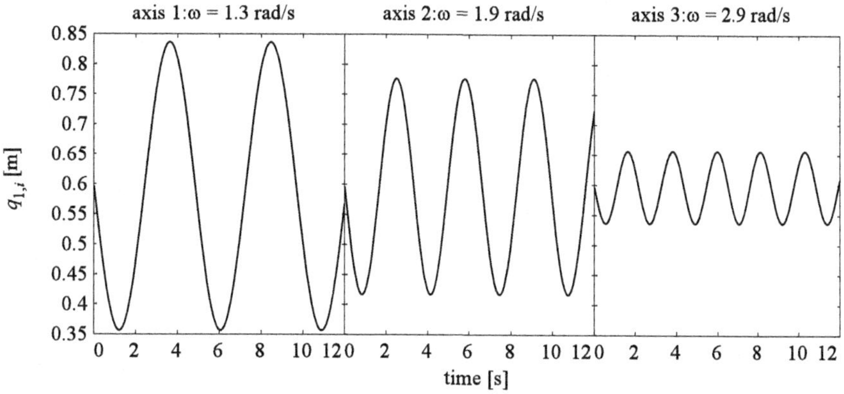

Figure 7. Harmonic displacement profiles used for model verification.

Such motion has been chosen in order to investigate a significant part of the robot workspace, pushing the machine to the physical limits given by the maximum velocity that the three motors are able to perform, which is 0.6 m s^{-1}.

Results of both virtual and mathematical models are used for computation of the relative difference subsisting between the two sets of forces. In particular, named the maximum absolute value of force recorded for motor i during multibody simulations, for the i-th axis it is defined the error ϵ_i as:

$$\epsilon_i = \frac{|\tau_{v,i} - \tau_{m,i}|}{|\tau_{v,i}|^{\max}} \tag{48}$$

where $\tau_{v,i}$ and $\tau_{m,i}$ are the instantaneous forces computed by multibody environment and mathematical model respectively. Equation (48) gives an idea of the deviation between the two methods and therefore it represents a sort of measure of the error introduced by the mathematical model. As visible in Fig. 8, this error never overcomes the 1.0 % of the maximum force during simulation, while the average error is always lower than 1 %.

According to Eq. (47), the actuation forces evaluated during the previous inverse dynamics simulation can be split in order to analyse the contribution of each part of the robot dynamics. Figure 9 shows the different contributions of the model on the total effort provided by each motor: as well visible, the most part of the force is due to the gravity acceleration acting on robot bodies while a negligible contribution is given by Coriolis and centripetal accelerations. This important information can be used during the realization of simplified mathematical models in which, the contribution of force vector τ_V can be ignored.

7 Simulations results

In this section, simulations are shown in order to test the reliability of the introduced model in actually reproducible conditions.

The first simulation approached is a linear trajectory inside the workspace. The robot, starting from its home position, moves to a given point in the space. The trajectory has been planned in order to obtain continuity on platform accelerations. The maximum velocity reached by the manipulator is 0.6 m s^{-1} (which corresponds to the maximum linear velocity available for the actuated joints), while the maximum acceleration is 1.40 m s^{-2}. The starting and the ending points of the motions are 0.5 m apart; it should be remarked that the robot workspace is a cube with a 0.6 m edge. Thus, the trajectory spans a relevant distance with respect to the maximum displacements that the manipulator is able to perform.

For this motion, the forces computed thanks to Eq. (47) are shown in Fig. 10. Also in this case, the most important contribution to motors total effort is given by the gravity acceleration, while the part of force due to Coriolis and centripetal acceleration is negligible.

A second simulation has been performed with a different trajectory in the space. In this case, the end-effector has been moved from its home configuration to a point in the space. From that point, a horizontal circular trajectory (with diameter equal to 0.3 m) centred on robot vertical axis has been performed. Also in this case, the trajectory planning has been performed in order to obtain triangular profiles of acceleration. In this case the maximum velocity reached by the moving platform is 0.6 m s^{-1}, with a maximum acceleration of 0.75 m s^{-2}.

Also for this motion, forces profiles are shown (see Fig. 11). Again, the gravity contribute to the total effort is prevailing with respect to τ_M and τ_V. The influence of τ_V, in particular, represents a negligible contribution to total actuation effort. Nevertheless, Fig. 11 shows that τ_M considerably contributes to the total effort being $|\tau_{M,i}|_{\max} \simeq 20\,\%|\tau_i|_{\max}$.

Given the results of the last simulation, it is interesting to investigate how much the term τ_M affects the dynamic behaviour of the manipulator when the motors are used at their maximum thrust. The robot I.Ca.Ro. is provided with brushless motors, which are able to feed the actuated joints with a maximum force of 420 N. To this aim a simulation has been performed with a circular trajectory, similar to the

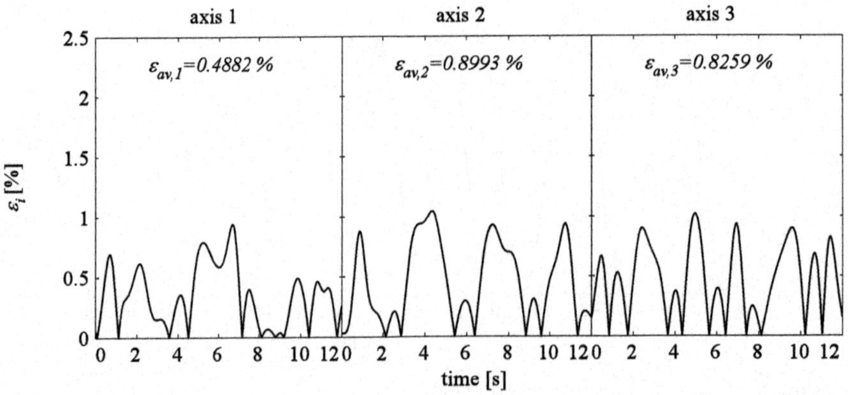

Figure 8. Difference between multibody model and mathematical model.

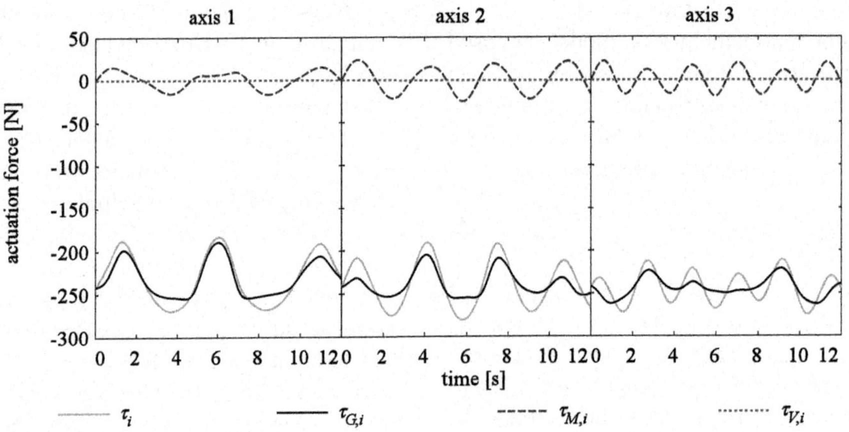

Figure 9. Different contributions to the total actuation efforts.

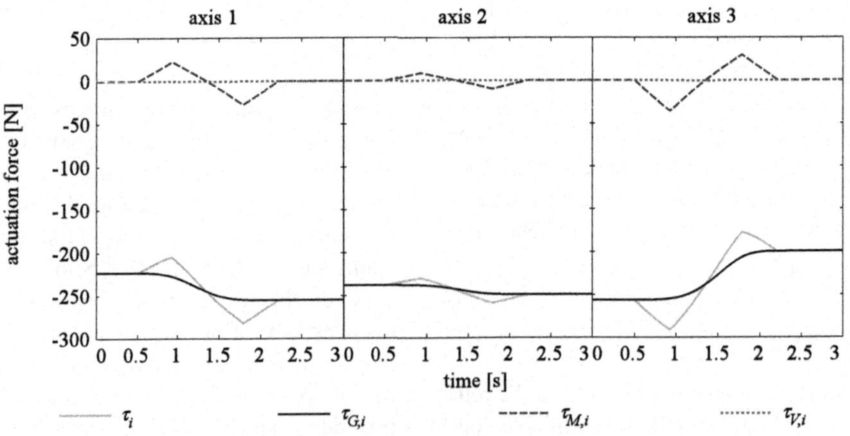

Figure 10. Different contributions to the total actuation efforts during a motion along a line.

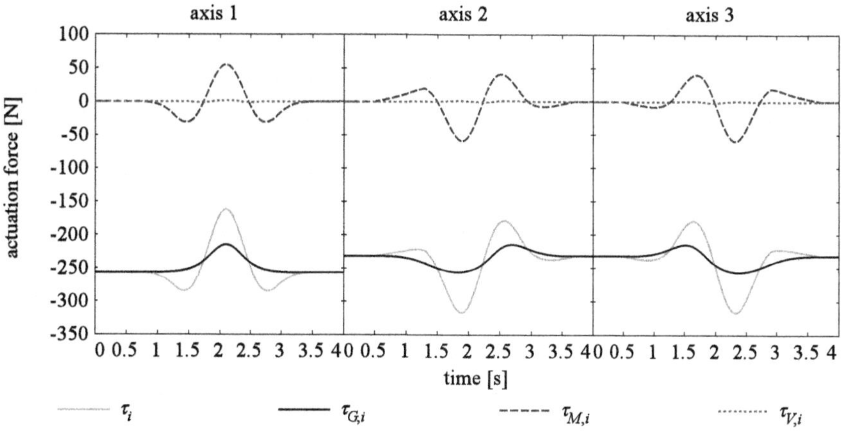

Figure 11. Different contributions to the total actuation efforts during a horizontal circular motion.

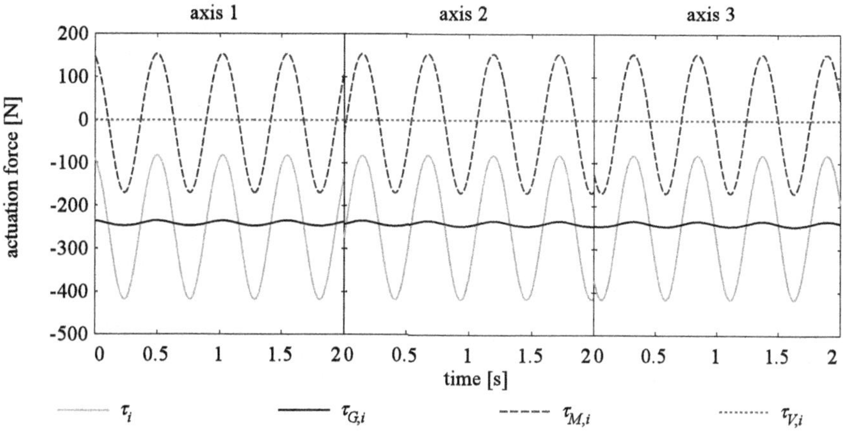

Figure 12. Different contributions to the total actuation efforts during a horizontal circular motion at maximum allowed thrust.

previous one. In this case the circle owns a diameter of 0.1 m and the manipulator is moved with a constant linear speed of $0.6\,\mathrm{m\,s^{-1}}$. Also the initial velocity has been taken equal to $0.6\,\mathrm{m\,s^{-1}}$ in order to overlook effects due to acceleration ramps. Figure 12 shows that forces τ_M are in this simulation comparable with τ_G demonstrating that a control based on the dynamic model can actually improve the performances of the 3-CPU robot. Moreover, Fig. 12 also confirms that the Coriolis terms are negligible and so they may be omitted in a simplified model (from Eq. 47, $\boldsymbol{\tau} \simeq -\mathbf{M}(\boldsymbol{q}_\mathrm{a})\ddot{\boldsymbol{q}}_\mathrm{a} - \boldsymbol{G}(\boldsymbol{q}_\mathrm{a})$).

Even though the elimination of the Coriolis terms significantly lightens the dynamics formulation, further simplifications can be carried on the matrix $\mathbf{M}$ itself. As an example, it is presented here the results obtained by the elimination of the terms out of the diagonal of such matrix. In particular, Fig. 13 shows the behaviour of τ_M and $\tau_{M'}$ for the circular motion just presented in Fig. 12; τ_M and $\tau_{M'}$ are computed as:

$$\tau_M = \mathbf{M}(\boldsymbol{q}_\mathrm{a})\,\ddot{\boldsymbol{q}}_\mathrm{a} \qquad \tau_{M'} = \mathbf{M}'(\boldsymbol{q}_\mathrm{a})\,\ddot{\boldsymbol{q}}_\mathrm{a} \tag{49}$$

where the matrix $\mathbf{M}'$ is the simplified $\mathbf{M}$ matrix:

$$\mathbf{M}' = \begin{bmatrix} M_{1,1} & 0 & 0 \\ 0 & M_{2,2} & 0 \\ 0 & 0 & M_{3,3} \end{bmatrix} \tag{50}$$

Figure 13 shows that the use of matrix $\mathbf{M}'$ overestimates the effect of the mass matrix on the robot dynamics of a maximum value of 30 N (see curve $\Delta\tau_{M,i}$). The error between τ_M and $\tau_{M'}$ is here estimated as:

$$\epsilon_{M,i} = \frac{|\tau_{M,i} - \tau_{M',i}|}{\max|\tau_{M,i}|} \tag{51}$$

Figure 14 shows the error ϵ_M for each motor: the graphic shows that the error does not overcome the value of 18 %.

8 Conclusions

The dynamic modelling of a pure translational PKM has been tackled in this work. Authors proposed a screw based approach for modelling the robot's kinematics, allowing a fast writing of the Jacobian matrices.

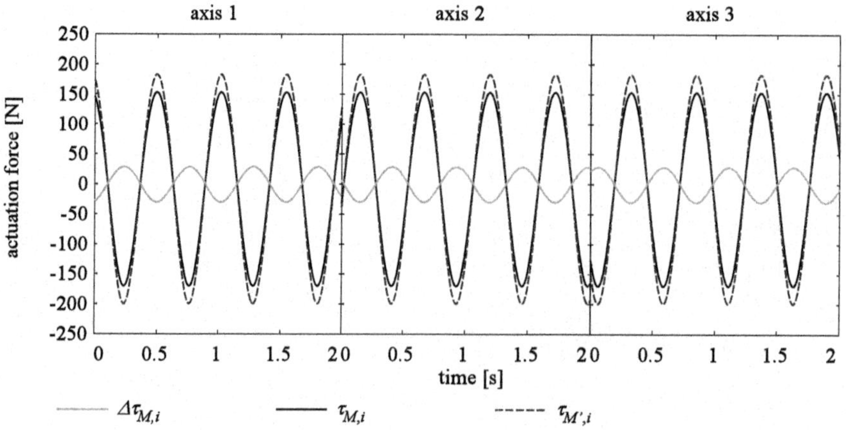

Figure 13. Mass matrix contribution to the dynamics in case of plain and simplified model.

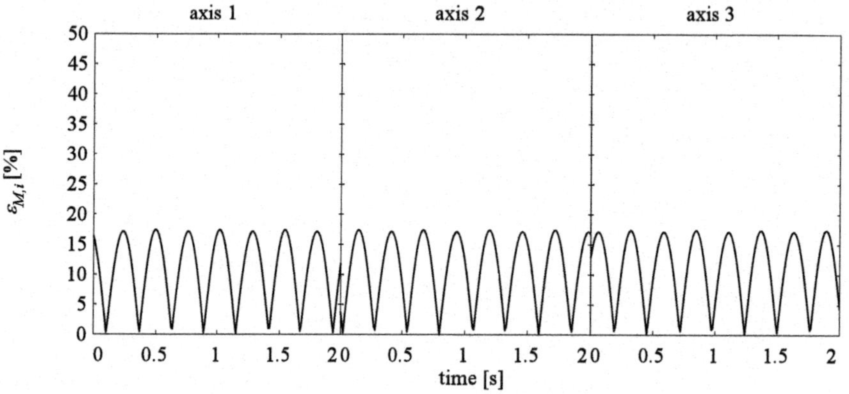

Figure 14. Difference between contributions of full mass matrix **M** and diagonalized mass matrix **M**′.

The dynamics of the I.Ca.Ro. manipulator was worked out by means of a virtual work principle approach. The resulting model has been verified through simple simulations, taking advance of a multibody model of the robot. Such verification pointed out that the error subsisting between the two virtual models never overcomes the 1.0 % of the maximum value of torque involved into the motion.

At last, two simulations have been performed on two trajectories with main aim of investigating the different contributions to the dynamics model. Observation of the results of such simulations yielded a further investigation on the contribution to the whole motors efforts. Such study pointed out that the robot is poorly affected by Coriolis and centrifugal forces while the influence of the mass matrix is not negligible. It is author thought that the compensation of this effect by means of a model based control may improve the performances of robot I.Ca.Ro.

The implementation of the mass matrix on a control algorithm may represent a low efficiency step of a control algorithm because of the heavy mathematical formulation. Due to this, authors also presented a simple simplification of the model based on a simplification of the mass matrix. Simulations demonstrated that such assumption yield an error that never overcomes the 18 % in a situation of high motors stress.

Edited by: A. Tasora

References

Abdellatif, H. and Heimann, B.: Computational efficient inverse dynamics of 6-DOF fully parallel manipulators by using the Lagrangian formalism, Mech. Mach. Theory, 44, 192–207, doi:10.1016/j.mechmachtheory.2008.02.003, 2009.

Caccavale, F., Siciliano, B., and Villani, L.: The Tricept robot: dynamics and impedance control, Mechatronics, IEEE/ASME Transactions on, 8, 263–268, doi:10.1109/TMECH.2003.812839, 2003.

Callegari, M. and Palpacelli, M.-C.: Prototype design of a translating parallel robot, Meccanica, 43, 133–151, doi:10.1007/s11012-008-9116-8, doi:10.1007/s11012-008-9116-8, 2008.

Carbonari, L.: Extended analysis of the 3-cpu reconfigurable class of parallel robotic manipulators, Ph.D. thesis, Polytechnic University of Marche, Ancona, Italy, 2012.

Carbonari, L. and Callegari, M.: The kinematotropic 3-CPU parallel robot: analysis of mobility and reconfigurability aspects, in: Latest Advances in Robot Kinematics, edited by: Lenarčič, J. and Husty, M., Springer, 2012.

Carbonari, L., Callegari, M., Palmieri, G., and Palpacelli, M.: A new class of reconfigurable parallel kinematics machines, Mech. Mach. Theory, in review, 2013.

Dasgupta, B. and Mruthyunjaya, T.: A Newton-Euler formulation for the inverse dynamics of the Stewart platform manipulator, Mech. Mach. Theory, 33, 1135–1152, doi:10.1016/S0094-114X(97)00118-3, 1998.

Daun, Q., Daun, B., and Daun, X.: Dynamics modelling and hybrid control of the 6-UPS platform, in: 2010 International Conference on Mechatronics and Automation (ICMA), 434 –439, doi:10.1109/ICMA.2010.5589105, 2010.

Di Gregorio, R. and Parenti-Castelli, V.: Dynamics of a Class of Parallel Wrists, J. Mech. Design, 126, 436–441, doi:10.1115/1.1737382, 2004.

Do, W. Q. D. and Yang, D. C. H.: Inverse dynamic analysis and simulation of a platform type of robot, J. Robotic Syst., 5, 209–227, doi:10.1002/rob.4620050304, 1988.

Gallardo, J., Rico, J., Frisoli, A., Checcacci, D., and Bergamasco, M.: Dynamics of parallel manipulators by means of screw theory, Mech. Mach. Theory, 38, 1113–1131, doi:10.1016/S0094-114X(03)00054-5, 2003.

Khalil, W. and Ibrahim, O.: General Solution for the Dynamic Modeling of Parallel Robots, J. Intell. Robotics Syst., 49, 19–37, doi:10.1007/s10846-007-9137-x, 2007.

Kovecses, J., Piedboeuf, J. C., and Lange, C.: Dynamics modeling and simulation of constrained robotic systems, Mechatronics, IEEE/ASME Transactions on, 8, 165–177, doi:10.1109/TMECH.2003.812827, 2003.

Kunquan, L. and Rui, W.: Closed-form Dynamic Equations of the 6-RSS Parallel Mechanism through the Newton-Euler Approach, in: 2011 Third International Conference on Measuring Technology and Mechatronics Automation (ICMTMA), 1, 712–715, doi:10.1109/ICMTMA.2011.180, 2011.

Lin, Y.-J. and Song, S.-M.: A comparative study of inverse dynamics of manipulators with closed-chain geometry, J. Robotic Syst., 7, 507–534, doi:10.1002/rob.4620070402, 1990.

Lopes, A. M.: Dynamic modeling of a Stewart platform using the generalized momentum approach, Commun. Nonlinear Sci., 14, 3389–3401, doi:10.1016/j.cnsns.2009.01.001, 2009.

Miller, K.: Optimal Design and Modeling of Spatial Parallel Manipulators, Int. J. Robot. Res., 23, 127–140, doi:10.1177/0278364904041322, 2004.

Moon, F.: Applied Dynamics: With Applications to Multibody and Mechatronic Systems, John Wiley & Sons, 2008.

Papastavridis, J.: Analytical Mechanics: A Comprehensive Treatise on the Dynamics of Constrained Systems, World Scientific, 2012.

Tsai, L.-W.: Solving the Inverse Dynamics of a Stewart-Gough Manipulator by the Principle of Virtual Work, J. Mech. Design, 122, 3–9, doi:10.1115/1.533540, 2000.

Wang, J. and Gosselin, C. M.: A New Approach for the Dynamic Analysis of Parallel Manipulators, Multibody Syst. Dyn., 2, 317–334, doi:10.1023/A:1009740326195, 1998.

Wang, J., Wu, J., Wang, L., and Li, T.: Simplified strategy of the dynamic model of a 6-UPS parallel kinematic machine for real-time control, Mech. Mach. Theory, 42, 1119–1140, doi:10.1016/j.mechmachtheory.2006.09.004, 2007.

Wronka, C. and Dunnigan, M.: Derivation and analysis of a dynamic model of a robotic manipulator on a moving base, Robot. Auton. Syst., 59, 758–769, doi:10.1016/j.robot.2011.05.010, 2011.

Yang, C., Han, J., Zheng, S., and Ogbobe Peter, O.: Dynamic modeling and computational efficiency analysis for a spatial 6-DOF parallel motion system, Nonlinear Dynam., 67, 1007–1022, doi:10.1007/s11071-011-0043-1, 2012.

Zhang, C.-D. and Song, S.-M.: An efficient method for inverse dynamics of manipulators based on the virtual work principle, J. Robotic Syst., 10, 605–627, doi:10.1002/rob.4620100505, 1993.

7

Prediction of railway induced ground vibration through multibody and finite element modelling

G. Kouroussis and O. Verlinden

Université de Mons – UMONS, Faculty of Engineering, Department of Theoretical Mechanics, Dynamics and Vibrations, Place du Parc 20, 7000 Mons, Belgium

Correspondence to: G. Kouroussis (georges.kouroussis@umons.ac.be)

Abstract. The multibody approach is now recognized as a reliable and mature computer aided engineering tool. Namely, it is commonly used in industry for the design of road or railway vehicles. The paper presents a framework developed for predicting the vibrations induced by railway transportation. Firstly, the vehicle/track subsystem is simulated, on the basis of the home-made C++ library EasyDyn, by mixing the multibody model of the vehicle and the finite element model of the track, coupled to each other through the wheel/rail contact forces. Only the motion in the vertical plane is considered, assuming a total symmetry between left and right rails. This first step produces the time history of the forces exerted by the ballast on the foundation, which are then applied to a full 3-D FEM model of the soil, defined under the commercial software ABAQUS. The paper points out the contribution of the pitch motion of the bogies and carbodies which were neglected in previous publications, as well as the interest of the so-called coupled-lumped mass model (CLM) to represent the influence of the foundation in the track model. The potentialities of the model are illustrated on the example of the Thalys high-speed train, riding at 300 km h^{-1} on the Belgian site of Mévergnies.

1 Introduction

After more than 40 yr of research and development, multibody dynamics simulation has now reached scientific and commercial maturity: several books exist describing well established methods to build and integrate the equations of motion (Géradin and Cardona, 2000; Garcia de Jalon and Bayo, 1993; Shabana, 2005; Bauchau, 2011), and commercial software's like MSC/ADAMS, SIMPACK or LMS/Virtual.Lab Motion are commonly used in robotics, car or railway industry along with other computer-aided engineering tools like finite element. The coupling of multibody systems with other disciplines offers nowadays a rich area of new developments. For example mechatronic systems which need to integrate specific equations related to controllers, actuators or sensors, or biomechanics where behaviour equations of tissues like muscles must be mixed with the ones of the mechanical system constituted by the skeleton and the limbs. In this paper, we will present a similar application: the multibody model of a vehicle and the finite element model of the track are merged in order to constitute a framework aiming at predicting the vibrations induced by railway vehicles. The model is used as a first step: it provides the time history of the forces exerted on the foundation, which are in turn used as inputs in a 3-D finite element of the soil. The complete process is performed in the time domain.

The focus of this study is to describe the approach, with a particular attention on the vehicle/track subsystem which involves the multibody model of the vehicle. The paper will first present a brief summary of the state of the art in terms of railway induced ground vibrations. The general organization of the global simulation framework will then be explained. The fourth section will detail the vehicle/track model with a focus on recent improvements either on the vehicle and track models. The potentialities of the approach will then be illustrated through the example of the Thalys high-speed train (HST). The paper is ended with some conclusions.

2 Railway-induced ground vibrations

It is largely admitted that railway transport constitutes a proper solution to traffic congestion and pollution observed in big cities. However, it also brings nuisances that must remain limited to avoid opposition of the dwellers. Noise and vibrations generated by the vehicle riding on the localized or distributed irregularities of the track are one of the usual reasons of complaint. It is then important to develop technical solutions to mitigate the railway vibrations, ideally from the beginning of the track and vehicle design. This requires reliable simulation tools, able to reproduce with a sufficient accuracy the propagation of vibrations, from the wheel-rail interface to the buildings, through the track and a medium, the soil, fundamentally inhomogeneous and infinite in three directions.

The problem of railway induced vibrations clearly involves 3 components: the vehicle, the track (rails, sleepers and ballast) and the soil.

Initially, the vehicle has often been reduced to a simple moving loaded mass (Wang and Zeng, 2004) or eventually a succession of the latter (Lefeuve-Mesgouez et al., 2002). With such a simplification, the origin of vibration lies in the irregular deflection of the track which induces up and down motion of the moving mass: the rail indeed offers a larger rigidity (and then a smaller displacement) above the sleepers than between the latter. This effect is sufficient to explain the so-called soil critical speed: when the speed of a train gets above the Rayleigh wave velocity of the superficial layer, it has a tendency to induce large vibrations. This phenomenon is generally observed for soft soils. Many works, as those proposed by Kaynia et al. (2000), Takemiya and Bian (2005) or Kouroussis et al. (2012c), have been conducted to reproduce by simulation this phenomenon. However, other sources contribute to the vibration content, among which the track and wheel irregularities and the vehicle dynamics. To reproduce this contribution, a more detailed model of the vehicle becomes necessary (Kouroussis et al., 2010). For example, Costa et al. (2011, 2012) have recently identified the influence and relevance of the mechanical properties of the train and have confirmed that the unsprung and semi-sprung masses must be included in the prediction model. In parallel, Kouroussis et al. (2012b) have analysed the vibratory effect of the unsprung masses in the specific case of the tramway of Brussels, showing that a modification of the resilient wheel stiffness notably reduces the ground vibrations when the vehicle is coming up against local rail defects.

The track is usually considered through a finite element model (Knothe and Grassie, 1993; Zhai and Sun, 1994). The rail is built from beam elements while lumped masses represent the sleepers, connected by springs and dampers to the rail and to the ground. One more layer of lumped masses can possibly be added to take into account the effect of the foundation in the track model.

Two principal approaches are used to simulate the wave propagation through the soil: the finite element method (FEM) and the boundary element method (BEM). Initially, BEM was preferably used due its natural ability to represent infinite domains and its good computational efficiency when the problem is formulated in the frequency domain (Do Rêgo Silva, 1994). However, the method becomes cumbersome when dealing with complex geometries, while frequency domain is limited to linear problems. In parallel, the continuously increasing power of computers and the development of infinite elements have opened the door to FEM models and it is presently possible to manage fully three-dimensional soil models, either in frequency (Wang et al., 2008) and, more recently in time domain (Kouroussis et al., 2011e). Indeed, Kouroussis et al. (2011d, 2009) have demonstrated that it is possible, in time domain, to alleviate the requirements in terms of domain and element size simulation. Let us also mention that some authors have developed combined BEM/FEM models (Galvín and Domínguez, 2009; François et al., 2009).

3 Railway vibration prediction model

A complete description of the model that we developed for predicting the vibrations induced by railway traffic can be found in Kouroussis et al. (2012a). Its main characteristics are the following:

- The simulation is performed in the time domain and in two successive steps (Fig. 1): firstly the simulation of the vehicle/track subsystem, whose result is the time history of the forces exerted by the track on the soil and, secondly, the simulation of the response of the soil to these forces through a finite element model.
- The vehicle/track subsystem is processed under the home-made framework EasyDyn and merges the non-linear equations of motion of the vehicle defined as a multibody model and the linear equations of a finite element model of the track. So far, a perfect symmetry has been assumed between left and right sides so that the motion is restricted to the vertical plane.
- The response of the ground is simulated under the commercial software ABAQUS. A particular care is given to the definition of the boundary conditions in order to get the best representation of the domain infinity and in particular to avoid wave reflection.

Although the track is modelled with finite elements, it is associated with the vehicle multibody model of the track instead of the finite element model of soil. The reasons are a good description of the contact location without additional artefacts (as for example wheel elements, Ju, 2009) and, for some cases (presence of a singular rail surface defect), the vehicle/track interaction (Kouroussis et al., 2010). Complete

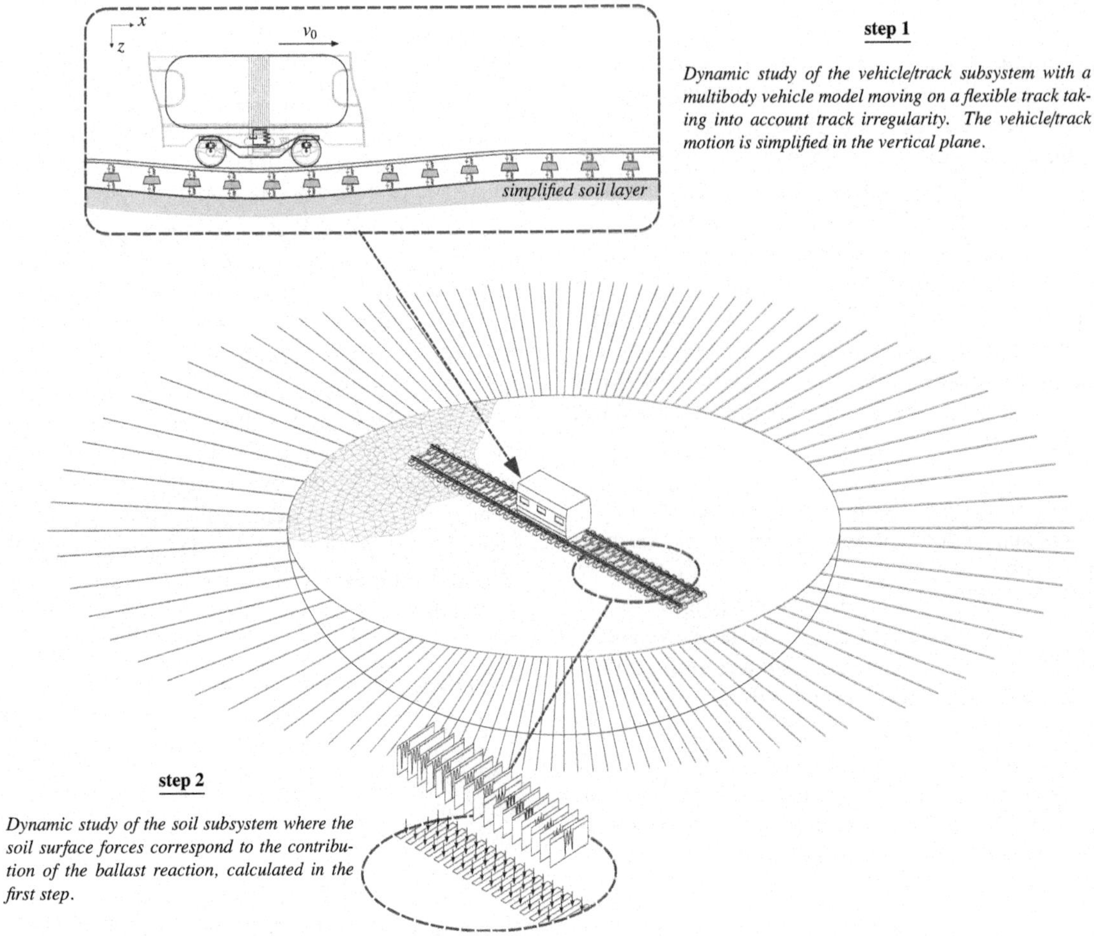
x
z
v_0
step 1
Dynamic study of the vehicle/track subsystem with a multibody vehicle model moving on a flexible track taking into account track irregularity. The vehicle/track motion is simplified in the vertical plane.
simplified soil layer
step 2
Dynamic study of the soil subsystem where the soil surface forces correspond to the contribution of the ballast reaction, calculated in the first step.

Figure 1. Vehicle/track/soil model, working in two successive steps.

models exist therefore combining multibody and finite element approaches in a single application for vehicle/track/soil analysis (see for example, Connolly et al., 2013).

The prediction model has been used in various practical cases and was successfully confronted to experimental results (Kouroussis et al., 2010, 2011e, 2012b,c, 2013). The main hypothesis in the approach relies in the two step strategy which assumes some decoupling between the track and the soil. However, it turns out that the hypothesis is reasonable as far as the mechanical impedance of the rail, as seen from the vehicle, is well represented, including the possible dynamic coupling between the sleepers through the soil. With classical track models where the soil is considered only through the stiffness under the sleepers (generally known as Winkler foundation, this term being used for continuously supported track and for discrete model as well as), the hypothesis is valid if the soil is sufficiently stiff, which is the case in most of railway lines (Kouroussis et al., 2011e, 2012a). With softer soils, the authors have proposed an enhanced track model, referred to as the coupled lumped mass model (CLM), which offers a faithful representation of soft soils impedance in the requested frequency range (Kouroussis et al., 2011b). In this paper, we will focus on the vehicle/track model.

4 Vehicle/track model

4.1 Global structure

The developed model merges the equations of motion of the vehicle, which have the form usually encountered in multibody system dynamics

$$\mathbf{M}_{\mathrm{v}}(\boldsymbol{q}_{\mathrm{v}}) \cdot \ddot{\boldsymbol{q}}_{\mathrm{v}} + \boldsymbol{h}_{\mathrm{v}}(\boldsymbol{q}_{\mathrm{v}}, \dot{\boldsymbol{q}}_{\mathrm{v}}, t) = \boldsymbol{f}_{\mathrm{v}}(\boldsymbol{q}_{\mathrm{v}}, \boldsymbol{q}_{\mathrm{t}}, t) \tag{1}$$

and the equations of the track, represented by a linear finite element model

$$\mathbf{M}_{\mathrm{t}} \cdot \ddot{\boldsymbol{q}}_{\mathrm{t}} + \mathbf{C}_{\mathrm{t}} \cdot \dot{\boldsymbol{q}}_{\mathrm{t}} + \mathbf{K}_{\mathrm{t}} \cdot \boldsymbol{q}_{\mathrm{t}} = \boldsymbol{f}_{\mathrm{t}}(\boldsymbol{q}_{\mathrm{v}}, \boldsymbol{q}_{\mathrm{t}}, t) \tag{2}$$

with

- $\boldsymbol{q}_{\mathrm{v}}$ and $\boldsymbol{q}_{\mathrm{t}}$ the vectors gathering the configuration parameters of the vehicle and track, respectively;

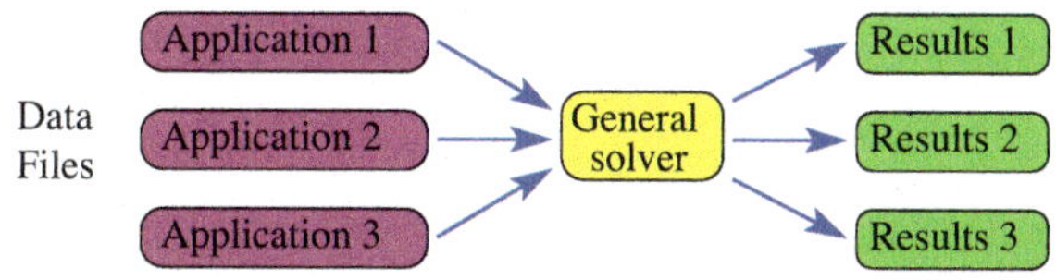

Figure 2. Computational organization with classical coordinates.

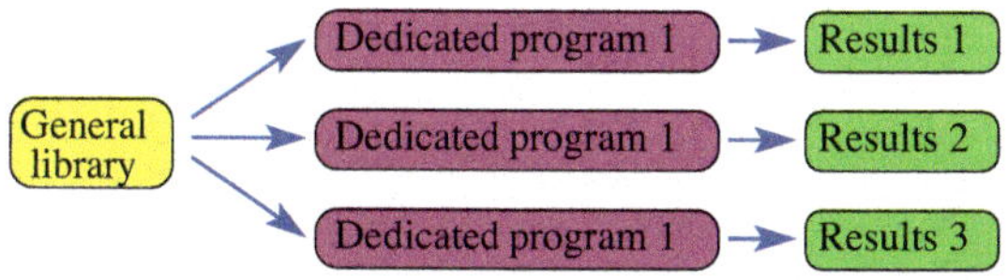

Figure 3. Computational organization with minimal coordinates.

- $\mathbf{M}_{\mathrm{v}}$ and $\mathbf{M}_{\mathrm{t}}$ the mass matrices of the vehicle and track, respectively;
- $\boldsymbol{h}_{\mathrm{v}}$ the term gathering the Coriolis, centrifugal and gyroscopic terms of the vehicle;
- $\mathbf{C}_{\mathrm{t}}$ and $\mathbf{K}_{\mathrm{t}}$ the damping and stiffness matrices of the track;
- $\boldsymbol{f}_{\mathrm{t}}$ and $\boldsymbol{f}_{\mathrm{v}}$ representing the external forces (gravity, suspensions, wheel/rail contact forces) exerted on the vehicle and the track respectively.

It appears that the coupling between the vehicle and the track is realized through external forces, and more precisely the contact forces. The latter are determined by considering an elastic contact between the wheel and the rail and then depend on both vehicle and track displacements.

4.2 Equations of motion of the vehicle

4.2.1 The choice of minimal coordinates

The construction of the equations of motion of the vehicle is based on the so-called *minimal coordinates* approach, developed by Anantharam and Hiller (1991); Hiller (1993) in the early 90's. With this approach, the configuration parameters used to express the kinematics of the multibody system are arbitrarily chosen but must be independent so that their number is equal to the number of degrees of freedom of the system. Compared to approaches like Cartesian or relative coordinates widely used in commercial products, the minimal coordinates approach has the major drawback to be less systematic as it requires to set up a specific kinematics of the considered system. However, it has the advantage to yield a system of pure ordinary differential equations, without constraint equations, which can be processed in a stable and robust way with standard numerical integration techniques. The approach proves anyway easy to use and efficient with open loop systems like the train model developed in this study.

It is worth to mention that the computational implementation with minimal coordinates is dramatically different from the one classically found with other coordinates. As illustrated in Fig. 2, a multibody simulation software classically consists of a main general solver able to simulate various mechanical systems, each of them being described in a specific data file. With minimal coordinates (Fig. 3), each application actually leads to a dedicated program, which is constructed with the help of a multibody library. The latter is expected to provide routines which facilitate the expression of kinematics and forces, and the construction and integration of the equations of motion.

When working with minimal coordinates, the equations of motion generally derive from the application of the d'Alembert's principle. If the system comprises n_B bodies and n_{cp} degrees of freedom, the n_{cp} differential equations governing the dynamic behaviour of the mechanical system are built according to

$$\sum_{i=1}^{n_B}\left[\boldsymbol{d}^{i,j}\cdot(\boldsymbol{R}_i - m_i\boldsymbol{a}_i) + \boldsymbol{\theta}^{i,j}\cdot(\boldsymbol{M}_{Gi} - \boldsymbol{\Phi}_{G_i}\dot{\boldsymbol{\omega}}_i - \boldsymbol{\omega}_i\times\boldsymbol{\Phi}_{G_i}\boldsymbol{\omega}_i)\right] = 0$$
$$j = 1,\ldots,n_{cp} \tag{3}$$

with

- m_i and $\boldsymbol{\Phi}_{G_i}$ the mass and the central inertia tensor of body i;
- $\boldsymbol{R}_i$ and $\boldsymbol{M}_{G_i}$ the resultant force and moment, at the centre of gravity G_i, of all applied forces exerted on body i;
- $\boldsymbol{a}_i$ the acceleration of the centre of gravity of body i;
- $\boldsymbol{d}^{i,j}$ and $\boldsymbol{\theta}^{i,j}$ the partial contributions of $\dot{q}_j$ in the velocity of the centre of gravity $\boldsymbol{v}_i$ and the rotational velocity $\boldsymbol{\omega}_i$ of body i, respectively, defined by

$$\boldsymbol{v}_i = \sum_{j=1}^{n_{cp}}\boldsymbol{d}^{i,j}\cdot\dot{q}_j \quad\leftrightarrow\quad \boldsymbol{d}^{i,j} = \frac{\partial\boldsymbol{v}_i}{\partial\dot{q}_j} \tag{4}$$

$$\boldsymbol{\omega}_i = \sum_{j=1}^{n_{cp}}\boldsymbol{\theta}^{i,j}\cdot\dot{q}_j \quad\leftrightarrow\quad \boldsymbol{\theta}^{i,j} = \frac{\partial\boldsymbol{\omega}_i}{\partial\dot{q}_j}\,. \tag{5}$$

The resulting equations of motion have the classical following form

$$\mathbf{M}(\boldsymbol{q})\cdot\ddot{\boldsymbol{q}} + \boldsymbol{h}(\boldsymbol{q},\dot{\boldsymbol{q}}) = \boldsymbol{g}(\boldsymbol{q},\dot{\boldsymbol{q}},t) \tag{6}$$

where, for example the mass matrix of dimension $n_{cp}\times n_{cp}$, is obtained by

$$\mathbf{M}_{jk} = \sum_{i=1}^{n_B}\left[m_i\boldsymbol{d}^{i,j}\cdot\boldsymbol{d}^{i,k} + \boldsymbol{\theta}^{i,j}\cdot(\boldsymbol{\Phi}_{G_i}\cdot\boldsymbol{\theta}^{i,k})\right] \tag{7}$$

while $\boldsymbol{h}$ represents the contribution of Coriolis, centrifugal and gyroscpic terms, and $\boldsymbol{f}$ the applied forces.

It turns out that the equations of motion can be constructed if the user provides

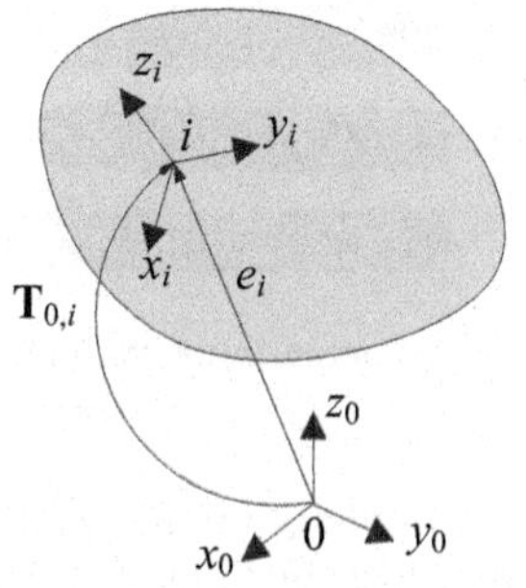

Figure 4. Formalism of homogeneous transformation matrix.

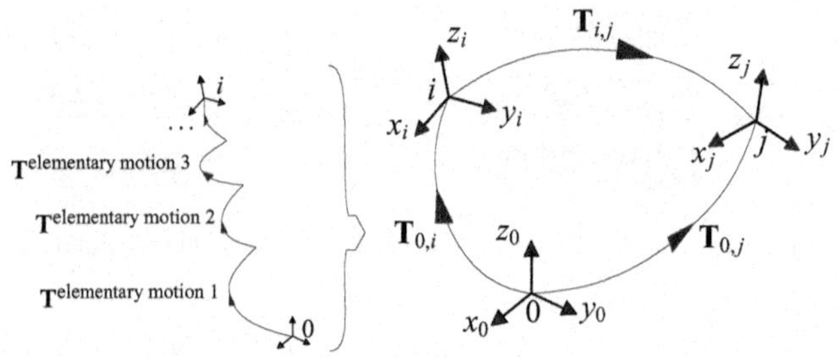

Figure 5. Homogeneous transformation matrices: illustration of the motion decomposition.

- the kinematics of the system, that's to say the expression of position, velocity, acceleration and partial velocities of each body i of the system in terms of the configuration parameters $\boldsymbol{q}$ and their first and second time derivatives;

- the applied efforts acting on each body i.

4.2.2 Kinematics

The expression of the complete kinematics, especially at acceleration level and for 3-D systems, is tricky, tedious and source of mistakes. To circumvent this difficulty, the EasyDyn framework provides a symbolic script which automatically generates the expressions of velocities and accelerations from only the position information.

To express the kinematics at position level, the formalism of homogeneous transformation matrices was retained. The position and orientation of each body i is expressed by means of the homogeneous transformation matrix $\mathbf{T}_{0,i}$ giving the situation of the local frame associated with body i with respect to the global reference frame 0 (Fig. 4). It is a 4×4 matrix of the following well-know form

$$\mathbf{T}_{0,i} = \begin{pmatrix} \mathbf{R}_{0,i} & \{\boldsymbol{e}_i\}_i \\ 0\,0\,0 & 1 \end{pmatrix} \tag{8}$$

where $\boldsymbol{e}_i$ is the coordinate vector of frame i with respect to the global coordinate system 0, and $\mathbf{R}_{0,i}$ is the rotation tensor describing the orientation of frame i with respect to frame 0.

Practically, the homogeneous transformation matrices have the deciding advantage to enjoy the following property

$$\mathbf{T}_{i,k} = \mathbf{T}_{i,j} \cdot \mathbf{T}_{j,k} \quad \forall i,j,k\,. \tag{9}$$

This means that a complex motion can be elegantly defined as a succession of elementary motions (Fig. 5) like for example

$$\mathbf{T}_{0,1} = \mathbf{T}^{\text{rot.}\,z}(\theta) \cdot \mathbf{T}^{\text{disp}}(0,l,0) \tag{10}$$

expressing a rotation about z of an angle θ followed by a displacement along the y-axis equal to l.

Translational and rotational velocities of each body i can then be easily obtained by differentiation of the homogeneous transformation matrix giving its situation

$$\{\boldsymbol{v}_i\}_0 = \frac{\mathrm{d}}{\mathrm{d}t}\{\boldsymbol{e}_i\}_0 \tag{11}$$

$$\{\tilde{\omega}_i\}_0 = \dot{\mathbf{R}}_{0,i} \cdot \mathbf{R}_{0,i}^T = \begin{pmatrix} 0 & -\omega_z & \omega_y \\ \omega_z & 0 & -\omega_x \\ -\omega_y & \omega_x & 0 \end{pmatrix} \tag{12}$$

from which accelerations and partial velocities can be determined by one more differentiation.

Under EasyDyn, a symbolic script called CAGeM, which stands for Computer-Aided generation of Motion, takes after the differentiation. The script can be run under MuPAD (SciFace Software GmbH & Co, 2012) or Xcas/Giac (Parisse and Graeve, 2010), the latter offering a completely open source environment.

4.2.3 Application to the vehicle modelling

In the present study, the vehicle is modelled by rigid bodies representing each inertial part of the train: car bodies, bogies, wheelsets and possibly wheel treads in case of compliant wheels (Kouroussis et al., 2012b), interconnected by the primary and secondary suspensions represented by springs and dampers.

To illustrate the use of homogeneous transformation matrices, let us consider the carriage of Fig. 6. The motion of the car body R_1 and bogies R_2 and R_3 is described by their vertical displacements and their pitch angle, involving a total of 6 degrees of freedom q_0 to q_5. The resulting homogeneous transformation matrix of the first bogie R_2 is then written

$$\begin{aligned}\mathbf{T}_{0,R_2} = {} & \mathbf{T}^{\text{disp}}(v_0 \cdot t, 0, q_0) \cdot \mathbf{T}^{\text{rot.}\,y}(q_1) \cdot \mathbf{T}^{\text{disp}}(l_b/2, 0, 0) \\ & \cdot \mathbf{T}^{\text{rot.}\,y}(-q_1) \cdot \mathbf{T}^{\text{disp}}(0,0,q_2) \cdot \mathbf{T}^{\text{rot.}\,y}(q_3)\end{aligned} \tag{13}$$

where $\mathbf{T}^{\text{disp}}(d_x, d_y, d_z)$ represents a translation and $\mathbf{T}^{\text{rot.}\,y}(\theta)$ a rotation about the y-axis.

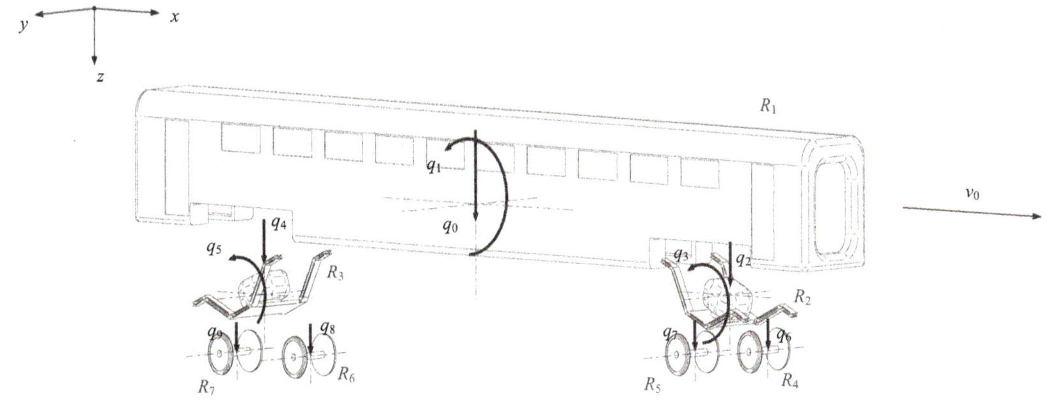

Figure 6. Vehicle modelling – kinematics in the *xz*-plane.

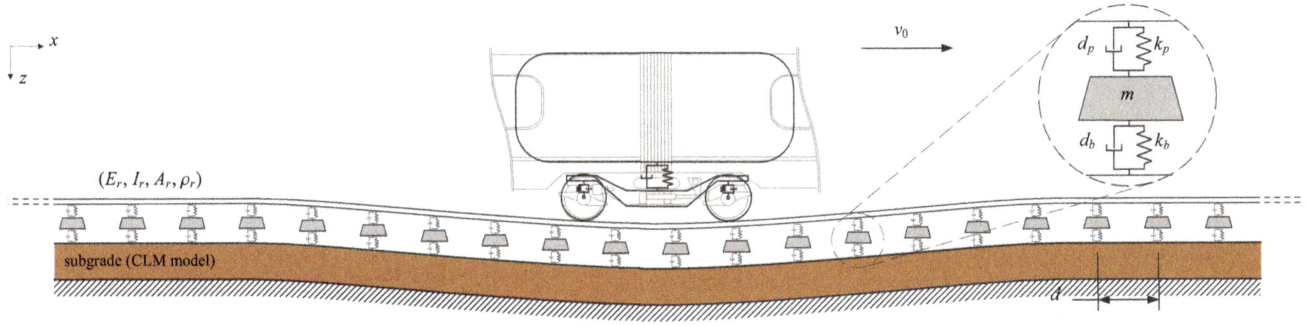

Figure 7. The vehicle/track/foundation model.

The following observations can be made

- the global progression of the vehicle is assured by the x coordinate of the form $v_0 \cdot t$ with v_0 the velocity and t the time;
- in this example, q_3 is desired to strictly correspond to the vertical relative displacement; that is why a backward rotation $-q_1$ is imposed between the car and the bogie to compensate the pitch angle q_1 of the carbody.

On the other hand, the motion of wheelsets R_4 to R_7 comes down to only the bounce motion, with respect to the bogie. For instance, the transformation matrix giving the situation of wheelset R_7 reads

$$\mathbf{T}_{0,R_7}=\mathbf{T}_{0,R_2}\cdot\mathbf{T}^{\text{disp}}(-l_{\text{w}}/2,0,0)\cdot\mathbf{T}^{\text{rot. }y}(-q_3)\cdot\mathbf{T}^{\text{disp}}(0,0,q_7) \quad (14)$$

where l_{b} and l_{w} represent the distances between bogies and wheelsets, respectively.

The formulation is not restrictive, and can be applied without any difficulty to more complex models and/or to three-dimensional kinematics.

4.2.4 Applied forces

The applied forces derive from gravity, suspensions and contact forces. The suspensions are classically defined as springs and dampers attached to specific points of the bodies. In the same manner, the contact forces are applied on the wheels. Their computation is explained further.

4.3 The track model

The track is represented by a finite element planar model, made up of 3 layers: the rail, the sleepers and the subgrade or foundation (Fig. 7). So far, only the vertical motion has been considered since the major contribution of ground vibration is induced by the vertical track deflection. Moreover, as symmetry is assumed, the track is condensed in the symmetry plane and embraces the 2 rails. The reduced central rail (Young modulus E_{r}, density ρ_{r}, geometrical moment of inertia I_{r} and section A_{r} for a single rail) consists of a succession of Euler beams, while the sleepers correspond to lumped masses of mass m, placed with a regular spacing L. The sleepers are connected by spring-damper systems to the rail (stiffness k_{p} and damping d_{p} representing a single railpad) and to the foundation (stiffness k_{b} and damping d_{b} representing the ballast).

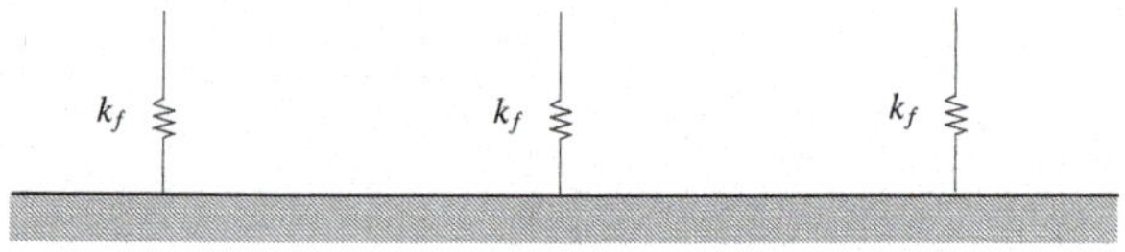

(a) Winkler model

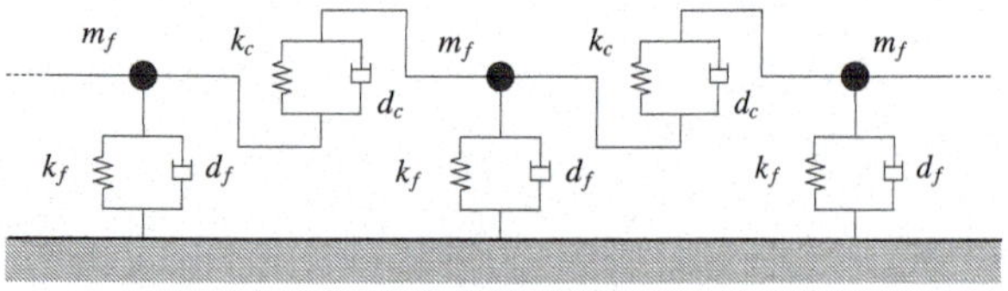

(b) CLM model

Figure 8. Foundation models (each discrete element is connected to each sleeper).

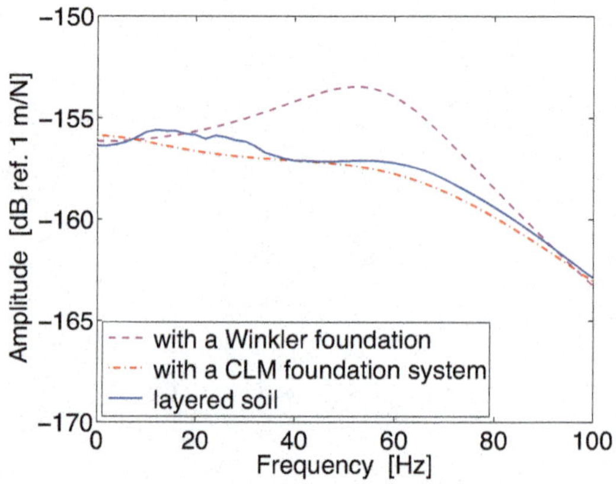

(a) Direct receptance

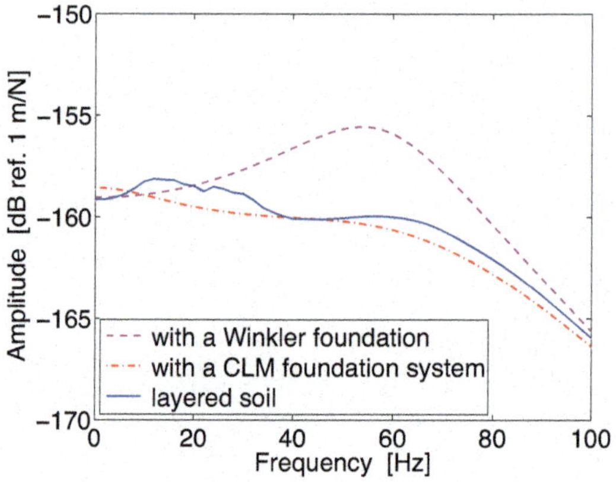

(b) Indirect receptance

Figure 9. Track receptances of the studied site (Mévernies – Belgium).

As indicated in the state of the art, the model retained to represent the track is the CLM (coupled lumped mass) model proposed in Kouroussis et al. (2011a,b). It permits to properly capture the track receptance even in presence of soft soils and therefore minimizes the error resulting from the hypothesis of track/soil decoupling. Compared to models based on the classical Winkler foundation representing one spring under each sleeper (Fig. 8a), it includes a better representation of the soil under the track through a layer of lumped masses of mass m_{f} placed below the sleepers (Fig. 8b), and viscoelastically connected to the bedrock (stiffness k_{f} and the damping d_{f}) and to the surrounding foundation masses (damping d_{c} and stiffness k_{c}). The values of the parameters of the CLM model ($m_{\mathrm{f}}, k_{\mathrm{f}}, d_{\mathrm{f}}, k_{\mathrm{c}}, d_{\mathrm{c}}$) are calculated, through simple analytical relations (Kouroussis et al., 2011b), so as to match soil impedances issued from the FEM model of the soil[1].

Figure 9 illustrates the accuracy of the CLM approach, by comparing the track receptances of a specific high-speed track in Belgium, lying on various foundation models, for the 3 following cases

- the track model is included in the three-dimensional FEM model of the soil, which can be considered as the reference solution;
- the proposed track model with a foundation represented by a Winkler foundation;
- the proposed track model where the CLM model is used for the foundation.

Presented direct receptances are defined as frequency response functions between the vertical displacement of the rail above a sleeper and the vertical force applied at the same point. Indirect receptances correspond to rail displacement above the adjacent sleeper.

It appears that the CLM model properly captures the track receptances between 0 Hz and 100 Hz, which is not the case of the Winkler foundation. A good agreement can be observed between the CLM and the reference case: the error remains below 1 dB which is well below the uncertainty range encountered in experimental receptances. The interest of this approach is certainly its ability to faithfully reproduce the track response with a limited number of degrees of freedom.

The size of the model depends on the track length required to get the requested level of accuracy, that is to say between 20 and 80 m (Kouroussis et al., 2010). Practically, this represents the flexible part of the track, surrounded by rigid parts, with a transition area along which the compliance progressively evolves. This assures a smooth loading of the soil and permits to manage vehicles longer than the considered track.

Concerning the degrees of freedom, the beam nodes introduce 2 degrees of freedom (vertical displacement and slope) while each lumped mass, representing either a sleeper or a foundation mass, introduces one degree of freedom. Knowing that 2 beam elements are defined between the sleepers,

[1] Impedances issued from experimental tests or from other models could be used equivalently.

this leads, roughly speaking, to 6 degrees of freedom per sleeper, and consequently a few hundreds degrees of freedom for the track.

5 Track interaction: wheel/rail contact forces

The contact forces between the wheels and the rail allow coupling between the vehicle and the track. We have chosen to calculate this force through the well-known Hertz contact theory, stating that the normal contact force N can be calculated from the penetration d according to

$$N = K_{\mathrm{Hz}} d^{3/2} , \tag{15}$$

where the coefficient K_{Hz} depends only on the radii of curvature of the wheel and the rail profiles, and the elastic properties of the material of both bodies.

The penetration for each wheel i depends on the vehicle and track states and is calculated from

$$d_{\mathrm{wheel},i} = z_{\mathrm{rail}}(x_{\mathrm{wheel},j}) - z_{\mathrm{wheel},i} + h(x_{\mathrm{wheel},i}) + R_{\mathrm{wheel},i} \tag{16}$$

with

- $x_{\mathrm{wheel},i}$ and $z_{\mathrm{wheel},i}$ the coordinates of the centre of wheel i, depending on the configuration parameters of the vehicle;
- $z_{\mathrm{rail}}(x_{\mathrm{wheel},j})$ the height of the rail below wheel i, depending on the degrees of freedom of the track model, through the shape functions of the beam elements;
- $R_{\mathrm{wheel},i}$ the radius of wheel i;
- $h(x_{\mathrm{wheel},i})$ the rail irregularity below wheel i, which can consist of local defects and/or overall track contribution like roughness (Kouroussis et al., 2010).

The action and reaction components of the contact force can then be applied to the multibody and track models. Let us note that, for the track, the force is transformed to nodal forces and torques, again through the shape functions of the beam element, calculated at the contact point.

Let us note that most of ground vibration models consider a linear relationship between the contact force and the penetration, neglecting the inherent non-linearity of the contact physics.

Numerical framework

The track/vehicle model explained in the previous sections has been implemented as a C++ program based on the home-made EasyDyn library (Kouroussis et al., 2011c; Verlinden et al., 2013).

The equations of motion and of the vehicle (Eq. 1) and of the track (Eq. 2), are rewritten in the following residual form

$$f(\boldsymbol{q},\dot{\boldsymbol{q}},\ddot{\boldsymbol{q}},t)=\left\{\begin{array}{c}\mathbf{M}_{\mathrm{v}}\cdot\ddot{\boldsymbol{q}}_{\mathrm{v}}+\boldsymbol{h}_{\mathrm{v}}(\boldsymbol{q}_{\mathrm{v}},\dot{\boldsymbol{q}}_{\mathrm{v}},t)-\boldsymbol{f}_{\mathrm{v}}(\boldsymbol{q}_{\mathrm{v}},\boldsymbol{q}_{\mathrm{t}},t)\\ \mathbf{M}_{\mathrm{t}}\cdot\ddot{\boldsymbol{q}}_{\mathrm{t}}+\mathbf{C}_{\mathrm{t}}\cdot\dot{\boldsymbol{q}}_{\mathrm{t}}+\mathbf{K}_{\mathrm{t}}\cdot\boldsymbol{q}_{\mathrm{t}}-\boldsymbol{f}_{\mathrm{t}}(\boldsymbol{q}_{\mathrm{v}},\boldsymbol{q}_{\mathrm{t}},t)\end{array}\right\}=0 \tag{17}$$

where the global vector of configuration parameters $\boldsymbol{q}$ results from the concatenation of the vehicle and track configuration parameters

$$\boldsymbol{q}=\left\{\begin{array}{c}\boldsymbol{q}_{\mathrm{v}}\\ \boldsymbol{q}_{\mathrm{t}}\end{array}\right\} . \tag{18}$$

When rewritten in this form, the equations can be integrated by the routines provided by the sim module, which implement the so-called Newmark-1/4 method. The latter is known to not introduce any numerical damping which is an advantage in the considered application. Numerical damping is anyway unnecessary as we deal with ordinary differential equations.

The vehicle is a multibody system and its equations of motion are computed from the routines offered by the mbs module. This assumes that the user provides two routines implementing on the one hand the kinematics of the multibody system and, on the other hand, the forces exerted on each body, corresponding in our case to the gravity, the suspension forces and the contact forces. Let us recall that concerning the kinematics, the velocities, accelerations and partial velocities are generated symbolically by the script CAGeM accompanying the C++ library of EasyDyn.

The equations of motion of the track (Eq. 2) are simple and are coded directly in C++ by using the classical assembly techniques.

The computer implementation is summarized in Fig. 10. Let us note that, thanks to the visu module, shapes can be attached to bodies, in order to visualize the motion of the system.

6 Soil simulation

The simulation of the vehicle/track subsystem provides the time history of the ground forces, defined as the visco-elastic action of the ballast on the subgrade. These forces are used in the second subproblem, managed under the finite element software ABAQUS which computes the free field response.

The finite element model of the soil is out of the scope of this paper, a detailed description of the finite element model being available in Kouroussis et al. (2010, 2012a). Let us mention anyway that

- only a half soil is considered due to the assumed left-right symmetry;
- the forces are not applied on nodes but on rigid surfaces corresponding to the area covered by the sleepers;
- the inner part of the model consists of one quarter of a sphere and defines the specific geometry of the considered track;
- a dedicated script generates the outer part consisting of a transition spherical slice with progressive element sizes, surrounded by the infinite elements.

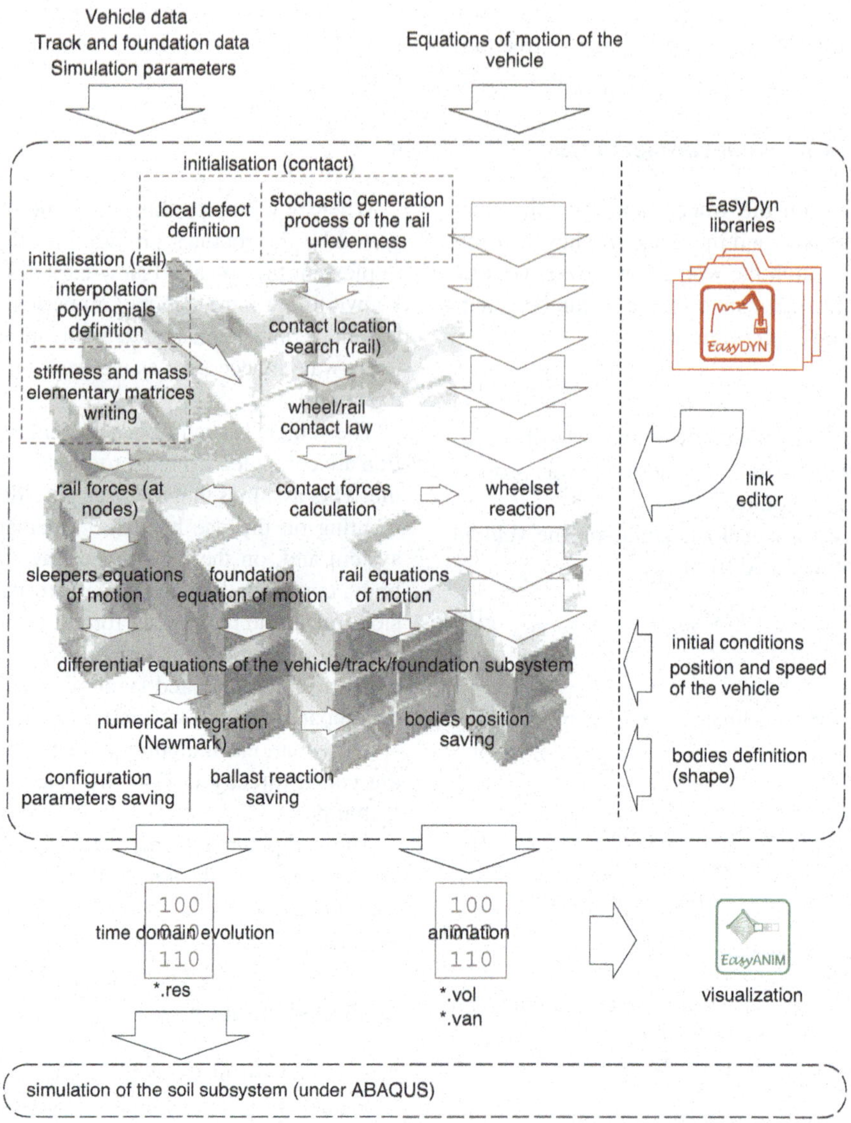

Figure 10. Computer implementation of the vehicle/track model.

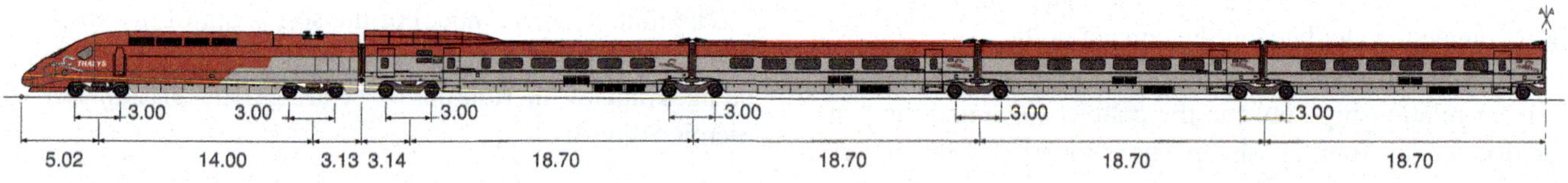

Figure 11. Thalys HST dimensions

7 Example: the Thalys high-speed train

7.1 Description of the model

Thalys trains are designed to operate over the French, Belgian, German and Dutch networks and therefore ensure the interconnection between the different high-speed lines. The high-speed vehicle studied in this work stems from the same generation as the French TGV Atlantique with some minor differences in the dynamical and geometrical parameters. The vehicle data were supplied by the Belgian railway operator. Figure 11 shows the configuration and the dimensions of these trainsets, consisting of 2 locomotives and 8 carriages, with a total length of 200 m. The two locomotives are supported by two bogies. Instead of the conventional bogie configuration of two-to-a-car, the carriage bogies are placed half under one car and half under the next, with the exception of

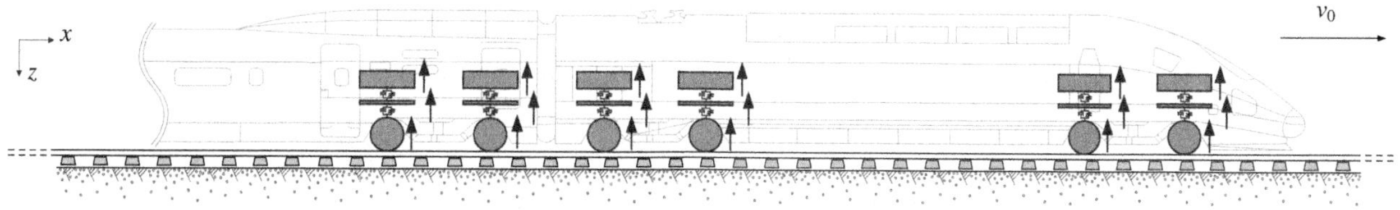

(a) Old model

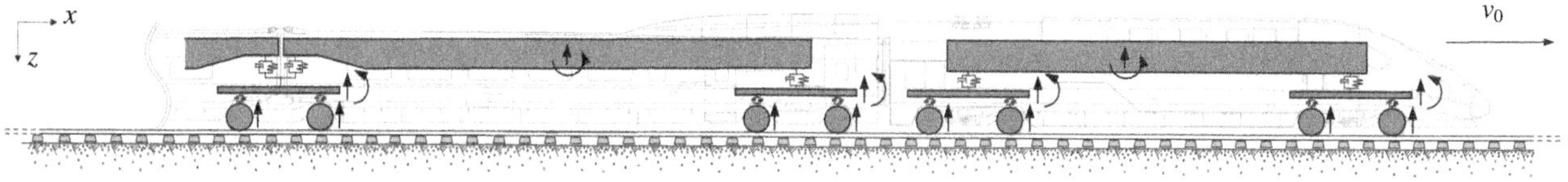

(b) New model

Figure 12. Studied multibody models for the vehicle.

Table 1. Dynamic parameters of Thalys HST – unladen weight.

	Bogie Y230A	Bogie Y237A	Bogie Y237B
Carbody mass m_c [kg]	26721	14250	20426
Carbody pitch moment of inertia I_c [$\times 10^6$ kg m^2]	1.15	0.61	0.88
Bogie mass m_b [kg]	3261	15650	8156
Bogie pitch moment of inertia I_b [kg m^2]	2870	13750	7185
Wheelset mass m_0 [kg]	2009	2050	2009
Primary suspension stiffness k_1 [MN m^{-1}]	2.09	1.63	2.09
Primary suspension damping d_1 [kN s m^{-1}]	40	40	40
Secondary suspension stiffness k_2 [MN m^{-1}]	2.45	0.93	2.45
Secondary suspension damping d_2 [kN s m^{-1}]	40	40	40

the side carriage bogies, which connect the power carriages (at the outer extremities of the train) to the main passenger carriages (in the centre of the train). The unladen mass is close to 386 tonnes, while the nominal loading is worth 439 tonnes.

Three bogie types are used in this vehicle:

- the Y230A motor bogie equipping the locomotives;
- the Y237 trailing bogie: variant A for the side carriages and variant B for the other ones.

An SR 10 pneumatic air-sprung suspension is used as the secondary suspension of the trailing bogies while the coil spring is preferred for the primary suspension. For the Y230A, classical rubber sandwich block (coil) spring is used for the primary (secondary) suspension. Table 1 summarizes the dynamic parameters of the bogies in terms of mass, stiffness and damping.

The Thalys HST has already been studied by the authors (Kouroussis et al., 2011e). However, due to a lack in the vehicle data that we were able to collect, the pitch motion of the bogie and carbodies was neglected. The bogie and the carbody were actually replaced by a front and a rear lumped mass whose only bounce motion was taken into account (Fig. 12a). The present model consists of a succession of carbodies, bogies and wheelsets involving 2 degrees of freedom (bounce and pitch) for each carbody/bogie and 1 degree of freedom (bounce) for each wheelset (Fig. 12b).

The Thalys HST, in its general configuration consists of a succession of one locomotive, one side carriage, six central carriages, one side carriage and finally one locomotive, with a total of 72 degrees of freedom. The relevant geometrical data of the train are specified in Table 2.

If the vehicle rides at velocity v_0, the position matrices relative to the first locomotive are written

$$\mathbf{T}_{0,\text{carbody}} = \mathbf{T}^{\text{disp}}(v_0 t, 0, q_{1,1}) \cdot \mathbf{T}^{\text{rot.}\,y}(q_{1,2}) \tag{19}$$
$$\mathbf{T}_{0,\text{front bogie}} = \mathbf{T}^{\text{disp}}(v_0 t + l_b/2, 0, q_{1,3}) \cdot \mathbf{T}^{\text{rot.}\,y}(q_{1,4}) \tag{20}$$
$$\mathbf{T}_{0,\text{rear bogie}} = \mathbf{T}^{\text{disp}}(v_0 t - l_b/2, 0, q_{1,5}) \cdot \mathbf{T}^{\text{rot.}\,y}(q_{1,6}) \tag{21}$$
$$\mathbf{T}_{0,\text{first wheel}} = \mathbf{T}^{\text{disp}}(v_0 t + l_b/2 + l_w/2, 0, q_{1,7}) \tag{22}$$
$$\mathbf{T}_{0,\text{second wheel}} = \mathbf{T}^{\text{disp}}(v_0 t + l_b/2 - l_w/2, 0, q_{1,8}) \tag{23}$$
$$\mathbf{T}_{0,\text{third wheel}} = \mathbf{T}^{\text{disp}}(v_0 t - l_b/2 + l_w/2, 0, q_{1,9}) \tag{24}$$
$$\mathbf{T}_{0,\text{fourth wheel}} = \mathbf{T}^{\text{disp}}(v_0 t - l_b/2 - l_w/2, 0, q_{1,10}) \tag{25}$$

where $q_{1,i}$ are the 10 configuration parameters of the locomotive, with $q_{1,1}$ and $q_{1,2}$ the bounce and pitch motions of the carbody, $q_{1,3}$ and $q_{1,4}$ ($q_{1,5}$ and $q_{1,6}$) the bounce and pitch

Table 2. Geometric parameters of Thalys HST.

Parameter	Symbol	Value (m)
Distance between bogies on locomotive	l_b	14
Bogie wheelbase	l_b	3
Distance between locomotive COM and side carriage COM	d_{lb}	23.12
Central carriage length	b	18.7

Table 3. Studied site parameters (Mévernies – Belgium).

Track parameters				
E_r	I_r	ρ_r	A_r	d
210 GPa	3055 cm^4	7850 kg m^{-3}	63.9 cm^2	0.6 m
k_p	d_p	k_b	d_b	m
120 MN m^{-1}	4 kNs m^{-1}	47 MN m^{-1}	72 kNs m^{-1}	150 kg
Soil parameters				
layer	d	E	ρ	ν
1	2.7 m	129 MPa	1600 kg m^{-3}	0.3
2	3.9 m	227 MPa	2000 kg m^{-3}	0.3
3	∞	659 MPa	2000 kg m^{-3}	0.3
m_f	k_f	d_f	k_c	d_c
460 kg	40 MN m^{-1}	426 kNs m^{-1}	63 MN m^{-1}	−73 kNs m^{-1}

motions of the front (rear) bogie, and $q_{l,7}$ to $q_{l,10}$ the bounce motions of the wheelsets.

For the side carriages, we get

$$\mathbf{T}_{0,\text{carbody}} = \mathbf{T}^{\text{disp}}(v_0 t - d_{lb}, 0, q_{s,1}) \cdot \mathbf{T}^{\text{rot. }y}(q_{s,2}) \quad (26)$$

$$\mathbf{T}_{0,\text{front bogie}} = \mathbf{T}^{\text{disp}}(v_0 t - d_{lb} + l_b/2, 0, q_{s,3}) \cdot \mathbf{T}^{\text{rot. }y}(q_{s,4}) \quad (27)$$

$$\mathbf{T}_{0,\text{rear bogie}} = \mathbf{T}^{\text{disp}}(v_0 t - d_{lb} - l_b/2, 0, q_{s,5}) \cdot \mathbf{T}^{\text{rot. }y}(q_{s,6}) \quad (28)$$

$$\mathbf{T}_{0,\text{first wheel}} = \mathbf{T}^{\text{disp}}(v_0 t - d_{lb} + l_b/2 + l_w/2, 0, q_{s,7}) \quad (29)$$

$$\mathbf{T}_{0,\text{second wheel}} = \mathbf{T}^{\text{disp}}(v_0 t - d_{lb} + l_b/2 - l_w/2, 0, q_{s,8}) \quad (30)$$

$$\mathbf{T}_{0,\text{third wheel}} = \mathbf{T}^{\text{disp}}(v_0 t - d_{lb} - l_b/2 + l_w/2, 0, q_{s,9}) \quad (31)$$

$$\mathbf{T}_{0,\text{fourth wheel}} = \mathbf{T}^{\text{disp}}(v_0 t - d_{lb} - l_b/2 - l_w/2, 0, q_{s,10}) \quad (32)$$

where $q_{s,i}$ ($i = 1 \mapsto 10$) have the same meaning as $q_{l,i}$ for the locomitve.

There is only one bogie per central carriage, which is the rear one, the front one being kinematically attached to the previous carriage. The corresponding position matrices of the j-th central carriage then read

$$\mathbf{T}_{0,\text{carbody j}} = \mathbf{T}^{\text{disp}}(v_0 t - d_{lb} - l_b/2 - (2j-1)b/2, 0, q_{cj,1}) \cdot \mathbf{T}^{\text{rot. }y}(q_{cj,2}) \quad (33)$$

$$\mathbf{T}_{0,\text{rear bog. j}} = \mathbf{T}^{\text{disp}}(v_0 t - d_{lb} - l_b/2 - jb, 0, q_{cj,3}) \cdot \mathbf{T}^{\text{rot. }y}(q_{cj,4}) \quad (34)$$

$$\mathbf{T}_{0,\text{1t wheel j}} = \mathbf{T}^{\text{disp}}(v_0 t - d_{lb} - l_b/2 - jb + l_w/2, 0, q_{cj,5}) \quad (35)$$

$$\mathbf{T}_{0,\text{2nd wheel j}} = \mathbf{T}^{\text{disp}}(v_0 t - d_{lb} - l_b/2 - jb - l_w/2, 0, q_{cj,6}) \quad (36)$$

with $q_{cj,i}$ ($i = 1 \mapsto 6$ and $j = 1 \mapsto 6$) the configuration parameters of the carriage, defined in the same manner as previously.

The track model involves 160 sleepers. The parameters of the track and of the CLM model used to represent the foundation are pointed out in Table 3. The CLM parameters issue from the identification of the foundation receptance with respect to a 3-D FEM model of the soil comprising 3 layers. For each layer, the depth d, the Young modulus E, the density ρ and the Poisson's number ν are also given in Table 3. Let us note that the damping d_c is negative, so as to properly capture the ground wave propagation delay, also called "tau effect".

7.2 Simulation results

7.2.1 Studied configurations

In the next sections, the results provided by three different models are compared

- The initial model Kouroussis et al. (2011e), without the pitch motion of bogies and carbodies and a Winkler foundation for the track subgrade. It is denoted by model A.

- An intermediary model, with the same vehicle as model A but where the CLM model has been adopted for the track subgrade. It is denoted by model B.

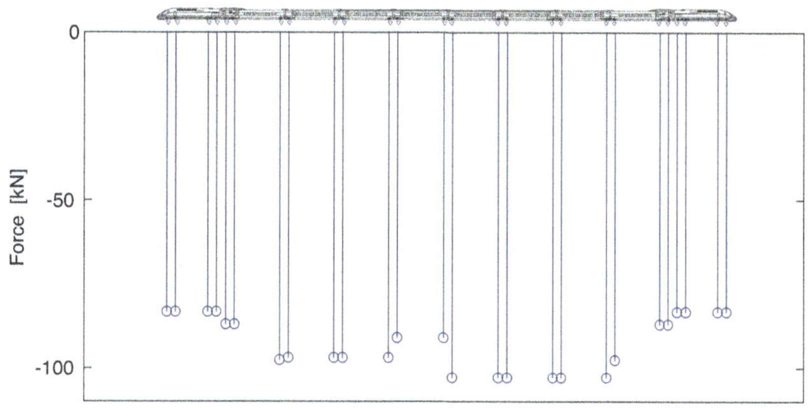

(a) Static axle load acting on the track

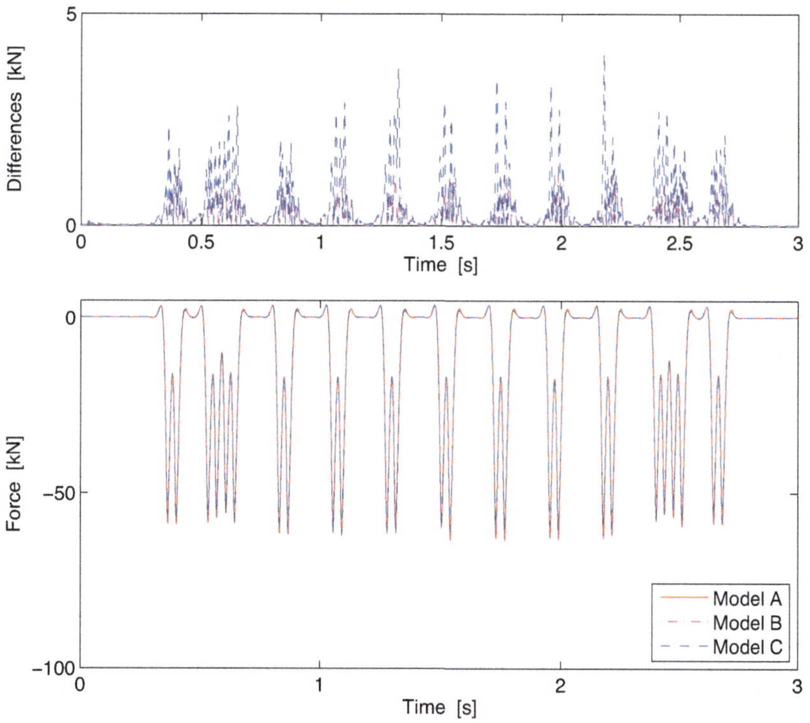

(b) Ballast reaction used as input in the soil model

Figure 13. Time history of track/foundation forces for each model.

- The complete model, denoted by model C, as described in this paper with the CLM model and the pitch motion of bogies and carbodies.

In all cases, the train speed is worth $v_0 = 300\,\mathrm{km\,h^{-1}}$ and the rail irregularity is calculated for a rail quality of 6 (Garg and Dukkipati, 1984; Kouroussis et al., 2012a).

7.2.2 Forces on the track

Figure 13b shows the time history of the force exerted by the track on the soil at the centre of the model, for each model. In addition to these curves, the difference of force magnitudes provided by models B and C, compared to model A, is presented. This force, along with the ones under the other sleepers, is used as input in the FEM model of the soil to study the wave propagation. The plot is to compare with the one of Fig. 13a which shows the static load on the track, in function of a pseudo-time corresponding to the distance divided by the velocity. The figures show how the track distributes the contact forces through the sleepers.

It turns out that there is no significant difference between the models, especially between model A and model B. The major difference appears when the deflection is maximum and reaches about 2 % between models A and B but more

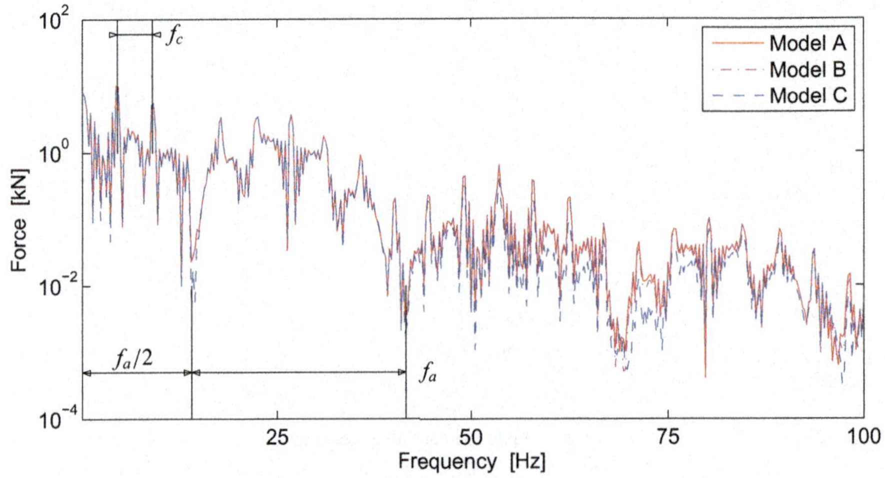

Figure 14. Frequency content of ballast reaction force for each model.

than 5 % between models A and C. This is confirmed by the corresponding frequency content, illustrated in Fig. 14. The latter reveals the usual peaks, related to the carriage passage excitation mechanisms at frequency $f_c = v_0/l_b = 4.5$ Hz modulated in amplitude by the fundamental axle passage frequency $f_a = v_0/l_w = 27.8$ Hz. The magnitudes at frequency $\frac{k}{2} f_a$ $(k = 1, 3, 5, \ldots)$ are completely suppressed.

The analysis shows that on one hand the benefit brought by the CLM model is not significant, due to the fact that the soil is relatively rigid. On the other hand, the differences observed with model C, although limited, indicate the importance of a careful vehicle modelling in the prediction of railway induced ground vibrations.

7.2.3 Ground vibrations

Figure 15 shows the time history of the vertical ground velocity at 9 m from the track, resulting from the application of the ballast reaction forces, obtained from the simulation of each vehicle/track model, on the 3-D FEM model of the soil. The figure also includes the experimental results presented in a previous work (Kouroussis et al., 2011e). The corresponding weighted severity can be found in Fig. 16, as defined in the DIN 4150 part 2 standard (Deutsches Institut für Normung, 1999). This indicator provides a quantification of the maximum vibratory dose felt by residents, and presents an interesting basis for drawing a parallel between discomfort and ground vibrations. Finally, Fig. 17 shows the frequency content of the ground velocity and indicates the maximum amplitude at 26 Hz. The latter is imposed by a resonance-like phenomenon, where the soil surface vibrates in phase with the vertical loading at a frequency corresponding to

$$f_{\text{layer}} = \frac{c_P}{4d} \tag{37}$$

with c_P the compression wave velocity of the first layer and d its depth.

The following observations are noteworthy:

- The comparison of the results obtained from models A, B and C leads to the same conclusions as in the previous section. The difference between models A/B and C is clearly observed on the weighted severity, at the beginning and end of the ground vibration. In the same way, the frequency contents differ in mid and high frequencies. Note that the experimental values at 23 m have greater amplitude than those at 25 and 18 m, whose origin is unfortunately unknown.

- Figures 15 and 16 show a good agreement between predicted and experimental ground vibrations, which validates the hypothesis made by the authors concerning the track/soil decoupling when the soil is sufficiently rigid with respect to the ballast (Kouroussis et al., 2012a). It must however be mentioned that the vibration peaks predicted in the frequency ranges 20–30 Hz and 50–60 Hz (Fig. 17) are larger than their experimental counterparts. At high frequencies, the gap is explained by the adopted material damping in the soil model: a time domain simulation imposes a viscous damping although the hysteretic damping better corroborates for soil motion since it does not significantly depend on the frequency of motion.

As a final result, Fig. 18 shows how the the peak particle velocity PPV and the $\text{KB}_{F,\max}$ indicators evolve with the distance from the track d. The second indicator is defined as the maximum of the weighted severity. The attenuation is identical for the three models, when fitted according to a simple

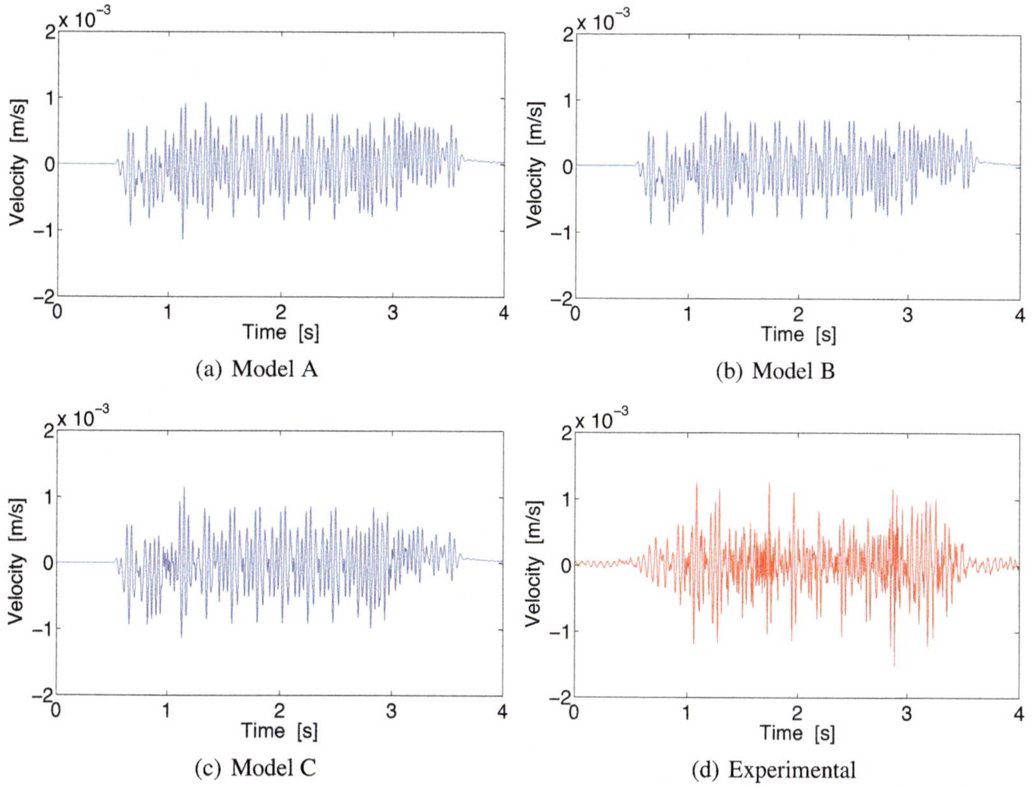

Figure 15. Predicted and measured time history of vertical ground velocity at 9 m from the track.

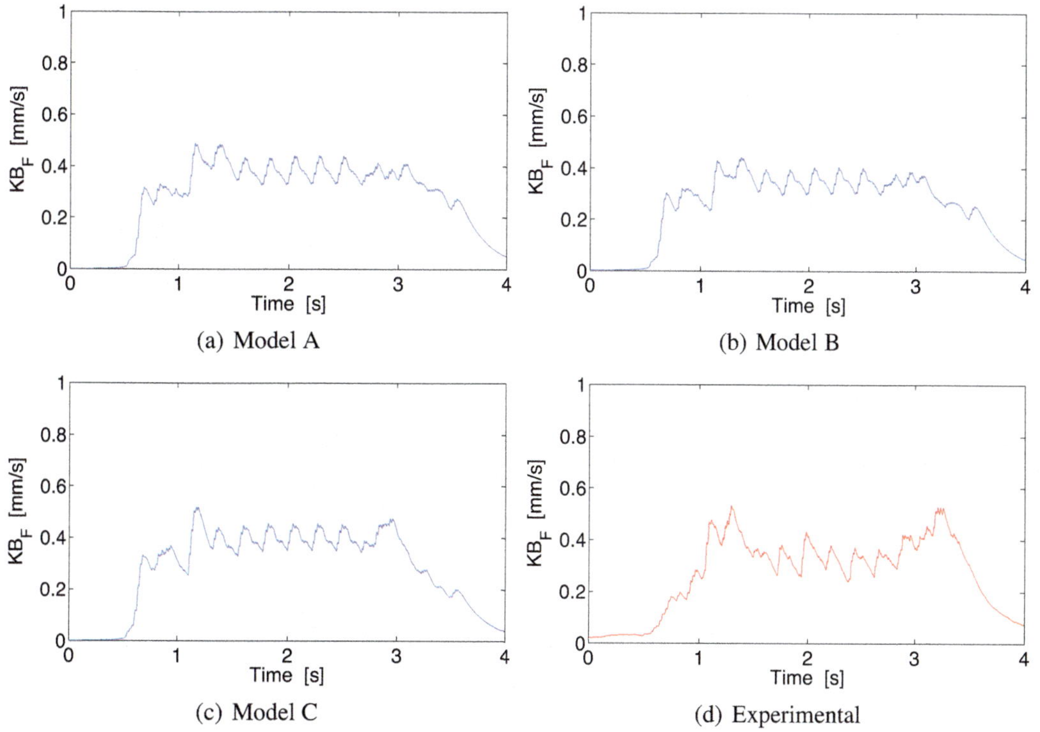

Figure 16. Predicted and measured weighted severity of vertical ground velocity at 9 m from the track.

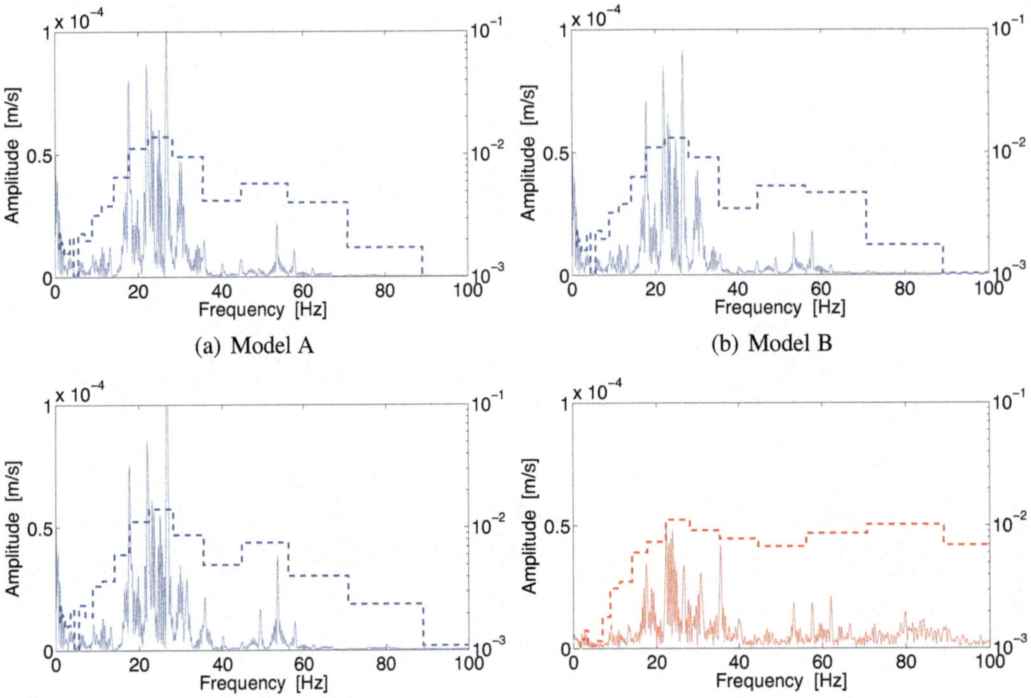

(a) Model A (b) Model B

(c) Model C (d) Experimental

Figure 17. Predicted and measured frequency content (spectra in solid line and one-third octave band in dashed line) of vertical ground velocity at 9 m from the track.

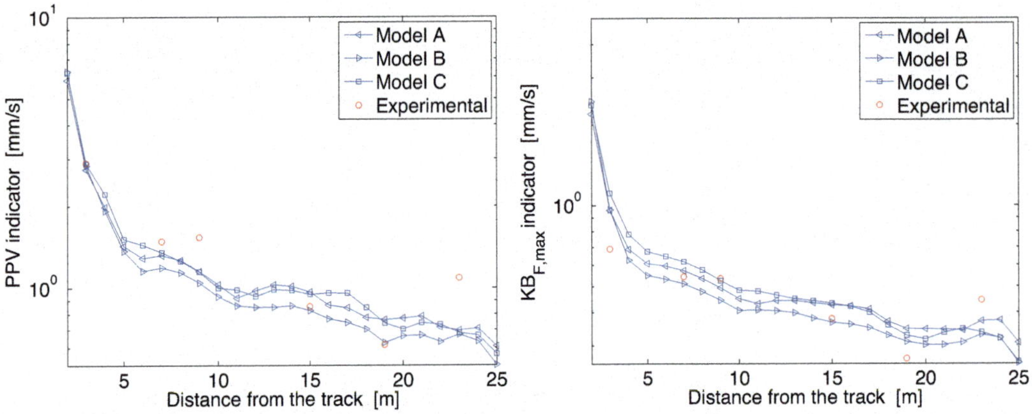

(a) Peak particle velocity comparison (b) $KB_{F,max}$ comparison

Figure 18. Overview of the ground vibration level difference for various distances from the track.

power-law function

$$\begin{aligned} \mathrm{PPV} &\propto d^{-0.7} \\ \mathrm{KB}_{F,\max} &\propto d^{-0.5}. \end{aligned}$$

The comparison with experimental results is not bad, with the exception of the point at 23 m from the track, where the experimental level is suspiciously greater than the ones at 15 and 18 m.

8 Conclusions

In this paper, we have presented the framework that we have developed for predicting the vibrations induced by railway transportation. The vibrations find their origin on the one hand on the nature of the track, discretely supported at the sleepers, and on the other hand in the irregularity of the rail surface. The proposed framework first considers the vehicle/track model mixing the multibody model of the vehicle and the finite element model of the track, coupled to each other through the wheel/rail contact. Only the motion in the vertical plane is considered, assuming a total symmetry between left and right rails. This first step is implemented on the basis of the home-made C++ library EasyDyn and produces a time history record of the forces exerted by the ballast on the foundation, which are then applied to a full 3-D FEM model of the soil. The latter is managed by the commercial software ABAQUS and involves finite elements to represent the unbounded nature of the considered domain. Again, symmetry is assumed betwen left and right parts so that only a half domain is defined.

With respect to previous publications, the model of the vehicle no longer consists of a succession of travelling masses but considers the carbodies and bogies as actual bodies, undergoing namely a pitch motion. The interest of the coupled lumped mass model has also been emphasized. The latter offers a better representation of the foundation contribution in the track model, and extends the application range of the approach, based on a decoupling of the vehicle/track and soil subsystems.

The potential applications of the model are illustrated on the example of the Thalys high-speed train, riding at $300\,\mathrm{km\,h^{-1}}$ on the Belgian site of Mévergnies. A good agreement is observed between experimental and predicted track and ground vibrations. In this particular case, the CLM model does not bring any significant improvement due to the relatively high stiffness of the soil. The new vehicle model, consisting essentially of the pitch motion of the bodies, introduces light changes in the response.

Although the model of the vehicle remains simple, the presented methodology does no longer suffer any limitations for the extension of the model to dissymetric cases or the effect of lateral contributions.

Edited by: O. Brüls

References

Anantharam, M. and Hiller, M.: Numerical Simulation of Mechanical Systems Using Methods for Differential-Algebraic Equations, Int. J. Num. Meth. Eng., 21, 1531–1542, 1991.

Bauchau, O.: Flexible Multibody Dynamics, Springer, Dordrecht, The Netherlands, 2011.

Connolly, D., Giannopoulos, A., and Forde, M. C.: Numerical modelling of ground borne vibrations from high speed rail lines on embankments, Soil Dyn. Earthq. Eng., 46, 13–19, 2013.

Costa, P., Calçada, R., and Cardoso, A. S.: Vibrations induced by railway traffic: influence of the mechanical properties of the train on the dynamic excitation mechanism, in: 8th European Conference on Structural Dynamics: Eurodyn 2011, edited by: De Roeck, G., Degrande, G., Lombaert, G., Müller, G., European Association for Structural Dynamics, Leuven (Belgium), 804–811, 2011.

Costa, P., Calçada, R., and Cardoso, A. S.: Influence of train dynamic modelling strategy on the prediction of track-ground vibrations induced by railway traffic, Proc. IMechE, Part F: J. Rail and Rapid Transit, 226, 434–450, 2012.

Deutsches Institut für Normung: DIN 4150-2: Structural vibrations – Part 2: Human exposure to vibration in buildings, 1999.

Do Rêgo Silva, J. J.: Acoustic and Elastic Wave Scattering using Boundary Elements, vol. 18, Computational Mechanics Publications, Southampton (UK), 1994.

François, S., Schevenels, M., Galvín, P., Degrande, G., and Lombaert, G.: Applications of a 2.5D coupled FE–BE methodology for the dynamic interaction between longitudinally invariant structures and the soil, in: 8th National Congress on Theoretical and Applied Mechanics, National Committee for Theoretical and Applied Mechanics, Brussels (Belgium), 2009.

Galvín, P. and Domínguez, J.: Experimental and numerical analyses of vibrations induced by high-speed trains on the Córdoba–Málaga line, Soil Dyn. Earthq. Eng., 29, 641–651, 2009.

Garcia de Jalon, J. and Bayo, E.: Kinematic and Dynamic Simulation of Multibody Systems: the Real-time challenge, Springer-Verlag, New York, 1993.

Garg, V. and Dukkipati, R.: Dynamics of Railway Vehicle Systems, Academic Press Canada, Don Mills, Ontario, 1984.

Géradin, M. and Cardona, A.: Flexible Multibody Dynamics. A Finite Element Approach, John Wiley & Sons, Chichester, England, 2000.

Hiller, M.: Dynamics of multibody system with minimal coordinates, in: Computer-Aided Analysis of Rigid and Flexible Mechanical Systems – Proceedings of the NATO Advanced Study Institute, edited by: Pereira, M. F. O. and Ambrosio, J. C., Tróia (Portugal), 119–163, 1993.

Ju, S.-H.: Finite element investigation of traffic induced vibrations, J. Sound Vib., 321, 837–853, 2009.

Kaynia, A. M., Madshus, C., and Zackrisson, P.: Ground vibration from high-speed trains: prediction and countermeasure, J. Geotech. Geoenviron., 126, 531–537, 2000.

Knothe, K. and Grassie, S. L.: Modelling of railway track and vehicle/track interaction at high frequencies, Vehicle Syst. Dyn., 22, 209–262, 1993.

Kouroussis, G., Verlinden, O., and Conti, C.: Ground propagation of vibrations from railway vehicles using a finite/infinite-element model of the soil, Proc. IMechE, Part F: J. Rail and Rapid Transit, 223, 405–413, 2009.

Kouroussis, G., Verlinden, O., and Conti, C.: On the interest of integrating vehicle dynamics for the ground propagation of vibrations: the case of urban railway traffic, Vehicle Syst. Dyn., 48, 1553–1571, 2010.

Kouroussis, G., Gazetas, G., Anastasopoulos, I., Conti, C., and Verlinden, O.: Lumped mass model of vertical dynamic coupling of a railway track on elastic homogeneous or layered halfspace, in: 8th European Conference on Structural Dynamics: Eurodyn 2011, edited by: De Roeck, G., Degrande, G., Lombaert, G., and Müller, G., European Association for Structural Dynamics, Leuven (Belgium), 676–683, 2011a.

Kouroussis, G., Gazetas, G., Anastasopoulos, I., Conti, C., and Verlinden, O.: Discrete modelling of vertical track-soil coupling for vehicle-track dynamics, Soil Dyn. Earthq. Eng., 31, 1711–1723, 2011b.

Kouroussis, G., Rustin, C., Bombled, Q., and Verlinden, O.: EasyDyn: multibody open-source framework for advanced research purposes, in: Multibody Dynamics 2011, edited by: J. C. Samin and Fisette, P., ECCOMAS Thematic Conference, Brussels (Belgium), 2011c.

Kouroussis, G., Van Parys, L., Conti, C., and Verlinden, O.: Prediction of environmental vibrations induced by railway traffic using a three-dimensional dynamic finite element analysis, in: Proceedings of the Thirteenth International Conference on Civil, Structural and Environmental Engineering Computing, edited by: Topping, B. H. V. and Tsompanakis, Y., Civil-Comp Press, Chania (Greece), 2011d.

Kouroussis, G., Verlinden, O., and Conti, C.: Free field vibrations caused by high-speed lines: measurement and time domain simulation, Soil Dynam. Earthq. Eng., 31, 692–707, 2011e.

Kouroussis, G., Verlinden, O., and Conti, C.: A two-step time simulation of ground vibrations induced by the railway traffic, J. Mech. Eng. Sci., 226, 454–472, 2012a.

Kouroussis, G., Verlinden, O., and Conti, C.: Efficiency of resilient wheels on the alleviation of railway ground vibrations, Proc. IMechE, Part F: J. Rail and Rapid Transit, 226, 381–396, 2012b.

Kouroussis, G., Verlinden, O., and Conti, C.: Influence of some vehicle and track parameters on the environmental vibrations induced by railway traffic, Vehicle Syst. Dyn., 50, 619–639, 2012c.

Kouroussis, G., Conti, C., and Verlinden, O.: Investigating the influence of soil properties on railway traffic vibration using a numerical model, Vehicle Syst. Dyn., 51, 421–442, 2013.

Lefeuve-Mesgouez, G., Peplow, A. T., and Houédec, D. L.: Surface vibration due to a sequence of high speed moving harmonic rectangular loads, Soil Dyn. Earthq. Eng., 22, 459–473, 2002.

Parisse, B. and Graeve, R. D.: Giac/Xcas (version 0.8.6), http://www-fourier.ujf-grenoble.fr/~parisse/giac_fr.html, last access: September 2010.

SciFace Software GmbH & Co: MuPAD, http://www.mupad.de, last access: July 2012.

Shabana, A. A.: Dynamics of Multibody Systems, Cambridge University Press, New York, USA, 2005.

Takemiya, H. and Bian, X.: Substructure simulation of inhomogeneous track and layered ground dynamic interaction under train passage, J. Eng. Mech., 131, 699–711, 2005.

Verlinden, O., Ben Fekih, L., and Kouroussis, G.: Symbolic generation of the kinematics of multibody systems in EasyDyn: from MuPAD to Xcas/Giac, Theoretical and Applied Mechanics Letters, 3, 013012, doi:10.1063/2.13013012, 2013.

Wang, J. and Zeng, X.: Numerical simulations of vibration attenuation of high-speed train foundations with varied trackbed underlayment materials, J. Vib. Control, 10, 1123–1136, 2004.

Wang, J., Zeng, X., and Gasparini, D. A.: Dynamic response of high-speed rail foundations using linear hysteretic damping and frequency domain substructuring, Soil Dyn. Earthq. Eng., 28, 258–276, 2008.

Zhai, W. and Sun, X.: A detailed model for investigating vertical interaction between railway vehicle and track, Vehicle Syst. Dyn., 23 (supplement), 603–615, 1994.

8

Multiple-task motion planning of non-holonomic systems with dynamics

A. Ratajczak and K. Tchoń

Institute of Computer Engineering, Control and Robotics, Wrocław University of Technology, ul. Janiszewskiego 11/17, 50–372 Wrocław, Poland

Correspondence to: A. Ratajczak (adam.ratajczak@pwr.wroc.pl)

Abstract. This paper addresses the motion planning problem in non-holonomic robotic systems. The system's kinematics and dynamics are represented as a control affine system with outputs. The problem is defined in terms of the end-point map of this system, using the endogenous configuration space approach. Special attention is paid to the multiple-task motion planning problem, i.e. a problem that beyond the proper motion planning task includes a number of additional tasks. For multiple-task motion planning two strategies have been proposed, called the egalitarian approach and the prioritarian approach. Also, two computational strategies have been launched of solving the motion planning problem: the parametric and the non-parametric. The motion planning and computational strategies have been applied to a motion planning problem of the trident snake robot. Performance of the motion planning algorithms is illustrated with computer simulations.

1 Introduction

The motion planning problem of a robotic system consists in determining an action in the configuration space that would drive the system along a desired trajectory or to a desired location in the task space. Frequently, reaching the goal is accompanied with avoiding obstacles and singularities, respecting constraints of motion, limitations of energy, etc. If the planning problem is decomposed into a number of tasks, it is called a multiple-task problem. Resolving multiple-task motion planning problems is enabled by the system's redundancy.

In the area of kinematic control of redundant manipulation robots the multiple-task problems have been addressed within the prioritized approach, initiated by Maciejewski and Klein (1985); Nakamura et al. (1987), and then refined in the works of Chiaverini (1997); Chiacchio et al. (1991); Choi et al. (2004); Antonelli (2009). An extension of these ideas towards mobile robotic systems, in particular mobile robots and mobile manipulators, has been promoted by the endogenous configuration space approach (Tchoń and Jakubiak, 2003; Tchoń and Zadarnowska, 2003). The endogenous configuration space approach employs a control system representation of the robotic system and focuses on the analysis of its end-point map. The central concept of endogenous configuration is identified with the system's control function, so singular controls become singular endogenous configurations, and controllability defines the dexterity of the robotic system. The concept of Jacobian of the robotic system relies on the linear approximation of the control system. Jacobian algorithms are introduced using the continuation method. On the kinematics level the endogenous configuration approach has been developed on the basis of the ideas of Sussmann (1993); Chitour and Sussmann (1998); Divelbiss et al. (1998). Founded on the end-point map of a control system, the endogenous configuration space approach extends in a natural way to robotic systems with dynamics (Zadarnowska and Tchoń, 2007; Ratajczak et al., 2010). Since the endogenous configuration space is infinite-dimensional, the mobile robotic systems have infinite redundancy, capable of accommodating an arbitrary big number of tasks. Using the endogenous configuration space approach, a prioritized approach to motion planning of underactuated robotic systems has been proposed in Ratajczak et al. (2010); Ratajczak (2012).

This paper addresses the multiple-task motion planning problem for control affine systems, that include the dynamics models of non-holonomic robotic systems. It is assumed that the proper motion planning task of reaching a desired point in the task space has been supplemented by a number of additional tasks, characterized by their specific task maps. Two methods of solving the problem have been proposed, called the egalitarian and the prioritarian approach. A motion planning algorithm has been derived by means of the endogenous configuration space approach, in the form of a functional differential equation for the control function. Furthermore, a parametric and a non-parametric strategy of computing numerically the control function have been launched.

Theoretical concepts are applied to the dynamics model of the trident snake robot (Ishikawa, 2004). The robot can be regarded as a realization of the undulatory locomotion principle and a demanding test bed of motion planning algorithms for non-holonomic systems. The design and kinematics analysis of the trident snake with passive wheels and active joints can be found in Ishikawa (2004); Ishikawa et al. (2010). Recently, this analysis has been extended to the case of active wheels in Paszuk et al. (2012) and complemented by a study of trident snake dynamics (Pietrowska, 2012). In both these cases the motion single-task planning algorithms have been derived from the endogenous configuration space approach.

Differently than in the references mentioned above, this paper concentrates on the multiple-task motion planning strategy for the trident snake robot. The robot's kinematics and dynamics are represented as a control affine system with outputs. The problem is defined in terms of the end-point map of this system. The control strategy of the system involves a preliminary state feedback. The motion planning problem includes two subtasks: the proper motion planning task of transferring the system to a desired task space location, and the singularity avoidance task guaranteeing well definiteness of the feedback transformation. Two motion planning strategies have been developed, referred to as the egalitarian and the prioritarian approach. The former strategy regards the component tasks as equivalent, the latter assigns the highest priority to the proper motion planning task. Simultaneously, two computational strategies have been proposed in order to solve the motion planning problem: the parametric and the non-parametric, depending on whether the computation of the control function utilizes a specific base in the endogenous configuration space or is base-independent. Computer simulations demonstrate the performance of the egalitarian non-parametric and the prioritarian parametric motion planning algorithms.

The paper is organized in the following way. Section 2 introduces a control system representation of the non-holonomic system, defines a Jacobian motion planning algorithm, and describes two computational strategies. The strategies of multiple-task motion planning are presented in Sect. 3. In Sect. 4 the motion planning strategies are specified to the trident snake robot. Results of numeric computations are included in Sect. 5. Section 6 concludes the paper. In order to not distract the reader's attention from the main thread of the paper, a derivation of the subtask Jacobian and its inverse as well as the dynamics model of the trident snake are placed in the Appendix.

2 Basic concepts

Since the dynamics of a non-holonomic robotic system can be represented as an affine control system with outputs, this system will define our *universe of discourse*. We begin with introducing basic concepts of the endogenous configuration space approach, and derive the Jacobian motion planning algorithm. This algorithm relies on the solution of a functional differential equation involving a Jacobian inverse operator. Depending on the method of solution of this equation, parametric or non-parametric motion planning algorithms are distinguished.

2.1 Endogenous configuration space approach

The basic concepts of the endogenous configuration space approach will be adopted to a general control affine system, of the form

$$\begin{cases} \dot{x} = f(x) + g(x)u = f(x) + \sum_{i=1}^{m} g_i(x)u_i, \\ y = k(x), \end{cases} \tag{1}$$

where $u \in \mathbb{R}^m$, $x \in \mathbb{R}^n$, $y \in \mathbb{R}^r$. All the functions and vector fields appearing in (1) will be assumed smooth. Let $T > 0$ denote a control time horizon. The admissible control functions entering system (1) will be assumed to belong to the space $L^2_m[0,T]$ of Lebesgue square integrable functions defined on the interval $[0,T]$. The space $L^2_m[0,T]$ is a Hilbert space with inner product

$$\langle u_1(\cdot), u_2(\cdot)\rangle_R = \int_0^T u_1^T(t)R(t)u_2(t)\mathrm{d}t, \tag{2}$$

where $R(t) = R^T(t) > 0$, and the corresponding norm $||u(\cdot)||_R = \langle u(\cdot), u(\cdot)\rangle_R^{1/2}$. For a control function $u(\cdot)$, let $x(t) = \varphi_{x_0,t}(u(\cdot))$ denote the state trajectory of (1), initialized at x_0. It will be assumed that this trajectory exists for every $t \in [0,T]$. The output y is identified with the vector of task variables.

Given an initial state x_0 of system (1) and the time horizon T, a general motion planning problem consists in defining a control $u(t)$ that drives the system's output at T to a prescribed point y_d, so that $y(T) = y_d$.

Our analysis of the motion planning problem will be based on the concept of the end-point map of system (1), defined as the value at T of the output function resulting from the application of a control function $u(\cdot)$,

$$K_{x_0,T}(u(\cdot)) = k(x(T)) = k(\varphi_{x_0,T}(u(\cdot))). \tag{3}$$

For bounded measurable control functions $u(\cdot) \in \mathcal{X}$ the end-point map $K : \mathcal{X} \longrightarrow \mathbb{R}^r$ is continuously differentiable ($\mathbb{C}^1$)

(Sontag, 1990). In the context of mobile robots or mobile manipulators the space $\mathcal{X}$ has been called the endogenous configurations space (Tchoń and Jakubiak, 2003). The derivative of the end-point map is computed by means of the linear approximation to system (1)

$$\dot{\xi}(t) = A(t)\xi(t) + B(t)v(t), \quad \eta(t) = C(t)\xi(t), \quad \xi(0) = 0, \tag{4}$$

along the (control,trajectory) pair $(u(t), x(t))$, where

$$\xi(t) = D\varphi_{x_0,t}(u(\cdot))v(\cdot), \tag{5}$$

and

$$A(t) = \frac{\partial(f(x(t)) + g(x(t))u(t))}{\partial x}, \quad B(t) = g(x(t)),$$
$$C(t) = \frac{\partial k(x(t))}{\partial x}. \tag{6}$$

Given the linear system (4), the derivative of the end-point map at $u(\cdot) \in \mathcal{X}$ is equal to

$$DK_{x_0,T}(u(\cdot))v(\cdot) = \eta(T) = C(T)\xi(T). \tag{7}$$

In compliance with the robotic terminology, the derivative (7) will be called the system's Jacobian,

$$DK_{x_0,T}(u(\cdot))v(\cdot) = J_{x_0,T}(u(\cdot))v(\cdot).$$

It follows that the computation of the Jacobian involves the integration of the differential equation (4) from 0 to T at zero initial condition. If $\Phi(t,s)$ denotes the transition matrix of (4),

$$\frac{\partial\Phi(t,s)}{\partial t} = A(t)\Phi(t,s), \quad \Phi(s,s) = I_n,$$

this means that the Jacobian $J_{x_0,T}(u(\cdot)) : \mathcal{X} \longrightarrow \mathbb{R}^r$ can be expressed as

$$J_{x_0,T}(u(\cdot))v(\cdot) = C(T)\int_0^T \Phi(T,s)B(s)v(s)\mathrm{d}s. \tag{8}$$

The Jacobian allows to distinguish regular and singular controls (endogenous configurations) of system (1). A control $u(\cdot) \in \mathcal{X}$ will be called regular, if the Jacobian is surjective onto $\mathbb{R}^r$, otherwise the control $u(\cdot)$ is referred to as singular. It can be shown that at regular controls the control affine system (1) is locally controllable.

Using the inner product (2) in the endogenous configuration space and the Euclidean structure of the output space, the dual Jacobian map $J^*_{x_0,T}(u(\cdot)) : \mathbb{R}^r \longrightarrow \mathcal{X}$ can be defined in the following way

$$\left(J^*_{x_0,T}(u(\cdot))\eta\right)(t) = R^{-1}(t)B^T(t)\Phi^T(T,t)C^T(T)\eta. \tag{9}$$

2.2 Jacobian motion planning

Using the end-point map, the general motion planning problem in system (1) is tantamount to computing a control function $u_d(\cdot)$, such that $K_{x_0,T}(u_d(\cdot)) = y_d$. The problem can be solved by means of a Jacobian motion planning algorithm whose derivation relies on the continuation (homotopy) method (Sussmann, 1993). Given the motion planning problem, we begin with any initial control $u_0(\cdot) \in \mathcal{X}$. If the initial choice does not solve the problem, i.e. $K_{x_0,T}(u_0(\cdot)) \neq y_d$, we choose in $\mathcal{X}$ a differentiable curve $u_\theta(\cdot)$, $\theta \in \mathbb{R}$, passing at $\theta = 0$ through $u_0(\cdot)$, and compute the task space error along this curve

$$e(\theta) = K_{x_0,T}(u_\theta(\cdot)) - y_d. \tag{10}$$

Next, we request that the error decrease exponentially along with θ, with a prescribed decay rate $\gamma > 0$, i.e.

$$\frac{\mathrm{d}e(\theta)}{\mathrm{d}\theta} = -\gamma e(\theta). \tag{11}$$

By differentiating the formula (10) with respect to θ, we arrive at the Ważewski-Davidenko equation

$$J_{x_0,T}(u_\theta(\cdot))\frac{\mathrm{d}u_\theta(\cdot)}{\mathrm{d}\theta} = -\gamma e(\theta), \tag{12}$$

involving the Jacobian (8). If $J^{\#}_{x_0,T}(u(\cdot)) : \mathbb{R}^r \longrightarrow \mathcal{X}$ denotes a right Jacobian inverse, such that $J_{x_0,T}(u(\cdot))J^{\#}_{x_0,T}(u(\cdot)) = I_r$, then (12) can be converted into a dynamic system

$$\frac{\mathrm{d}u_\theta(\cdot)}{\mathrm{d}\theta} = -\gamma J^{\#}_{x_0,T}(u_\theta(\cdot))e(\theta) \tag{13}$$

evolving in $\mathcal{X}$. Finally, a solution of the motion planning problem is obtained as the limit

$$u_d(t) = \lim_{\theta\to+\infty} u_\theta(t).$$

A frequently used right Jacobian inverse is the Moore–Penrose inverse derived from minimizing the square norm $||v(\cdot)||^2_R$ under the equality constraint (a Jacobian equation)

$$J_{x_0,T}(u(\cdot))v(\cdot) = \eta, \tag{14}$$

$\eta \in \mathbb{R}^r$. The resulting formula is (Tchoń and Jakubiak, 2003),

$$\left(J^{\#}_{x_0,T}(u(\cdot))\eta\right)(t) = R^{-1}(t)B^T(t)\Phi^T(T,t)C^T(T)\mathcal{G}^{-1}_{x_0,T}(u(\cdot))\eta, \tag{15}$$

where

$$\mathcal{G}_{x_0,T}(u(\cdot)) = C(T)\int_0^T \Phi(T,s)B(s)R^{-1}(s)B^T(s)\Phi^T(T,s)\mathrm{d}sC^T(T).$$

denotes the Gram matrix associated with the linear system (4), that in robotics context is referred to as the dexterity or mobility matrix of system (1) (Tchoń and Zadarnowska, 2003). At regular control functions the Gram matrix is full rank, making the inverse (15) well defined. The Gram matrix can be conveniently computed by integrating the Lyapunov differential equation

$$\dot{M}(t) = B(t)R^{-1}(t)B^T(t) + A(t)M(t) + M(t)A^T(t), \tag{16}$$

at zero initial condition $M(0) = 0$, and then substituting $\mathcal{G}_{x_0,T}(u(\cdot)) = C(T)M(T)C^T(T)$. By invoking the definition of dual Jacobian (9), it is easily checked that the Gram matrix equals the composition

$$\mathcal{G}_{x_0,T}(u(\cdot)) = J_{x_0,T}(u(\cdot))J^*_{x_0,T}(u(\cdot)).$$

Moreover, it follows that the Jacobian inverse (15) can be written as

$$\left(J^\#_{x_0,T}(u(\cdot))\eta\right)(t) = \left(J^*_{x_0,T}(u(\cdot))\mathcal{G}^{-1}_{x_0,T}(u(\cdot))\eta\right)(t),$$

what justifies the identity

$$J^\#_{x_0,T}(u(\cdot)) = J^*_{x_0,T}(u(\cdot))\mathcal{G}^{-1}_{x_0,T}(u(\cdot)). \tag{17}$$

2.3 Computations

The Jacobian motion planning algorithm derived in the previous subsection exploits a solution of a functional differential equation (13) in the endogenous configuration space $\mathcal{X}$. In order to make this equation tractable, we need to discretize the θ variable, and pass to a discrete control updating scheme

$$u_{\theta+1}(t) = u_\theta(t) - \gamma R^{-1}(t)B^T_\theta(t)\Phi^T_\theta(T,t)C^T_\theta(T)\mathcal{G}^{-1}_{x_0,T}(u_\theta(\cdot))e(\theta), \tag{18}$$

where $B_\theta(t)$, $\Phi_\theta(t,s)$, $C_\theta(t)$ are computed along $(u_\theta(t), x_\theta(t))$ in agreement with (6), and $e(\theta) = K_{x_0,T}(u_\theta(\cdot)) - y_d$, for $\theta = 0, 1, \ldots$.

Given $u_\theta(t)$, a basic step of the updating (18) consists in solving simultaneously a system of differential equations

$$\begin{cases} \dot{x}_\theta(t) = f(x_\theta(t)) + g(x_\theta(t))u_\theta(t), \\ \frac{\partial\Phi_\theta(T,t)}{\partial t} = -\Phi_\theta(T,t)A_\theta(t), \\ \dot{M}_\theta(t) = B_\theta(t)R^{-1}(t)B^T_\theta(t) + A_\theta(t)M_\theta(t) + M_\theta(t)A^T_\theta(t) \end{cases}$$

with respective boundary conditions $x_\theta(0) = x_0$, $\Phi_\theta(T,T) = I_n$ and $M_\theta(0) = 0$.

Alternatively, we can expand the control functions into a truncated orthogonal series that leads to a finite parametrization of controls. More specifically, we select a finite base of orthogonal functions $\{\varphi_0(t), \varphi_1(t), \ldots, \varphi_k(t)\}$, and assume that

$$u_\lambda(t) = \Psi(t)\lambda, \quad u_{\lambda i}(t) = \sum_{j=0}^{k} \varphi_j(t)\lambda_{ij}, \; i = 1, 2, \ldots, m,$$

where $\lambda \in \mathbb{R}^s$, $s = m(k+1)$, and matrix $\Psi(t)$ of dimension $m \times s$ contains the suitably arranged basic functions. By orthogonality of the base $\int_0^T \Psi^T(t)\Psi(t)dt = I_s$. Under assumption that $v(t) = \Psi(t)\mu$, the Jacobian (8) takes the matrix form

$$\hat{J}_{x_0,T}(\lambda) = C_\lambda(T)\int_0^T \Phi_\lambda(T,s)B_\lambda(s)\Psi(s)ds,$$

where the matrices (6) need to be computed for $u(t) = u_\lambda(t)$ and $x(t) = \varphi_{x_0,t}(u_\lambda(\cdot))$. It follows that this Jacobian can be obtained by the integration of the matrix differential equation

$$\dot{N}_\lambda(t) = B_\lambda(t)\Psi(t) + A_\lambda(t)N_\lambda(t)$$

with initial condition $N_\lambda(0) = 0$, followed by the substitution $\hat{J}_{x_0,T}(\lambda) = C_\lambda(T)N_\lambda(T)$. For the matrix Jacobian, the Jacobian equation (14) assumes the form $\hat{J}_{x_0,T}(\lambda)\mu = \eta$. It can be shown that the Moore–Penrose Jacobian inverse assumes the matrix form

$$\hat{J}^\#_{x_0,T}(\lambda) = S^{-1}\hat{J}_{x_0,T}(\lambda)\left(\hat{J}_{x_0,T}(\lambda)S^{-1}\hat{J}^T_{x_0,T}(\lambda)\right)^{-1}, \tag{19}$$

where $S = \int_0^T \Psi^T(t)R(t)\Psi(t)dt$. Consequently, the updating scheme of the control function (18) converts to

$$\lambda_{\theta+1} = \lambda_\theta - \gamma\hat{J}^\#_{x_0,T}(\lambda_\theta)e(\theta). \tag{20}$$

The representation of the control functions by means of a truncated orthogonal series will be called parametric. By contrast, the representation that does not use any base will be referred to as non-parametric. This terminology extends to the corresponding motion planning algorithms.

3 Multiple-task motion planning

If the motion planning problem consists only of the motion planning task, it will be called single-task. When the motion planning task is augmented with additional tasks, the motion planning problem will be referred to as a multiple-task problem. We shall assume that the additional tasks are defined by task maps ${}^iK_{x_0,T} : \mathcal{X} \longrightarrow R\; i = 1, 2, \ldots, p$, operating in the endogenous configuration space of (1). Most frequently, these task maps have the form

$${}^iK_{x_0,T}(u(\cdot)) = \int_0^T F_i(x(t), u(t))dt, \tag{21}$$

for a non-negative function $F_i(x,u) \geq 0$ that is assumed differentiable wherever defined. Given p additional tasks, the multiple-task motion planning problem consists in determining a control function $u(\cdot) \in \mathcal{X}$ minimizing the collective error

$$\mathbf{e}(\theta) = \left({}^0e(\theta), {}^1e(\theta), \ldots, {}^pe(\theta)\right), \tag{22}$$

where ${}^0e(\theta) = K_{x_0,T}(u_\theta(\cdot)) - y_d$ and ${}^ie(\theta) = {}^iK_{x_0,T}(u_\theta(\cdot))$ $i = 1,2,\ldots,p$ are errors corresponding to subsequent subtasks. We associate with the task maps the Jacobians: the Jacobian for the task number 0 is given by (8), while for the i-th subtask the Jacobian

$$ {}^iJ_{x_0,T}(u(\cdot))v(\cdot) = \int_0^T \left(\frac{\partial F_i(x(t),u(t))}{\partial x}\xi(t) + \frac{\partial F_i(x(t),u(t))}{\partial u}v(t) \right) \mathrm{d}t, \tag{23} $$

where $\xi(t)$ denotes the solution of (4), see Appendix for a derivation. The Moore–Penrose inverse for ${}^0J_{x_0,T}(u(\cdot))$ is defined by expression (15). In Appendix we have shown that the Moore–Penrose inverse of the ith Jacobian, $i = 1,2,\ldots,p$, takes the following form

$$ \left({}^iJ^{\#}_{x_0,T}(u(\cdot))\eta\right)(t) = \frac{b_i(t)+c_i(t)}{\|b_i(\cdot)+c_i(\cdot)\|^2_R}\eta, \tag{24} $$

$\eta \in \mathbb{R}$, where

$$ b_i(t) = R^{-1}(t)B^T(t)\int_t^T \Phi^T(s,t)\left(\frac{\partial F_i(x(s),u(s))}{\partial x}\right)^T \mathrm{d}s, \tag{25} $$

$$ c_i(t) = R^{-1}(t)\left(\frac{\partial F_i(x(t),u(t))}{\partial u}\right)^T. \tag{26} $$

The task Jacobians can be arranged into a collective Jacobian

$$ \mathbf{J}_{x_0,T}(u_\theta(\cdot))v(\cdot) = \left({}^0J_{x_0,T}(u(\cdot))v(\cdot), {}^1J_{x_0,T}(u(\cdot))v(\cdot),\ldots,{}^pJ_{x_0,T}(u(\cdot))v(\cdot)\right). \tag{27} $$

These subtasks can either be treated as equivalent and solved simultaneously or ordered according to their importance and solved sequentially. These two approaches will be further referred to as egalitarian and prioritarian. In the following two subsections we shall derive an egalitarian and a prioritarian Jacobian motion planning algorithm.

3.1 Egalitarian approach

In accordance with the egalitarian approach all subtasks need to be solved simultaneously. Imposing the exponential decrease of the error (22) with a decay rate $\gamma > 0$, and using the collective Jacobian (27), we arrive at the Ważewski-Davidenko equation

$$ \mathbf{J}_{x_0,T}(u_\theta(\cdot))\frac{\mathrm{d}u_\theta(\cdot)}{\mathrm{d}\theta} = -\gamma\mathbf{e}(\theta). \tag{28} $$

Finally, similarly to (13), the equation (28) can be solved by means of the Moore–Penrose Jacobian inverse of the collective Jacobian,

$$ \frac{\mathrm{d}u_\theta(\cdot)}{\mathrm{d}\theta} = -\gamma\mathbf{J}^{\#}_{x_0,T}(u_\theta(\cdot))\mathbf{e}(\theta). \tag{29} $$

The egalitarian solution of the multiple-task motion planning problem is obtained as the limit $u_d = \lim_{\theta\to+\infty} u_\theta(t)$ of the trajectory of dynamic system (29).

3.2 Prioritarian approach

Differently to the egalitarian algorithm, now the subtasks will be ordered with decreasing priorities. The essential assumption of the prioritarian approach is that the solution of a lower priority task is sought in a space rendered accessible by the higher priority tasks, so it should not affect the solution of any task with higher priority. Given the i-th error (22) and the i-th Jacobian (27), we start the derivation from a statement of the Jacobian equation for the i-th subtask,

$$ \frac{\mathrm{d}\,{}^ie(\theta)}{\mathrm{d}\theta} = {}^iJ_{x_0,T}(u_\theta(\cdot))\frac{\mathrm{d}u_\theta(\cdot)}{\mathrm{d}\theta} = -{}^i\gamma\,{}^ie(\theta), \tag{30} $$

with ${}^i\gamma > 0$ and $i = 0,1,\ldots,p$. A general solution of this equation involves the Jacobian inverse with projection (Tchoń and Jakubiak, 2003),

$$ \frac{\mathrm{d}u_\theta(\cdot)}{\mathrm{d}\theta} = -{}^i\gamma\,{}^iJ^{\#}_{x_0,T}(u_\theta(\cdot))\,{}^ie(\theta) + {}^iP_{x_0,T}(u_\theta(\cdot))\,{}^i\zeta_\theta(\cdot), \tag{31} $$

where

$$ {}^iP_{x_0,T}(u_\theta(\cdot)) = \mathrm{id}_{\mathcal{X}} - {}^iJ^{\#}_{x_0,T}(u_\theta(\cdot))\,{}^iJ_{x_0,T}(u_\theta(\cdot)) $$

denotes the projection onto $\ker {}^iJ_{x_0,T}(u_\theta(\cdot))$, and ${}^i\zeta_\theta(\cdot) \in \mathcal{X}$ is any element of the endogenous configuration space. For the highest priority task (number 0) the equality (31) is required to hold in the whole endogenous configuration space,

$$ \frac{\mathrm{d}u_\theta(\cdot)}{\mathrm{d}\theta} = -{}^0\gamma\,{}^0J^{\#}_{x_0,T}(u_\theta(\cdot))\,{}^0e(\theta) + {}^0P_{x_0,T}(u_\theta(\cdot))\,{}^0\zeta_\theta(\cdot). \tag{32} $$

For the lower priority task (number 1) this equality needs to be satisfied only within the kernel of ${}^0J^{\#}_{x_0,T}(u_\theta(\cdot))$, i.e.

$$ {}^0P_{x_0,T}(u_\theta(\cdot))\frac{du_\theta(\cdot)}{d\theta} = -{}^1\gamma{}^0P_{x_0,T}(u_\theta(\cdot))\,{}^1J^{\#}_{x_0,T}(u_\theta(\cdot))\,{}^1e(\theta) + {}^0P_{x_0,T}(u_\theta(\cdot)){}^1P_{x_0,T}(u_\theta(\cdot))\,{}^1\zeta_\theta(\cdot). \tag{33} $$

Having projected (32) onto $\ker {}^0J_{x_0,T}(u_\theta(\cdot))$, we get

$$ {}^0P_{x_0,T}(u_\theta(\cdot))\frac{du_\theta(\cdot)}{d\theta} = {}^0P_{x_0,T}(u_\theta(\cdot))\,{}^0\zeta_\theta(\cdot), \tag{34} $$

where we have used the identity ${}^iP_{x_0,T}(u_\theta(\cdot))\,{}^iJ^{\#}_{x_0,T}(u_\theta(\cdot)) = 0$, and the idempotency of the projection. Now, a combination of (33) and (34) results in

$$ {}^0P_{x_0,T}(u_\theta(\cdot))\,{}^0\zeta_\theta(\cdot) = -{}^1\gamma{}^0P_{x_0,T}(u_\theta(\cdot))\,{}^1J^{\#}_{x_0,T}(u_\theta(\cdot))\,{}^1e(\theta) + {}^0P_{x_0,T}(u_\theta(\cdot)){}^1P_{x_0,T}(u_\theta(\cdot))\,{}^1\zeta_\theta(\cdot). $$

Finally, a substitution of the above identity into (32) yields the following prioritarian Jacobian motion planning algorithm for two subtasks

$$ \frac{du_\theta(\cdot)}{d\theta} = -{}^0\gamma\,{}^0J^{\#}_{x_0,T}(u_\theta(\cdot))\,{}^0e(\theta) - {}^1\gamma\,{}^0P_{x_0,T}(u_\theta(\cdot))\,{}^1J^{\#}_{x_0,T}(u_\theta(\cdot))\,{}^1e(\theta) + {}^0P_{x_0,T}(u_\theta(\cdot)){}^1P_{x_0,T}(u_\theta(\cdot))\,{}^1\zeta_\theta(\cdot) \tag{35} $$

If there are only two subtasks, the last term ${}^1\zeta_\theta(\cdot)$ is zero, otherwise it could be used to define the prioritarian algorithm for the remaining lower priority tasks.

In this way the presented derivation extends to $p+1$ subtasks, resulting in the following motion planning algorithm (Ratajczak et al., 2010; Ratajczak, 2012),

$$\frac{\mathrm{d}u_\theta(t)}{\mathrm{d}\theta} = -\sum_{i=0}^{p} {}^i\gamma \left(\prod_{j=0}^{i} {}^{j-1}P_{x_0,T}(u_\theta(\cdot)) \right) {}^i\mathcal{J}^{\#}_{x_0,T}(u_\theta(\cdot))\, {}^ie(\theta), \quad (36)$$

where ${}^{-1}P_{x_0,T}(u_\theta(\cdot)) = \mathrm{id}_X$. In the terminology of Antonelli (2009) this algorithm belongs to the successive inverse-based projection methods. Again, the solution of the multiple-task motion planning problem is obtained as the limit $u_d(t) = \lim_{\theta\to+\infty} u_\theta(t)$ of the trajectory of system (36).

Let us take a closer examination of the control formula (35). By a substitution of this formula into (32), it is easily seen that the first error

$$\frac{\mathrm{d}^0e(\theta)}{\mathrm{d}\theta} = -{}^0\gamma\, {}^0e(\theta)$$

decreases exponentially, as requested in (11). However, because of the task-priority assumption, the second error

$$\frac{\mathrm{d}^1e(\theta)}{\mathrm{d}\theta} = -{}^1\gamma\, {}^1e(\theta) - {}^0\gamma {}^1J_{x_0,T}(u_\theta(\cdot))\, {}^0\mathcal{J}^{\#}_{x_0,T}(u_\theta(\cdot))\, {}^0e(\theta)$$
$$+{}^1\gamma {}^1J_{x_0,T}(u_\theta(\cdot))\, {}^0\mathcal{J}^{\#}_{x_0,T}(u_\theta(\cdot))\, {}^0J_{x_0,T}(u_\theta(\cdot))\, {}^1\mathcal{J}^{\#}_{x_0,T}(u_\theta(\cdot))\, {}^1e(\theta),$$

behaves differently. The error ${}^1e(\theta)$ will converge toward zero exponentially, provided that it does not affect convergence of the error ${}^0e(\theta)$. This happens when ${}^1J_{x_0,T}(u_\theta(\cdot))\, {}^0\mathcal{J}^{\#}_{x_0,T}(u_\theta(\cdot)) = 0$. Taking into account the form of (17), the last equality will be satisfied whenever

$${}^1J_{x_0,T}(u_\theta(\cdot))\, {}^0\mathcal{J}^{*}_{x_0,T}(u_\theta(\cdot)) = 0,$$

what may be interpreted as a sort of orthogonality of Jacobians ${}^0J_{x_0,T}(u(\cdot))$ and ${}^1J_{x_0,T}(u(\cdot))$.

4 Motion planning of trident snake

The presented algorithms will be utilized in order to solve a motion planning problem for the trident snake mobile robot (Ishikawa, 2004; Ishikawa et al., 2010; Paszuk et al., 2012). In the following subsections we are going to present a control-theoretic model, state a multiple-task motion planning problem, and introduce the egalitarian and the prioritarian motion planning algorithm for the trident snake robot.

4.1 Model

The trident snake is a non-holonomic mobile robot. It consists of a triangular-shape body and three links able to rotate around the attachment points (see Fig. 1). Each link is ended with a passive wheel which is subjected to non-holonomic constraints. The control inputs of the robot are the driving torques exerted at the joints between the body and the links. The angles $\alpha_1 = -\frac{2}{3}\pi$, $\alpha_2 = 0$, $\alpha_3 = \frac{2}{3}\pi$ are constant geometric parameters of the robot, l denotes the length of each link, and r stands for the distance between the center of the robot body and the joints.

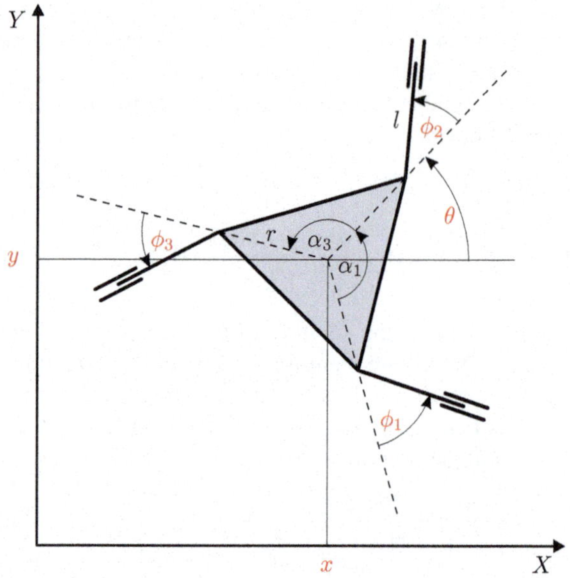

Figure 1. Trident snake robot: conceptual model.

4.1.1 Kinematics

According to Fig. 1, the vector of generalized coordinates is defined as $q = (x, y, \theta, \phi_1, \phi_2, \phi_3) = (x, y, \theta, \phi) \in \mathbb{R}^6$. For every wheel the non-holonomic constraints arise from the assumption of non-slipping laterally. The corresponding constraints matrix takes the form (Ishikawa et al., 2010; Paszuk et al., 2012)

$$\mathsf{A}(q) = \begin{bmatrix} \mathsf{A}_1(q)\mathrm{Rot}^T(Z,\theta) & -lI_3 \end{bmatrix},$$

where $\mathrm{Rot}(Z,\theta)$ is the rotation matrix around Z-axis by θ angle, and $\mathsf{A}_1(q)$ is defined as

$$\mathsf{A}_1(q) = \begin{bmatrix} \sin(\alpha_1+\phi_1) & -\cos(\alpha_1+\phi_1) & -l-r\cos\phi_1 \\ \sin(\alpha_2+\phi_2) & -\cos(\alpha_2+\phi_2) & -l-r\cos\phi_2 \\ \sin(\alpha_3+\phi_3) & -\cos(\alpha_3+\phi_3) & -l-r\cos\phi_3 \end{bmatrix}.$$

Using the relation $\mathsf{A}(q)\mathsf{G}(q) = 0$, we find

$$\mathsf{G}(q) = \begin{bmatrix} \cos\theta & -\sin\theta & 0 \\ \sin\theta & \cos\theta & 0 \\ 0 & 0 & 1 \\ \frac{\sin(\alpha_1+\phi_1)}{l} & -\frac{\cos(\alpha_1+\phi_1)}{l} & -1-\frac{r\cos\phi_1}{l} \\ \frac{\sin(\alpha_2+\phi_2)}{l} & -\frac{\cos(\alpha_2+\phi_2)}{l} & -1-\frac{r\cos\phi_2}{l} \\ \frac{\sin(\alpha_3+\phi_3)}{l} & -\frac{\cos(\alpha_3+\phi_3)}{l} & -1-\frac{r\cos\phi_3}{l} \end{bmatrix} = \begin{bmatrix} \mathsf{G}_1(\theta) \\ \mathsf{G}_2(\phi) \end{bmatrix}, \quad (37)$$

so the trident snake kinematics can be represented by the driftless control system

$$\dot{q} = \mathsf{G}(q)v,$$

$v \in \mathbb{R}^3$ denoting the control variable.

4.1.2 Dynamics

The dynamics equations of the trident snake can be obtained from the Lagrange formalism together with the d'Alembert principle. A detailed derivation can be found in Pietrowska (2012). After the elimination of the traction forces, the resulting equations of motion take the form

$$\begin{cases} \dot{q} = \mathsf{G}(q)v \\ \dot{v} = \mathsf{N}(q,v) + \mathsf{P}(q)w, \\ y = k(q,v). \end{cases} \tag{38}$$

The matrix $\mathsf{G}(q)$ is given by (37), the terms $\mathsf{N}(q,v)$ and $\mathsf{P}(q)$ can be reconstructed from the data provided in the Sect. A2. It follows that the trident snake has $\dim v = \dim w = 3$, so it is fully controlled. Therefore, it is possible to apply a partially linearizing feedback transformation

$$w = \mathsf{P}^{-1}(q)(u - \mathsf{N}(q,v)), \tag{39}$$

u denoting a new control variable, that simplifies the control system representation substantially, and produces a kinematically reduced system (Lewis, 1999),

$$\begin{cases} \dot{q} = \mathsf{G}(q)v \\ \dot{v} = u, \\ y = k(q,v). \end{cases} \tag{40}$$

In Paszuk et al. (2012), it has been shown that the feedback (39) is well defined provided that the lower submatrix of (37) is non-singular, i.e. $\det \mathsf{G}_2(\phi) \neq 0$.

To obtain a control affine system (1), we need to introduce a state vector $x = (q,v) \in \mathbb{R}^9$. Then, the kinematics and dynamics of the trident snake may be rewritten as

$$\begin{cases} \dot{x} = f(x) + g(x)u, \\ y = k(x), \end{cases} \tag{41}$$

where $u \in \mathbb{R}^3$ and

$$f(x) = \begin{pmatrix} \mathsf{G}(q)v \\ 0_{3\times1} \end{pmatrix}, \quad g(x) = \begin{bmatrix} 0_{3\times3} \\ I_3 \end{bmatrix}.$$

Assuming that all state variables x are subject to motion planning, we get the output function $y = k(x) = x$.

4.2 Motion planning

The motion planning problem for the trident snake will consist in defining a control function $u(t)$ in system (41) that drives the system from an initial state to a desired terminal output, over a prescribed time horizon $T > 0$. In order to make the feedback (39) well defined, the control $u(t)$ should also prevent matrix $\mathsf{G}_2(\phi)$ from getting singular. Clearly, the motion planning problem includes two subtasks, of which the former will be called the proper motion planning task (index 0), whereas the latter will be named the singularity avoidance task (index 1). The task map ${}^0K_{x_0,T}(u(\cdot))$ is just the end-point map of system (41). The singularity avoidance task will be assigned the task map ${}^1K_{x_0,T}(u(\cdot))$ of the form (21) with function $F_1(x,u) = \det^{-2}(\mathsf{G}_2(\phi))$. The collective error $\mathbf{e}(\theta)$ and the collective Jacobian $\mathbf{J}_{x_0,T}(u(\cdot))$ can be derived directly from (22) and (27).

4.3 Algorithms

The motion planning problem for the trident snake robot will be solved by means of the multiple-task motion planning algorithms introduced in Sect. 3. For the egalitarian motion planning the collective error is defined as

$$\mathbf{e}(\theta) = ({}^0e(\theta), \delta\,{}^1e(\theta)),$$

where the second component has been multiplied by a scaling factor δ in order to maintain a dynamic balance of the two subtasks. The egalitarian motion planning algorithm takes the form (29). The solution is the limit $u_d = \lim_{\theta\to+\infty} u_\theta(t)$ of the trajectory of (29).

In the prioritarian approach, the proper motion planning subtask will be assigned a higher priority, while the singularity avoidance subtask will get a lower priority. Having defined two errors: ${}^0e(\theta)$, ${}^1e(\theta)$ and two Jacobian inverses: ${}^0J^{\#}_{x_0,T}(u(\cdot))$ and ${}^1J^{\#}_{x_0,T}(u(\cdot))$ given, respectively, by (15) and (24), we can provide the following formula for the prioritarian motion planning algorithm (36)

$$\frac{\mathrm{d}u_\theta(\cdot)}{\mathrm{d}\theta} = -{}^0\gamma\, {}^0J^{\#}_{x_0,T}(u_\theta(\cdot))\, {}^0e(\theta) - {}^1\gamma\, {}^0P_{x_0,T}(u_\theta(\cdot))\, {}^1J^{\#}_{x_0,T}(u_\theta(\cdot))\, {}^1e(\theta),$$

where ${}^0P_{x_0,T}(u(\cdot))$ denotes the projection into $\ker {}^0J_{x_0,T}(u(\cdot))$. In the next section the performance of both these algorithms will be illustrated by means of numerical computations.

5 Implementation and computations

Potentially, in the egalitarian as well as in the prioritarian motion planning algorithm, both the parametric and non-parametric computational approaches introduced in Sect. 2.3 are applicable. In order to demonstrate this possibility, two numerical solutions of the same motion planning problem for the trident snake will be presented. The problem involves the proper motion planning task and the singularity avoidance task. In the first case, the solution will be delivered by the non-parametric egalitarian algorithm. In the second case, the parametric prioritarian algorithm is used. The

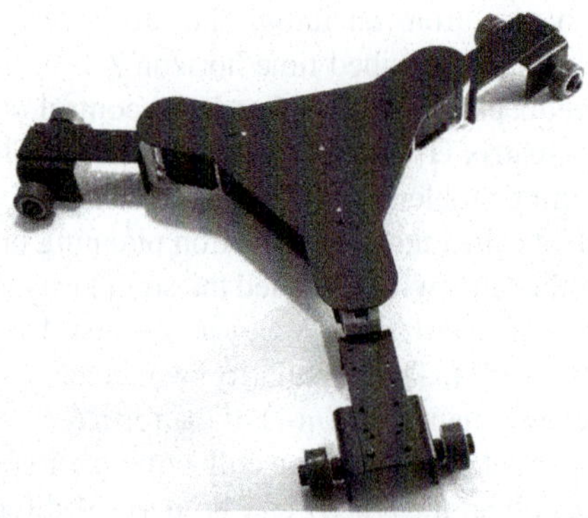

Figure 2. Trident snake robot: physical model.

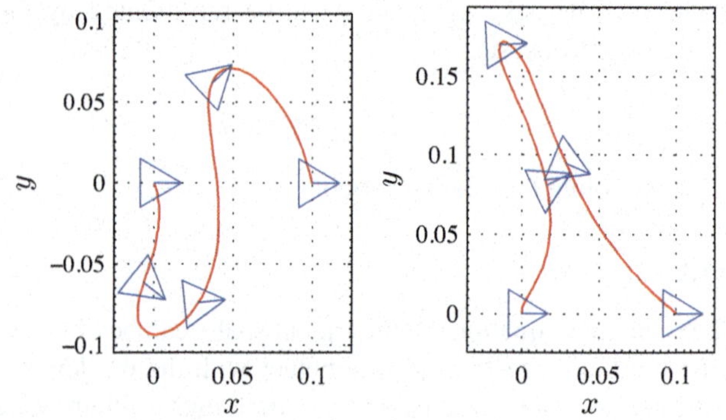

Figure 3. Motion path in XY plane: egalitarian solution (left) and prioritarian solution (right).

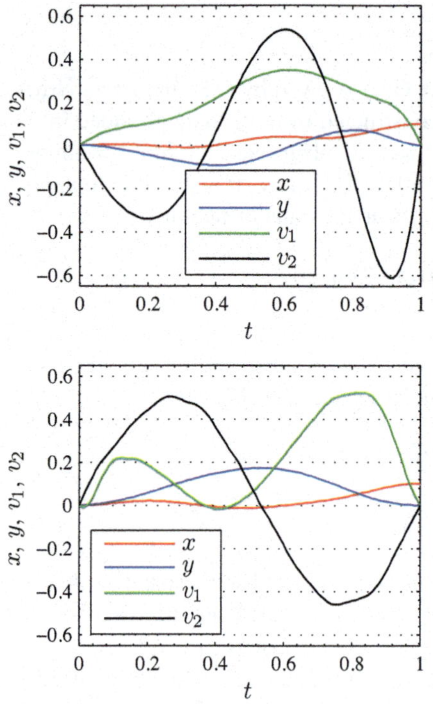

Figure 4. Position and linear velocity: egalitarian solution (top) and prioritarian solution (bottom).

proper motion planning task consists in driving the system (41) from the initial state $x_0 = (0,0,0,0,0,0,0,0,0)$ to the desired $y_d = x_d = (0.1,0,0,0,0,0,0,0,0)$, over the time horizon $T = 1$. Taking into account geometric parameters of the trident snake $r = l = 0.12$, this task could be interpreted as an elementary a rest-to-rest move forward. For both these simulations, the motion planning problem is regarded as solved, when the norm $||^0e(\theta)||$ drops below 10^{-4}, and the determinant $\det(\mathsf{G}_2(\phi)) \neq 0$ during the whole motion time. The initial control functions for both cases are chosen constant, $u_0 = (2,1,-1)$. In the egalitarian algorithm the error decay rate $\gamma = 0.1$, and the scaling factor $\delta = 10^{-4}$. The prioritarian algorithm uses decay coefficients ${}^0\gamma = 0.5$ and ${}^1\gamma = 10^{-3}$. In the parametric representation a truncated Fourier series of length $k = 20$ has been chosen, so the vector of control parameters $\lambda \in \mathbb{R}^{63}$. The dynamics model of the trident snake robot has been borrowed from Pietrowska (2012). The model corresponds to the trident snake robot designed at our laboratory, Gospodarek (2011), displayed in Fig. 2. Its dynamic parameters are the following:

- body mass $m_0 = 0.52$kg,
- wheel mass $m_{\mathrm{w}} = 0.03$kg,
- wheel radius $r_{\mathrm{w}} = 0.02$m,
- wheel thickness $d = 0.001$m,
- moment of inertia of the body around its center of mass $I_0 = \frac{m_0 r^2}{4}$,
- moment of inertia of the wheel around the center of mass of the robot body $I_{0\mathrm{w}} = m_{\mathrm{w}}(\frac{r_{\mathrm{w}}^2}{4} + \frac{d^2}{12})$,
- link mass $m_{\mathrm{l}} = 0.07$kg,
- motor mass $m_{\mathrm{m}} = 0.055$kg,
- total mass $m_c = m_0 + 3(m_{\mathrm{w}} + m_{\mathrm{m}} + m_{\mathrm{l}})$.

The results of computations are depicted in Figs. 3–11. SI units of measure have been employed. The motion paths corresponding to the solution $u_d(t)$ of the motion planning problem provided by the algorithms are presented in Fig. 3, that also displays the robot body every 0.2 s. As it can be seen, these paths are different for the two algorithms. Figures 4–6 show the state space trajectories. One can observe that the proper motion planning task has been solved correctly. The robot position and orientation (x,y,θ) as well as the joint variables ϕ have reached the desired values at T. Also, all velocities v have become equal to zero. Figure 7 presents the control $u(t)$ in the linearized model. Comparing the control

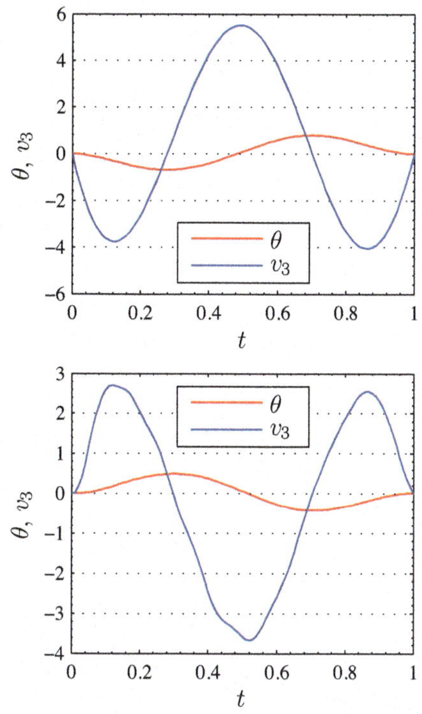

Figure 5. Orientation and angular velocity: egalitarian solution (top) and prioritarian solution (bottom).

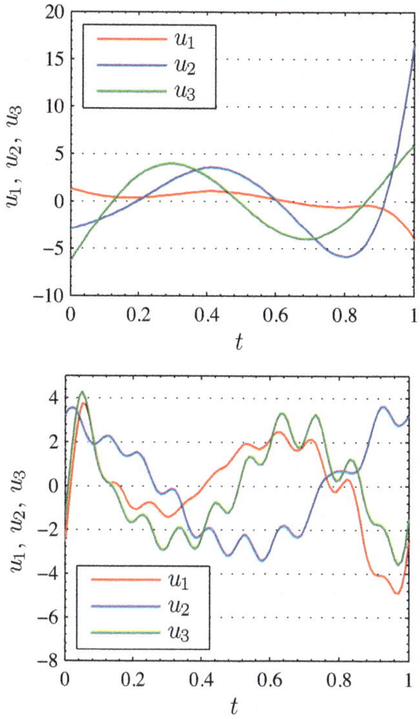

Figure 7. Linearized control $u(t)$, egalitarian solution (top) and prioritarian solution (bottom).

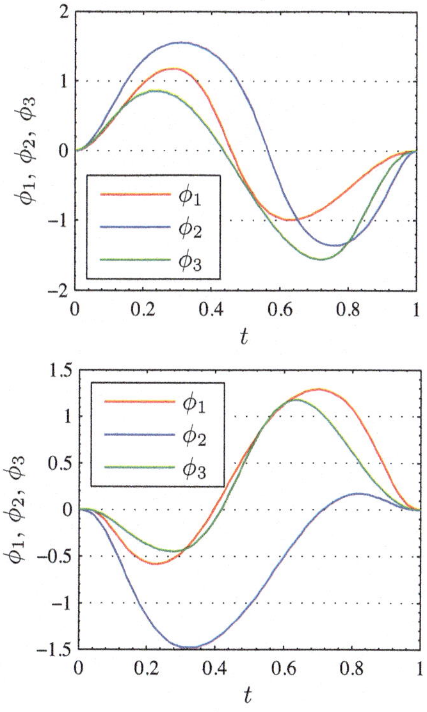

Figure 6. Joint angles: egalitarian solution (top) and prioritarian solution (bottom).

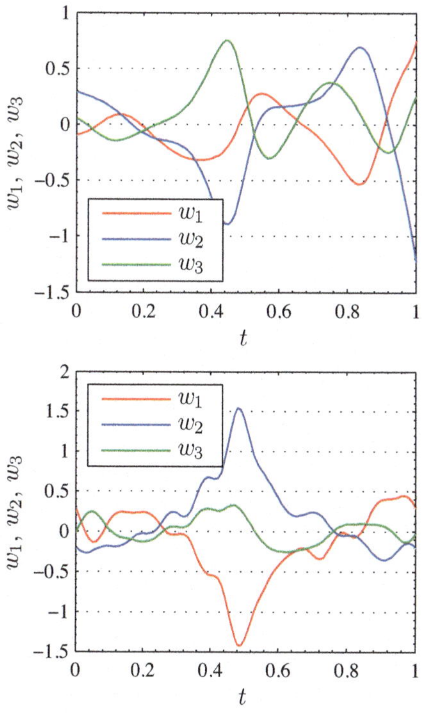

Figure 8. Original control $w(t)$: egalitarian solution (top) and prioritarian solution (bottom).

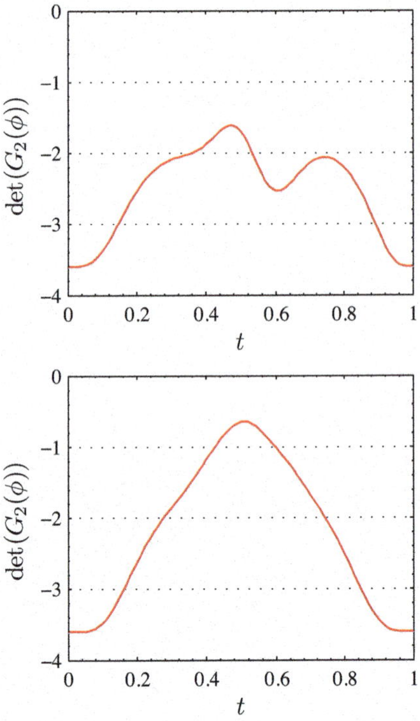

Figure 9. Determinant $\det(G_2(\phi))$: egalitarian solution (top) and prioritarian solution (bottom).

functions obtained from the non-parametric egalitarian and the parametric prioritarian approaches, one can see that the non-parametric controls are smoother. The original control $w(t)$ representing the torques in the joints is depicted in Fig. 8. Again, as in the case of $u(t)$ control, the function $w(t)$ produced by the non-parametric algorithm is smoother. Figure 9 refers to the singularity avoidance subtask. The plots present the value of the determinant $\det(\mathsf{G}_2(\phi))$ during the motion time. In both algorithms the determinant stays quite far away from zero, what means that the egalitarian and the prioritarian algorithms have solved the second task in a satisfactory way. This conclusion is confirmed by Fig. 10 showing three-dimensional plots in the joint space ϕ. The surface $\det(\mathsf{G}_2(\phi)) = 0$ represents singularities. Joint trajectories $\phi(t)$ (starting with the black color and going towards the light green) remain safely inside the regular set. Finally, the last Fig. 11 illustrates the convergence of the algorithms. It follows that the non-parametric egalitarian algorithm needs more steps in order to fulfill the stop condition $||^0e(\theta)|| < 10^{-4}$. In the egalitarian algorithm the total error $\mathbf{e}(\theta)$ decreases exponentially. In the case of the prioritarian algorithm the error of the second task $|^1e(\theta)|$ has been forced to increase by the higher priority task.

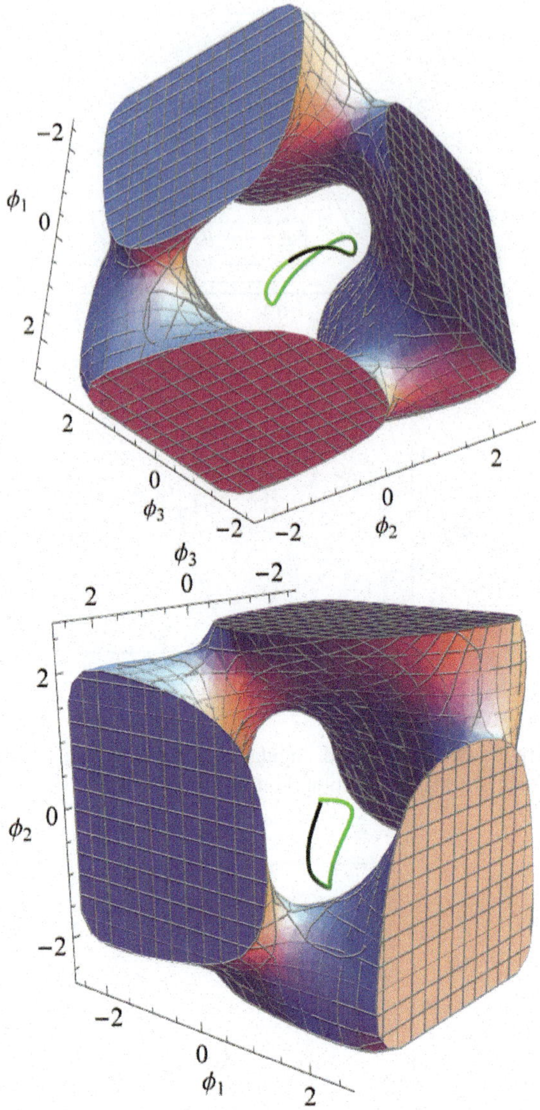

Figure 10. Motion in joint space ϕ: egalitarian solution (top) and prioritarian solution (bottom).

6 Conclusions

We have designed two multiple-task motion planning algorithms for non-holonomic systems with dynamics. Conceptually, their derivation is rooted in the endogenous configuration space approach, so the solution of the problem is a control function in the control system representation of the dynamics. The egalitarian algorithm treats all the component tasks as equivalent. The prioritarian algorithm arranges the tasks according to decreasing priorities. Computations of the control function resulting from these algorithms can either use a finite-dimensional expansion of control functions at a prescribed base in the endogenous configuration space or be base-independent. This gives rise to either the parametric or the non-parametric computational technique. The

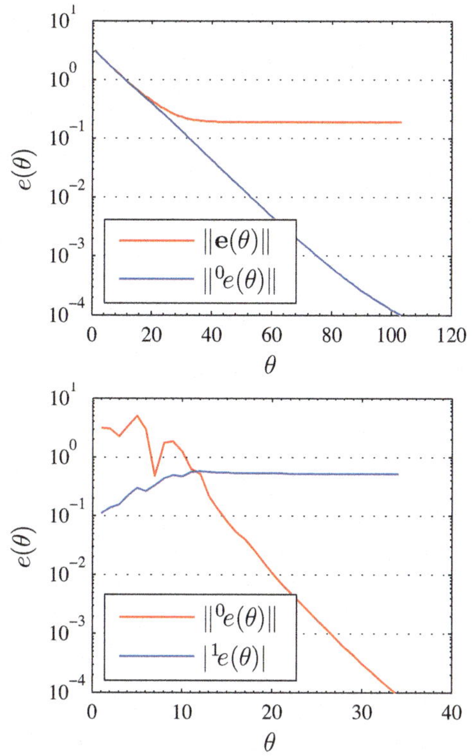

Figure 11. Convergence: egalitarian solution (top) and prioritarian solution (bottom)top.

algorithms have been applied to solve a multiple-task motion planning problem for the trident snake robot, that consists of the proper motion planning task and the singularity avoidance task. Numerical computations have shown that both the algorithms provide correct, although different, solutions to the problem. Also, the algorithms behave differently. In the case of egalitarian algorithm, when all the tasks are solved simultaneously, there are only two possibilities: either both tasks are solved correctly or none of the tasks is solved whatsoever. Contrary to that, in the prioritarian algorithm there is a risk that only the highest priority tasks will be solved. This observation has been confirmed by the plots of error convergence. In the egalitarian algorithm the total error decreases. It follows from the derivation of the prioritarian algorithm that the error of a lower priority task can even increase to enable the decrease of a higher priority task error. The parametric computations of solution of the motion planning problem are base-dependent, what usually appears to be quite restrictive. The non-parametric representation of control functions does not depend on the base choice, and is free from the limitations of the parametric approach. The computational effort in the parametric approach depends on the dimension of the parameter space. If the number of parameters is small, the parametric computations are much more efficient than the non-parametric. This, however, reverts, when the number of parameters grows up. In the example presented in this paper (63 control parameters), the parametric computations have been about 1.5 times more time consuming than the non-parametric. This being so, the choice of the motion planning strategy should depend on the complexity of the non-holonomic system subject to motion planning as well as on the type of the tasks defining the problem.

Appendix A

In this section we shall make a derivation of the Jacobian (23) and its Moore–Penrose inverse (24), and present the dynamics model of the trident snake robot used in Sect. 4.

A1 Subtask Jacobian and its inverse

The i-th subtask Jacobian is equal to

$$^iJ_{x_0,T}(u(\cdot))v(\cdot) = D\,^iK_{x_0,T}(u(\cdot))v(\cdot)$$

$$= \frac{\mathrm{d}}{\mathrm{d}\alpha}\bigg|_{\alpha=0}\int_0^T F_i(\varphi_{x_0,t}(u(\cdot)+\alpha v(\cdot)), u(t)+\alpha v(t))\mathrm{d}t,$$

where $x(t) = \varphi_{x_0,t}(u(\cdot))$ denotes the state trajectory of (1) driven by the control function $u(\cdot)$. The differentiation gives

$$^iJ_{x_0,T}(u(\cdot))v(\cdot)$$

$$= \int_0^T \left(\frac{\partial F_i(x(t),u(t))}{\partial x} D\varphi_{x_0,t}(u(\cdot))v(\cdot) + \frac{\partial F_i(x(t),u(t))}{\partial u} v(t)\right)\mathrm{d}t.$$

Finally, a substitution for $\xi(t) = D\varphi_{x_0,t}(u(\cdot))v(\cdot)$ from (5) yields (23).

In order to find a formula for the Moore–Penrose Jacobian inverse, we begin with a Jacobian equation

$$^iJ_{x_0,T}(u(\cdot))v(\cdot) = \eta,$$

where $\eta \in \mathbb{R}$. A solution of this equation will be sought by minimizing the squared norm of the control function,

$$\min_{v(\cdot)}\left(\|v(\cdot)\|_R^2 = \int_0^T v^T(t)Rv(t)\mathrm{d}t\right),$$

with the equality constraint. After the substitution

$$\xi(t) = \int_0^t \Phi(t,s)B(s)v(s)\mathrm{d}s,$$

the corresponding Lagrange function becomes

$$L(v(\cdot),\lambda) = \int_0^T v^T(t)Rv(t)\mathrm{d}t + \lambda\int_0^T\left(\frac{\partial F_i(x(t),u(t))}{\partial x}\right.$$

$$\left.\int_0^t \Phi(t,s)B(s)v(s)\mathrm{d}s + \frac{\partial F_i(x(t),u(t))}{\partial u}v(t)\mathrm{d}t\right),$$

$\lambda \in \mathbb{R}$. Now, the differentiation of the Lagrange function with respect to $v(\cdot)$

$$DL(v(\cdot),\lambda)w(\cdot) = \left.\frac{\mathrm{d}}{\mathrm{d}\alpha}\right|_{\alpha=0} L(v(\cdot)+\alpha w(\cdot),\lambda),$$

use of the identity

$$\int_0^T \int_0^t f(t,s)\mathrm{d}s\mathrm{d}t = \int_0^T \int_s^T f(t,s)\mathrm{d}t\mathrm{d}s,$$

and then equating the derivative to 0 yield

$$v_i(t) = -\frac{1}{2}\lambda R^{-1}(t)\left(B^T(t) \int_t^T \Phi^T(s,t)\left(\frac{\partial F_i(x(s),u(s))}{\partial x}\right)^T \mathrm{d}s + \left(\frac{\partial F_i(x(t),u(t))}{\partial u}\right)^T \right),$$

where the subscript i refers to the i-th subtask. To simplify the notations, let us define a pair of functions

$$b_i(t) = R^{-1}(t)B^T(t) \int_t^T \Phi^T(s,t)\left(\frac{\partial F_i(x(s),u(s))}{\partial x}\right)^T \mathrm{d}s$$

and

$$c_i(t) = R^{-1}(t)\left(\frac{\partial F_i(x(t),u(t))}{\partial u}\right)^T,$$

so that

$$v_i(t) = -\frac{1}{2}\lambda(b_i(t)+c_i(t)).$$

The Lagrange multiplier λ can be eliminated by inserting the control $v_i(t)$ into the Jacobian equation, that implies

$$-\frac{1}{2}\lambda \int_0^T \left(b_i^T(t)R(t) + c_i^T(t)R(t)\right)(b_i(t)+c_i(t))\mathrm{d}t = -\frac{1}{2}\lambda \|b_i(\cdot)+c_i(\cdot)\|_R^2 = \eta.$$

From the last identity we compute λ, and conclude that

$$v_i(t) = \frac{b_i(t)+c_i(t)}{\|b_i(\cdot)+c_i(\cdot)\|_R^2}\eta,$$

what is just (24).

A2 Dynamics model of trident snake

A standard derivation based on the Lagrangian mechanics and d'Alembert principle leads to the following definition of terms appearing in the equations of motion (38) of the trident snake robot

$$\mathsf{P}(q) = \mathsf{M}^{-1}(q)\mathsf{G}^T(q)\mathsf{B}(q),$$
$$\mathsf{M}(q) = \mathsf{G}^T(q)\mathsf{Q}(q)\mathsf{G}(q),$$
$$\mathsf{N}(q,v) = -\mathsf{M}^{-1}(q)\mathsf{G}^T(q)\left(\mathsf{Q}(q)\dot{\mathsf{G}}(q) + \mathsf{C}(q,\mathsf{G}(q)v)\mathsf{G}(q)\right)v,$$

where $\mathsf{G}(q)$ is given by (37),

$$\mathsf{B}(q) = \begin{bmatrix} 0_{3\times3} \\ I_3 \end{bmatrix}$$

denotes a control matrix, $\mathsf{Q}(q)$ is the inertia matrix defined below, and $\mathsf{C}(q,\dot{q})$ is the matrix of Coriolis and centripetal forces whose entries

$$\mathsf{C}_{kj}(q,\dot{q}) = \sum_{i=1}^{6} c_{ij}^k(q)\dot{q}_i$$

are determined by the Christoffel's symbols of the first kind associated with $\mathsf{Q}(q)$

$$c_{ij}^k(q) = \frac{1}{2}\left(\frac{\partial \mathsf{Q}_{ik}(q)}{\partial q_j} + \frac{\partial \mathsf{Q}_{jk}(q)}{\partial q_i} - \frac{\partial \mathsf{Q}_{ij}(q)}{\partial q_k}\right), \quad i,j,k = 1,\ldots,6.$$

The following form of the inertia matrix for the trident snake robot can be found in Pietrowska (2012)

$$\mathsf{Q}(q) = \begin{bmatrix} m_{11} & 0 & m_{13} & m_{14} & m_{15} & m_{16} \\ 0 & m_{22} & m_{23} & m_{24} & m_{25} & m_{26} \\ m_{13} & m_{23} & m_{33} & m_{34} & m_{35} & m_{36} \\ m_{14} & m_{24} & m_{34} & m_{44} & 0 & 0 \\ m_{15} & m_{25} & m_{35} & 0 & m_{55} & 0 \\ m_{16} & m_{26} & m_{36} & 0 & 0 & m_{66} \end{bmatrix},$$

where

$$m_{11} = m_{22} = m_c,$$
$$m_{33} = I_0 + 3I_{0w} + 3m_w(r^2 + l^2) + 2m_w rl\cos\phi_1$$
$$+2m_w rl\cos\phi_2 + 2m_w rl\cos\phi_3 + m_l(l^2 + 3r^2 + lr\cos\phi_1$$
$$+lr\cos\phi_2 + lr\cos\phi_3) + 6m_m r^2,$$
$$m_{44} = m_{55} = m_{66} = I_{0w} + m_w l^2 + \frac{1}{3}m_l l^2,$$
$$m_{13} = -m_w l\sin(\alpha_1 + \phi_1 + \theta) - m_w r\sin(\alpha_1 + \theta)$$
$$-m_w l\sin(\alpha_2 + \phi_2 + \theta) - m_w r\sin(\alpha_2 + \theta)$$
$$-m_w l\sin(\alpha_3 + \phi_3 + \theta) - m_w r\sin(\alpha_3 + \theta)$$
$$-\frac{1}{2}m_l(2r\sin(\alpha_1 + \theta) + l\sin(\alpha_1 + \phi_1 + \theta) + 2r\sin(\alpha_2 + \theta)$$
$$+l\sin(\alpha_2 + \phi_2 + \theta) + 2r\sin(\alpha_3 + \theta) + l\sin(\alpha_3 + \phi_3 + \theta))$$
$$-m_m r\sin(\theta + \alpha_1) - m_m r\sin(\theta + \alpha_2) - m_m r\sin(\theta + \alpha_3),$$
$$m_{14} = -m_w l\sin(\alpha_1 + \phi_1 + \theta) - \frac{1}{2}m_l l\sin(\alpha_1 + \phi_1 + \theta),$$
$$m_{15} = -m_w l\sin(\alpha_2 + \phi_2 + \theta) - \frac{1}{2}m_l l\sin(\alpha_2 + \phi_2 + \theta),$$
$$m_{16} = -m_w l\sin(\alpha_3 + \phi_3 + \theta) - \frac{1}{2}m_l l\sin(\alpha_3 + \phi_3 + \theta),$$
$$m_{23} = m_w l\cos(\alpha_1 + \phi_1 + \theta) + m_w r\cos(\alpha_1 + \theta)$$
$$+m_w l\cos(\alpha_2 + \phi_2 + \theta) + m_w r\cos(\alpha_2 + \theta)$$
$$+m_w l\cos(\alpha_3 + \phi_3 + \theta) + m_w r\cos(\alpha_3 + \theta)$$
$$+\frac{1}{2}m_l(2r\cos(\alpha_1 + \theta) + l\cos(\alpha_1 + \phi_1 + \theta) + 2r\cos(\alpha_2 + \theta)$$
$$+l\cos(\alpha_2 + \phi_2 + \theta) + 2r\cos(\alpha_3 + \theta) + l\cos(\alpha_3 + \phi_3 + \theta))$$
$$+m_m r\cos(\theta + \alpha_1) + m_m r\cos(\theta + \alpha_2) + m_m r\cos(\theta + \alpha_3),$$
$$m_{24} = m_w l\cos(\alpha_1 + \phi_1 + \theta) + \frac{1}{2}m_l l\cos(\alpha_1 + \phi_1 + \theta),$$
$$m_{25} = m_w l\cos(\alpha_2 + \phi_2 + \theta) + \frac{1}{2}m_l l\cos(\alpha_2 + \phi_2 + \theta),$$
$$m_{26} = m_w l\cos(\alpha_3 + \phi_3 + \theta) + \frac{1}{2}m_l l\cos(\alpha_3 + \phi_3 + \theta),$$
$$m_{34} = I_{0w} + m_w l^2 + m_w rl\cos\phi_1 + \frac{1}{6}m_l l(2l + 3r\cos\phi_1),$$
$$m_{35} = I_{0w} + m_w l^2 + m_w rl\cos\phi_2 + \frac{1}{6}m_l l(2l + 3r\cos\phi_2),$$
$$m_{36} = I_{0w} + m_w l^2 + m_w rl\cos\phi_3 + \frac{1}{6}m_l l(2l + 3r\cos\phi_3).$$

Acknowledgements. This research was supported by the Wrocław University of Technology under a statutory grant.

Edited by: A. Müller

References

Antonelli, G.: Stablity analysis for prioritized closed-loop inverse kinematic algorithms for redundant robotic systems, IEEE Trans. Robotics, 25, 985–994, 2009.

Chiacchio, P., Chiaverini, S., Sciavicco, L., and Siciliano, B.: Closed loop inverse kinematics schemes for constrained redundant manipulators with the task space augmentation and task priority strategy, Int. J. Robotics Res., 10, 410–425, 1991.

Chiaverini, S.: Singularity-robust task-priority redundancy resolution for real time kinematic control of robot manipulators, IEEE Trans. Robotics Autom., 13, 398–410, 1997.

Chitour, Y. and Sussmann, H. J.: Motion planning using the continuation method, in: Essays on Mathematical Robotics, edited by: Baillieul, J., Sastry, S. S., and Sussmann, H. J., Springer-Verlag, New York, 91–125, 1998.

Choi, Y., Oh, Y., Oh, S., Park, J., and Chung, W. K.: Multiple task manipulation for a robotic manipulator, Adv. Robotics, 18, 637–653, 2004.

Divelbiss, A., Seereeram, S., and Wen, J. T.: Kinematic path planning for robots with holonomic and nonholonomic constraints, in: Essays on Mathematical Robotics, edited by: Baillieul, J., Sastry, S. S., and Sussmann, H. J., Springer-Verlag, New York, 127–150, 1998.

Gospodarek, S.: Design and modelling of the trident snake type mobile robot, Master's thesis, Wrocław University of Technology, 2011.

Ishikawa, M.: Trident snake robot: Locomotion analysis and control, in: Proc 6th IFAC Symp. NOLCOS, Stuttgart, Germany, 1169–1174, 2004.

Ishikawa, M., Minati, Y., and Sugie, T.: Development and control experiment of the trident snake robot, IEEE/ASME Trans. Mechatronics, 15, 9–16, 2010.

Lewis, A.: When is a mechanical control system kinematic?, in: Proc. 38th IEEE CDC, Phenix, Arizona, 1162–1167, 1999.

Maciejewski, A. A. and Klein, C. A.: Obstacle avoidance for kinematically redundant manipulators in dynamically varying environments, Int. J. Robotics Res., 4, 109–117, 1985.

Nakamura, Y., Hanafusa, H., and Yoshikawa, T.: Task-priority based redundancy control of robot manipulators, Int. J. Robotics Res., 6, 3–15, 1987.

Paszuk, D., Tchoń, K., and Pietrowska, Z.: Motion planning of the trident snake robot equipped with passive or active wheels, in: Bull. Pol. Acad. Sci., Tech. Sci., 60, 547–554, 2012.

Pietrowska, Z.: Kinematics, dynamics, and control of a trident snake nonholonomic system, Master's thesis, Wrocław University of Technology, 2012.

Ratajczak, A.: Motion planning of underactuated robotic systems, Ph.D. thesis, Wrocław University of Technology, 2012.

Ratajczak, A., Karpińska, J., and Tchoń, K.: Task-priority motion planning of underactuated systems: An endogenous configuration space approach, Robotica, 28, 885–892, 2010.

Sontag, E. D.: Mathematical Control Theory, Springer-Verlag, New York, 1990.

Sussmann, H. J.: A continuation method for non-holonomic path finding problems, in: Proc. 32nd IEEE CDC, San Antonio, Texas, 2718–2723, 1993.

Tchoń, K. and Jakubiak, J.: Endogenous configuration space approach to mobile manipulators: A derivation and performance assessment of Jacobian inverse kinematics algorithms, Int. J. Control, 76, 1387–1419, 2003.

Tchoń, K. and Zadarnowska, K.: Kinematic dexterity of mobile manipulators: An endogenous configuration space approach, Robotica, 21, 521–530, 2003.

Zadarnowska, K. and Tchoń, K.: A control theory framework for performance evaluation of mobile manipulators, Robotica, 25, 703–715, 2007.

9

A novel self-aligning mechanism to decouple force and torques for a planar exoskeleton joint

J. F. Schorsch[1], A. Q. L. Keemink[2], A. H. A. Stienen[2], F. C. T. van der Helm[1], and D. A. Abbink[1]

[1]Department of BioMechanical Engineering, Technical University Delft, Delft, the Netherlands
[2]Laboratory of Biomechanical Engineering, University of Twente, Enschede, the Netherlands

Correspondence to: J. F. Schorsch (j.f.schorsch@tudelft.nl)

Abstract. The design of exoskeletons is a popular and promising area of research both for restoring lost function and rehabilitation, and for augmentation in military and industrial applications. A major practical challenge to the comfort and usability for exoskeletons is the need to avoid misalignment of the exoskeletal joint with the underlying human joint. Alignment mismatches are difficult to prevent due to large inter-user variability, and can create large stresses on the attachment system and underlying human anatomy. Previous self-aligning systems have been proposed in literature, which can compensate for muscle forces, but leave large residual forces passed directly to the skeletal system. In this paper we propose a new mechanism to reduce misalignment complications. A decoupling approach is proposed which allows large forces to be carried by the exoskeletal system while allowing both the muscle and skeletal joint force presented to the user to be compensated to any desired degree.

1 Introduction

Exoskeletons have been proposed to augment strength (Makinson, 1971; Zoss et al., 2006), enhance endurance (Lockheed Martin, 2012; Raytheon, 2012), and restore lost abilities (Berkeley Bionics, 2012; Jezernik et al., 2003; Stienen et al., 2007). These applications span military, (Zoss et al., 2006; Lockheed Martin, 2012; Raytheon, 2012), industrial (Makinson, 1971), and medical (Argo, 2012; Berkeley Bionics, 2012; Jezernik et al., 2003; Stienen et al., 2007) fields – each with their own challenges; but there exists a common subset of design challenges inherent to all exoskeleton systems. One of these challenges is how to fit the exoskeleton system to the operator. This problem of fit is of particular difficulty for anthropomorphic exoskeletons (Schiele and van der Helm, 2006; Stienen et al., 2009), as they are typically attached to each limb segment of the user. A misalignment between a joint in a rigid exoskeleton and the corresponding biomechanical joint in the human operator produces unexpected and potentially dangerous internal joint forces on the human as well as potential forces on the human-robot physical interface (Schiele and van der Helm, 2006) that couple the operator to the mechanism. Furthermore it has been shown that the attachment pressure has a large effect on comfort, mental load, physical demand, and effort (Schiele, 2009). There have been many different approaches to solving this issue which can be subdivided into three major classes; manual adjustment, compliant mechanisms, and kinematic redundancy. The first class of approach is simply manual linkage/patient adjustment (Evryon, 2012) – this is the most commonly used approach. In this approach, a technician will manually adjust the exoskeleton linkage lengths and restraints in an attempt to align the exoskeleton joint with the joint of the specific user. This approach can give good results, provided the degree of motion through the joint closely approximates a one degree of freedom hinge (Schiele and van der Helm, 2006; Colombo et al., 2000), and the skill of the technician is high. One of the main downsides is that this method can be time consuming, and has to be repeated each time a different user employs the exoskeleton. Additionally, choosing to use manual linkage adjustment requires careful consideration for the type of control and the placement of force and torque sensors (De Rossi et al., 2001); in that admittance can be controlled at the locations of the force sensors, while impedance control requires an accurate knowledge of the robot configuration and from that, an accurate dynamic predictive model (Kooij et al., 2006). The manual adjustment type systems may be considered a kinematically redundant system, in that the points of manual adjustment to represent a kinematic redundancy, but while in operation those degrees of freedom are locked, and thus should be considered separate from the true kinematically redundant systems.

The second approach commonly used, especially in rehabilitation applications (Argo, 2012; Stienen et al., 2007; Colombo et al., 2000) is not to avoid misalignment itself, but to make the restraint and attachment system compliant. This built-in compliance is assumed to be non-detrimental to performance and at least adequate for the comfort of the user. However, the comfort is not optimal due to resulting forces, and more importantly; reduced device stiffness is closely related to reduced force feedback bandwidth (Pratt et al., 1997) which may have implications for the use of high fidelity haptic force feedback, which may be necessary for the learning or rehabilitation process. For example, Wildenbeest et al. (2013) have shown that a reduction in haptic fidelity can impact both motor learning and skill generalization – both critical aspects of exoskeleton usage in an application.

A third approach is to use a kinematically redundant system such as in Lockheed Martin (2012), Argo (2012), Schiele and van der Helm (2006) and Colombo et al. (2000). In such a system degrees of freedom exist in the exoskeleton that are not present in the human operator. This allows the exoskeleton, when coupled in parallel to the human user, to be properly constrained. The kinematically redundant systems can be further subdivided into self-aligning systems and alignment free systems. A self-aligning mechanism, such used by Schiele (2009) provides excellent fitting ability; however, passive linear-motion joints do not allow for large forces to be transferred through the exoskeleton, and can introduce non-uniform inertial properties throughout the usable workspace. Conversely, in alignment free systems an additional actuator and sensors capable of controlling the DOF are required, with the associated mass, power, and control penalties inherent to such a solution. Further, such a system typically does not share a per-segment kinematic relationship with the limb segments of the operator, making multi-point contact with the operator contraindicated.

This creates a design constraint, where redundant DOF systems that need to provide high levels of force through the exoskeleton must make use of powerful actuators at each and every exoskeleton joint, and can require significant engineering design to maintain stability across the entire range of movements and joint geometries.

We propose a self-aligning joint structure that may allow for the high force transfer potential of completely powered systems with the lower actuator count and simpler control design of passive DOF systems.

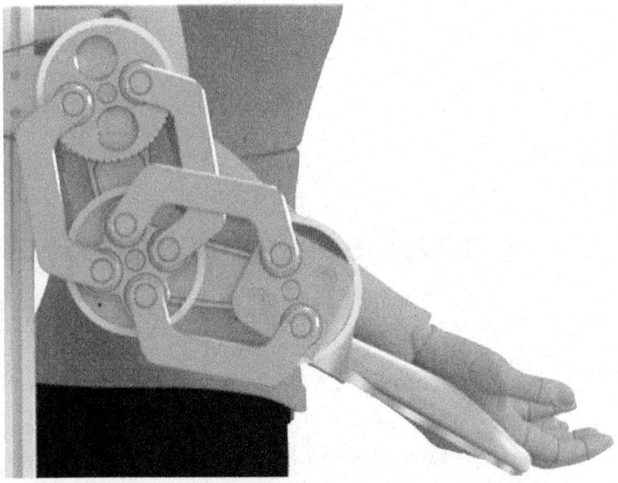

Figure 1. CAD concept of gravity compensation and torque assistance mechanism used as an elbow support exoskeleton.

2 Design and analysis

2.1 Design

In this paper, we discuss a type of planar mechanical linkage, for use in exoskeleton devices (Fig. 1), which allows both forces and torques to be passed across a virtual center of rotation, which is not known a priori. This center of rotation can exist arbitrarily at a point described within the working area of the linkage. We will show that by providing independent torque and force compensation components it is possible to both provide augmented force output to the desired load and to maintain physiologically appropriate loading of the underlying joint. In contrast to a purely passive alignment scheme, this mechanism allows for transfer of large forces and high working loads due to the novel configuration of the mechanism.

An additional advantage of the mechanism described in this paper is that forces and torques are completely decoupled from the local mechanical configuration of the mechanism. This eliminates the need for position and force sensing *within* the mechanism, and furthermore, does not require knowledge of the location of operator's joint, relative to the mechanism, as long as the operator's joint lies within the working range of the mechanism.

Forces and torques applied to the end effector of the linkage can also be partially decoupled. Forces which are applied through this mechanism create an apparent moment about the supported physiological joint, which need to be taken into account, while moments applied do not create additional force requirements. The degree to which the applied force is decoupled from applied torque is dependant entirely on the ratio of the amount of misalignment accommodated by the mechanism to the length of the proximal limb supported. This torque and force decoupling characteristic allows the device to be tailored for specific loading conditions,

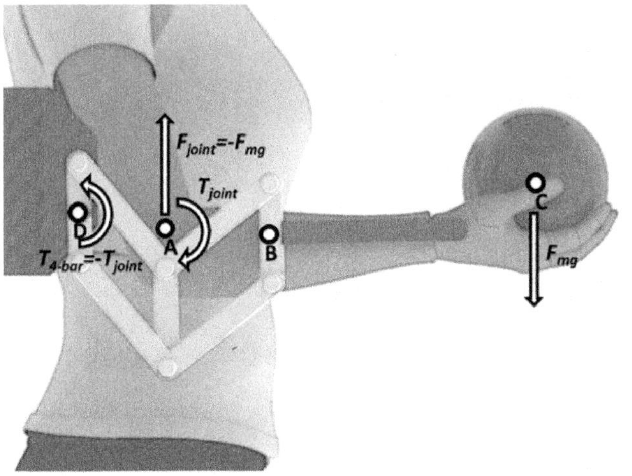

Figure 2. Example of serial 4-bar mechanism used for an elbow support exoskeleton. The user's elbow join is located at A, a load is applied at point C, and point B is located arbitrarily along the line AC. The load at C creates a torque, T_{joint} about point A. The 4-bar mechanism is anchored at point D. F_{joint} is the resultant force needed to support the load.

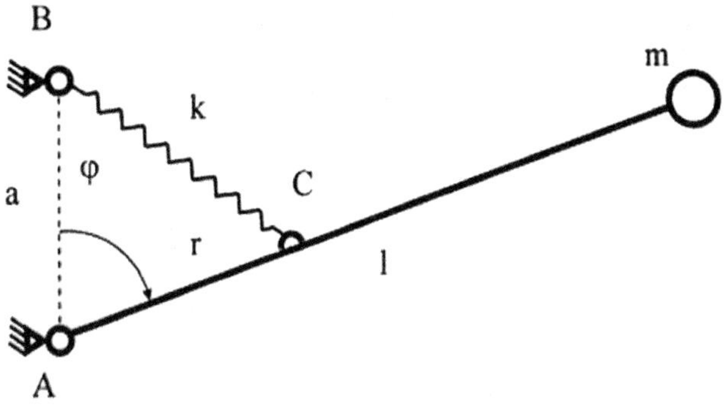

Figure 3. akr type spring gravity compensation mechanism, as introduced in Tuijthof and Herder (2000), reprinted with permission. Mass m, supported by bar l, acted on by gravity, g, can be exactly balanced at any point in its rotation about point A by a zero free length spring with stiffness k attached to ground reference at a vertical distance a, and at point r along bar l. For practical implementations spring k either must act through some type of cable structure, to allow the approximation of zero free length, or some other mechanical construct must be used to allow for zero free length approximation. The design of such structure can be non-trivial, especially in constrained workspace.

such as pure weight compensation, combined weight and inertia compensation, or targeted joint loading scenarios, independent of the basic mechanical design. This allows the same mechanical design to be generally applied for strength enhancement, endurance improvement, or for rehabilitative purposes.

2.2 Predecessors

The mechanism as discussed in this paper can be considered a novel combination of two existing, well understood, types of mechanisms:

1. a double four-bar parallelogram mechanism, such as used in the LIMPACT rehabilitation system (Stienen et al., 2009),
2. an energy-free gravity compensation mechanism as described by Tuijthof and Herder (2000).

2.2.1 Predicate 1

This type of mechanism can transfer torques from the ground or actuator through the linkage to a proximal end effector. If not attached to a user, the linkage is under-constrained. When the mechanism end effector is attached to a user through a restraint, the assembly becomes properly constrained.

For evaluating the loading of the human elbow during static equilibrium, the net forces and moments on the system must both sum to zero, and it can be seen that in the case (Eq. 1a) of the serial arrangement of a pair of four-bar parallel mechanisms, (see Fig. 2) the mechanism is capable of reducing the torque created by the muscles about the elbow to zero. There is a zero sum relationship between the torque borne by the mechanism and the torque at the human joint, such that the torques created by the applied load can be distributed as desired to the human joint by increasing or reducing the amount of torque provided by the exoskeleton.

$$T_{4-\text{bar}} = -m_{\text{object}} \times g \times l_{\overline{\text{AC}}} \tag{1a}$$

$$F_{\text{joint}} = -m_{\text{object}} \times g \tag{1b}$$

By examining the balancing of forces (Eq. 1b) it is shown that a double four-bar mechanism which reduces the residual moment about the elbow to zero will leave a residual force on the elbow that is antiparallel and proportional to the weight of the load. This is contrary to the expected loading pattern experienced during unaugmented lifting, and could result in injury or unexpected control activity by the human operator.

During unaugmented lifting, the force created on the humeroulnar joint is compressive and of significantly larger magnitude than the weight of the object being lifted (Stormont et al., 1985). When changing both the direction and magnitude of the force experienced at the elbow there are two potentially important physiological effects. The first effect is the potential to increase the rates of musculoskeletal injuries. Kumar clearly makes the link that musculoskeletal injuries occur at a higher rate when the natural mechanical order of a task is disrupted (Kumar, 2001). It is unknown to the extent that disrupting the expected loading pattern by use of a double four-bar mechanism would result in increased injuries. The second effect experienced is in creating a scaled force in the opposite direct to what is expected during normal movement, and will require a new muscle activation pattern

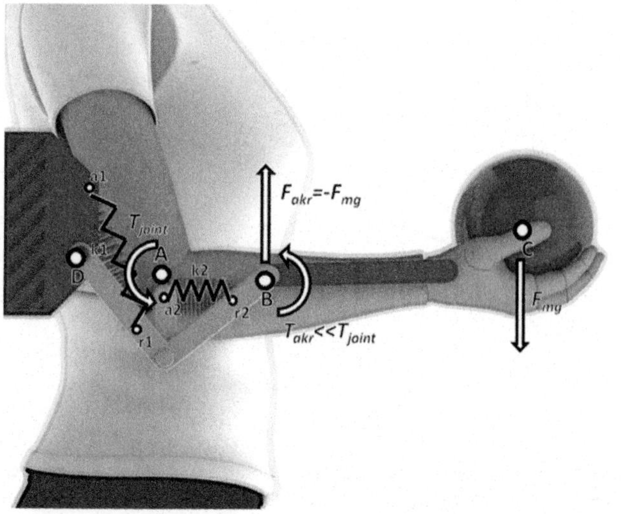

Figure 4. Example of serial akr mechanism used for an elbow support exoskeleton. The user's elbow join is located at A, a load is applied at point C, and point B is located arbitrarily along the line AC.

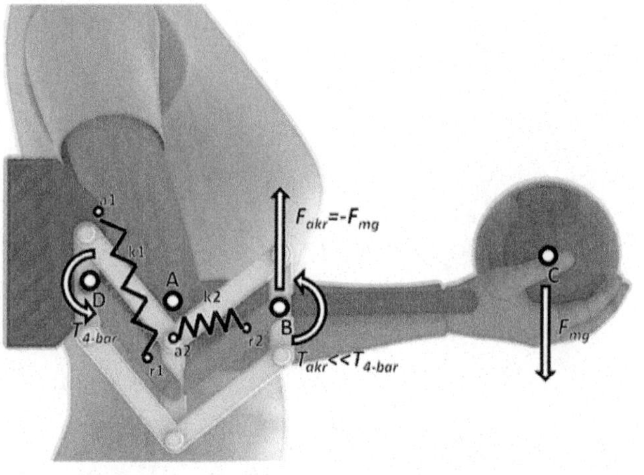

Figure 5. Example of combined mechanism used as an elbow support exoskeleton. The user's elbow join is located at A, a load is applied at point C, and point B is located arbitrarily along the line AC. All forces supplied by the mechanism are explicit. Compensation case illustrated is for 100 % compensation. Residual forces left to the operator at point A can be controlled to appropriate level.

in musculoskeletal joints more proximally located along the kinematic chain, in order to produce a specific desired movement.

In addition to the physiological complications, there are mechanical and control considerations to take into account. The moment supplied by the actuator through the serial four-bar mechanism is purely dependant on the location and magnitude of the load, relative to the elbow joint of the user, not the current mechanical configuration of the parallel mechanism. This makes it difficult to estimate the actual mass of the load based on torques or forces perceived at the mechanism, and requires direct sensing of the load, or precise knowledge of the user's anatomy, which may be difficult to obtain.

2.2.2 Predicate 2

The second predecessor mechanism is the so-called "akr" type of energy free gravity compensation mechanism as described by Tuijthof and Herder (2000) (Fig. 3). This type of mechanism relies on ideal linear spring behaviour to precisely compensate a fixed mass at a given distance. As the mass m translates along the arc described by l, it stretches the spring k attached at point r in such a way to allow the system to remain energy neutral; e.g., the system provides a force perpendicular to l and through m with the component aligned parallel to AB remaining constant. It is possible to connect multiple single-link akr type mechanisms in series (see Fig. 4) to give an improved range of motion, and eliminate any force vector components that are not parallel to AB. In this type of serial connection, the endpoint of the proximal akr mechanism acts as the reference grounding point for the more distal akr mechanism. An important consideration in the design of serial akr mechanisms is the cancelation of force components perpendicular to AB, allowing for a zero net energy cost for displacements in the entire working plane. Potential energy lost or gained via vertical displacement is stored or recovered simultaneously in the springs for both akr mechanisms, while the energy is only transferred between springs when the system is displaced horizontally. The key feature which allows for the novelty of the combined mechanism is that the theoretical working area for two akr type compensation mechanisms working in series is the same as for a serial pair of four-bar parallelogram mechanisms. The static behaviour of a realization for such system with two serial connected akr mechanisms is described by Eq. (2a–b)

$$T_{\text{joint}} = -m_{\text{object}} \times g \times \left(l_{\overline{AC}} - l_{\overline{AB}}\right) \tag{2a}$$

$$F_{\text{joint}} \equiv 0 \tag{2b}$$

This mechanism is capable of creating a consistently aligned force vector through the supported limb segment regardless of the attachment point of the mechanism, and also provides a degree of moment reduction, depending on the relative positions of the attachment point, relative to the physiological joint it is supporting (Eq. 2b).

2.3 Implementation

Each mechanism discussed previously is capable of providing one component of the solution for equilibrium; e.g. forces and moments respectively for the serial akr mechanisms and double four-bar mechanisms. The equations for equilibrium could be satisfied while controlling both the residual joint and muscle loads, if the endpoints of both mechanisms could be coupled through a joint that allowed rotational freedom, but no translational freedom. A simple rotational bearing provides exactly such a connection. This creates a parallel arrangement of the double four-bar mechanism and the serial akr compensation mechanism, (Fig. 5). Equations 3a–l show the solution for the static equilibrium of such a system.

$$\sum F_A = 0 \tag{3a}$$

$$F_{joint} + F_{4-bar} + F_{akr} + W_{object} = 0 \tag{3b}$$

$$W_{object} = m_{object} \times g \tag{3c}$$

$$F_{4-bar} \equiv 0 \tag{3d}$$

$$F_{akr} \equiv -m_{object} \times g \tag{3e}$$

$$F_{joint} = 0 \tag{3f}$$

$$\sum T_A = 0 \tag{3g}$$

$$T_{joint} + T_{4-bar} + T_{akr} + T_{object} = 0 \tag{3h}$$

$$T_{object} = m_{object} \times g \times l_{\overline{AC}} \tag{3i}$$

$$T_{akr} = m_{object} \times g \times l_{\overline{AB}} \tag{3j}$$

$$T_{joint} \equiv 0 \tag{3k}$$

$$F_{4-bar} = m_{object} \times g \times \left(l_{\overline{AC}} - l_{\overline{AB}}\right) \tag{3l}$$

3 Discussion

The mechanism described in this paper has many potential applications for use in exoskeletal systems, such as transferring large forces to the ground in load bearing applications, providing reduced loading of damaged joints for rehabilitation orthotics, or for providing physiologically appropriate increased loading of bones to counter disuse atrophy. It provides a mechanism for use as a prismatic self-aligning exoskeleton joint within a described planar area. Decoupling the application of force and moments from each other creates a method to design a planar self-aligning exoskeleton joint in which both the forces in the muscles and at the joint of the user can be compensated or controlled independently. It is important to note that in the derivation of the equations of static equilibrium, the complete reduction of forces and moments at the joint was chosen for illustrative purposes. Any combination of residual joint force and torques can be designed for, depending on a particular application or experimental purpose.

There are some practical limitations to implementation of the parallel 4-bar linkage used for torque transmission. If an actual mechanical four-bar mechanism is used, when any pair of links become collinear, a mechanical singularity is reached, which leads to kinematic uncertainty. As this mechanical singularity is approached, internal forces become very high, relative to the amount of torque transmitted from the input to the output. This limits the practical working area to a smaller subset of the theoretical maximum, dependent on the initial configuration of the parallel four-bar mechanisms. By using a pseudo four-bar mechanism, such as a capstan-and-cable or a belted pulley (Morita et al., 2003) anchored through rotational joints to the akr subassemblies, this mechanical singularity can be avoided.

It was noted that both the double four-bar mechanism and the serial akr mechanisms have identical working areas, as long as the effective linkage length is the same. Further, by keeping the individual linkage segments equal for both mechanisms, each point in one mechanism segment will remain in exactly the same position relative to the other. This allows further coupling of the endpoints of parallel linkage segments through rotational bearings. This coupling of the intermediate endpoints is not strictly necessary from a kinematic standpoint, however, it would allow the use of each 4-bar parallelogram as member l of the akr compensation mechanism or vice versa, allowing for a reduction in the number of parts, as well as reducing the volume of the mechanism for a given compensation weight.

If the parallelogram linkages are used as a component of an akr mechanism that makes use of a pulley system (Morita et al., 2003) to transfer force to the more distal akr unit, care must be taken to maintain proper pretension of the cable or belting solution. Improper design of the relative linkages can result in overstress of the load carrying belt, which can lead to breakage, loss of load carrying ability, and premature wear.

If the 4-bar parallelograms are not used as members in the akr mechanism, and the two systems are attached together directly at the intermediate pivot point through rotational bearings, an over-constrained system is created. This can result in very high internal forces and large asymmetric loads on the coupling bearings. The magnitudes of the internal forces are purely dependent on the design of the connecting linkages and the relative precision of the rotational center distances. It is recommended that care be taken in the design of the akr mechanism to reduce the internal forces if intermediate coupling is attempted.

This mechanism was realized in hardware, according to the design shown in Fig. 6. This realization makes use of an intermediate endpoint coupling arrangement for the four-bar parallelograms and akr mechanisms. This implementation was chosen to allow use of a synchronous belt transmission to transmit force to the second of the akr linkages while still maintaining proper pretension on the synchronous belt. A compliant bushing is used around the bearing connection between the akr linkages and the double-parallelogram linkage to reduce the internal forces produced on the bearings by tolerance induced misalignment. The design decision to use a synchronous belt as part of the akr mechanism was to make the effective attachment point of both akr mechanisms coaxial. This allows the springs which drive both akr mechanisms to be physically remote from the actual joint placement, with use of a cable routing system to transmit the spring force to the akr mechanisms.

Assembly of the described mechanism is challenging. Bearings in the 4-bar mechanism were pre-compressed via heat shrink fitting of the mounting structure around the bearing race. This allows for precise control of the bearing

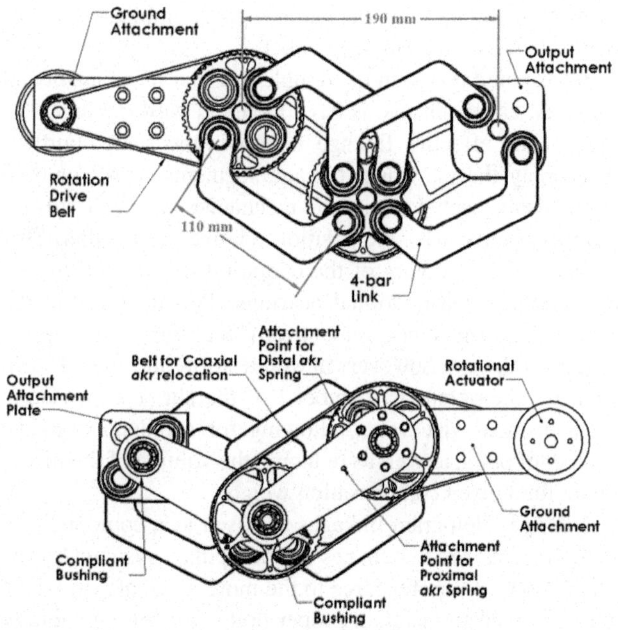

Figure 6. Technical Illustration for the mechanism as constructed. Illustrated from both sides. Top illustrates primarily the 4-bar component of the mechanism, while bottom illustrates the akr mechanism. This iteration will accommodate up to 10cm of misalignment, without restriction of motion. Mechanism as designed to compensate up to 100 kg of applied load. Many structural members are significantly oversize to simplify manufacture and construction. All rotational joints for 4-bar linkages are supported by cylindrical roller bearings (SKF Bearing) rated to 13kN dynamic loading. Rotational joints for akr mechanism are supported by standard ball bearings (SKF Bearing). Teflon impregnated delrin is used as the material for the compliant bushing, due to its relatively high strength and desirable deformation characteristics. Synchronous belts are Gates GT type.

preload, and significantly increased off-axial stiffness, at a cost of high assembly and disassembly difficulty. The authors do not recommend this approach unless absolutely required by application. Testing of the akr gravity compensation scheme was performed using a fixed pulley routing system, and substitution of varying stiffness springs. The fixed pulley system allows for a slight and expected variation of the output load, due to a small change in the effective distance "a" as the system moves through various kinematic constraints. This variation may not be desirable, and additionally, the necessary pulley routing systems in addition to the synchronous belt to collocate the point of rotation for both akr systems significantly complicate the mechanical design. Future researchers interested in pursuing this type of mechanism are encouraged to pursue alternate methods of routing forces to the akr mechanism, such as Bowden cables (Schiele, 2008).

The mechanism described in this paper can be kinematically expanded through straightforward conceptual alterations. By using a parallelipiped in place of the one of the serial four-bar mechanisms, with a similar expansion for the akr type gravity compensation mechanism, it would allow accommodation of joint misalignment along all three Cartesian axis, as opposed to the purely planar mechanism described in this paper. This kinematic expansion is a non-trivial engineering challenge, and should be carefully approached.

4 Conclusions

This mechanism provides a new mechanical solution that allows for a planar exoskeleton joint with the ability to carry high forces and torques, independently of the load applied to the user. It allows the transmission of very high forces through an exoskeletal joint that requires minimal effort to align. It allows the use of physical human-robot interfaces that are appropriate to the application, without need to accommodate for kinematic misalignment. The basic mechanism as described can be tailored to accommodate any desired degree of Cartesian misalignment by varying a single design parameter. Further, forces and torques can be partially decoupled, the degree of which is dependent on the relative degree of misalignment accommodated.

Such a mechanism(-s) can be integrated into an exoskeleton to allow for many users to quickly don and comfortably use the system, and allows for the introduction of exoskeletal systems to a much broader user audience.

Author Contributions.

J. F. Schorsch contributed the initial design concept, development of the technical design requirements, analysis of the mechanism, and preparation of the manuscript, with assistance from all co-authors. A. Q. L. Keemink provided analysis of the self-alignment mathematics, and many contributions to the manuscript preparation. A. H. A. Stienen assisted with the development of the concept, development of the requirements, review of prior literature, as well as technical recommendations for implementation and construction. D. A. Abbink provided many contributions to the preparation of the manuscript.

Acknowledgements. We would like to thank J. Herder, Emile Rosenberg, and Guiseppe Radaelli for their insights into the function of spring compensated mechansisms.

Edited by: D. Dodou

References

Argo: Rewalk, available at: http://www.argomedtec.coml (last access: 17 September 2012), 2012.

Berkeley Bionics: eLegs, available at: http://berkeleybionics.comlexoskeletons-rehabmobility/ (last access: 17 September 2012), 2012.

Colombo, G., Joerg, M., Schreier, R., and Dietz, V.: Treadmill training of paraplegic patients using a robotic orthosis, J. Rehabil. Res. Dev., 37, 693–700, 2000.

De Rossi, S. M. M., Vitiello, N., Lenzi, T., Ronsse, R., Koopman, B., Persichetti, A., Vecchi, F., Ijspeert, A. J., van der Kooij, H., and Carrozza, M. C.: Sensing pressure distribution on a lower-limb exoskeleton physical human-machine interface, Sensors, 11 207–227, 2011.

Evryon: Evryon, available at: www.evryon.eu/ (last access: 17 September 2012), 2009.

Jezernik, S., Colombo, G., Keller, T., Frueh, H., and Morari, M.: Robotic orthosis lokomat: a rehabilitation and research tool, Neuromodulation, 6, 108–115, 2003.

Kumar, S.: Theories of musculoskeletal injury causation, Ergonomics, 44, 17–47, 2001.

Lockheed Martin: HULC, available at: http://www.lockheedmartin.com/us/products/hulc.html (last access: 17 September 2012), 2012.

Makinson, B. J.: Research and Development Prototype for Machine Augmentation of Human Strength and Endurance, Hardiman I Project, General Electric Report S-71-1056, Schenectady, NY, 1971.

Morita, T., Kuribara, F., Yuki, S., and Sugano, S.: A Novel Mechanism Design for Gravity Compensation in Three Dimensional Space, Proceedings of the 2003 IEEE/ASME International Conference on Advanced Intelligent Mechatronics, 163–168, 2003.

Pratt, G. A., Williamson, M. M., Dillworth, P., Pratt, J., and Wright, A.: Stiffness Isn't Everything, Experimental Robotics IV, edited by: Salisbury J. K. and Khatib O., Springer, Berlin, 263–262, 1997.

Schiele, A.: Performance difference of Bowden Cable relocated and non-relocated master actuators in virtual environment applications, 2008 IEEE/RSJ International Conference on Intelligent Robots and Systems, IROS, 3507–3512, 2008.

Schiele, A.: Ergonomics of Exoskeletons: Subjective Performance Metrics, The 2009 IEEE/RSJ International Conference on Intelligent Robots and Systems, St. Louis, USA, 480–485, 2009.

Schiele, A. and van der Helm, F. C. T.: Kinematic design to improve ergonomics in human machine interaction, IEEE Trans. Neural Syst. Rehabil. Eng., 14 456–469, 2006.

Stienen, A. H. A., Hekman, E. E. G., Van der Helm, F. C. T., Prange, G. B., Jannink, M. J. A., Aalsma, A. M. M., and Van der Kooij, H.: Dampace: dynamic force-coordination trainer for the extremities, Proceedings of the 2007 IEEE Conference on Rehabilitation Robotics, Noordwijk, The Netherlands, 2007.

Stienen, A. H. A., Hekman, E. E. G., van der Helm, F. C. T., and van der Kooij, and H.: Self-Aligning Exoskeleton Axes Through Decoupling of Joint Rotations and Translations, IEEE Trans. Robot., 25, 628–633, 2009.

Stormont, T. J., An, K. N., Morrey, B. F., and Chao, E. Y.: Elbow joint contact study: comparison of techniques, J. Biomech., 18, 329–336, 1985.

Tuijthof, G. J. M., and Herder, J. L.: Design, actuation and control of an anthropomorphic robot arm, Mech. Mach. Theory, 35, 945–962, 2000.

Van Der Kooij, H., Veneman, J. F., and Ekkelenkamp, R.: Compliant Actuation of Exoskeletons, Mobile Robots: towards New Applications, edited by: Lazinica, A., 129–148, 2006.

Wildenbeest, J. G. W., Abbink, D. A., and Schorsch, J. F.: Haptic transparency increases the generalizability of telemanipulated motor learning, IEEE World Haptics Conference, Daejeon, Korea, 707–712, 2013.

Raytheon: XOS 2, available at: http://www.raytheon.com/newsroom/technology/rtn08_exoskeleton/ (last access: 17 September 2012), 2012.

Zoss, A. B., Kazerooni, H., and Chu, A.: Biomechanical Design of the Berkeley Lower Limb Exoskeleton (BLEEX), IEEE/ASME Trans. Mechatronics, 11, 128–138, 2006.

10

A recursive multibody formalism for systems with small mass and inertia terms

M. Arnold

Martin Luther University Halle-Wittenberg, NWF II – Institute of Mathematics, 06099 Halle (Saale), Germany

Correspondence to: M. Arnold (martin.arnold@mathematik.uni-halle.de)

Abstract. Complex multibody system models that contain bodies with small mass or nearly singular inertia tensor may suffer from high frequency solution components that deteriorate the solver efficiency in time integration. Singular perturbation theory suggests to neglect these small mass and inertia terms to allow a more efficient computation of the smooth solution components. In the present paper, a recursive multibody formalism is developed to evaluate the equations of motion for a tree structured N body system with $O(N)$ complexity even if isolated bodies have a rank-deficient body mass matrix. The approach is illustrated by some academic test problems in 2-D.

1 Introduction

Classical time integration methods in technical simulation are tailored to problems with smooth solution. Small system parameters in a mathematical model may introduce rapidly oscillating or strongly damped solution components that cause problems in time integration. Singular perturbation theory gives much insight in the analytical background of these phenomena and allows furthermore an efficient approximation of *smooth* solutions neglecting all terms that contain small parameters, see, e.g., Hairer and Wanner (1996).

The application of these classical results to multibody dynamics is non-trivial since the numerical algorithms for evaluating the equations of motion efficiently (multibody formalisms) are based on regularity assumptions that may be violated if small mass and inertia terms are neglected. A modified multibody formalism for chain structured multibody systems with an isolated "zero mass" body was developed in Arnold et al. (2010).

The present paper is a revised and extended version of the author's contribution to this conference paper. A recursive multibody formalism is developed for tree structured systems with bodies that suffer from a rank-deficient body mass matrix and may be considered as limit case of systems with bodies of (very) small mass or nearly singular inertia tensor. This research is guided by well known results from general singular perturbation theory, see Hairer and Wanner (1996), and its extensions to singularly perturbed problems in the context of multibody dynamics by Lubich (1993) and Stumpp (2008).

Related problems are, e.g., the modelling of serial spring-damper elements using an auxiliary zero mass body between spring and damper (Eich-Soellner and Führer, 1998, Section 1.3.4), the modification of inertia forces for high-frequency eigenmodes of flexible bodies in multibody system models for the analysis of elastohydrodynamic bearing coupling in Schönen (2003) and recently proposed methods from FE contact mechanics in Hager and Wohlmuth (2009).

For real-time applications in multibody dynamics, the neglection of inertia forces for small mass bodies was studied by Eichberger and Rulka (2004). A more detailed analysis shows, that this neglection of inertia forces is straightforward if all small mass bodies of the multibody system are leaf bodies in the kinematic tree. In numerical experiments for the model of a walking mobile robot (mobot) with stiff contact forces between lightweight legs and ground floor, the numerical effort was reduced by a factor of 4, see Weber et al. (2012).

The singular perturbation analysis is technically more complicated for multibody system models with small mass bodies having in the kinematic tree a successor of substantially larger mass since classical multibody formalisms and topological solvers are not applicable to the limit case of zero mass bodies in a kinematic chain. In Burgermeister et al. (2011), a smoothed velocity approximation for small mass bodies was proposed as a work-around.

In the present paper, we discuss an alternative approach that extends the recursive multibody formalism directly to kinematic trees containing bodies of vanishing mass or rank-deficient inertia tensor. The classical set of second order equations of motion in the joint coordinates $\boldsymbol{q}(t)$ is substituted by a suitable combination of second and first order ODEs describing the system dynamics in the limit case of a zero mass body. These results extend the previous analysis for chain structured systems in Arnold et al. (2010) and may be considered as a next step to extend advanced multibody formalisms for flexible multibody systems to models with bodies of (very) small mass.

The remaining part of the paper is organized as follows: Basic results of singular perturbation theory and its application to multibody system dynamics are recalled in Sect. 2. A recursive multibody formalism and the resulting mixed set of first and second order equations of motion for a body of rank-deficient body mass matrix are derived in Sect. 3. Finally, two simple test problems are discussed in Sect. 4. Some basic information on Moore-Penrose pseudo-inverses is provided in Appendix A.

2 Singular perturbations in multibody system models

After a short introduction to singularly perturbed ODEs we consider in the present section singular perturbations in multibody system models that are caused by stiff potential forces, see Sect. 2.1. Furthermore, some typical problems in the dynamical simulation of multibody system models with small mass bodies are illustrated by the analysis of two coupled oscillators in Sect. 2.2.

2.1 Time integration of singularly perturbed problems in multibody dynamics

The generic form of singularly perturbed ODEs are partitioned systems

$$\dot{\boldsymbol{y}}_\varepsilon = \boldsymbol{\varphi}(\boldsymbol{y}_\varepsilon, \boldsymbol{z}_\varepsilon), \tag{1a}$$
$$\varepsilon \dot{\boldsymbol{z}}_\varepsilon = \boldsymbol{\gamma}(\boldsymbol{y}_\varepsilon, \boldsymbol{z}_\varepsilon) \tag{1b}$$

with a small perturbation parameter $\varepsilon > 0$ that are considered at a finite time interval $[0, t_e]$, see Hairer and Wanner (1996, Chapter VI) and the references therein.

For any given initial value $\boldsymbol{y}^0$, the singularly perturbed system (1) has a smooth solution $(\boldsymbol{y}_\varepsilon^0(t), \boldsymbol{z}_\varepsilon^0(t))$ with $\boldsymbol{y}_\varepsilon^0(0) = \boldsymbol{y}^0$ if the right hand side of the singularly perturbed subsystem (1b) has a Jacobian with eigenvalues satisfying along each solution trajectory $(\boldsymbol{y}_\varepsilon(t), \boldsymbol{z}_\varepsilon(t))$ the condition

$$\operatorname{Re} \lambda_i [(\partial \boldsymbol{\gamma} / \partial \boldsymbol{z}_\varepsilon)(\boldsymbol{y}_\varepsilon, \boldsymbol{z}_\varepsilon)] \le -\beta < 0$$

for some positive constant $\beta > 0$. This smooth solution remains in an $O(\varepsilon)$-neighbourhood of the solution $(\boldsymbol{y}_0(t), \boldsymbol{z}_0(t))$ of the reduced problem that results from setting formally the perturbation parameter in (1) to $\varepsilon := 0$. We get

$$\dot{\boldsymbol{y}}_0 = \boldsymbol{\varphi}(\boldsymbol{y}_0, \boldsymbol{z}_0), \tag{2a}$$
$$\boldsymbol{0} = \boldsymbol{\gamma}(\boldsymbol{y}_0, \boldsymbol{z}_0) \tag{2b}$$

with $\boldsymbol{y}_0(0) = \boldsymbol{y}^0$ and $\boldsymbol{z}_0(0)$ being implicitly defined by (2b). The general solution of the singularly perturbed problem (1) has the form

$$\boldsymbol{y}_\varepsilon(t) = \boldsymbol{y}_\varepsilon^0(t) + \varepsilon \boldsymbol{\eta}_\varepsilon(t/\varepsilon) = \boldsymbol{y}_0(t) + O(\varepsilon), \tag{3a}$$
$$\boldsymbol{z}_\varepsilon(t) = \boldsymbol{z}_\varepsilon^0(t) + \boldsymbol{\zeta}_\varepsilon(t/\varepsilon) = \boldsymbol{z}_0(t) + O(\varepsilon) + \boldsymbol{\zeta}_\varepsilon(t/\varepsilon) \tag{3b}$$

with smooth functions $\boldsymbol{\eta}_\varepsilon(t/\varepsilon)$, $\boldsymbol{\zeta}_\varepsilon(t/\varepsilon)$ that decay like $\exp(-\bar{\beta}t/\varepsilon)$ for some positive constant $\bar{\beta} \in (0, \beta)$, see Hairer and Wanner (1996, Theorem VI.3.2).

For many stiff integrators, the numerical solution of (1) may be decomposed as well in a smooth part and a rapidly decaying part reflecting the transient behaviour for initial values $(\boldsymbol{y}_\varepsilon(0), \boldsymbol{z}_\varepsilon(0))$ that do not belong to a smooth solution. If there is no particular interest in this transient phase, an approximate numerical solution may be obtained much more efficiently solving for given initial values $\boldsymbol{y}_0(0) := \boldsymbol{y}_\varepsilon(0)$ the DAE (2) by appropriate time integration methods (Hairer and Wanner, 1996, Chapter VI). Note, that the initial values $\boldsymbol{z}_0(0)$ in DAE (2) are not free but have to satisfy the consistency condition $\boldsymbol{\gamma}(\boldsymbol{y}_0(0), \boldsymbol{z}_0(0)) = \boldsymbol{0}$.

Lubich (1993) extended these classical results to a class of singular singularly perturbed problems

$$\mathbf{M}(\boldsymbol{q})\ddot{\boldsymbol{q}} = \boldsymbol{\psi}(\boldsymbol{q}, \dot{\boldsymbol{q}}) - \frac{1}{\varepsilon^2} \nabla U(\boldsymbol{q}) \tag{4}$$

with $\boldsymbol{q}(t)$ denoting the position coordinates of a multibody system. Matrix $\mathbf{M}(\boldsymbol{q})$ is the symmetric positive definite mass matrix and $\boldsymbol{\psi}(\boldsymbol{q}, \dot{\boldsymbol{q}})$ denotes a vector of forces and momenta. The crucial term in (4) are the (very) stiff potential forces $-\varepsilon^{-2} \nabla U(\boldsymbol{q})$ that depend on the perturbation parameter ε with $0 < \varepsilon \ll 1$. If the potential $U(\boldsymbol{q})$ attains a local minimum on a manifold $\mathcal{U}$ and is strongly convex along all directions that are non-tangential to $\mathcal{U}$, then the smooth solution of (4) may be approximated up to $O(\varepsilon^2)$ by the solution $\boldsymbol{q}_0(t)$ of a constrained (differential-algebraic) system with $\boldsymbol{q}_0(t) \in \mathcal{U}$, $(t \ge 0)$. In general, this constrained system may be solved much more efficiently than the original singularly perturbed problem (4), see Hairer and Wanner (1996).

In Stumpp (2008), these results were extended to mechanical systems with strong damping forces $-\varepsilon^{-1}\mathbf{D}(\boldsymbol{q})\dot{\boldsymbol{q}}$ in the right hand side of (4). In the limit case $\varepsilon \to 0$, the solution of the singularly perturbed problem is again approximated by the solution of a DAE problem that can be obtained in a robust and efficient way by standard DAE time integration methods, see Hairer and Wanner (1996).

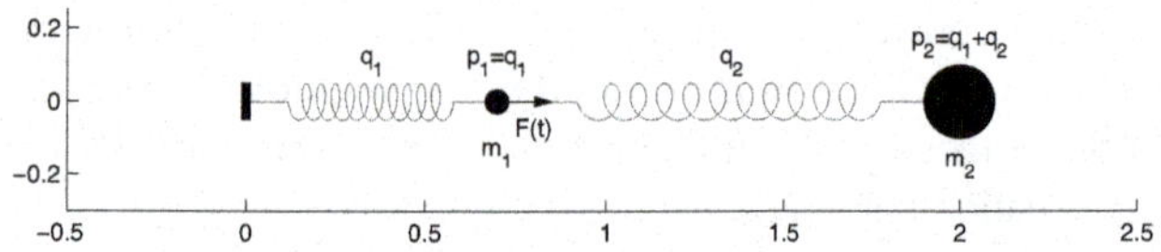

Figure 1. Example problem: two coupled oscillators with a fast oscillating small mass m_1, see Burgermeister et al. (2011).

2.2 The small mass oscillator as singularly perturbed problem

The eigenfrequency of a harmonic oscillator is given by $\omega = \sqrt{c/m}$ with mass m and spring constant c. High frequency oscillations in a mechanical system may not only be introduced by (very) stiff potential forces but also by potential forces of moderate size that act on a body with (very) small mass.

In Burgermeister et al. (2011), this phenomenon was studied for the simple model problem in Fig. 1. In two coupled oscillators, a small mass m_1 is connected to a large mass m_2 and the reference system by stiff springs with constants c_1, c_2 and damping with damping parameters d_1, d_2. Both bodies can only move along the x-axis. Additional forces $F(t)$ are only acting on m_1. In absolute coordinates $\boldsymbol{p}(t) = (p_1(t), p_2(t))^\top$, the equations of motion are given by

$$m_1\ddot{p}_1 = F(t) - d_1\dot{p}_1 - c_1 p_1 + d_2(\dot{p}_2 - \dot{p}_1) + c_2(p_2 - p_1), \tag{5a}$$

$$m_2\ddot{p}_2 = -d_2(\dot{p}_2 - \dot{p}_1) - c_2(p_2 - p_1). \tag{5b}$$

The small mass can oscillate very fast depending on the ratio of the masses and spring parameters. If the perturbation parameter $\varepsilon := m_1$ gets smaller, the frequency of the oscillations increases and a time integration method with stepsize control would choose very small stepsizes to resolve these oscillations and meet the integration tolerances. The number of time steps and the computing time increase significantly, see Fig. 2. In the limit case $\varepsilon = m_1 = 0$, the inertia forces of the first mass point are neglected and (5a) results in an implicit first order differential equation

$$0 = F(t) - d_1\dot{p}_1 - c_1 p_1 + d_2(\dot{p}_2 - \dot{p}_1) + c_2(p_2 - p_1). \tag{6}$$

The fast oscillations of the small mass disappear and the integrator can use large stepsizes, see Fig. 2.

In a system description by relative coordinates $\boldsymbol{q}(t) = (q_1(t), q_2(t))^\top$ with $p_1(t) =: q_1(t)$ and $p_2(t) - p_1(t) =: q_2(t)$, the limit process $\varepsilon \to 0$ causes substantial problems since the equations of motion (5b) of the *large* mass depend on $\ddot{p}_2 = \ddot{q}_1 + \ddot{q}_2$ but $\ddot{q}_1 = \ddot{p}_1$ does not appear in (6). Furthermore, the differentiation of (6) w.r.t. time t shows that $\ddot{q}_1(t)$ depends on the time derivative of $F(t)$ in the limit case $\varepsilon = m_1 = 0$.

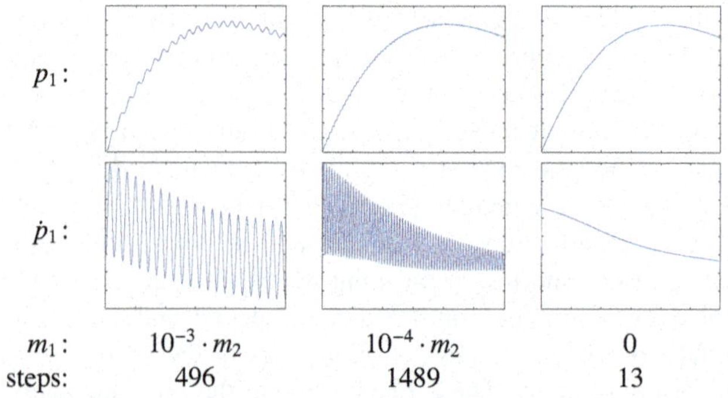
p_1:
$\dot{p}_1$:
m_1: $10^{-3} \cdot m_2$ $10^{-4} \cdot m_2$ 0
steps: 496 1489 13

Figure 2. Oscillations and number of time steps for decreasing mass ratios m_1/m_2 (BDF solver DASSL with tolerances 10^{-6} for absolute and relative errors), see Burgermeister et al. (2011).

3 Mixed coordinate formulation of the equations of motion

(Very) small masses or (nearly) singular inertia tensors in a multibody system model may be interpreted as small perturbations. The analysis of Sect. 2.1 suggests to study a reduced system that neglects these perturbations. Therefore, we consider now the limit case of multibody systems that have one or more bodies with zero mass or singular inertia tensor, i.e., with a rank-deficient body mass matrix.

In that case, classical recursive multibody formalisms ("$O(N)$-formalisms") like the ones by Brandl et al. (1988), Lubich et al. (1992) and Eichberger (1994) fail since they use the inverse of projected body mass matrices. Udwadia and Phohomsiri (2006) consider multibody systems with singular mass matrix but do not exploit the system's topology to evaluate the equations of motion efficiently by a sequence of forward and backward recursions in the kinematic tree.

In this section, we describe the multibody system by the absolute position and orientation of all N bodies relative to the inertial frame and use joint coordinates as generalized coordinates for a tree structured system ("mixed coordinates"), see also Schiehlen (1997). As in Lubich et al. (1992) and Eich-Soellner and Führer (1998), the recursive multibody formalism is interpreted as a block Gaussian elimination for a sparse, (very) large block-structured system of linear equations with the block structure being determined by the topology of the multibody system.

Using generalized inverses, the recursive formalism of Lubich et al. (1992) is modified to skip bodies with rank-deficient body mass matrix in the second forward recursion, see Sects. 3.2 and 3.3. The resulting equations of motion are not longer explicit but form a (linearly) implicit set of first and second order differential equations (Sect. 3.4). Finally, we show in Sect. 3.5 by some analytical transformations that the equations of motion form a first order index-1 DAE if a certain regularity assumption is satisfied.

3.1 Tree structured multibody systems: kinematics

Recursive multibody formalisms are tailored to tree structured systems. Here, the term *tree structure* corresponds to the structure of the labelled graph being associated to the multibody system model. In this graph, each (rigid or flexible) body of the system is represented by a vertex. Two vertices of the graph are connected by an edge if and only if the corresponding bodies in the multibody system model are connected by a joint restricting their relative motion.

The graph of a tree structured multibody system is acyclic, i.e., it is free of loops. Then, the multibody system has a *root* body $(\bullet)^{(0)}$ that corresponds to the root vertex of the tree structured graph and is supposed to be inertially fixed. All other bodies $(\bullet)^{(i)}$ have a uniquely defined predecessor $(\bullet)^{(\pi_i)}$ in the kinematic tree. Each body $(\bullet)^{(i)}$ may have successors $(\bullet)^{(j)}$ being characterized by $\pi_j = i$ or, equivalently, by $j \in I_i := \{k : \pi_k = i\}$ with an index set I_i that represents the set of all successors of a given body $(\bullet)^{(i)}$ in the multibody system model. Bodies without successors ($I_i = \emptyset$) correspond to leafs of the kinematic tree and are therefore called "leaf bodies".

We suppose that position and orientation of body $(\bullet)^{(i)}$ may be characterized by (absolute) position coordinates $\boldsymbol{p}_i(t) \in \mathbb{R}^d$ with $d = 6$ for 3-D models and $d = 3$ for 2-D models (the position of point masses may be characterized by $d = 3$ absolute coordinates in 3-D and by $d = 2$ absolute coordinates in 2-D, see Sect. 4 below). The *relative* position and orientation of body $(\bullet)^{(i)}$ w.r.t. its predecessor $(\bullet)^{(\pi_i)}$ is characterized by joint coordinates $\boldsymbol{q}_i(t) \in \mathbb{R}^{n_i}$ representing the n_i degrees of freedom of the joint connecting $(\bullet)^{(i)}$ with $(\bullet)^{(\pi_i)}$:

$$\mathbf{0} = \boldsymbol{k}_i(\boldsymbol{p}_i, \boldsymbol{p}_{\pi_i}, \boldsymbol{q}_i, t). \tag{7}$$

Here and in the following we suppose that (7) is locally uniquely solvable w.r.t. $\boldsymbol{p}_i$ and that the Jacobian $\mathbf{K}_i = \partial \boldsymbol{k}_i / \partial \boldsymbol{p}_i$ is non-singular along the solution. In its most simple form, Eq. (7) defines $\boldsymbol{p}_i$ explicitly by $\boldsymbol{p}_i(t) = \boldsymbol{r}_i(\boldsymbol{p}_{\pi_i}(t), \boldsymbol{q}_i(t), t)$ resulting in $\mathbf{K}_i = \mathbf{I}_d$.

The kinematic relations (7) at the level of position coordinates imply relations at the level of velocity and acceleration coordinates that may formally be obtained by (total) differentiation of (7) w.r.t. time t:

$$\begin{aligned} \mathbf{0} &= \frac{\mathrm{d}}{\mathrm{d}t} \boldsymbol{k}_i(\boldsymbol{p}_i(t), \boldsymbol{p}_{\pi_i}(t), \boldsymbol{q}_i(t), t) \\ &= \mathbf{K}_i \dot{\boldsymbol{p}}_i + \mathbf{H}_i \dot{\boldsymbol{p}}_{\pi_i} + \mathbf{J}_i \dot{\boldsymbol{q}}_i + \boldsymbol{k}_i^{(\mathrm{I})}(\boldsymbol{p}_0, \boldsymbol{p}, \boldsymbol{q}, t), \end{aligned} \tag{8}$$

$$\mathbf{0} = \mathbf{K}_i \ddot{\boldsymbol{p}}_i + \mathbf{H}_i \ddot{\boldsymbol{p}}_{\pi_i} + \mathbf{J}_i \ddot{\boldsymbol{q}}_i + \boldsymbol{k}_i^{(\mathrm{II})}(\boldsymbol{p}_0, \dot{\boldsymbol{p}}_0, \boldsymbol{p}, \dot{\boldsymbol{p}}, \boldsymbol{q}, \dot{\boldsymbol{q}}, t) \tag{9}$$

with

$$\mathbf{K}_i := \frac{\partial \boldsymbol{k}_i}{\partial \boldsymbol{p}_i} \in \mathbb{R}^{d \times d}, \quad \mathbf{H}_i := \frac{\partial \boldsymbol{k}_i}{\partial \boldsymbol{p}_{\pi_i}} \in \mathbb{R}^{d \times d}, \quad \mathbf{J}_i := \frac{\partial \boldsymbol{k}_i}{\partial \boldsymbol{q}_i} \in \mathbb{R}^{d \times n_i}.$$

It is supposed that the joint coordinates $\boldsymbol{q}_i(t)$ are defined such that all Jacobians $\mathbf{J}_i$ have full column rank: $\operatorname{rank} \mathbf{J}_i = n_i \le d$.

Functions $\boldsymbol{k}_i^{(\mathrm{I})} := \partial \boldsymbol{k}_i / \partial t$ and $\boldsymbol{k}_i^{(\mathrm{II})}$ summarize partial time derivatives and all lower order terms in the first and second time derivative of (7), respectively. They may depend on the (absolute) coordinates $\boldsymbol{p}_0$ of the root body, on the absolute coordinates $\boldsymbol{p} := (\boldsymbol{p}_1, \ldots, \boldsymbol{p}_N)$ of the remaining N bodies in the system, on the corresponding joint coordinates $\boldsymbol{q} := (\boldsymbol{q}_1, \ldots, \boldsymbol{q}_N)$ and on $\dot{\boldsymbol{p}}_0$, $\dot{\boldsymbol{p}}$ and $\dot{\boldsymbol{q}}$.

Note, that (9) is simplified substantially for all bodies $(\bullet)^{(i)}$ that follow directly the root body ($\pi_i = 0$) since the root body is inertially fixed ($\ddot{\boldsymbol{p}}_{\pi_i} = \mathbf{0}$):

$$\mathbf{0} = \mathbf{K}_i \ddot{\boldsymbol{p}}_i + \mathbf{J}_i \ddot{\boldsymbol{q}}_i + \boldsymbol{k}_i^{(\mathrm{II})}(\boldsymbol{p}_0, \dot{\boldsymbol{p}}_0, \boldsymbol{p}, \dot{\boldsymbol{p}}, \boldsymbol{q}, \dot{\boldsymbol{q}}, t) \quad \text{if} \quad \pi_i = 0. \tag{10}$$

In recursive multibody formalisms, it is supposed that position and velocity of the root body $(\boldsymbol{p}_0(t), \dot{\boldsymbol{p}}_0(t))$ and all joint coordinates $\boldsymbol{q}_i(t)$, $\dot{\boldsymbol{q}}_i(t)$, $(i = 1, \ldots, N)$, at a current time t are known. Starting from the root body, the absolute position and velocity coordinates $\boldsymbol{p}_i(t)$, $\dot{\boldsymbol{p}}_i(t)$ of all N bodies $(\bullet)^{(i)}$, $(i = 1, \ldots, N)$, may then be computed recursively using (7) and (8), respectively, (*forward recursion*).

3.2 Tree structured multibody systems: equilibrium conditions

The equations of motion of a multibody system with N bodies may be obtained from the equilibrium conditions for forces and momenta for each individual body that are formulated in absolute coordinates $\boldsymbol{p}_i$:

$$\mathbf{M}_i \ddot{\boldsymbol{p}}_i + \mathbf{K}_i^\top \boldsymbol{\mu}_i + \sum_{j \in I_i} \mathbf{H}_j^\top \boldsymbol{\mu}_j = \boldsymbol{f}_i, \; (i = 1, \ldots, N). \tag{11}$$

The body mass matrix $\mathbf{M}_i \in \mathbb{R}^{d \times d}$ contains mass and inertia tensor of body $(\bullet)^{(i)}$ and is supposed to be symmetric, positive semi-definite. The equilibrium conditions contain the reaction forces of the joints connecting body $(\bullet)^{(i)}$ with its predecessor ($\mathbf{K}_i^\top \boldsymbol{\mu}_i$) and with its successors in the kinematic tree ($\mathbf{H}_j^\top \boldsymbol{\mu}_j$, $j \in I_i$). All remaining forces and momenta acting on body $(\bullet)^{(i)}$ are summarized in the force vector $\boldsymbol{f}_i = \boldsymbol{f}_i(\boldsymbol{p}, \dot{\boldsymbol{p}}, \boldsymbol{q}, \dot{\boldsymbol{q}}, t) \in \mathbb{R}^d$.

The specific structure of the joint reaction forces with Lagrange multipliers $\boldsymbol{\mu}_i(t) \in \mathbb{R}^d$ that satisfy

$$\mathbf{J}_i^\top \boldsymbol{\mu}_i = \mathbf{0}, \; (i = 1, \ldots, N), \tag{12}$$

results from the joint equations (7) and from d'Alembert's principle since the virtual work of constraint forces vanishes for all (virtual) displacements being compatible with (7). In (12), matrix $\mathbf{J}_i$ denotes the Jacobian of the constraint function $\boldsymbol{k}_i$ w.r.t. joint coordinates $\boldsymbol{q}_i \in \mathbb{R}^{n_i}$, see Sect. 3.1 above.

For leaf bodies $(\bullet)^{(i)}$, the equilibrium conditions (11) get a simpler form since $I_i = \{j : \pi_j = i\} = \emptyset$. We obtain

$$\bar{\mathbf{M}}_i \mathbf{K}_i \ddot{\boldsymbol{p}}_i + \boldsymbol{\mu}_i = \bar{\boldsymbol{f}}_i \tag{13}$$

with $\bar{\boldsymbol{f}}_i := \mathbf{K}_i^{-\top} \boldsymbol{f}_i$, $\mathbf{K}_i^{-\top} := (\mathbf{K}_i^\top)^{-1}$ and the symmetric, positive semi-definite mass matrix $\bar{\mathbf{M}}_i := \mathbf{K}_i^{-\top} \mathbf{M}_i \mathbf{K}_i^{-1}$.

One of the basic components of recursive multibody formalisms are algorithms to transform the equilibrium conditions (11) recursively for *all* bodies $(\bullet)^{(i)}$ to the simpler form (13) with suitable $\bar{\mathbf{M}}_i$ and $\bar{\boldsymbol{f}}_i$. With the common assumption that all body mass matrices $\mathbf{M}_i$ are non-singular, $\boldsymbol{\mu}_i$ may be expressed in terms of $\ddot{\boldsymbol{p}}_{\pi_i}$, $\bar{\mathbf{M}}_i$, $\bar{\boldsymbol{f}}_i$ and $\boldsymbol{k}_i^{(\mathrm{II})}$ by block Gaussian elimination applied to (9), (12) and (13), see, e.g., Lubich et al. (1992).

It is an important observation that this *backward recursion* may be generalized to multibody systems with rank-deficient body mass matrices $\mathbf{M}_i$ as long as all body mass matrices $\mathbf{M}_i$ are symmetric, positive semi-definite. In the following, this will be shown by mathematical induction: let us suppose that the equilibrium conditions of all successors $(\bullet)^{(j)}$ of body $(\bullet)^{(i)}$ are given in form (13), i.e.,

$$\bar{\mathbf{M}}_j\mathbf{K}_j\ddot{\boldsymbol{p}}_j + \boldsymbol{\mu}_j = \bar{\boldsymbol{f}}_j,\ (j \in I_i). \tag{14}$$

Since $\boldsymbol{\mu}_j$ belongs to the null space of $\mathbf{J}_j^\top$, see (12), we get

$$\begin{aligned}\boldsymbol{\mu}_j &= (\mathbf{I}_d - \bar{\mathbf{M}}_j\mathbf{J}_j(\mathbf{J}_j^\top\bar{\mathbf{M}}_j\mathbf{J}_j)^+\mathbf{J}_j^\top)\boldsymbol{\mu}_j \\ &= (\mathbf{I}_d - \bar{\mathbf{M}}_j\mathbf{J}_j(\mathbf{J}_j^\top\bar{\mathbf{M}}_j\mathbf{J}_j)^+\mathbf{J}_j^\top)(\bar{\boldsymbol{f}}_j - \bar{\mathbf{M}}_j\mathbf{K}_j\ddot{\boldsymbol{p}}_j)\end{aligned}$$

with the Moore-Penrose pseudo-inverse $(\mathbf{J}_j^\top\bar{\mathbf{M}}_j\mathbf{J}_j)^+$ of the projected body mass matrix $\mathbf{J}_j^\top\bar{\mathbf{M}}_j\mathbf{J}_j \in \mathbb{R}^{n_j\times n_j}$, see Remark 1 in Appendix A. Because of $\pi_j = i$, the term $\mathbf{K}_j\ddot{\boldsymbol{p}}_j$ is given by

$$\mathbf{K}_j\ddot{\boldsymbol{p}}_j = -\mathbf{H}_j\ddot{\boldsymbol{p}}_i - \mathbf{J}_j\ddot{\boldsymbol{q}}_j - \boldsymbol{k}_j^{(\mathrm{II})}, \tag{15}$$

see (9), and we obtain finally

$$\begin{aligned}\boldsymbol{\mu}_j &= (\bar{\mathbf{M}}_j - \bar{\mathbf{M}}_j\mathbf{J}_j(\mathbf{J}_j^\top\bar{\mathbf{M}}_j\mathbf{J}_j)^+\mathbf{J}_j^\top\bar{\mathbf{M}}_j)\mathbf{H}_j\ddot{\boldsymbol{p}}_i \\ &\quad + (\mathbf{I}_d - \bar{\mathbf{M}}_j\mathbf{J}_j(\mathbf{J}_j^\top\bar{\mathbf{M}}_j\mathbf{J}_j)^+\mathbf{J}_j^\top)(\bar{\boldsymbol{f}}_j + \bar{\mathbf{M}}_j\boldsymbol{k}_j^{(\mathrm{II})})\end{aligned} \tag{16}$$

since

$$\begin{aligned}&(\bar{\mathbf{M}}_j - \bar{\mathbf{M}}_j\mathbf{J}_j(\mathbf{J}_j^\top\bar{\mathbf{M}}_j\mathbf{J}_j)^+\mathbf{J}_j^\top\bar{\mathbf{M}}_j)\mathbf{J}_j\ddot{\boldsymbol{q}}_j \\ &= \bar{\mathbf{M}}_j^{1/2}\mathbf{C}(\mathbf{I}_{n_j} - (\mathbf{C}^\top\mathbf{C})^+(\mathbf{C}^\top\mathbf{C}))\ddot{\boldsymbol{q}}_j = \bar{\mathbf{M}}_j^{1/2}\mathbf{0}_{d\times n_j}\ddot{\boldsymbol{q}}_j = \mathbf{0}\end{aligned}$$

with $\mathbf{C} := \bar{\mathbf{M}}_j^{1/2}\mathbf{J}_j$, see Lemma 2a in Appendix A.

Multiplying (11) from the left by $\mathbf{K}_i^{-\top}$ and inserting $\boldsymbol{\mu}_j$ from (16) for all $j \in I_i$, the equilibrium conditions for body $(\bullet)^{(i)}$ get the more compact form (13) with

$$\begin{aligned}\bar{\mathbf{M}}_i &:= \mathbf{K}_i^{-\top}\mathbf{M}_i\mathbf{K}_i^{-1} \\ &\quad + \sum_{j\in I_i}\mathbf{K}_i^{-\top}\mathbf{H}_j^\top(\bar{\mathbf{M}}_j - \bar{\mathbf{M}}_j\mathbf{J}_j(\mathbf{J}_j^\top\bar{\mathbf{M}}_j\mathbf{J}_j)^+\mathbf{J}_j^\top\bar{\mathbf{M}}_j)\mathbf{H}_j\mathbf{K}_i^{-1},\end{aligned} \tag{17a}$$

$$\begin{aligned}\bar{\boldsymbol{f}}_i &:= \mathbf{K}_i^{-\top}\boldsymbol{f}_i \\ &\quad - \sum_{j\in I_i}\mathbf{K}_i^{-\top}\mathbf{H}_j^\top(\mathbf{I}_d - \bar{\mathbf{M}}_j\mathbf{J}_j(\mathbf{J}_j^\top\bar{\mathbf{M}}_j\mathbf{J}_j)^+\mathbf{J}_j^\top)(\bar{\boldsymbol{f}}_j + \bar{\mathbf{M}}_j\boldsymbol{k}_j^{(\mathrm{II})}).\end{aligned} \tag{17b}$$

Lemma 1 in Appendix A shows that matrix $\bar{\mathbf{M}}_i$ in (17a) is symmetric, positive semi-definite since it is a finite sum of symmetric, positive semi-definite matrices. Starting from the leaf bodies and following all branches of the kinematic tree to the root, the compact form (13) of the equilibrium conditions may be obtained recursively for all N bodies $(\bullet)^{(i)}$ of the multibody system. The condensed mass matrices $\bar{\mathbf{M}}_i \in \mathbb{R}^{d\times d}$. are symmetric, positive semi-definite.

The backward recursion with $\bar{\mathbf{M}}_i$, $\bar{\boldsymbol{f}}_i$ being defined by (17) is well-defined whenever all body mass matrices $\mathbf{M}_i$, $(i = 1,\dots,N)$, are symmetric, positive semi-definite and matrices $\mathbf{K}_i$ are non-singular. Assuming additionally that the body mass matrices $\mathbf{M}_i \in \mathbb{R}^{d\times d}$ are positive definite, matrices $\bar{\mathbf{M}}_i \in \mathbb{R}^{d\times d}$ and $\mathbf{J}_j^\top\bar{\mathbf{M}}_j\mathbf{J}_j \in \mathbb{R}^{n_i\times n_i}$ in (17) are non-singular and $(\mathbf{J}_j^\top\bar{\mathbf{M}}_j\mathbf{J}_j)^+$ may be substituted by $(\mathbf{J}_j^\top\bar{\mathbf{M}}_j\mathbf{J}_j)^{-1}$. In that special case, the recursive definitions (17) are well known from classical multibody formalisms, see, e.g., Lubich et al. (1992).

3.3 Forward recursion: Absolute coordinates

The second time derivative (9) of the kinematic relations (7) defines the acceleration $\ddot{\boldsymbol{p}}_i$ of body $(\bullet)^{(i)}$ in terms of the acceleration $\ddot{\boldsymbol{p}}_{\pi_i}$ of its predecessor $(\bullet)^{(\pi_i)}$ and in terms of the corresponding joint coordinates $\ddot{\boldsymbol{q}}_i$. Eliminating the joint coordinates, all (absolute) accelerations $\ddot{\boldsymbol{p}}_i$, $(i = 1,\dots,N)$, may be computed recursively starting at the root body since $\ddot{\boldsymbol{p}}_0 \equiv \mathbf{0}$ (*forward recursion*).

Left multiplication of (9) by $(\mathbf{J}_i^\top\bar{\mathbf{M}}_i\mathbf{J}_i)^+\mathbf{J}_i^\top\bar{\mathbf{M}}_i$ results in

$$\begin{aligned}&(\mathbf{J}_i^\top\bar{\mathbf{M}}_i\mathbf{J}_i)^+(\mathbf{J}_i^\top\bar{\mathbf{M}}_i\mathbf{J}_i)\ddot{\boldsymbol{q}}_i \\ &= -(\mathbf{J}_i^\top\bar{\mathbf{M}}_i\mathbf{J}_i)^+\mathbf{J}_i^\top\bar{\mathbf{M}}_i(\mathbf{H}_i\ddot{\boldsymbol{p}}_{\pi_i} + \boldsymbol{k}_i^{(\mathrm{II})}) - (\mathbf{J}_i^\top\bar{\mathbf{M}}_i\mathbf{J}_i)^+\mathbf{J}_i^\top\bar{\boldsymbol{f}}_i\end{aligned} \tag{18}$$

since

$$\mathbf{J}_i^\top\bar{\mathbf{M}}_i\mathbf{K}_i\ddot{\boldsymbol{p}}_i = \mathbf{J}_i^\top\bar{\boldsymbol{f}}_i - \mathbf{J}_i^\top\boldsymbol{\mu}_i = \mathbf{J}_i^\top\bar{\boldsymbol{f}}_i,$$

see (12) and (13). If $\bar{\mathbf{M}}_i \in \mathbb{R}^{d\times d}$ is symmetric, positive definite then $\mathbf{J}_i^\top\bar{\mathbf{M}}_i\mathbf{J}_i \in \mathbb{R}^{n_i\times n_i}$ is symmetric, positive definite as well and (18) defines an explicit expression for $\ddot{\boldsymbol{q}}_i$ because of $(\mathbf{J}_i^\top\bar{\mathbf{M}}_i\mathbf{J}_i)^+(\mathbf{J}_i^\top\bar{\mathbf{M}}_i\mathbf{J}_i) = \mathbf{I}_{n_i}$ in that case. In general, however, Eq. (18) determines only $r_i := \mathrm{rank}(\mathbf{J}_i^\top\bar{\mathbf{M}}_i\mathbf{J}_i) \le n_i$ components of $\ddot{\boldsymbol{q}}_i \in \mathbb{R}^{n_i}$, see also Remark 1b in Appendix A.

Substituting (18) in (9), we get the expression

$$\begin{aligned}\ddot{\boldsymbol{p}}_i &= -\bar{\mathbf{H}}_i\ddot{\boldsymbol{p}}_{\pi_i} - \bar{\boldsymbol{k}}_i^{(\mathrm{II})} \\ &\quad - \mathbf{K}_i^{-1}\mathbf{J}_i\left(\mathbf{I}_{n_i} - (\mathbf{J}_i^\top\bar{\mathbf{M}}_i\mathbf{J}_i)^+(\mathbf{J}_i^\top\bar{\mathbf{M}}_i\mathbf{J}_i)\right)\ddot{\boldsymbol{q}}_i\end{aligned} \tag{19}$$

with

$$\bar{\mathbf{H}}_i := \mathbf{K}_i^{-1}(\mathbf{I}_d - \mathbf{J}_i(\mathbf{J}_i^\top\bar{\mathbf{M}}_i\mathbf{J}_i)^+\mathbf{J}_i^\top\bar{\mathbf{M}}_i)\mathbf{H}_i, \tag{20a}$$

$$\begin{aligned}\bar{\boldsymbol{k}}_i^{(\mathrm{II})} &:= \mathbf{K}_i^{-1}(\mathbf{I}_d - \mathbf{J}_i(\mathbf{J}_i^\top\bar{\mathbf{M}}_i\mathbf{J}_i)^+\mathbf{J}_i^\top\bar{\mathbf{M}}_i)\boldsymbol{k}_i^{(\mathrm{II})} \\ &\quad - \mathbf{K}_i^{-1}\mathbf{J}_i(\mathbf{J}_i^\top\bar{\mathbf{M}}_i\mathbf{J}_i)^+\mathbf{J}_i^\top\bar{\boldsymbol{f}}_i\end{aligned} \tag{20b}$$

that proves to be useful in the forward recursion if $\mathbf{J}_i^\top\bar{\mathbf{M}}_i\mathbf{J}_i$ is rank-deficient. The main difference between the full rank and the rank-deficient case is the additional non-zero term $\left(\mathbf{I}_{n_i} - (\mathbf{J}_i^\top\bar{\mathbf{M}}_i\mathbf{J}_i)^+(\mathbf{J}_i^\top\bar{\mathbf{M}}_i\mathbf{J}_i)\right)\ddot{\boldsymbol{q}}_i$ in the right hand side of (19) if $r_i = \mathrm{rank}(\mathbf{J}_i^\top\bar{\mathbf{M}}_i\mathbf{J}_i) < n_i$.

It is an important (and non-trivial) observation that this additional term does not affect the successors of body $(\bullet)^{(i)}$ in the kinematic tree (if there are any). Suppose $I_i \neq \emptyset$ and consider a (direct) successor $(\bullet)^{(j)}$ of body $(\bullet)^{(i)}$, $\pi_j = i$. In the condensed equilibrium conditions (14), the reaction force between bodies $(\bullet)^{(j)}$ and $(\bullet)^{(i)}$ is represented by $\boldsymbol{\mu}_j \in \mathbb{R}^d$ that depends on $\ddot{\boldsymbol{p}}_i$, $\bar{\boldsymbol{f}}_j$ and $\bar{\boldsymbol{k}}_j^{(\mathrm{II})}$, see (16):

$$\boldsymbol{\mu}_j = \bar{\boldsymbol{f}}_j + \bar{\mathbf{M}}_j \mathbf{K}_j (\bar{\mathbf{H}}_j \ddot{\boldsymbol{p}}_i + \bar{\boldsymbol{k}}_j^{(\mathrm{II})}). \tag{21a}$$

From (19), we see that the right hand side of (21a) contains the product of matrices

$$\bar{\mathbf{J}}_j := \bar{\mathbf{M}}_j \mathbf{K}_j \bar{\mathbf{H}}_j \mathbf{K}_i^{-1} \mathbf{J}_i \left(\mathbf{I}_{n_i} - (\mathbf{J}_i^\top \bar{\mathbf{M}}_i \mathbf{J}_i)^+ (\mathbf{J}_i^\top \bar{\mathbf{M}}_i \mathbf{J}_i) \right)$$

that is rewritten as

$$\bar{\mathbf{J}}_j = (\bar{\mathbf{M}}_j - \bar{\mathbf{M}}_j \mathbf{J}_j (\mathbf{J}_j^\top \bar{\mathbf{M}}_j \mathbf{J}_j)^+ \mathbf{J}_j^\top \bar{\mathbf{M}}_j)^{1/2} \cdot \\ \cdot \mathbf{B}_j \mathbf{J}_i (\mathbf{I}_{n_i} - (\mathbf{C}^\top \mathbf{C})^+ (\mathbf{C}^\top \mathbf{C}))$$

with

$$\begin{aligned} \mathbf{B}_l &:= (\bar{\mathbf{M}}_l - \bar{\mathbf{M}}_l \mathbf{J}_l (\mathbf{J}_l^\top \bar{\mathbf{M}}_l \mathbf{J}_l)^+ \mathbf{J}_l^\top \bar{\mathbf{M}}_l)^{1/2} \mathbf{H}_l \mathbf{K}_i^{-1}, \ (l \in I_i), \\ \mathbf{B}_0 &:= \mathbf{M}_i^{1/2} \mathbf{K}_i^{-1}, \\ \mathbf{C} &:= \Big(\mathbf{B}_0^\top \mathbf{B}_0 + \sum_{l \in I_i} \mathbf{B}_l^\top \mathbf{B}_l \Big)^{1/2} \mathbf{J}_i = \bar{\mathbf{M}}_i^{1/2} \mathbf{J}_i, \end{aligned}$$

see (17a). Lemma 3 from Appendix A proves $\bar{\mathbf{J}}_j = \mathbf{0}_{d \times n_i}$ resulting in

$$\boldsymbol{\mu}_j = \bar{\boldsymbol{f}}_j + \bar{\mathbf{M}}_j \mathbf{K}_j (-\bar{\mathbf{H}}_j \bar{\mathbf{H}}_i \ddot{\boldsymbol{p}}_{\pi_i} - \bar{\mathbf{H}}_j \bar{\boldsymbol{k}}_i^{(\mathrm{II})} + \bar{\boldsymbol{k}}_j^{(\mathrm{II})}), \tag{21b}$$

see (21a) and (19).

The two alternative representations of $\boldsymbol{\mu}_j$ in (21a,b) provide two different ways to evaluate $\ddot{\boldsymbol{p}}_j$ by forward recursion, see (14):

$$\begin{aligned} \mathbf{0} &= \bar{\mathbf{M}}_j \mathbf{K}_j (\ddot{\boldsymbol{p}}_j + \bar{\mathbf{H}}_j \ddot{\boldsymbol{p}}_i + \bar{\boldsymbol{k}}_j^{(\mathrm{II})}), & (22a) \\ \mathbf{0} &= \bar{\mathbf{M}}_j \mathbf{K}_j (\ddot{\boldsymbol{p}}_j - \bar{\mathbf{H}}_j \bar{\mathbf{H}}_i \ddot{\boldsymbol{p}}_{\pi_i} - \bar{\mathbf{H}}_j \bar{\boldsymbol{k}}_i^{(\mathrm{II})} + \bar{\boldsymbol{k}}_j^{(\mathrm{II})}). & (22b) \end{aligned}$$

To keep the presentation compact, we assume in the following that the bodies with rank-deficient body mass matrix $\mathbf{M}_i$ are isolated in the kinematic tree, i.e., the predecessor of a body $(\bullet)^{(i)}$ with rank-deficient $\mathbf{M}_i$ is either the root body ($\pi_i = 0$) or a body with non-singular body mass matrix:

$$\operatorname{rank} \mathbf{M}_i < d \quad \Rightarrow \quad (\ \pi_i = 0 \quad \text{or} \quad \operatorname{rank} \mathbf{M}_{\pi_i} = d\ ). \tag{23}$$

For $\pi_i \neq 0$, this assumption implies that $\bar{\mathbf{M}}_{\pi_i}$ is non-singular as well since a symmetric, positive semi-definite matrix $\mathbf{M}_{\pi_i}$ with $\operatorname{rank} \mathbf{M}_{\pi_i} = d$ is positive definite and $\bar{\mathbf{M}}_{\pi_i}$ is defined by a sum of $\mathbf{K}_{\pi_i}^{-\top} \mathbf{M}_{\pi_i} \mathbf{K}_{\pi_i}^{-1}$ and a finite number of symmetric, positive semi-definite matrices, see (17a).

The technical assumption (23) allows to evaluate recursively $\ddot{\boldsymbol{p}}_j$ for all bodies with non-singular $\bar{\mathbf{M}}_j$ starting at the root body $(\bullet)^{(0)}$ that was supposed to be inertially fixed ($\ddot{\boldsymbol{p}}_0 = \mathbf{0}$). Let a body $(\bullet)^{(j)}$ be given with $\operatorname{rank} \bar{\mathbf{M}}_j = d$ and denote its predecessor by $i := \pi_j$. If $\pi_j = 0$ or $i = \pi_j \neq 0$ and the condensed mass matrix $\bar{\mathbf{M}}_i$ of the predecessor is non-singular then $\ddot{\boldsymbol{p}}_i$ has been computed before and $\ddot{\boldsymbol{p}}_j$ may be obtained from (22a) since $\mathbf{K}_j$ was supposed to be non-singular:

$$\ddot{\boldsymbol{p}}_j = -\bar{\mathbf{H}}_j \ddot{\boldsymbol{p}}_i - \bar{\boldsymbol{k}}_j^{(\mathrm{II})}. \tag{24a}$$

Otherwise, $i = \pi_j \neq 0$ and $\bar{\mathbf{M}}_i$ is rank-deficient and the corresponding body mass matrix $\mathbf{M}_i$ has to be rank-deficient as well. Then, body $(\bullet)^{(i)}$ may be skipped in the forward recursion since assumption (23) guarantees that $\ddot{\boldsymbol{p}}_{\pi_i}$ has been computed before and $\ddot{\boldsymbol{p}}_j$ may be obtained from (22b):

$$\ddot{\boldsymbol{p}}_j = \bar{\mathbf{H}}_j \bar{\mathbf{H}}_i \ddot{\boldsymbol{p}}_{\pi_i} + \bar{\mathbf{H}}_j \bar{\boldsymbol{k}}_i^{(\mathrm{II})} - \bar{\boldsymbol{k}}_j^{(\mathrm{II})}. \tag{24b}$$

The proposed forward recursion algorithm evaluates $\ddot{\boldsymbol{p}}_j$ for all bodies $(\bullet)^{(j)}$ with non-singular $\bar{\mathbf{M}}_j$ provided that the technical assumption (23) is satisfied and the body mass matrices $\mathbf{M}_i$ are symmetric, positive semi-definite for all N bodies of the multibody system model, ($i = 1, \ldots, N$). It provides furthermore the algorithmic basis to evaluate the equations of motion as mixed second and first order system for the joint coordinates $\boldsymbol{q} = (\boldsymbol{q}_1, \ldots, \boldsymbol{q}_N)$.

3.4 Equations of motion: mixed second and first order system of differential equations

Let us consider again a body $(\bullet)^{(j)}$ with non-singular $\bar{\mathbf{M}}_j$ and denote $i = \pi_j$. Multiplying (15) from the left by $\mathbf{J}_j^\top \bar{\mathbf{M}}_j$, we get

$$(\mathbf{J}_j^\top \bar{\mathbf{M}}_j \mathbf{J}_j) \ddot{\boldsymbol{q}}_j = -\mathbf{J}_j^\top \bar{\mathbf{M}}_j (\mathbf{H}_j \ddot{\boldsymbol{p}}_i + \boldsymbol{k}_j^{(\mathrm{II})}) - \mathbf{J}_j^\top \bar{\boldsymbol{f}}_j \tag{25a}$$

since $\mathbf{J}_j^\top \bar{\mathbf{M}}_j \mathbf{K}_j \ddot{\boldsymbol{p}}_j = \mathbf{J}_j^\top \bar{\boldsymbol{f}}_j - \mathbf{J}_j^\top \boldsymbol{\mu}_j = \mathbf{J}_j^\top \bar{\boldsymbol{f}}_j$, see (14) and (12).

As before, we get an alternative expression from substituting in the right hand side of (25a) the term $\ddot{\boldsymbol{p}}_i$ according to (19):

$$\begin{aligned} &(\mathbf{J}_j^\top \bar{\mathbf{M}}_j \mathbf{J}_j) \ddot{\boldsymbol{q}}_j - \mathbf{J}_j^\top \bar{\mathbf{M}}_j \mathbf{H}_j \mathbf{K}_i^{-1} \mathbf{J}_i \left(\mathbf{I}_{n_i} - (\mathbf{J}_i^\top \bar{\mathbf{M}}_i \mathbf{J}_i)^+ (\mathbf{J}_i^\top \bar{\mathbf{M}}_i \mathbf{J}_i) \right) \ddot{\boldsymbol{q}}_i \\ &= \mathbf{J}_j^\top \bar{\mathbf{M}}_j (\mathbf{H}_j \bar{\mathbf{H}}_i \ddot{\boldsymbol{p}}_{\pi_i} + \mathbf{H}_j \bar{\boldsymbol{k}}_i^{(\mathrm{II})} - \boldsymbol{k}_j^{(\mathrm{II})}) - \mathbf{J}_j^\top \bar{\boldsymbol{f}}_j. \qquad (25b) \end{aligned}$$

The simpler expression (25a) can be used whenever $\pi_j = 0$ or $i = \pi_j \neq 0$ and $\bar{\mathbf{M}}_i$ is non-singular since $\ddot{\boldsymbol{p}}_i$ has been evaluated by forward recursion in that case. Eq. (25b) shows that the situation is substantially more complicated if $i = \pi_j \neq 0$ and $\bar{\mathbf{M}}_i$ is rank-deficient since in that case body $(\bullet)^{(i)}$ and the corresponding acceleration term $\ddot{\boldsymbol{p}}_i$ have been skipped in the forward recursion and $\ddot{\boldsymbol{p}}_{\pi_i}$ has to be used instead.

In the latter case, the technical assumption (23) guarantees that $\ddot{\boldsymbol{p}}_{\pi_i}$ is really available from the forward recursion since either $\pi_i = 0$ or $\pi_i \neq 0$ and $\bar{\mathbf{M}}_{\pi_i}$ is non-singular. Therefore, the right hand side of (25a) with (j, i) being substituted by (i, π_i) may be evaluated straightforwardly:

$$(\mathbf{J}_i^\top \bar{\mathbf{M}}_i \mathbf{J}_i) \ddot{\boldsymbol{q}}_i = -\mathbf{J}_i^\top \bar{\mathbf{M}}_i (\mathbf{H}_i \ddot{\boldsymbol{p}}_{\pi_i} + \boldsymbol{k}_i^{(\mathrm{II})}) - \mathbf{J}_i^\top \bar{\boldsymbol{f}}_i. \tag{26}$$

In (26), the coefficient $\mathbf{J}_i^\top \bar{\mathbf{M}}_i \mathbf{J}_i$ of $\ddot{\boldsymbol{q}}_i$ is symmetric, positive semi-definite and may therefore be diagonalized, see Remark 1 in Appendix A:

$$\mathbf{J}_i^\top \bar{\mathbf{M}}_i \mathbf{J}_i =: \mathbf{A}_i = \mathbf{X}_i \boldsymbol{\Lambda}_i \mathbf{X}_i^\top = \mathbf{X}_i \begin{pmatrix} \bar{\boldsymbol{\Lambda}}_i & \mathbf{0} \\ \mathbf{0} & \mathbf{0} \end{pmatrix} \mathbf{X}_i^\top \in \mathbb{R}^{n_i \times n_i}.$$

Here, $\bar{\boldsymbol{\Lambda}}_i \in \mathbb{R}^{r_i \times r_i}$ with $r_i = \operatorname{rank}(\mathbf{J}_i^\top \bar{\mathbf{M}}_i \mathbf{J}_i) \le n_i$ is a positive diagonal matrix containing the non-zero eigenvalues of $\mathbf{J}_i^\top \bar{\mathbf{M}}_i \mathbf{J}_i$. Matrix $\mathbf{X}_i \in \mathbb{R}^{n_i \times n_i}$ is orthogonal. The projector $\mathbf{I}_{n_i} - (\mathbf{J}_i^\top \bar{\mathbf{M}}_i \mathbf{J}_i)^+ (\mathbf{J}_i^\top \bar{\mathbf{M}}_i \mathbf{J}_i)$ in (25b) may be expressed as

$$\begin{aligned} \mathbf{I}_{n_i} - (\mathbf{J}_i^\top \bar{\mathbf{M}}_i \mathbf{J}_i)^+ (\mathbf{J}_i^\top \bar{\mathbf{M}}_i \mathbf{J}_i) &= \mathbf{X}_i (\mathbf{I}_{n_i} - \boldsymbol{\Lambda}_i^+ \underbrace{\mathbf{X}_i^\top \mathbf{X}_i}_{= \mathbf{I}_{n_i}} \boldsymbol{\Lambda}_i) \mathbf{X}_i^\top \\ &= \mathbf{X}_i \Big(\mathbf{I}_{n_i} - \begin{pmatrix} \mathbf{I}_{r_i} & \mathbf{0} \\ \mathbf{0} & \mathbf{0} \end{pmatrix} \Big) \mathbf{X}_i^\top = \mathbf{X}_i \begin{pmatrix} \mathbf{0} & \mathbf{0} \\ \mathbf{0} & \mathbf{I}_{n_i - r_i} \end{pmatrix} \mathbf{X}_i^\top . \end{aligned} \quad (27)$$

Multiplying (26) from the left by $\mathbf{X}_i^\top$, we end up with a decoupled system of r_i linearly implicit second order differential equations

$$\begin{aligned} & \bar{\boldsymbol{\Lambda}}_i (\, \mathbf{I}_{r_i} \;\; \mathbf{0}_{r_i \times (n_i - r_i)} \,) \mathbf{X}_i^\top \ddot{\boldsymbol{q}}_i & (28a) \\ = & -(\, \mathbf{I}_{r_i} \;\; \mathbf{0}_{r_i \times (n_i - r_i)} \,) \mathbf{X}_i^\top \mathbf{J}_i^\top (\bar{\mathbf{M}}_i (\mathbf{H}_i \ddot{\boldsymbol{p}}_{\pi_i} + \boldsymbol{k}_i^{(\mathrm{II})}) + \bar{\boldsymbol{f}}_i) \end{aligned}$$

and $n_i - r_i$ additional equations that do not contain $\ddot{\boldsymbol{q}}_i$ and may further be simplified to

$$\mathbf{0}_{n_i - r_i} = (\, \mathbf{0}_{(n_i - r_i) \times r_i} \;\; \mathbf{I}_{n_i - r_i} \,) \mathbf{X}_i^\top \mathbf{J}_i^\top \bar{\boldsymbol{f}}_i \quad (28b)$$

since $(\, \mathbf{0}_{(n_i - r_i) \times r_i} \;\; \mathbf{I}_{n_i - r_i} \,) \mathbf{X}_i^\top \mathbf{J}_i^\top \bar{\mathbf{M}}_i (\mathbf{H}_i \ddot{\boldsymbol{p}}_{\pi_i} + \boldsymbol{k}_i^{(\mathrm{II})})$ vanishes because (27) and Lemma 2b with $\mathbf{C} := \bar{\mathbf{M}}_i^{1/2} \mathbf{J}_i \in \mathbb{R}^{d \times n_i}$ imply

$$\begin{aligned} & (\, \mathbf{0}_{(n_i - r_i) \times r_i} \;\; \mathbf{I}_{n_i - r_i} \,) \mathbf{X}_i^\top \mathbf{J}_i^\top \bar{\mathbf{M}}_i = \\ = & (\, \mathbf{0}_{(n_i - r_i) \times r_i} \;\; \mathbf{I}_{n_i - r_i} \,) \underbrace{\mathbf{X}_i^\top \mathbf{X}_i}_{= \mathbf{I}_{n_i}} \begin{pmatrix} \mathbf{0} & \mathbf{0} \\ \mathbf{0} & \mathbf{I}_{n_i - r_i} \end{pmatrix} \mathbf{X}_i^\top \mathbf{J}_i^\top \bar{\mathbf{M}}_i \\ = & (\, \mathbf{0}_{(n_i - r_i) \times r_i} \;\; \mathbf{I}_{n_i - r_i} \,) \mathbf{X}_i^\top \left(\mathbf{I}_{n_i} - (\mathbf{J}_i^\top \bar{\mathbf{M}}_i \mathbf{J}_i)(\mathbf{J}_i^\top \bar{\mathbf{M}}_i \mathbf{J}_i)^+ \right) \mathbf{J}_i^\top \bar{\mathbf{M}}_i \\ = & (\, \mathbf{0}_{(n_i - r_i) \times r_i} \;\; \mathbf{I}_{n_i - r_i} \,) \mathbf{X}_i^\top (\mathbf{I}_{n_i} - (\mathbf{C}^\top \mathbf{C})(\mathbf{C}^\top \mathbf{C})^+) \mathbf{C}^\top \bar{\mathbf{M}}_i^{1/2} \\ = & \mathbf{0}_{n_i \times d} . \end{aligned}$$

The equations of motion for the multibody system model are given by (25b) for all bodies $(\bullet)^{(j)}$ with $i := \pi_j \neq 0$ and $\operatorname{rank} \bar{\mathbf{M}}_i < d$ and by (25a) for the remaining bodies, $(j = 1, \ldots, N)$. They are composed of $\sum_i r_i$ linearly implicit second order differential equations (25b), (28a) and $\sum_i (n_i - r_i)$ additional equations (28b) to define $(\, \mathbf{0}_{(n_i - r_i) \times r_i} \;\; \mathbf{I}_{n_i - r_i} \,) \mathbf{X}_i^\top \dot{\boldsymbol{q}}_i$, $(i = 1, \ldots, N)$, see also the detailed discussion in Sect. 3.5 below. Similar to a classical residual formalism, see Eichberger (1994), the residuals in (25) may be used to integrate the equations of motion by general purpose DAE solvers like DASSL, see Brenan et al. (1996).

3.5 Equations of motion: formal analysis

For a formal analysis of equations of motion (25), we introduce velocity coordinates $\boldsymbol{v}_0 := \dot{\boldsymbol{p}}_0$,

$$\boldsymbol{v}_j := \begin{cases} \mathbf{X}_j^\top \dot{\boldsymbol{q}}_j - \mathbf{T}_j \dot{\boldsymbol{q}}_i & \text{if } \; i := \pi_j \neq 0 \text{ and } \operatorname{rank} \bar{\mathbf{M}}_i < d, \\ \mathbf{X}_j^\top \dot{\boldsymbol{q}}_j & \text{otherwise}, \end{cases} \quad (29)$$

$(j = 1, \ldots, N)$, with

$$\mathbf{T}_j := \boldsymbol{\Lambda}_j^{-1} \mathbf{X}_j^\top \mathbf{J}_j^\top \bar{\mathbf{M}}_j \mathbf{H}_j \mathbf{K}_i^{-1} \mathbf{J}_i \left(\mathbf{I}_{n_i} - (\mathbf{J}_i^\top \bar{\mathbf{M}}_i \mathbf{J}_i)^+ (\mathbf{J}_i^\top \bar{\mathbf{M}}_i \mathbf{J}_i) \right).$$

Because of (29), the joint coordinates $\dot{\boldsymbol{q}}_j$ may also be expressed in terms of $\boldsymbol{v} := (\boldsymbol{v}_1, \ldots, \boldsymbol{v}_N)$:

$$\dot{\boldsymbol{q}}_j = \boldsymbol{\varphi}_j^{[1]}(\boldsymbol{q}, \boldsymbol{v}, t), \;\; (j = 1, \ldots, N),$$

with

$$\boldsymbol{\varphi}_j^{[1]} = \begin{cases} \mathbf{X}_j (\boldsymbol{v}_j + \mathbf{T}_j \mathbf{X}_i \boldsymbol{v}_i) & \text{if } \; i := \pi_j \neq 0 \text{ and } \operatorname{rank} \bar{\mathbf{M}}_i < d, \\ \mathbf{X}_j \boldsymbol{v}_j & \text{otherwise} \end{cases} \quad (30)$$

since $\operatorname{rank} \bar{\mathbf{M}}_i < d$ and the technical assumption (23) imply $\dot{\boldsymbol{q}}_i = \mathbf{X}_i \boldsymbol{v}_i$. For all bodies $(\bullet)^{(i)}$ with $\operatorname{rank} \bar{\mathbf{M}}_i < d$, vector $\boldsymbol{v}_i$ is split according to

$$\boldsymbol{v}_i = \mathbf{X}_i^\top \dot{\boldsymbol{q}}_i = \begin{pmatrix} \boldsymbol{\eta}_i \\ \boldsymbol{\zeta}_i \end{pmatrix} \quad (31)$$

with

$$\begin{aligned} \boldsymbol{\eta}_i &:= \left(\mathbf{I}_{r_i} \;\; \mathbf{0}_{r_i \times (n_i - r_i)} \right) \mathbf{X}_i^\top \dot{\boldsymbol{q}}_i \in \mathbb{R}^{r_i}, \\ \boldsymbol{\zeta}_i &:= \left(\mathbf{0}_{(n_i - r_i) \times r_i} \;\; \mathbf{I}_{n_i - r_i} \right) \mathbf{X}_i^\top \dot{\boldsymbol{q}}_i \in \mathbb{R}^{n_i - r_i}. \end{aligned}$$

In the full rank case ($\operatorname{rank} \bar{\mathbf{M}}_i = d$), we set $\boldsymbol{\eta}_i := \boldsymbol{v}_i \in \mathbb{R}^{n_i}$ and leave $\boldsymbol{\zeta}_i$ "empty" since $r_i = \operatorname{rank}(\mathbf{J}_i^\top \bar{\mathbf{M}}_i \mathbf{J}_i) = n_i$, i.e., $n_i - r_i = 0$. In the rank-deficient case ($\operatorname{rank} \bar{\mathbf{M}}_i < d \Rightarrow r_i < n_i$), the technical assumption (23) guarantees that $\bar{\mathbf{M}}_j$ is nonsingular resulting in $\boldsymbol{\eta}_j = \boldsymbol{v}_j$. With (29) and the diagonalized projector in (27), we see that $\dot{\boldsymbol{q}}_j$ may be written as a linear combination of $\boldsymbol{\eta}_j$ and $\boldsymbol{\zeta}_i$ that is independent of $\boldsymbol{\eta}_i$:

$$\dot{\boldsymbol{q}}_j = \mathbf{X}_j (\boldsymbol{v}_j + \mathbf{T}_j \dot{\boldsymbol{q}}_i) = \mathbf{X}_j \boldsymbol{\eta}_j + \mathbf{X}_j \boldsymbol{\Lambda}_j^{-1} \mathbf{X}_j^\top \mathbf{J}_j^\top \bar{\mathbf{M}}_j \mathbf{H}_j \mathbf{K}_i^{-1} \mathbf{J}_i \mathbf{X}_i \begin{pmatrix} \mathbf{0}_{r_i} \\ \boldsymbol{\zeta}_i \end{pmatrix}.$$

The time derivative of $\boldsymbol{v}_j$ in (29) depends on time derivatives of

$$\mathbf{X}_j^\top = \mathbf{X}_j^\top (\boldsymbol{p}_0(t), \boldsymbol{p}(t), \boldsymbol{q}(t), t), \quad \mathbf{T}_j = \mathbf{T}_j (\boldsymbol{p}_0(t), \boldsymbol{p}(t), \boldsymbol{q}(t), t)$$

with $\boldsymbol{p} = \boldsymbol{p}(\boldsymbol{p}_0, \boldsymbol{q}, t)$, see (7). Let $\dot{\mathbf{X}}_j^\top = \dot{\mathbf{X}}_j^\top (\boldsymbol{p}_0, \boldsymbol{v}_0, \boldsymbol{q}, \dot{\boldsymbol{q}}, t)$ and $\dot{\mathbf{T}}_j = \dot{\mathbf{T}}_j (\boldsymbol{p}_0, \boldsymbol{v}_0, \boldsymbol{q}, \dot{\boldsymbol{q}}, t)$ be defined such that

$$\dot{\mathbf{X}}_j^\top \boldsymbol{w} = \frac{\mathrm{d}}{\mathrm{d}t} (\mathbf{X}_j^\top \boldsymbol{w}), \; (\boldsymbol{w} \in \mathbb{R}^{n_j}), \quad \dot{\mathbf{T}}_j \boldsymbol{w} = \frac{\mathrm{d}}{\mathrm{d}t} (\mathbf{T}_j \boldsymbol{w}), \; (\boldsymbol{w} \in \mathbb{R}^{n_{\pi_j}}).$$

For bodies $(\bullet)^{(j)}$ with $i := \pi_j \neq 0$ and $\operatorname{rank}\bar{\mathbf{M}}_i < d$, we get from (29), (30) and from the product rule

$$\begin{aligned}\dot{\boldsymbol{v}}_j &= \mathbf{X}_j^{\top}\ddot{\boldsymbol{q}}_j - \mathbf{T}_j\ddot{\boldsymbol{q}}_i + \dot{\mathbf{X}}_j^{\top}\dot{\boldsymbol{q}}_j - \dot{\mathbf{T}}_j\dot{\boldsymbol{q}}_i \\ &= \mathbf{X}_j^{\top}\ddot{\boldsymbol{q}}_j - \mathbf{T}_j\ddot{\boldsymbol{q}}_i + \dot{\mathbf{X}}_j^{\top}\mathbf{X}_j(\boldsymbol{v}_j + \mathbf{T}_j\mathbf{X}_i\boldsymbol{v}_i) - \dot{\mathbf{T}}_j\mathbf{X}_i\boldsymbol{v}_i\,.\end{aligned}$$

Multiplying the equations of motion (25b) from the left by $\boldsymbol{\Lambda}_j^{-1}\mathbf{X}_j^{\top}$, we observe

$$\boldsymbol{\Lambda}_j^{-1}\mathbf{X}_j^{\top}(\mathbf{J}_j^{\top}\bar{\mathbf{M}}_j\mathbf{J}_j) = \boldsymbol{\Lambda}_j^{-1}\mathbf{X}_j^{\top}\mathbf{X}_j\boldsymbol{\Lambda}_j\mathbf{X}_j^{\top} = \mathbf{X}_j^{\top}$$

and end up with

$$\begin{aligned}\dot{\boldsymbol{\eta}}_j = \dot{\boldsymbol{v}}_j &= \dot{\mathbf{X}}_j^{\top}\mathbf{X}_j(\boldsymbol{v}_j + \mathbf{T}_j\mathbf{X}_i\boldsymbol{v}_i) - \dot{\mathbf{T}}_j\mathbf{X}_i\boldsymbol{v}_i \\ &\quad + \boldsymbol{\Lambda}_j^{-1}\mathbf{X}_j^{\top}\mathbf{J}_j^{\top}(\bar{\mathbf{M}}_j(\mathbf{H}_j\bar{\mathbf{H}}_i\ddot{\boldsymbol{p}}_{\pi_i} + \mathbf{H}_j\bar{\boldsymbol{k}}_i^{(\mathrm{II})} - \boldsymbol{k}_j^{(\mathrm{II})}) - \bar{\boldsymbol{f}}_j)\,.\end{aligned} \tag{32}$$

Since $\ddot{\boldsymbol{p}}_{\pi_i}$ has been evaluated by forward recursion and the joint coordinates $\dot{\boldsymbol{q}}$ are given in terms of $\boldsymbol{q}$, $\boldsymbol{v}$, $\boldsymbol{p}_0(t)$, $\boldsymbol{v}_0(t)$ and t, see (30), the system of n_j first order differential equations (32) may be written as

$$\dot{\boldsymbol{\eta}}_j = \boldsymbol{\varphi}_j^{[2]}(\boldsymbol{q},\boldsymbol{v},t) \quad \text{if} \quad i := \pi_j \neq 0 \text{ and } \operatorname{rank}\bar{\mathbf{M}}_i < d\,.$$

In the same way, (28a) is seen to imply a system of $r_i \leq n_i$ first order differential equations

$$\begin{aligned}\dot{\boldsymbol{\eta}}_i = \boldsymbol{\varphi}_i^{[2]}(\boldsymbol{q},\boldsymbol{v},t) &:= \\ &-\bar{\boldsymbol{\Lambda}}_i^{-1}(\,\mathbf{I}_{r_i}\;\; \mathbf{0}_{r_i\times(n_i-r_i)}\,)\mathbf{X}_i^{\top}\mathbf{J}_i^{\top}(\bar{\mathbf{M}}_i(\mathbf{H}_i\ddot{\boldsymbol{p}}_{\pi_i} + \boldsymbol{k}_i^{(\mathrm{II})}) + \bar{\boldsymbol{f}}_i) \\ &\qquad + (\,\mathbf{I}_{r_i}\;\; \mathbf{0}_{r_i\times(n_i-r_i)}\,)\dot{\mathbf{X}}_i^{\top}\boldsymbol{v}_i\,.\end{aligned}$$

Finally the $n_i - r_i$ equations (28b) are written as algebraic equations

$$\mathbf{0}_{n_i-r_i} = \boldsymbol{\gamma}_i(\boldsymbol{q},\boldsymbol{v},t)\,.$$

With these transformations, the equations of motion are re-formulated as differential-algebraic system (2) with $\sum_i(n_i+r_i)$ differential variables $y_0 = (\boldsymbol{q}_1,\ldots,\boldsymbol{q}_N,\boldsymbol{\eta}_1,\ldots,\boldsymbol{\eta}_N)$ satisfying (2a) with right hand sides $\boldsymbol{\varphi}_i^{[1]}$ of dimension n_i and right hand sides $\boldsymbol{\varphi}_i^{[2]}$ of dimension r_i, $(i = 1,\ldots,N)$, and $\sum_i(n_i-r_i)$ algebraic variables $z_0 = \boldsymbol{\zeta} := (\boldsymbol{\zeta}_1,\ldots,\boldsymbol{\zeta}_N)$ satisfying (2b) with functions $\boldsymbol{\gamma}_i$, $(i = 1,\ldots,N)$.

The algebraic equations (2b) define implicitly the "algebraic" velocity components $z_0 = \boldsymbol{\zeta}$, if the Jacobian $\partial\boldsymbol{\gamma}/\partial\boldsymbol{\zeta}$ is non-singular along the solution. In practical applications, this regularity assumption will typically be satisfied if $\partial\boldsymbol{\gamma}_i/\partial\boldsymbol{\zeta}_i$ is non-singular for all bodies $(\bullet)^{(i)}$ with $r_i = \operatorname{rank}\bar{\mathbf{M}}_i < n_i$, $(i = 1,\ldots,N)$, which may be achieved by appropriate damping terms in the force vector $\bar{\boldsymbol{f}}_i$ that should depend on the velocity coordinates $\boldsymbol{\zeta}_i = \left(\,\mathbf{0}_{(n_i-r_i)\times r_i}\;\; \mathbf{I}_{n_i-r_i}\,\right)\mathbf{X}_i^{\top}\dot{\boldsymbol{q}}_i$, see (28b). A more detailed analysis of the regularity of Jacobian $\partial\boldsymbol{\gamma}/\partial\boldsymbol{\zeta}$ is subject of further research.

4 Neglecting inertia forces in multibody systems: two examples

The theoretical analysis of Sect. 3 generalizes the results of Arnold et al. (2010) from chain structured systems to general tree structured systems. In this section, we recall two academic test problems from Arnold et al. (2010) to illustrate the basic steps of these investigations. We consider a chain of two mass points $(\bullet)^{(i)}$, $(\bullet)^{(j)}$ in 2-D with $i = \pi_j$ and $\pi_i = 0$. I.e., body $(\bullet)^{(i)}$ follows in the kinematic chain directly the inertial system ("root") and is the predecessor of body $(\bullet)^{(j)}$. In Sect. 3, there are no specific physical assumptions on the joints between bodies $(\bullet)^{(0)}$ and $(\bullet)^{(i)}$ and between bodies $(\bullet)^{(i)}$ and $(\bullet)^{(j)}$, respectively. Therefore, the resulting set of r_i explicit second order differential equations (28a) and $n_i - r_i$ implicit first order differential equations (28b) for body $(\bullet)^{(i)}$ describes arbitrary joint configurations and has a substantially more complex mathematical structure than the corresponding equations of motion in a classical multibody formalism.

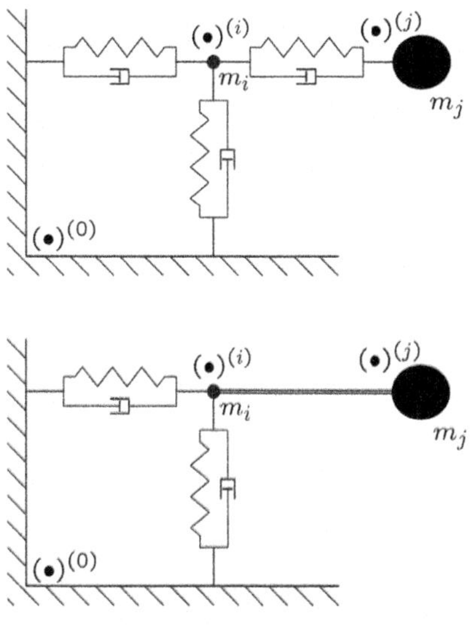

Figure 3. Two planar configurations illustrating the analysis of Sect. 3, see Arnold et al. (2010).

In the test problems, bodies $(\bullet)^{(i)}$ and $(\bullet)^{(j)}$ are represented by point masses with $d = 2$ degrees of freedom, see Fig. 3. The root body $(\bullet)^{(0)}$ is inertially fixed resulting in $\boldsymbol{p}_{\pi_i}(t) \equiv \mathbf{0}$. The absolute coordinates of bodies $(\bullet)^{(i)}$ and $(\bullet)^{(j)}$ are denoted by $\boldsymbol{p}_i = (p_{i,x}, p_{i,y})^{\top}$, $\boldsymbol{p}_j = (p_{j,x}, p_{j,y})^{\top} \in \mathbb{R}^2$. In this simplified setting, the diagonal mass matrices $\mathbf{M}_i$, $\mathbf{M}_j$ have format 2×2 and the joint Jacobians satisfy $\mathbf{J}_i \in \mathbb{R}^{2\times n_i}$ and $\mathbf{J}_j \in \mathbb{R}^{2\times n_j}$.

Bodies $(\bullet)^{(0)}$ and $(\bullet)^{(i)}$ are connected by two linear spring-damper elements acting parallel to the x-axis and y-axis, see Fig. 3. The $n_i = 2$ degrees of freedom in this joint are represented by joint coordinates $\boldsymbol{q}_i(t) = \boldsymbol{p}_i(t) \in \mathbb{R}^2$ such that the functions $\boldsymbol{k}_i^{(\mathrm{I})}$, $\boldsymbol{k}_i^{(\mathrm{II})}$ in (8), (9) and (10) vanish identically, $\mathbf{K}_i = \mathbf{I}_2$, $\mathbf{H}_i = \mathbf{J}_i = -\mathbf{I}_2$. The free motion of body $(\bullet)^{(j)}$ in y-direction is represented by the joint coordinate $q_{j,y}(t) := p_{j,y}(t) - p_{i,y}(t)$.

For the configuration in the upper plot of Fig. 3, bodies $(\bullet)^{(i)}$ and $(\bullet)^{(j)}$ are connected by another linear spring-damper element acting parallel to the x-axis.

The joint has $n_j = 2$ degrees of freedom with joint coordinates $\boldsymbol{q}_j = (q_{j,x}, q_{j,y})^\top \in \mathbb{R}^2$ and $q_{j,x}(t) := p_{j,x}(t) - p_{i,x}(t)$. Functions $\boldsymbol{k}_j^{(\mathrm{I})}$, $\boldsymbol{k}_j^{(\mathrm{II})}$ in (8) and (9) vanish identically, $\mathbf{K}_j = \mathbf{I}_2$, $\mathbf{H}_j = \mathbf{J}_j = -\mathbf{I}_2$.

In absolute coordinates, the equations of motion are given by

$$\begin{aligned} m_i\ddot{p}_{i,x} &= -d_{i,x}\dot{p}_{i,x} - c_{i,x}p_{i,x}\,, && (33a)\\ &= \quad + d_{j,x}(\dot{p}_{j,x} - \dot{p}_{i,x}) + c_{j,x}(p_{j,x} - p_{i,x})\,, \\ m_i\ddot{p}_{i,y} &= -d_{i,y}\dot{p}_{i,y} - c_{i,y}p_{i,y}\,, && (33b)\\ m_j\ddot{p}_{j,x} &= -d_{j,x}(\dot{p}_{j,x} - \dot{p}_{i,x}) - c_{j,x}(p_{j,x} - p_{i,x})\,, && (33c)\\ m_j\ddot{p}_{j,y} &= 0 && (33d) \end{aligned}$$

because (12) with $\mathbf{J}_i = \mathbf{J}_j = -\mathbf{I}_2$ implies $\boldsymbol{\mu}_i = \boldsymbol{\mu}_j = \mathbf{0}$.

With $\mathbf{J}_j = -\mathbf{I}_2$, Eqs. (17a) and (17b) simplify to $\bar{\mathbf{M}}_i = \mathbf{M}_i$ and $\bar{\boldsymbol{f}}_i = \boldsymbol{f}_i$. If the mass m_i of body $(\bullet)^{(i)}$ vanishes, we get $\bar{\mathbf{M}}_i = \mathbf{M}_i = m_i\mathbf{I}_2 = \mathbf{0}_{2\times 2}$ and $r_i = \operatorname{rank}(\mathbf{J}_i^\top\bar{\mathbf{M}}_i\mathbf{J}_i) = 0$, $\mathbf{X}_i = \mathbf{I}_2$. The equations of motion for coordinates $\boldsymbol{q}_i(t)$ are composed of $r_i = 0$ second order differential equations (28a) and the $n_i - r_i = 2$ implicit first order differential equations

$$\mathbf{0} = \boldsymbol{f}_i = \begin{pmatrix} -d_{i,x}\dot{q}_{i,x} - c_{i,x}q_{i,x} + d_{j,x}\dot{q}_{j,x} + c_{j,x}q_{j,x} \\ -d_{i,y}\dot{q}_{i,y} - c_{i,y}q_{i,y} \end{pmatrix} \qquad (34)$$

see (28b). This result is in perfect agreement with (33a,b) in the limit case $m_i = 0$.

The joint coordinates $\boldsymbol{q}_j = \boldsymbol{p}_j - \boldsymbol{q}_i$ are not defined explicitly if $\bar{\mathbf{M}}_i = \mathbf{0}$. The equations of motion (25b) yield

$$\begin{aligned} m_j(\ddot{q}_{j,x} + \ddot{q}_{i,x}) &= -d_{j,x}\dot{q}_{j,x} - c_{j,x}q_{j,x}\,, && (35a)\\ m_j(\ddot{q}_{j,y} + \ddot{q}_{i,y}) &= 0\,. && (35b) \end{aligned}$$

In the lower plot of Fig. 3, the relative motion of body $(\bullet)^{(j)}$ w.r.t. body $(\bullet)^{(i)}$ is restricted in x-direction by the scalar constraint $p_{j,x}(t) = p_{i,x}(t) + l_i$. The joint has only $n_j = 1$ degree of freedom $\boldsymbol{q}_j(t) = q_{j,y}(t)$ with a joint Jacobian $\mathbf{J}_j = -(0,1)^\top$, $\mathbf{K}_j = \mathbf{I}_2$, $\mathbf{H}_j = -\mathbf{I}_2$. The update formula (17a) with $\mathbf{M}_i = m_i\mathbf{I}_2$, $\mathbf{M}_j = m_j\mathbf{I}_2$ results in

$$\begin{aligned} \bar{\mathbf{M}}_i &= \mathbf{M}_i + (\bar{\mathbf{M}}_j - \bar{\mathbf{M}}_j\begin{pmatrix}0\\1\end{pmatrix}((0\ 1)\bar{\mathbf{M}}_j\begin{pmatrix}0\\1\end{pmatrix})^+(0\ 1)\bar{\mathbf{M}}_j) \\ &= \begin{pmatrix} m_i + m_j & 0 \\ 0 & m_i \end{pmatrix}. \end{aligned}$$

In (17b), we have $\bar{\boldsymbol{f}}_i = \boldsymbol{f}_i$ since $\boldsymbol{k}_j^{(\mathrm{I})} = \boldsymbol{k}_j^{(\mathrm{II})} = \mathbf{0}$ and $\boldsymbol{f}_j = \mathbf{0}$. The equations of motion in absolute coordinates are

$$\begin{aligned} m_i\ddot{p}_{i,x} - \mu_{j,x} &= -d_{i,x}\dot{p}_{i,x} - c_{i,x}p_{i,x}\,, && (36a)\\ m_i\ddot{p}_{i,y} &= -d_{i,y}\dot{p}_{i,y} - c_{i,y}p_{i,y}\,, && (36b)\\ m_j\ddot{p}_{j,x} + \mu_{j,x} &= 0\,, && (36c)\\ m_j\ddot{p}_{j,y} &= 0\,. && (36d) \end{aligned}$$

Because of $\ddot{p}_{j,x}(t) = \ddot{p}_{i,x}(t)$, we obtain $\mu_{j,x}(t) = -m_j\ddot{p}_{j,x}(t) = -m_j\ddot{p}_{i,x}(t)$ and the equations of motion (36a,b) get the form

$$\begin{aligned} (m_i + m_j)\ddot{p}_{i,x} &= -d_{i,x}\dot{p}_{i,x} - c_{i,x}p_{i,x}\,, \\ m_i\ddot{p}_{i,y} &= -d_{i,y}\dot{p}_{i,y} - c_{i,y}p_{i,y}\,. \end{aligned}$$

In the limit case $m_i = 0$, a combined set of $r_i = 1$ second order differential equation for $q_{i,x}(t)$ and $n_i - r_i = 1$ first order differential equation for $q_{i,y}(t)$ is obtained, see also (28a) and (28b) with $\mathbf{J}_i = -\mathbf{I}_2$, $\mathbf{X}_i = \mathbf{I}_2$ and $\bar{\boldsymbol{\Lambda}}_i = m_j$:

$$\begin{aligned} m_j\ddot{q}_{i,x} &= -d_{i,x}\dot{q}_{i,x} - c_{i,x}q_{i,x}\,, \\ 0 &= -d_{i,y}\dot{q}_{i,y} - c_{i,y}q_{i,y}\,. \end{aligned}$$

For this second test problem, we have a scalar joint coordinate $\boldsymbol{q}_j = q_{j,y} \in \mathbb{R}$ that is again not explicitly defined but has to satisfy the linearly implicit second order differential equation

$$m_j(\ddot{q}_{j,y} + \ddot{q}_{i,y}) = 0\,,$$

see (25b) and (35b).

5 Conclusions

Motivated by results from singular perturbation theory, multibody system models with bodies of small mass or nearly singular inertia terms are analysed considering the limit case of systems with rank-deficient body mass matrices. Replacing in a classical recursive multibody formalism the inverse of condensed body mass matrices by their Moore-Penrose pseudo-inverse, the backward recursion phase may be adapted to the rank-deficient case.

The crucial point in the analysis is the evaluation of accelerations for successors of bodies with rank-deficient body mass matrix in the forward recursion phase. It was shown that bodies with rank-deficient body mass matrix may simply be skipped in forward recursion. The acceleration coordinates of joints leaving such bodies to one of its successors are not given in explicit form but satisfy a linearly implicit equation that may be handled conveniently by common general purpose DAE solvers.

For each body with rank-deficient body mass matrix, a mixed system of first and second order differential equations is obtained resulting in a first order DAE that has index 1 for multibody system models with appropriate damping terms in the force elements acting at the "zero mass" body. Further investigations will be necessary to analyse practical aspects of this index-1 assumption in more detail.

In future research, the basic framework that has been developed in the present paper for tree structured rigid multibody system models will be extended to flexible systems and to multibody system models with (holonomic) constraints. These additional results will provide the algorithmic basis for a reference implementation in industrial multibody system simulation software.

Appendix A

Useful results from numerical linear algebra

To make the paper self-contained, we summarize in this appendix some basics of numerical linear algebra. A comprehensive discussion of these topics is given, e.g., by Golub and van Loan (1996).

Remark 1 Any symmetric, positive semi-definite matrix $\mathbf{A} \in \mathbb{R}^{m\times m}$ with $r := \operatorname{rank}\mathbf{A} \leq m$ has an orthonormal basis of eigenvectors $\boldsymbol{x}_1, \boldsymbol{x}_2, \dots, \boldsymbol{x}_r, \boldsymbol{x}_{r+1}, \dots, \boldsymbol{x}_m$ corresponding to its eigenvalues $\lambda_1 \geq \lambda_2 \geq \dots \geq \lambda_r > \lambda_{r+1} = \dots = \lambda_m = 0$. Summarizing the eigenvectors in the orthogonal matrix $\mathbf{X} := [\boldsymbol{x}_1, \dots, \boldsymbol{x}_m] \in \mathbb{R}^{m\times m}$, we get $\mathbf{AX} = \mathbf{X\Lambda}$ with the diagonal matrix $\mathbf{\Lambda} := \operatorname{diag}_{1\leq k\leq m} \lambda_k$.

(a) With $\mathbf{\Lambda}^{1/2} := \operatorname{diag}_{1\leq k\leq m} \sqrt{\lambda_k}$, matrix $\mathbf{A}^{1/2} := \mathbf{X\Lambda}^{1/2}\mathbf{X}^\top$ is well defined and independent of the specific choice of orthogonal eigenvectors $\boldsymbol{x}_1, \dots, \boldsymbol{x}_m$. Matrix $\mathbf{A}^{1/2}$ is symmetric, positive semi-definite and satisfies $\mathbf{A}^{1/2}\mathbf{A}^{1/2} = (\mathbf{X\Lambda}^{1/2}\mathbf{X}^\top)(\mathbf{X\Lambda}^{1/2}\mathbf{X}^\top) = \mathbf{X\Lambda}^{1/2}\mathbf{\Lambda}^{1/2}\mathbf{X}^\top = \mathbf{X\Lambda X}^\top = \mathbf{A}$.

(b) With $\bar{\mathbf{\Lambda}} := \operatorname{diag}_{1\leq k\leq r} \lambda_k \in \mathbb{R}^{r\times r}$, the *Moore-Penrose pseudo-inverse* of $\mathbf{A}$ is given by

$$\mathbf{A}^+ = \mathbf{X\Lambda}^+\mathbf{X}^\top \text{ with } \mathbf{\Lambda}^+ := \begin{pmatrix} \bar{\mathbf{\Lambda}}^{-1} & \mathbf{0}_{r\times(m-r)} \\ \mathbf{0}_{(m-r)\times r} & \mathbf{0}_{(m-r)\times(m-r)} \end{pmatrix}. \tag{A1}$$

It defines orthogonal projectors $\mathbf{AA}^+$ and $\mathbf{I}_m - \mathbf{A}^+\mathbf{A}$ projecting on the range of $\mathbf{A}$ and on the null space of $\mathbf{A}$, respectively:

$$\mathbf{AA}^+\mathbf{A} = \mathbf{A}, \ (\mathbf{I}_m - \mathbf{AA}^+)\mathbf{A} = \mathbf{0}_{m\times m}, \ \mathbf{A}(\mathbf{I}_m - \mathbf{A}^+\mathbf{A}) = \mathbf{0}_{m\times m}. \tag{A2}$$

If $\mathbf{A}$ is not only positive semi-definite, but even positive definite, the Moore-Penrose pseudo-inverse $\mathbf{A}^+$ coincides with the classical inverse $\mathbf{A}^{-1}$ since $\mathbf{A}$ is non-singular and $r = \operatorname{rank}\mathbf{A} = m$ in that case, $\mathbf{\Lambda}^+ = \mathbf{\Lambda}^{-1}$.

Lemma 1 *Consider a symmetric, positive semi-definite matrix* $\mathbf{M} \in \mathbb{R}^{d\times d}$, *(square) matrices* $\mathbf{H}, \mathbf{K} \in \mathbb{R}^{d\times d}$ *of the same format and a (rectangular) matrix* $\mathbf{J} \in \mathbb{R}^{d\times n}$ *with* $0 < n \leq d$. *If* $\mathbf{K}$ *is non-singular then matrix*

$$\mathbf{K}^{-\top}\mathbf{H}^\top(\mathbf{M} - \mathbf{MJ}(\mathbf{J}^\top\mathbf{MJ})^+\mathbf{J}^\top\mathbf{M})\mathbf{HK}^{-1}$$

with $\mathbf{K}^{-\top} = (\mathbf{K}^{-1})^\top$ *is symmetric and positive semi-definite.*

Proof The argument $\mathbf{J}^\top\mathbf{MJ} \in \mathbb{R}^{n\times n}$ of the Moore-Penrose pseudo-inverse may be written as $\mathbf{J}^\top\mathbf{MJ} = \mathbf{C}^\top\mathbf{C}$ with $\mathbf{C} := \mathbf{M}^{1/2}\mathbf{J} \in \mathbb{R}^{d\times n}$. The singular value decomposition of $\mathbf{C}$ has the form $\mathbf{C} = \mathbf{P\Sigma Q}^\top$ with orthogonal matrices $\mathbf{P} \in \mathbb{R}^{d\times d}$, $\mathbf{Q} \in \mathbb{R}^{n\times n}$ and the $(d\times n)$-matrix

$$\mathbf{\Sigma} = \begin{pmatrix} \mathbf{\Sigma}_0 \\ \mathbf{0}_{(d-n)\times n} \end{pmatrix} \text{ with } \mathbf{\Sigma}_0 = \begin{pmatrix} \bar{\mathbf{\Sigma}}_0 & \mathbf{0}_{r\times(n-r)} \\ \mathbf{0}_{(n-r)\times r} & \mathbf{0}_{(n-r)\times(n-r)} \end{pmatrix} \in \mathbb{R}^{n\times n},$$

$\bar{\mathbf{\Sigma}}_0 := \operatorname{diag}_{1\leq k\leq r} \sigma_k$, that summarizes the positive singular values $\sigma_1 \geq \sigma_2 \geq \dots \geq \sigma_r > 0$ of matrix $\mathbf{C}$. Here, $r \leq n$ denotes the rank of $\mathbf{C}$ and the remaining singular values vanish identically: $\sigma_{r+1} = \dots = \sigma_n = 0$.

Applying (A1) to

$$\mathbf{A} := \mathbf{C}^\top\mathbf{C} = (\mathbf{P\Sigma Q}^\top)^\top(\mathbf{P\Sigma Q}^\top) = \mathbf{Q\Sigma}^\top\mathbf{\Sigma Q}^\top,$$

we get $m = n$, $\mathbf{X} = \mathbf{Q}$ and $\mathbf{\Lambda} = \mathbf{\Sigma}^\top\mathbf{\Sigma} = \mathbf{\Sigma}_0^\top\mathbf{\Sigma}_0$, i.e., $\bar{\mathbf{\Lambda}} = \bar{\mathbf{\Sigma}}_0^\top\bar{\mathbf{\Sigma}}_0$. The assertion of the Lemma follows from

$$\begin{aligned}&\mathbf{K}^{-\top}\mathbf{H}^\top(\mathbf{M} - \mathbf{MJ}(\mathbf{J}^\top\mathbf{MJ})^+\mathbf{J}^\top\mathbf{M})\mathbf{HK}^{-1}\\ &= (\mathbf{M}^{1/2}\mathbf{HK}^{-1})^\top(\mathbf{I}_d - \mathbf{C}(\mathbf{C}^\top\mathbf{C})^+\mathbf{C}^\top)(\mathbf{M}^{1/2}\mathbf{HK}^{-1})\end{aligned}$$

and

$$\begin{aligned}&\mathbf{I}_d - \mathbf{C}(\mathbf{C}^\top\mathbf{C})^+\mathbf{C}^\top\\ &= \mathbf{P}\Big(\mathbf{I}_d - \begin{pmatrix}\mathbf{\Sigma}_0\\ \mathbf{0}\end{pmatrix}\underbrace{\mathbf{Q}^\top\mathbf{Q}}_{=\mathbf{I}_n}(\mathbf{\Sigma}_0^\top\mathbf{\Sigma}_0)^+\underbrace{\mathbf{Q}^\top\mathbf{Q}}_{=\mathbf{I}_n}(\,\mathbf{\Sigma}_0^\top \ \ \mathbf{0}\,)\Big)\mathbf{P}^\top\\ &= \mathbf{P}\left(\mathbf{I}_d - \begin{pmatrix}\mathbf{I}_r & \mathbf{0}\\ \mathbf{0} & \mathbf{0}\end{pmatrix}\right)\mathbf{P}^\top = \mathbf{P}\begin{pmatrix}\mathbf{0} & \mathbf{0}\\ \mathbf{0} & \mathbf{I}_{d-r}\end{pmatrix}\mathbf{P}^\top,\end{aligned}$$

see (A1).

Lemma 2 *For any matrix* $\mathbf{C} \in \mathbb{R}^{d\times n}$ *with* $0 < n \leq d$ *the Moore-Penrose pseudo-inverse of* $(\mathbf{C}^\top\mathbf{C}) \in \mathbb{R}^{n\times n}$ *satisfies*

(a) $\mathbf{C}(\mathbf{I}_n - (\mathbf{C}^\top\mathbf{C})^+(\mathbf{C}^\top\mathbf{C})) = \mathbf{0}_{d\times n}$,

(b) $(\mathbf{I}_n - (\mathbf{C}^\top\mathbf{C})(\mathbf{C}^\top\mathbf{C})^+)\mathbf{C}^\top = \mathbf{0}_{n\times d}$.

Proof (a) Eq. (A2) with $\mathbf{A} := \mathbf{C}^\top\mathbf{C} \in \mathbb{R}^{n\times n}$ yields

$$\begin{aligned}&(\mathbf{C}^\top\mathbf{C})(\mathbf{I}_n - (\mathbf{C}^\top\mathbf{C})^+(\mathbf{C}^\top\mathbf{C}))\\ &= \mathbf{C}^\top\mathbf{C} - (\mathbf{C}^\top\mathbf{C})(\mathbf{C}^\top\mathbf{C})^+(\mathbf{C}^\top\mathbf{C}) = \mathbf{C}^\top\mathbf{C} - \mathbf{C}^\top\mathbf{C} = \mathbf{0}_{n\times n},\end{aligned}$$

i.e., $\mathbf{C}^\top\mathbf{C}z_k = \mathbf{0}_n$ for the column vectors $z_k \in \mathbb{R}^n$, $(k = 1, \dots, n)$, of matrix $\mathbf{I}_n - (\mathbf{C}^\top\mathbf{C})^+(\mathbf{C}^\top\mathbf{C})$. Since $\mathbf{C}^\top\mathbf{C}z_k = \mathbf{0}_n$ implies $z_k^\top\mathbf{C}^\top\mathbf{C}z_k = 0$ and $\|\mathbf{C}z_k\|_2^2 = 0$, we get $\mathbf{C}z_k = \mathbf{0}_d$, $(k = 1, \dots, n)$, and see that all column vectors of $\mathbf{I}_n - (\mathbf{C}^\top\mathbf{C})^+(\mathbf{C}^\top\mathbf{C})$ belong to the null space of $\mathbf{C}$.

(b) Assertion (b) follows in the same way from

$$(\mathbf{I}_n - (\mathbf{C}^\top\mathbf{C})(\mathbf{C}^\top\mathbf{C})^+)(\mathbf{C}^\top\mathbf{C}) = \mathbf{C}^\top\mathbf{C} - \mathbf{C}^\top\mathbf{C} = \mathbf{0}_{n\times n}.$$

Lemma 3 *Consider a finite index set* $I \subset \mathbb{N}_+$ *and matrices* $\mathbf{B}_0 \in \mathbb{R}^{d\times d}$, $\mathbf{B}_j \in \mathbb{R}^{d\times d}$, $(j \in I)$, $\mathbf{J} \in \mathbb{R}^{d\times n}$ *with* $0 < n \leq d$. *Then, matrix* $\mathbf{B}_0^\top\mathbf{B}_0 + \sum_{j\in I}\mathbf{B}_j^\top\mathbf{B}_j$ *is symmetric, positive semi-definite and*

$$\mathbf{C} := \Big(\mathbf{B}_0^\top\mathbf{B}_0 + \sum_{j\in I}\mathbf{B}_j^\top\mathbf{B}_j\Big)^{1/2}\mathbf{J} \in \mathbb{R}^{d\times n}$$

is well-defined and satisfies

$$\mathbf{B}_j\mathbf{J}(\mathbf{I}_n - (\mathbf{C}^\top\mathbf{C})^+(\mathbf{C}^\top\mathbf{C})) = \mathbf{0}_{d\times n}, \ (j \in I).$$

Proof For any vector $z \in \mathbb{R}^n$, we get

$$z^\top(\mathbf{B}_0^\top \mathbf{B}_0 + \sum_{j \in I} \mathbf{B}_j^\top \mathbf{B}_j)z = \|\mathbf{B}_0 z\|_2^2 + \sum_{j \in I} \|\mathbf{B}_j z\|_2^2 \geq 0\,,$$

i.e., the symmetric matrix $\mathbf{B}_0^\top \mathbf{B}_0 + \sum_j \mathbf{B}_j^\top \mathbf{B}_j$ is positive semi-definite and matrix $\mathbf{C} \in \mathbb{R}^{d \times n}$ is well-defined.

From the proof of Lemma 2 we know that $\mathbf{C} z_k = \mathbf{0}_d$ for all column vectors z_k of matrix $\mathbf{I}_n - (\mathbf{C}^\top \mathbf{C})^+(\mathbf{C}^\top \mathbf{C})$. Therefore,

$$\begin{aligned} 0 &= (\mathbf{C} z_k)^\top (\mathbf{C} z_k) = z_k^\top (\mathbf{C}^\top \mathbf{C}) z_k \\ &= z_k^\top \mathbf{J}^\top (\mathbf{B}_0^\top \mathbf{B}_0 + \sum\nolimits_j \mathbf{B}_j^\top \mathbf{B}_j) \mathbf{J} z_k \\ &= z_k^\top \mathbf{J}^\top \mathbf{B}_0^\top \mathbf{B}_0 \mathbf{J} z_k + \sum\nolimits_j z_k^\top \mathbf{J}^\top \mathbf{B}_j^\top \mathbf{B}_j \mathbf{J} z_k \\ &= \|\mathbf{B}_0 \mathbf{J} z_k\|_2^2 + \sum\nolimits_j \|\mathbf{B}_j \mathbf{J} z_k\|_2^2 . \end{aligned}$$

This sum of non-negative numbers may vanish only, if $\|\mathbf{B}_0 \mathbf{J} z_k\|_2 = 0$ and $\|\mathbf{B}_j \mathbf{J} z_k\|_2 = 0$, $(j \in I,\ k = 1, \ldots, n)$. Therefore, $\mathbf{B}_j \mathbf{J} z_k = \mathbf{0}_d$ for all $j \in I$ and all column vectors z_k, $(k = 1, \ldots, n)$, of matrix $\mathbf{I}_n - (\mathbf{C}^\top \mathbf{C})^+(\mathbf{C}^\top \mathbf{C})$.

Acknowledgements. This research on recursive multibody formalisms for tree structured systems is funded by the German Federal Ministry of Education and Research (BMBF grant 05M10NHB). It extends previous results for chain structured systems that were obtained in a project "Quasistatic methods for numerical analysis of flexible multibody systems" that was funded by the German Research Foundation (grant AR 243/2-1). The fruitful cooperation with Bernhard Burgermeister (SIMPACK AG), Steffen Weber (Martin Luther University Halle-Wittenberg) and Michael Valášek (CTU Prague) is gratefully acknowledged. Figures 1, 2 and 3 were adapted from Arnold et al. (2010) and were provided by Bernhard Burgermeister and Steffen Weber, respectively.

Edited by: A. Tasora

References

Arnold, M., Burgermeister, B., and Weber, S.: Improved time integration of multibody system models using methods from singular perturbation theory, in: Proc. of The 1st Joint International Conference on Multibody System Dynamics, 25–27 May 2010, Lappeenranta, Finland, 2010.

Brandl, H., Johanni, R., and Otter, M.: A very efficient algorithm for the simulation of robots and similar multibody systems without inversion of the mass matrix, in: Theory of Robots, edited by: Kopacek, P., Troch, I., and Desoyer, K., 95–100, Pergamon Press, Oxford, 1988.

Brenan, K., Campbell, S., and Petzold, L.: Numerical solution of initial-value problems in differential-algebraic equations, SIAM, Philadelphia, 2nd Edn., 1996.

Burgermeister, B., Arnold, M., and Eichberger, A.: Smooth velocity approximation for constrained systems in real-time simulation, Multibody Syst. Dyn., 26, 1–14, doi:10.1007/s11044-011-9243-1, 2011.

Eich-Soellner, E. and Führer, C.: Numerical Methods in Multibody Dynamics, Teubner-Verlag, Stuttgart, 1998.

Eichberger, A.: Transputer-based multibody system dynamic simulation: Part I. The residual algorithm – a modified inverse dynamic formulation, Mech. Struct. Mach., 22, 211–237, 1994.

Eichberger, A. and Rulka, W.: Process save reduction by macro joint approach: The key to real time and efficient vehicle simulation, Vehicle Syst. Dyn., 41, 401–413, 2004.

Golub, G. and van Loan, C.: Matrix Computations, The Johns Hopkins University Press, Baltimore London, 3rd Edn., 1996.

Hager, C. and Wohlmuth, B.: Analysis of a space-time discretization for dynamic elasticity problems based on mass-free surface elements, SIAM J. Numer. Anal., 47, 1863–1885, 2009.

Hairer, E. and Wanner, G.: Solving Ordinary Differential Equations. II. Stiff and Differential-Algebraic Problems, Springer-Verlag, Berlin Heidelberg New York, 2nd Edn., 1996.

Lubich, C.: Integration of stiff mechanical systems by Runge–Kutta methods, Z. Angew. Math. Phys., 44, 1022–1053, 1993.

Lubich, C., Nowak, U., Pöhle, U., and Engstler, C.: MEXX – Numerical software for the integration of constrained mechanical multibody systems, Tech. Rep. SC 92–12, ZIB Berlin, 1992.

Schiehlen, W.: Multibody system dynamics: roots and perspectives, Multibody Syst. Dyn., 1, 149–188, 1997.

Schönen, R.: Strukturdynamische Mehrkörpersimulation des Verbrennungsmotors mit elastohydrodynamischer Grundlagerkopplung, Fortschrittsberichte Strukturanalyse und Tribologie, Band 15, Kassel University Press, Kassel, 2003.

Stumpp, T.: Asymptotic expansions and attractive invariant manifolds of strongly damped mechanical systems, ZAMM – Z. Angew. Math. Mech., 88, 630–643, 2008.

Udwadia, F. and Phohomsiri, P.: Explicit equations of motion for constrained mechanical systems with singular mass matrices and applications to multi-body dynamics, Proc. R. Soc. A, 462, 2097–2117, doi:10.1098/rspa.2006.1662, 2006.

Weber, S., Arnold, M., and Valášek, M.: Quasistatic approximations for stiff second-order differential equations, Appl. Numer. Math., 62, 1579–1590, doi:10.1016/j.apnum.2012.06.030, 2012.

11

Geometrically exact Cosserat rods with Kelvin–Voigt type viscous damping

J. Linn[1], H. Lang[2], and A. Tuganov[1]

[1]Fraunhofer Institute for Industrial Mathematics, Fraunhofer Platz 1, 67633 Kaiserslautern, Germany
[2]Chair of Applied Dynamics, Univ. Erlangen-Nürnberg, Konrad-Zuse-Str. 3–5, 91052 Erlangen, Germany

Correspondence to: J. Linn (joachim.linn@itwm.fraunhofer.de)

Abstract. We present the derivation of a simple viscous damping model of Kelvin–Voigt type for geometrically exact Cosserat rods from three-dimensional continuum theory. Assuming moderate curvature of the rod in its reference configuration, strains remaining small in its deformed configurations, strain rates that vary slowly compared to internal relaxation processes, and a homogeneous and isotropic material, we obtain explicit formulas for the damping parameters of the model in terms of the well known stiffness parameters of the rod and the retardation time constants defined as the ratios of bulk and shear viscosities to the respective elastic moduli. We briefly discuss the range of validity of the Kelvin–Voigt model and illustrate its behaviour for large bending deformations with a numerical example.

1 Introduction

Simulation models for computing the transient response of structural members to dynamic excitations should contain a good approach to account for *dissipative effects* in order to be useful in realistic applications. If the structure considered may be treated within the range of *linear* dynamics with small vibration amplitudes, there is a well established set of standard approaches, e.g. Rayleigh damping, or a more general modal damping ansatz, to add such effects on the level of discretized versions of linear elastic structural models (see e.g. Craig and Kurdila, 2006). In the case of *geometrically exact* structure models for rods and shells (Antman, 2005), such linear approaches are not applicable. Geometrically exact rods, in particular, have a wide range of applications in *flexible multibody dynamics*. We refer to the brief introduction given in ch. 6 of Géradin and Cardona (2001) for a summary of the related work published before 2000, and to ch. 15 of Bauchau (2011) for a more recent account on this subject. Here the proper way to model viscous damping requires the inclusion of a *frame-indifferent viscoelastic constitutive model* into the continuum formulation of the structure model that is capable of dealing with *large displacements* and *finite rotations* (see Bauchau et al., 2008).

1.1 Viscous Kelvin–Voigt damping for Cosserat rods

In our recent work (Lang et al., 2011), we suggested the possibly simplest model of this kind to introduce *viscous material damping* in our quaternionic reformulation of Simo's dynamic continuum model for Cosserat rods (Simo, 1985). Following general considerations of Antman (2005) about the functional form of viscoelatic constitutive laws for Cosserat rods, we simply added viscous contributions, which we assumed to be proportional to the *rates* of the *material strain measures* $\boldsymbol{U}(s,t)$ and $\boldsymbol{V}(s,t)$ of the rod, to the *material stress resultants* $\boldsymbol{F}(s,t)$ and *stress couples* $\boldsymbol{M}(s,t)$, resulting in a constitutive model of *Kelvin–Voigt* type:

$$\boldsymbol{F} = \hat{\mathbb{C}}_F \cdot (\boldsymbol{V} - \boldsymbol{V}_0) + \hat{\mathbb{V}}_F \cdot \partial_t \boldsymbol{V}, \; \boldsymbol{M} = \hat{\mathbb{C}}_M \cdot (\boldsymbol{U} - \boldsymbol{U}_0) + \hat{\mathbb{V}}_M \cdot \partial_t \boldsymbol{U}. \quad (1)$$

A detailed presentation of the kinematical quantities and dynamic equilibrium equations of a Cosserat rod is given in Sect. 2 (see Figs. 1 and 2 for a compact summary).

In the material constitutive equations (1) the elastic properties of the rod are determined by the *effective stiffness parameters* contained in the symmetric 3×3 matrices $\hat{\mathbb{C}}_F$ and $\hat{\mathbb{C}}_M$. For homogeneous isotropic materials, both matrices are diagonal and given by:

$$\hat{\mathbb{C}}_F = \mathrm{diag}(GA, GA, EA), \; \hat{\mathbb{C}}_M = \mathrm{diag}(EI_1, EI_2, GI_3), \quad (2)$$

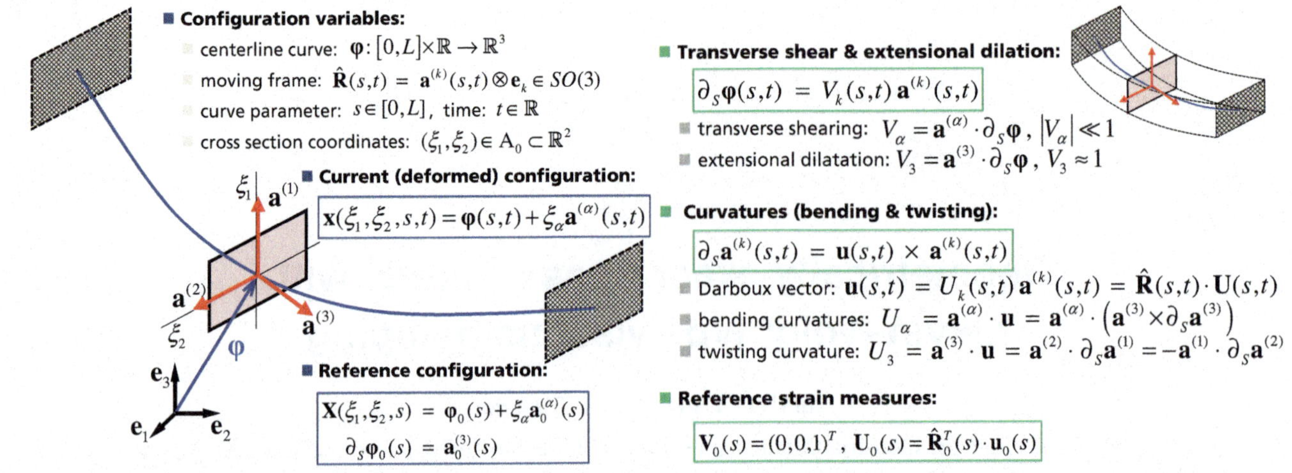

Figure 1. Left: kinematic quantities for the (deformed) current and (undeformed) reference configurations of a Cosserat rod. Right: strain measures of a Cosserat rod for transverse shearing, extensional dilatation, bending and twisting.

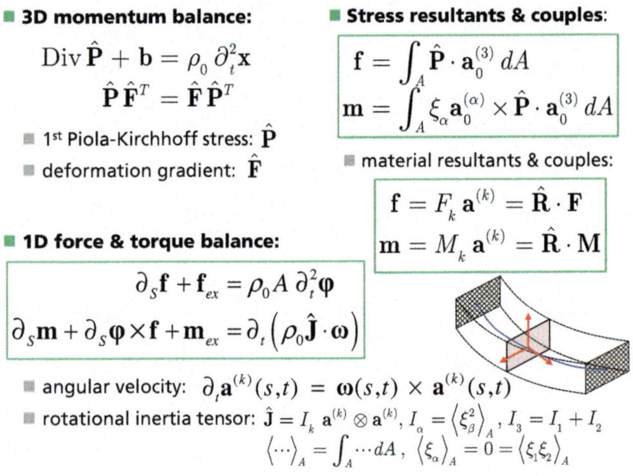

Figure 2. Dynamic equilibrium equations of a Cosserat rod.

with stiffness parameters given by the elastic moduli E and G and geometric parameters (area A, geometric moments I_k) of the cross section. In Lang et al. (2011) we assumed a similar structure for the matrices $\hat{\mathbb{V}}_F$ and $\hat{\mathbb{V}}_M$, which determine the viscous response:

$$\hat{\mathbb{V}}_F = \mathrm{diag}(\gamma_{S1},\gamma_{S2},\gamma_E)\ , \quad \hat{\mathbb{V}}_M = \mathrm{diag}(\gamma_{B1},\gamma_{B2},\gamma_T)\ . \qquad (3)$$

The set of six *effective viscosity parameters* γ_{xx} introduced in Eq. (3) represents the *integrated cross-sectional viscous damping behaviour* associated to the basic deformation modes (bending, twisting, transverse shearing and extension) of the rod, in the same way as the well known set of stiffness parameters given above determines the corresponding elastic response.

1.2 Effective damping parameter formulas

However, in Lang et al. (2011) the damping parameters γ_{xx} remained undetermined w.r.t. their specific dependence on material and geometric properties. Considering the special case of homogeneous and isotropic material properties, they certainly cannot be independent, but rather should be mutually related in a similar way as the stiffness parameters of the rod in terms of two material parameters (E, G) and the geometrical quantities (A, I_k) associated to the cross section. Assuming moderate curvature of the rod in its reference configuration, strains remaining small in its deformed configurations, strain rates that vary slowly compared to internal relaxation processes within the material, and a homogeneous and isotropic material, we will show that they are given by

$$\frac{\gamma_{S1/2}}{A} = \frac{\gamma_T}{I_3} = \eta,\ \frac{\gamma_E}{A} = \frac{\gamma_{B1/2}}{I_{1/2}} = \zeta(1-2\nu)^2 + \frac{4}{3}\eta(1+\nu)^2, \qquad (4)$$

where ζ and η are the *bulk and shear viscosities* of a viscoelastic *Kelvin–Voigt solid* (Lemaitre and Chaboche, 1990) with elastic moduli G and $E = 2G(1+\nu)$. While the viscous damping of the deformation modes of pure shear type is solely affected by shear viscosity η, extensional and bending deformations are both associated to normal stresses in the direction orthogonal to the cross section, which are damped by a specific combination of both bulk and shear viscosity that depends on the compressibilty of the material and may be interpreted as *extensional viscosity* parameter

$$\eta_E := \zeta(1-2\nu)^2 + \frac{4}{3}\eta(1+\nu)^2\ . \qquad (5)$$

Introducing the *retardation time* constants $\tau_S = \eta/G$ and $\tau_B = \zeta/K$, which relate the viscosities η and ζ to the shear and bulk moduli G and $3K = E/(1-2\nu)$, as well as the time constant $\tau_E := \eta_E/E = \frac{1}{3}[(1-2\nu)\tau_B + 2(1+\nu)\tau_S]$ relating extensional viscosity to Young's modulus, the formulas (4) may be rewritten equivalently as

$$\frac{\gamma_{S\,1/2}}{GA} = \frac{\gamma_T}{GI_3} = \tau_S\,,\ \frac{\gamma_E}{EA} = \frac{\gamma_{B1/2}}{EI_{1/2}} = \tau_E \tag{6}$$

in terms of the stiffness parameters of the rod and the retardation time constants. Interesting special cases of Eq. (6) are the simplified expressions $\eta_E = \zeta + \frac{4}{3}\eta$, $\tau_E = \frac{1}{3}(\tau_B + 2\tau_S)$ for completely compressible materials ($\nu = 0$), and $\eta_E = 3\eta$, $\tau_E = \tau_S$ for incompressible materials ($\nu = \frac{1}{2}$). The relation $\eta_E/\eta = 3$ between shear and extensional viscosity is well known as *Trouton's ratio* for incompressible Newtonian fluids (Trouton, 1906) and holds more generally for viscoelastic fluids in the limit of very small strain rates (Petrie, 2006). If $\zeta/\eta = K/G \Leftrightarrow \tau_B = \tau_S$ holds, one obtains $\tau_E = \tau_{B/S}$ as extensional retardation time constant (independent of ν).

Effective parameters modified by shear correction factors

It is well known that the stiffness parameters GA and GI_3 related to *shearing type* deformation modes systematically overestimate the actual stiffness of the structure for cross section geometries that display non-negligible warping. In the case of transverse shearing, this is accounted for via a modification of the corresponding stiffness parameter $GA \rightarrow GA_\alpha := GA\varkappa_\alpha$ by introducing dimensionless *shear correction factors* $\varkappa_\alpha \leq 1$ depending on the cross section geometry (see Cowper, 1966; Gruttmann and Wagner, 2001). Likewise, the *torsional rigidity* $C_T = GJ_T$ of a rod exactly equals GI_3 in the case of (annular) circular cross sections only, but is smaller than this value otherwise due to the presence of out-of-plane warping of cross sections. The replacement $GI_3 \rightarrow C_T$ correcting this deficieny corresponds to the introduction of another dimensionless correction factor $\varkappa_3 = J_T/I_3 \leq 1$ depending on the cross section geometry[1] which modifies the torsional stiffness according to the replacement rule $GI_3 \rightarrow GJ_T = GI_3\varkappa_3$. Altogether the various shear corrections mentioned above yield the corrected set of stiffness parameter values[2]

$$\hat{\mathbb{C}}_F = \mathrm{diag}(GA_1, GA_2, EA)\,,\ \hat{\mathbb{C}}_M = \mathrm{diag}(EI_1, EI_2, GJ_T)\,. \tag{7}$$

[1]In the case of an *elliptic* cross section with half axes a and b, the area moments are given by $I_1 = \frac{\pi}{4}a^3b$ and $I_2 = \frac{\pi}{4}ab^3$, while $C_T/G = J_T = \pi a^3b^3/(a^2+b^2) = 4I_1I_2/(I_1+I_2)$, such that $\varkappa_3 = J_T/I_3 = 4I_1I_2/(I_1+I_2)^2 \leq 1$ in this case. Equality ($\varkappa_3 = 1$) holds in the case of a circular cross section with $a = b = r \Rightarrow I_{1/2} = \frac{\pi}{4}r^4 = \frac{1}{2}I_3$ only. According to *Nikolai's inequality* $C_T \leq 4GI_1I_2/(I_1+I_2)$ the special case of an elliptic cross section maximes torsional rigidity among all asymmetric cross section geometries, and the value $GI_3 = 2GI$ valid for circular cross sections provides the absolute maximum of torsional rigidity (Berdichevsky, 1981).

[2]The stiffness parameters EA and EI_α are not affected by shear warping effects. However, they already account for *uniform lateral contraction*, which is a simple specific type of *in plane* cross section warping. This topic is discussed further in Sect. 3.4 below.

We argue that the analogously modified damping parameters

$$\gamma_{S\,1/2} = GA_{1/2}\tau_S\,,\ \gamma_T = GJ_T\tau_S \tag{8}$$

associated to shearing type rod deformations likewise provide a corresponding improvement of the formulas (6), which accounts for the influence of cross section warping on effective viscous dissipation, such that the *effective viscosity matrices* $\hat{\mathbb{V}}_F$ and $\hat{\mathbb{V}}_M$ introduced in Eq. (3) may be rewritten as

$$\hat{\mathbb{V}}_F = \hat{\mathbb{C}}_F \cdot \mathrm{diag}(\tau_S, \tau_S, \tau_E)\,,\ \hat{\mathbb{V}}_M = \hat{\mathbb{C}}_M \cdot \mathrm{diag}(\tau_E, \tau_E, \tau_S) \tag{9}$$

in terms of the effective stiffness matrices and retardation time constants given above.

1.3 Related work on viscoelastic rods

While there is a rather large number of articles considering various kinds of damping terms (also of Kelvin–Voigt type) added to *linear* Euler–Bernoulli or Timoshenko beam models (usually assumed to have a straight reference geometry), one hardly finds any work on viscous damping models for *geometrically nonlinear* beams or rods in the literature.

One notable exception is Antman's work (2003), where a damping model as given by Eq. (1) with positive, but otherwise undetermined parameters (3) is suggested from a completely different, mathematically motivated viewpoint, namely: as a simple possibility to introduce dissipative terms (denoted as *artificial viscosity*) into the dynamic balance equations of a Cosserat rod, which constitute a nonlinear coupled hyperbolic system of PDEs (see also Weiss, 2002a), and thereby achieve a *regularization* effect in view of the possible formation of shock waves that might appear in the *undamped* hyperbolic equations.

The recent article of Abdel-Nasser and Shabana (2011) is another relevant work for our topic. By inserting a 3-D Kelvin–Voigt model into a geometrically nonlinear beam given in *absolute nodal coordinate formulation* (ANCF), the authors obtain a viscous damping model for such ANCF beams which (by construction) is closely related, but conceptually quite different from our approach proposed for Cosserat rods. Later we briefly discuss the relation of both damping models (see Sect. 4.3). We refer othwise to the article of Romero (2008) for a comparison of the geometrically exact and ANCF approaches to nonlinear rods.

Mata et al. (2008) model the inelastic constitutive behaviour of composite beam structures under dynamic loading, using a Cosserat model as kinematical basis. However, they evaluate inelastic stresses by *numerical integration* of 3-D Piola–Kirchhoff stresses *over 2-D discretizations of the local cross sections* to obtain the stress resultants and couples of Simo's model. This differs from our approach aiming at a *direct* formulation of frame-indifferent inelastic constitutive laws in terms of $\boldsymbol{F}$ and $\boldsymbol{M}$, as achieved e.g. by Simo et al. (1984) for viscoplastic rods. The viscous model proposed in Sect. 3.2 of their paper is likewise of Kelvin–Voigt (KV) type, but formulated in terms of a vectorial strain measure related to the *Biot strain* (see also Sect. A2) and defined *pointwise* within the cross section. Moreover, they set up their model using only a *single* viscosity parameter.

Although there seems to be no further work on viscoelastic Cosserat rods made from solid material, *viscoelastic flow* in domains with rod-like geometries has been discussed in a number of articles. In his work on the coiling of viscous jets, Ribe (2004) presents a reduction of the three-dimensional Navier–Stokes equations to the dynamic equilibrium equations of a Kirchhoff/Love rod, endowed with Maxwell type constitutive equations for the viscous forces and moments which govern the finite resistance of the jet axis to stretching, bending and twisting. Although the derivation approach is different from ours, it represents its fluid-mechanical counterpart, as it likewise provides effective damping parameters[3] as given in Eq. (4), in the special case of an incompressible viscous fluid ($\nu = \frac{1}{2}$) with extensional viscosity given by Trouton's relation $\eta_E = 3\eta$, which in turn confirms our derivation of this special result.

A systematic derivation and mathematical investigation of viscous string and rod models in the context of Ribe's work is given by Panda et al. (2008) and Marheineke and Wegener (2009). Klar et al. (2009) and Arne et al. (2011) likewise use Ribe's Maxwell type constitutive law in their related work on the simulation of viscous fibers aiming at applications in the area of textile and nonwoven production. Lorenz et al. (2012) extend constitutive modelling for viscous strings by deriving an *upper convected Maxwell* model using mathematical methods of asymptotic analysis.

In the same context we finally mention the discrete modelling approach for viscous threads presented by Bergou et al. (2010), which extends earlier work of Bergou et al. (2008) on discrete elastic rods that, similar to our own approach as briefly presented in Linn et al. (2008) (see also Jung et al., 2011), relies on geometrically exact rod kinematics based on the *discrete differential geometry* of framed curves.

1.4 Overview of the remaining sections of the paper

After collecting a few basics of Cosserat rod theory in the following Sect. 2, we proceed with our derivation of the formulas (4) in of a two-step procedure: in Sect. 3 we start with the derivation of the elastic (*stored*) energy function

$$W_e(t) = \int_0^L ds \frac{1}{2} \Big[\Delta \boldsymbol{V}(s,t)^T \cdot \hat{\mathbb{C}}_F \cdot \Delta \boldsymbol{V}(s,t) + \Delta \boldsymbol{U}(s,t)^T \cdot \hat{\mathbb{C}}_M \cdot \Delta \boldsymbol{U}(s,t) \Big] \tag{10}$$

of a Cosserat rod, which is a quadratic functional of the terms $\Delta\boldsymbol{U}(s,t) = \boldsymbol{U}(s,t) - \boldsymbol{U}_0(s)$ and $\Delta\boldsymbol{V}(s,t) = \boldsymbol{V}(s,t) - \boldsymbol{V}_0(s)$ measuring the *change of the strain measures* w.r.t their reference values, from three-dimensional continuum theory.

This sets the notational and conceptional framework for the subsequent derivation of the viscous part of our damping model given in Sect. 4 by an analogous procedure, which yields the *dissipation function*

$$D_v = \int_0^L ds \frac{1}{2} \Big[\partial_t \boldsymbol{V}^T \cdot \hat{\mathbb{V}}_F \cdot \partial_t \boldsymbol{V} + \partial_t \boldsymbol{U}^T \cdot \hat{\mathbb{V}}_M \cdot \partial_t \boldsymbol{U} \Big] \tag{11}$$

of a Cosserat rod introduced[4] in Lang et al. (2011). The dissipation function (11), deduced from the three-dimensional (volumetric) continuum version of the dissipation function of a *Kelvin–Voigt solid* (Landau and Lifshitz, 1986; Lemaitre and Chaboche, 1990), corresponds to one half of the volume-integrated *viscous stress power* of a rod-shaped Kelvin–Voigt solid, such that $2D_v$ yields the rate at which the rod dissipates mechanical energy.

Having completed our derivation of the Kelvin–Voigt model, we proceed by a discussion of a seemingly straightforward, but, as it turns out, erroneous approach to derive the viscous parts of the forces and moments as given by Eq. (1) as resultants in analogy to the elastic counterparts. This shows that our energy-based approach to derive viscous damping is the proper one. After that, we briefly comment on the relation of our continuum model to the Kelvin–Voigt type model recently proposed by Abdel-Nasser and Shabana (2011) within their alternative ANCF approach to geometrically nonlinear rods, and conclude Sect. 4 by a short discussion of the validity of the Kelvin–Voigt model w.r.t. a more general viscoelastic model of generalized Maxwell type.

In Sect. 5, we illustrate the behaviour of our viscous damping model (1) by some simple numerical experiments with a clamped cantilever beam subject to bending with large deflections. We conclude our article with a short summary.

2 Basic Cosserat rod theory

The configuration variables of a Cosserat rod (see Antman, 2005) are its *centerline* curve $\boldsymbol{\varphi}(s,t) = \varphi_k(s,t)\, \boldsymbol{e}_k$ with cartesian component functions $\varphi_k(s,t)$ w.r.t. the fixed global ONB

[3] In the case of viscous flow in a rod-shaped domain, the area $A(s)$ of the (circular) cross section as well as its geometric area moment $I(s)$ vary along the centerline curve in accordance with mass conservation modeled by a divergence-free velocity field of an extensional flow with uniform lateral contraction.

[4] In Lang et al. (2011) we absorbed the prefactor 1/2 into the definition (3) of the damping parameters (see Eqs. 9 and 10 in Sect. 2.2). This leads to an additional factor of 2 multiplying $\mathbb{V}_F$ and $\mathbb{V}_M$ in the constitutive equations (1) of the rod model.

$\{\boldsymbol{e}_1, \boldsymbol{e}_2, \boldsymbol{e}_3\}$ of Euclidian space and "*moving frame*" $\hat{\mathbf{R}}(s,t) = \boldsymbol{a}^{(k)}(s,t) \otimes \boldsymbol{e}_k \in \mathsf{SO(3)}$ of orthonormal director vectors, both smooth functions of the curve parameter s and the time t, with the pair $\{\boldsymbol{a}^{(1)}, \boldsymbol{a}^{(2)}\}$ of directors spanning the local cross sections with normals $\boldsymbol{a}^{(3)}$ along the rod (see Fig. 1).

2.1 Material strain measures

The material strain measures associated to the configuration variables are given by (i) the components $V_k = \boldsymbol{a}^{(k)} \cdot \partial_s \boldsymbol{\varphi}$ of the tangent vector in the local frame (i.e.: $\boldsymbol{V} = \hat{\mathbf{R}}^T \cdot \partial_s \boldsymbol{\varphi} = V_k \boldsymbol{e}_k$), with V_1, V_2 measuring *transverse shear* deformation and V_3 measuring *extensional dilatation*, and (ii) the *material Darboux vector* $\boldsymbol{U} = \hat{\mathbf{R}}^T \cdot \boldsymbol{u} = U_k \boldsymbol{e}_k$, obtained from its spatial counterpart $\boldsymbol{u} = U_k \boldsymbol{a}^{(k)}$ governing the Frénet equations $\partial_s \boldsymbol{a}^{(k)} = \boldsymbol{u} \times \boldsymbol{a}^{(k)}$ of the frame directors, with U_1, U_2 measuring *bending curvature* w.r.t. the director axes $\{\boldsymbol{a}^{(1)}, \boldsymbol{a}^{(2)}\}$, and U_3 measuring *torsional twist* around the cross section normal.

In general, the *reference configuration* of the rod, given by its centerline $\boldsymbol{\varphi}_0(s)$ and frame $\hat{\mathbf{R}}_0(s) = \boldsymbol{a}_0^{(k)}(s) \otimes \boldsymbol{e}_k$, may have non-zero curvature and twist (i.e.: $\boldsymbol{U}_0 \neq \boldsymbol{0}$). However we may assume zero initial shear ($V_{01} = V_{02} = 0$), such that all cross sections of the reference configuration are orthogonal to the centerline tangent vector, which coincides with the cross section normal (i.e.: $\partial_s \boldsymbol{\varphi}_0 = \boldsymbol{a}_0^{(3)} \Rightarrow V_{03} = 1$) if we choose the *arc–length* of the reference centerline as curve parameter s.

2.2 Dynamic equilibrium equations

The constitutive equations (1) – or more general ones of viscoelastic type (see ch. 8.2 in Antman, 2005) – are required to close the system of dynamic equilibrium equations

$$\partial_s \boldsymbol{f} + \boldsymbol{f}_{\text{ext}} = (\rho_0 A)\, \partial_t^2 \boldsymbol{\varphi} \tag{12}$$

$$\partial_s \boldsymbol{m} + \partial_s \boldsymbol{\varphi} \times \boldsymbol{f} + \boldsymbol{m}_{\text{ext}} = \partial_t \left(\rho_0 \hat{\mathbf{J}} \cdot \omega\right) \tag{13}$$

(see Fig. 2) which has to be satisfied by the *spatial* stress resultants $\boldsymbol{f} = \hat{\mathbf{R}} \cdot \boldsymbol{F}$ and stress couples $\boldsymbol{m} = \hat{\mathbf{R}} \cdot \boldsymbol{M}$ with appropriate boundary conditions (see Simo, 1985). The *inertial terms* appearing on the r.h.s. of the equations of the *balance of forces* (linear momentum) (12) and the *balance of moments* (angular momentum) (13) depend parametrically on the local *mass density* $\rho_0(s)$ along the rod as well as on geometrical parameters of the local cross section (area $A(s)$ and area moment tensor $\hat{\mathbf{J}}(s,t) = \hat{\mathbf{R}} \cdot \hat{\mathbf{J}}_0(s) \cdot \hat{\mathbf{R}}^T$) and contain the accelerations of the centerline positions $\partial_t^2 \boldsymbol{\varphi}(s,t)$ as well as the *angular velocity* vector $\omega(s,t)$, which is implicitely defined by the the temporal evolution equations $\partial_t \boldsymbol{a}^{(k)} = \omega \times \boldsymbol{a}^{(k)}$ of the frame in close analogy to the Darboux vector, and its time derivative $\partial_t \omega(s,t)$ as dynamical variables (see Simo, 1985; Antman, 2005; Lang et al., 2011 for details).

Although we implemented Kelvin–Voigt type viscous damping given by Eq. (1) for our discrete[5] Cosserat model formulated with unit quaternions as explained in detail by Lang et al. (2011) and investigated further in Lang and Arnold (2012) w.r.t. numerical aspects, we do not make use of this particular formulation here, as it is more practical to work with the directors associated to SO(3) frames for the vector-algebraic calculations which we have to carry out within our derivations of one-dimensional rod functionals from three-dimensional continuum formulation.

2.3 Spatial configurations of a Cosserat rod

Introducing cartesian coordinates (ξ_1, ξ_2) w.r.t. the director basis $\{\boldsymbol{a}_0^{(1)}(s), \boldsymbol{a}_0^{(2)}(s)\}$ of the cross section located at the centerline point $\boldsymbol{\varphi}_0(s)$, the spatial positions of material points in the reference configuration of the rod are given by[6]

$$\boldsymbol{X}(\xi_1, \xi_2, s) = \boldsymbol{\varphi}_0(s) + \xi_\alpha\, \boldsymbol{a}_0^{(\alpha)}(s)\,. \tag{14}$$

The positions of the same material points in the current (deformed) configuration are then given by

$$\boldsymbol{x}(\xi_1, \xi_2, s, t) = \boldsymbol{\varphi}(s,t) + \xi_\alpha\, \boldsymbol{a}^{(\alpha)}(s,t) + \boldsymbol{w}(\xi_1, \xi_2, s, t) \tag{15}$$

in terms of the deformed centerline curve $\boldsymbol{\varphi}(s,t)$, the rotated orthonormal cross section basis vectors $\{\boldsymbol{a}^{(1)}(s,t), \boldsymbol{a}^{(2)}(s,t)\}$, the same pair of cartesian cross section coordinates (ξ_1, ξ_2), and an additional displacement vector field $\boldsymbol{w}(\xi_1, \xi_2, s, t)$, which by definition describes the (in-plane and out-of-plane) *warping* deformations of the cross sections along the deformed rod.

The kinematic assumption that the cross sections of a rod remain *plane and rigid* in a configuration is equivalent to the assumption that the displacement field $\boldsymbol{w}$ vanishes identically. Although we will initially adhere to this very common assumption for rod models, we will later admit some specific form of in-plane deformation of cross sections – namely: a *uniform lateral contraction* – to correct a deficiency w.r.t. artificial in-plane normal stresses caused by the excessively rigid kinematical ansatz (15) with $\boldsymbol{w} \equiv \boldsymbol{0}$.

For simplicity we assume the rod to be *prismatic*, such that all cross sections along the rod are identical, and the domain of the cartesian coordinates (ξ_1, ξ_2) coincides with one fixed domain $\mathcal{A} \subset \mathbb{R}^2$. As usual we choose the geometrical center of the domain $\mathcal{A}$ to coincide with the origin of $\mathbb{R}^2$

[5]Practical applications of our Cosserat rod model with Kelvin–Voigt damping in Multibody System Dynamics are reported in our recent collaboration with Schulze et al. (2012). We refer to the article of Zupan et al. (2009) for fundamental aspects of Cosserat rods with rotational d.o.f. represented by unit quaternions, as well as to the recent work (2012B) of the same authors discussing the *undamped* dynamics of quaternionic Cosserat rods with various time integration approaches. Appendix B contains additional remarks related to alternative discretization approaches and model variants.

[6]Within this paper we make use of Einstein's summation convention – as the reader may have observed already – w.r.t. all indices occuring twice within *product* terms, with greek indices $\alpha, \beta, \ldots$ running from 1 to 2 and latin ones $i, j, k, \ldots$ from 1 to 3.

such that $\langle \xi_\alpha \rangle_{\mathcal{A}} = 0$ holds, where we introduced the shorthand notation $\langle f \rangle_{\mathcal{A}} := \int_{\mathcal{A}} f(\xi_1, \xi_2)\, d\xi_1 d\xi_2$ for the *cross section integral* of functions. In addition we choose the orientation of the orthonormal director pairs $\{\boldsymbol{a}_0^{(1)}(s), \boldsymbol{a}_0^{(2)}(s)\}$ as well as $\{\boldsymbol{a}^{(1)}(s,t), \boldsymbol{a}^{(2)}(s,t)\}$ to coincide with the principle geometrical axes of $\mathcal{A}$, such that $\langle \xi_1 \xi_2 \rangle_{\mathcal{A}} = 0$ holds. The quantities that characterize the geometric properties of the cross section in the Cosserat rod model are the *cross section area* $A = \langle 1 \rangle_{\mathcal{A}}$, the two *area moments* $I_1 = \left\langle \xi_2^2 \right\rangle_{\mathcal{A}}$, $I_2 = \left\langle \xi_1^2 \right\rangle_{\mathcal{A}}$ and the *polar area moment* $I_3 = \left\langle \xi_1^2 + \xi_2^2 \right\rangle_{\mathcal{A}} = I_1 + I_2$. With these definitions we obtain the centerline of the reference configuration as the average position $\boldsymbol{\varphi}_0(s) = \langle \boldsymbol{X} \rangle_{\mathcal{A}} / A$ of all material points of the cross section located at fixed s. The same relation $\boldsymbol{\varphi}(s,t) = \langle \boldsymbol{x} \rangle_{\mathcal{A}} / A$ holds for deformed configurations provided that the warping field $\boldsymbol{w}(\xi_1, \xi_2, s, t)$ satisfies $\langle \boldsymbol{w} \rangle_{\mathcal{A}} = \boldsymbol{0}$.

3 The stored energy function of a Cosserat rod

In order to set the notational and conceptional framework for the derivation of the viscous part of our damping model, we first give a brief account of the derivation of its elastic part, i.e.: the stored energy function (10) of a Cosserat rod. Within this derivation we will encounter a variety of smallness assumptions w.r.t. the curvatures describing the reference geometry of the rod as well as the local strains occuring in its deformed configurations. In our subsequent derivation of the viscous dissipation function (11) we will use the same assumptions and thereby remain consistent with the derivation of the elastic part.

3.1 Three-dimensional strain measures

In the first step we compute the *deformation gradient* $\hat{\mathbf{F}} = \boldsymbol{g}_k \otimes \boldsymbol{G}^k$, the *right Cauchy–Green tensor* $\hat{\mathbf{C}} = \hat{\mathbf{F}}^T \cdot \hat{\mathbf{F}}$ and the *Green–Lagrange strain tensor* $\hat{\mathbf{E}} = \frac{1}{2}(\hat{\mathbf{C}} - \hat{\mathbf{I}})$ from the basis vectors $\boldsymbol{G}_k = \partial_k \boldsymbol{X}$ and $\boldsymbol{g}_k = \partial_k \boldsymbol{x}$ associated to the curvilinear coordinates of the rod configurations given by Eqs. (14) and (15), with $\partial_k = \frac{\partial}{\partial \xi_k}$ for $k = 1,2$ and $\partial_3 = \partial_s$ for $\xi_3 = s$.

The dual basis vectors $\boldsymbol{G}^j$ and $\boldsymbol{g}^j$ are defined by the relations $\boldsymbol{G}_i \cdot \boldsymbol{G}^j = \delta_{ij}$ and $\boldsymbol{g}_i \cdot \boldsymbol{g}^j = \delta_{ij}$, respectively. Proceeding in this way we obtain the basis vectors of the reference configuration (14) as $\boldsymbol{G}_\alpha = \boldsymbol{a}_0^{(\alpha)}(s)$ and $\boldsymbol{G}_3 = \boldsymbol{a}_0^{(3)}(s) + \xi_\alpha U_{0\alpha}(s) \boldsymbol{a}_0^{(\alpha)}(s)$. Their duals may be computed from the general formula $\boldsymbol{G}^i = \boldsymbol{G}_j \times \boldsymbol{G}_k / J_0$ with $J_0 := (\boldsymbol{G}_1 \times \boldsymbol{G}_2) \cdot \boldsymbol{G}_3$, where (ijk) is a cyclic permutation of the indices (123), resulting in: $\boldsymbol{G}^1 = \boldsymbol{a}_0^{(1)} + \xi_2 \frac{U_{03}}{J_0} \boldsymbol{a}_0^{(3)}$, $\boldsymbol{G}^2 = \boldsymbol{a}_0^{(2)} - \xi_1 \frac{U_{03}}{J_0} \boldsymbol{a}_0^{(3)}$, and $\boldsymbol{G}^3 = \frac{1}{J_0} \boldsymbol{a}_0^{(3)}$.

The inital curvatures $U_{0\alpha}(s)$ contained in the determinant $J_0(s) = 1 + \xi_2 U_{01}(s) - \xi_1 U_{02}(s)$ and the initial twist $U_{03}(s)$ of the reference configuration (14) influence the deviation of the dual vectors $\boldsymbol{G}^k$ from the frame directors $\boldsymbol{a}_0^{(k)}(s)$ within the cross section. Both vectors coincide if the reference configuration of the rod is straight and untwisted (i.e.: $\boldsymbol{U}_0 = \boldsymbol{0}$). We have approximate coincidence $\boldsymbol{G}^k \approx \boldsymbol{a}_0^{(k)}(s)$ if curvature and twist of the reference configuration are sufficiently weak, in the sense that for the curvature radii given by $R_k = 1/|U_{0k}|$ the estimates $|\xi_\alpha|/R_3 \ll 1$ and $|\xi_\alpha|/R_\beta \ll 1 \Rightarrow J_0 \approx 1$ hold throughout each cross section along the rod, such that all initial curvature radii R_α are large compared to the cross section diameter. The geometric approximation $J_0(s) \approx 1$ will occur repeatedly and therefore play an important role in the derivation of the elastic energy and dissipation function of a Cosserat rod. To compute the deformation gradient we also need the basis vectors $\boldsymbol{g}_\alpha = \boldsymbol{a}^{(\alpha)}(s,t)$ and $\boldsymbol{g}_3 = \boldsymbol{a}^{(3)}(s,t) + \xi_\alpha U_\alpha(s,t) \boldsymbol{a}^{(\alpha)}(s,t)$ of the deformed configuration (15) with vanishing gradient of the warping vector field $(\partial_k \boldsymbol{w} = \boldsymbol{0})$. For the dual vectors $\boldsymbol{g}^k$ one obtaines analogous expressions as those for the dual vectors $\boldsymbol{G}^k$ given above, which we omit here.

For the special kinematical relations of a Cosserat rod, the deformation gradient $\hat{\mathbf{F}} = \boldsymbol{g}_k \otimes \boldsymbol{G}^k$ may be expressed in terms of a *pseudo-polar decomposition* (see Géradin and Cardona, 2001) by a factorization of the *relative rotation* $\hat{\mathbf{R}}_{\text{rel}}(s,t) := \hat{\mathbf{R}}(s,t) \cdot \hat{\mathbf{R}}_0^T(s) = \boldsymbol{a}^{(k)}(s,t) \otimes \boldsymbol{a}_0^{(k)}(s)$ connecting the moving frames of the reference and deformed configurations of the rod. The resulting formula

$$\hat{\mathbf{F}}(\xi_1, \xi_2, s, t) = \hat{\mathbf{R}}_{\text{rel}}(s,t) \left[\hat{\mathbf{I}} + \frac{1}{J_0(s)} \boldsymbol{H}(\xi_1, \xi_2, s, t) \otimes \boldsymbol{a}_0^{(3)}(s) \right] \tag{16}$$

depends on the *absolute values* of the curvatures of the reference configuration (14) through $J_0(s)$, and on the *change of the strain measures* of the Cosserat rod given by the difference vectors $\boldsymbol{U}(s,t) - \boldsymbol{U}_0(s)$ and $\boldsymbol{V}(s,t) - \boldsymbol{V}_0 = (V_1(s,t), V_2(s,t), V_3(s,t) - 1)^T$ in terms of the *material strain vector* $\boldsymbol{H}(\xi_1, \xi_2, s, t) = H_k(\xi_1, \xi_2, s, t) \boldsymbol{a}_0^{(k)}(s)$ with components

$$\begin{aligned} H_1(\xi_2, s, t) &= V_1(s,t) - \xi_2 \left[U_3(s,t) - U_{03}(s) \right], \\ H_2(\xi_1, s, t) &= V_2(s,t) + \xi_1 \left[U_3(s,t) - U_{03}(s) \right], \\ H_3(\xi_1, \xi_2, s, t) &= \left[V_3(s,t) - 1 \right] + \xi_2 \left[U_1(s,t) - U_{01}(s) \right] \\ &\quad - \xi_1 \left[U_2(s,t) - U_{02}(s) \right], \end{aligned} \tag{17}$$

which can be written more compactly[7] in the form of a cartesian vector $\hat{\mathbf{R}}_0^T \cdot \boldsymbol{H} = (\boldsymbol{V} - \boldsymbol{V}_0) - \xi_\alpha \boldsymbol{e}_\alpha \times (\boldsymbol{U} - \boldsymbol{U}_0) = H_k \boldsymbol{e}_k$ w.r.t. the fixed global frame $\{\boldsymbol{e}_1, \boldsymbol{e}_2, \boldsymbol{e}_3\}$.

Computing the right Cauchy–Green tensor $\hat{\mathbf{C}} = \hat{\mathbf{F}}^T \cdot \hat{\mathbf{F}}$ with the deformation gradient given by Eq. (16) results in the following kinematically exact expression for the Green–Lagrange strain tensor:

[7] Our derivation generalizes the one given by Géradin and Cardona (2001) for the simpler case of a straight and untwisted reference configuration of the rod (i.e. $\boldsymbol{U}_0 = 0$). Apart from using a slightly different and more compact notation, the kinematically exact expression of the deformation gradient given by Eqs. (16) and (17) is algebraically equivalent to the one given by Kapania and Li (2003) in eq. (47) of their paper. We note that the difference terms $\boldsymbol{U} - \boldsymbol{U}_0$ and $\boldsymbol{V} - \boldsymbol{V}_0$ appear already in the kinematically exact expression (18) *before* discarding second order terms. This shows that our approach is more general than the one chosen by Weiss (2002a).

$$\hat{\mathbf{E}} = \frac{1}{2J_0}\left[\boldsymbol{H}\otimes\boldsymbol{a}_0^{(3)} + \boldsymbol{a}_0^{(3)}\otimes\boldsymbol{H}\right] + \frac{\boldsymbol{H}^2}{2J_0^2}\,\boldsymbol{a}_0^{(3)}\otimes\boldsymbol{a}_0^{(3)}\,. \tag{18}$$

The approximate expression[8]

$$\hat{\mathbf{E}} \approx \frac{1}{2}\left[\boldsymbol{H}\otimes\boldsymbol{a}_0^{(3)} + \boldsymbol{a}_0^{(3)}\otimes\boldsymbol{H}\right] \tag{19}$$

may be obtained from Eq. (18) by the geometric approximation $J_0 \approx 1$ assumed to hold for the reference geometry and the additional assumption $||\boldsymbol{H}|| \ll 1$ of a *small material strain vector*. Later we will make use of the approximate strain tensor (19), which is *linear* in the vector field $\boldsymbol{H}$ and therefore also in the change of the strain measures of the rod, to obtain the stored energy function (10), which then becomes a *quadratic form* in the change of the strain measures. Likewise we will use Eq. (19) to obtain an approximation of the strain rate $\partial_t\boldsymbol{E}$ in terms of the rate $\partial_t\boldsymbol{H}$ of the strain vector.

3.2 Validity of the small strain approximation

For deformed configurations of a slender rod one observes large displacements and rotations, but local strains remain small. To estimate the size of the strain tensor it is useful to compute its components $E_{ij} = \boldsymbol{a}_0^{(i)}\cdot(\hat{\mathbf{E}}\cdot\boldsymbol{a}_0^{(j)})$ w.r.t. the tensor basis $\boldsymbol{a}_0^{(i)}\otimes\boldsymbol{a}_0^{(j)}$ obtained from the directors of the reference frame $\hat{\mathbf{R}}_0(s)$. From Eqs. (18) and (19) we obtain identically vanishing in-plane components ($E_{\alpha\beta} = E_{\beta\alpha} \equiv 0$), as well as the exact and approximate expressions

$$E_{\alpha 3} = E_{3\alpha} = \frac{H_\alpha}{2J_0} \approx \frac{H_\alpha}{2}\,,\quad E_{33} = \frac{H_3}{J_0} + \frac{\boldsymbol{H}^2}{2J_0^2} \approx H_3 \tag{20}$$

of the components related to out-of-plane deformations of the local cross section. Introducing the the quantity $|\xi|_{\max} := \max_{(\xi_1,\xi_2)\in\mathcal{A}}(|\xi_1|,|\xi_2|)$ to estimate the maximal linear extension of the cross section $\mathcal{A}$, one may estimate the deviation of the determinant $J_0(s)$ from unity by $|J_0(s)-1| \le |\xi|_{\max}(1/R_1+1/R_2)$ as a coarse check of the validity of the approximation $J_0 \approx 1$. Otherwise the smallness of the components of $\hat{\mathbf{E}}$ is implied by the smallness of the components H_k of the strain vector. According to Eq. (17) these components in turn become small if the change of the strain measures of the Cosserat rod is small, i.e. if the estimates $|V_\alpha| \ll 1$, $|V_3-1| \ll 1$, $|U_k-U_{0k}| \ll 1/|\xi|_{\max}$ hold. For slender rods with moderately curved undeformed geometry these estimates are obviously easily satisfiable, except for extreme deformations of the rod that produce large curvatures or twists of the order of the inverse cross section diameter. In this case, the assumption of small strains obviously would be invalid.

[8] We note that Eq. (19) may alternatively be interpreted as an *approximation of the Biot strain* (see Sect. A1 of the Appendix).

3.3 Elastic constitutive behaviour of rods at small strains

If we assume the rod material to behave hyperelastically with a stored energy density function $\Psi_e(\hat{\mathbf{E}})$, a simple Taylor expansion argument[9] shows that the behaviour of the energy density within the range of small strains may be well approximated by the quadratic function $\Psi_e(\hat{\mathbf{E}}) \approx \frac{1}{2}\hat{\mathbf{E}}:\mathbb{H}:\hat{\mathbf{E}}$, where $\mathbb{H} = \partial^2_{\hat{\mathbf{E}}}\Psi_e(\hat{\mathbf{0}})$ is the fourth order *Hookean material tensor* known from linear elasticity. This quadratic approximation yields a well defined frame-indifferent elastic energy density that is suitable for structure deformations at small local strains, but arbitrary large displacements and rotations, and therefore serves as a proper basis for the derivation of the stored energy function of a Cosserat rod.

The corresponding approximation of the stress-strain relation yields the 2nd *Piola–Kirchhoff stress* tensor $\hat{\mathbf{S}} = \partial_{\hat{\mathbf{E}}}\Psi_e(\hat{\mathbf{E}}) \approx \mathbb{H}:\hat{\mathbf{E}}$ for small strains. The 1st *Piola–Kirchhoff stress* tensor $\hat{\mathbf{P}}$, which is used to define the stress resultants and stress couples of the Cosserat rod model (see Simo, 1985, for details), is obtained by the transformation $\hat{\mathbf{P}} = \hat{\mathbf{F}}\cdot\hat{\mathbf{S}}$ using the deformation gradient, and the Cauchy stress tensor as the inverse Piola transformation $\hat{\sigma} = J^{-1}\hat{\mathbf{P}}\cdot\hat{\mathbf{F}}^T$ depending also on $J = \det(\hat{\mathbf{F}})$. If we approximate the strain tensor $\hat{\mathbf{E}}$ by Eq. (19) and consistently discard all terms that are of second order in $||\boldsymbol{H}||$ in accordance with our assumption of small strains, we have to use the approximation $\hat{\mathbf{F}} \approx \hat{\mathbf{R}}_{\mathrm{rel}}(s)$ (which implies $J \approx 1$) for the deformation gradient in all stress tensor transformations. This means that all pull back or push forward transformations are carried out approximately as simple relative rotations connecting corresponding frames $\hat{\mathbf{R}}_0(s)$ and $\hat{\mathbf{R}}(s,t)$ of the undeformed and deformed configurations of a Cosserat rod. Alltogether we obtain the approximate expressions[10]

$$\hat{\mathbf{S}} \approx \mathbb{H}:\hat{\mathbf{E}} \;\Rightarrow\; \hat{\mathbf{P}} \approx \hat{\mathbf{R}}_{\mathrm{rel}}\cdot\hat{\mathbf{S}}\,,\; \hat{\sigma} \approx \hat{\mathbf{R}}_{\mathrm{rel}}\cdot\hat{\mathbf{S}}\cdot\hat{\mathbf{R}}_{\mathrm{rel}}^T \tag{21}$$

for the various stress tensors, which are valid for the specific type of small strain assumptions encountered for Cosserat rods, as discussed above.

In the case of a *homogeneous and isotropic* material, the Hookean tensor acquires the special form of an isotropic fourth order tensor $\mathbb{H}_{\mathrm{SVK}} = \lambda\hat{\mathbf{I}}\otimes\hat{\mathbf{I}} + 2\mu\mathbb{I}$ depending on two constant elastic moduli: the *Lamé parameters* λ and μ. Here $\hat{\mathbf{I}}$ and $\mathbb{I}$ are the second and fourth order identity tensors, which act on (symmetric) second order tensors $\hat{\mathbf{Q}}$ by double contraction as $\mathbb{I}:\hat{\mathbf{Q}} = \hat{\mathbf{Q}}$ and $\hat{\mathbf{I}}:\hat{\mathbf{Q}} = \mathrm{Tr}(\hat{\mathbf{Q}})$, such that one obtains $\hat{\mathbf{Q}}:(\hat{\mathbf{I}}\otimes\hat{\mathbf{I}}):\hat{\mathbf{Q}} = \mathrm{Tr}(\hat{\mathbf{Q}})^2$ and $\hat{\mathbf{Q}}:\mathbb{I}:\hat{\mathbf{Q}} = \hat{\mathbf{Q}}:\hat{\mathbf{Q}} = \mathrm{Tr}(\hat{\mathbf{Q}}^2) = ||\hat{\mathbf{Q}}||_F^2$, where $||\ldots||_F$ is the Frobenius norm. The corresponding energy function is the *Saint–Venant Kirchhoff*

[9] Additional assumptions are the vanishing of the elastic energy density at zero strain ($\Psi_e(\hat{\mathbf{0}}) = 0$), as well as the absence of initial stresses in the undeformed configuration (i.e.: $\hat{\mathbf{S}}_0 = \partial_{\hat{\mathbf{E}}}\Psi_e(\hat{\mathbf{0}}) = \hat{\mathbf{0}}$).

[10] An alternative interpretation of Eq. (21) in terms of the Biot stress tensor is briefly discussed in Sect. A3 of the Appendix.

potential

$$\Psi_{\text{SVK}}(\hat{\mathbf{E}}) = \frac{1}{2}\hat{\mathbf{E}} : \mathbb{H}_{\text{SVK}} : \hat{\mathbf{E}} \quad (22)$$
$$= \frac{\lambda}{2}\text{Tr}(\hat{\mathbf{E}})^2 + \mu\|\hat{\mathbf{E}}\|_F^2 = \frac{K}{2}\text{Tr}(\hat{\mathbf{E}})^2 + \mu\|\mathbb{P} : \hat{\mathbf{E}}\|_F^2 ,$$

where $\mathbb{P} = \mathbb{I} - \frac{1}{3}\hat{\mathbf{I}} \otimes \hat{\mathbf{I}}$ is the orthogonal projector on the subspace of traceless second order tensors, such that $\mathbb{P} : \hat{\mathbf{E}} = \hat{\mathbf{E}} - \frac{1}{3}\text{Tr}(\hat{\mathbf{E}})\hat{\mathbf{I}}$ yields the traceless (deviatoric) part of the strain tensor, and $K = \lambda + \frac{2}{3}\mu$ is the bulk modulus.

3.4 Modified strain tensor including lateral contraction

The stress-strain relation obtained from (22) is given by

$$\hat{\mathbf{S}}_{\text{SVK}} = \lambda\,\text{Tr}(\hat{\mathbf{E}})\,\hat{\mathbf{I}} + 2\mu\,\hat{\mathbf{E}} = K\,\text{Tr}(\hat{\mathbf{E}})\,\hat{\mathbf{I}} + 2\mu\,\mathbb{P} : \hat{\mathbf{E}} . \quad (23)$$

Inserting the approximate expressions (19) and (20) of the strain tensor and its components into Eq. (23) yields the small strain approximation $\hat{\mathbf{S}}_{\text{SVK}} \approx \lambda H_3\hat{\mathbf{I}} + \mu[\boldsymbol{H} \otimes \boldsymbol{a}_0^{(3)} + \boldsymbol{a}_0^{(3)} \otimes \boldsymbol{H}]$ of the stress tensor $\hat{\mathbf{S}}_{\text{SVK}}$ for Cosserat rods. The computation of the stress components w.r.t. the basis of $\hat{\mathbf{R}}_0(s)$ directors yields normal stress components $S_{\alpha\alpha} \approx \lambda H_3$ and $S_{33} \approx (\lambda + 2\mu)H_3$, and the shear stress components are given by $S_{12} = S_{21} = 0$ and $S_{\alpha 3} = S_{3\alpha} \approx \mu H_\alpha$, respectively.

As both elastic moduli $\lambda = 2\mu\nu/(1-2\nu)$ and $\lambda + 2\mu = 2\mu(1-\nu)/(1-2\nu)$ appearing in the expressions for the normal stress components, expressed in terms of the shear modulus $\mu = G$ and Poisson's ratio given by $2\nu = \lambda/(\lambda+\mu)$, diverge in the incompressible limit $\nu \to \frac{1}{2}$ (just as the bulk modulus $K = \frac{2}{3}\frac{1+\nu}{1-2\nu}G$ does), the normal stresses would become infinitely large whenever the normal strain $E_{33} \approx H_3$ becomes nonzero. This unphysical behaviour is a direct consequence of the kinematical assumption of plain and *rigid* cross section, which prevents any lateral contraction of the cross section in the case of a longitudinal extension. Therefore the assumption of a *perfectly rigid* cross section, as well as the expressions (18) and (19) derived under this assumption, are strictly compatible only with *perfectly compressible* materials (i.e.: in the special case $\nu = 0$).

The standard procedure to fix this deficiency (see e.g. Weiss, 2002a) is based on the plausible requirement that *all* in-plane stress components $S_{\alpha\beta}$ (including the normal stresses $S_{\alpha\alpha}$), which for rods in practice are very small compared to the out of plain normal and shear stresses $S_{\alpha 3}$ and S_{33}, should *vanish* completely. This may be achieved by imposing a *uniform lateral contraction* with in-plane normal strain components $E_{\alpha\alpha} = -\nu E_{33}$ upon the cross section. Although this procedure seems to be rather ad hoc, it may be justified by an asymptotic analysis[11] of the local strain field for rods, e.g. in the way as presented by Love (1927) in the paragraph §256 on the "*Nature of the strain in a bent and twisted rod*" in ch. XVIII of his book. Following Love's analysis, we obtain the in-plane normal strains to leading order as $E_{\alpha\alpha} = \partial_\alpha w_\alpha = -\nu E_{33}$ with the additional requirement that $E_{12} = E_{21} = \partial_1 w_2 + \partial_2 w_1 = 0$, which determines the in-plane components w_α of the the warping field $\boldsymbol{w}$ corresponding to the lateral contraction in terms of E_{33}.

To obtain the modified value of $E_{\alpha\alpha} = -\nu E_{33}$ one has to add an additional term $-\nu E_{33}\,\boldsymbol{a}_0^{(\alpha)} \otimes \boldsymbol{a}_0^{(\alpha)}$ to the exact expression (18) of the strain tensor. Using the identity $\hat{\mathbf{I}} = \boldsymbol{a}_0^{(k)} \otimes \boldsymbol{a}_0^{(k)}$, we obtain the modified expression

$$\hat{\mathbf{E}}' = \hat{\mathbf{E}} - \nu E_{33}\left[\hat{\mathbf{I}} - \boldsymbol{a}_0^{(3)} \otimes \boldsymbol{a}_0^{(3)}\right] \quad (24)$$

for the strain tensor, with $E_{33} \approx H_3$ as small strain approximation according to Eq. (19). Inserting the modified strain tensor (24) into the stress-strain equation of the Saint–Venant–Kirchhoff material with $\text{Tr}(\hat{\mathbf{E}}') = (1-2\nu)E_{33} \approx (1-2\nu)H_3$, and using the relation $\lambda(1-2\nu) = \frac{\nu}{1+\nu}E$ that relates the Lamé parameter λ to Young's modulus E, we obtain the following modified expression for the stress of a Cosserat rod:

$$\hat{\mathbf{S}}'_{\text{SVK}} \approx \frac{E\nu}{1+\nu}H_3\,\boldsymbol{a}_0^{(3)} \otimes \boldsymbol{a}_0^{(3)} + G\left[\boldsymbol{H} \otimes \boldsymbol{a}_0^{(3)} + \boldsymbol{a}_0^{(3)} \otimes \boldsymbol{H}\right] . \quad (25)$$

By construction, we now obtain vanishing in-plane stress components $S'_{12} = S'_{21} = S'_{\alpha\alpha} \equiv 0$, while the transverse shear stresses remain unaffected by the modification (i.e.: $S'_{\alpha 3} = S'_{3\alpha} \approx GH_\alpha$ with $G = \mu$). As $2G = E/(1+\nu)$, we likewise obtain the modified expression $S'_{33} \approx E\,H_3$ for the normal stress component orthogonal to the cross section, which corresponds to the familiar expression from elementary linear beam theory, with Young's modulus E replacing $\lambda + 2\mu$.

3.5 Elastic energy of a Cosserat rod

Next we demonstrate briefly that the modified expressions (24) and (25) immediately lead to the known stored energy function (10) mentioned in the introduction.

In the case of a hyperelastic material with an elastic (stored) energy density Ψ_e the elastic potential energy of a body is given by the volume integral $\int_{V_0} \mathrm{d}V\,\Psi_e$ of the energy density over the volume V_0 of the reference configuration of the body. In the case of a rod shaped body parametrized by the coordinates (ξ_1, ξ_2, s) of the reference configuration (14), the volume measure of V_0 is given by $\mathrm{d}V = J_0\mathrm{d}s\mathrm{d}\xi_1\mathrm{d}\xi_2$, where J_0 is the Jacobian of the reference configuration (see Sect. 3.1). Using the geometric approximation $J_0 \approx 1$, the stored energy function of a rod shaped body is obtained as the integral $\int_{V_0} \mathrm{d}V\,\Psi_e \approx \int_0^L \mathrm{d}s\,\langle\Psi_e\rangle_{\mathcal{A}}$ of the density over the cross sections and along the centerline of the reference configuration of the rod.

In the special case of the energy density (22) this leads to the stored energy function $W_e = \int_0^L \mathrm{d}s\,\left\langle\Psi_{\text{SVK}}(\hat{\mathbf{E}}')\right\rangle_{\mathcal{A}}$, using the modified strain tensor $\hat{\mathbf{E}}'$ from Eq. (24). Applying our previously introduced approximations of small strains and

[11] See Berdichevsky (1981) and ch. 15 of Berdichevsky (2009) for a modern comprehensive analysis within Berdichevsky's variational asymptotic approach.

small initial curvature, we obtain the approximate expression

$$\Psi_{\rm SVK}(\hat{\mathbf{E}}') = \frac{1}{2}\hat{\mathbf{S}}'_{\rm SVK} : \hat{\mathbf{E}}' \approx \frac{1}{2}\left[EH_3^2 + G(H_1^2 + H_2^2)\right] \quad (26)$$

for the energy density. Its cross section integral $\left\langle \Psi_{\rm SVK}(\hat{\mathbf{E}}') \right\rangle_{\mathcal{A}}$ may be evaluated in terms of the integrals

$$\begin{aligned} \left\langle H_1^2 + H_2^2 \right\rangle_{\mathcal{A}} &= A(V_1^2 + V_2^2) + I_3(U_3 - U_{03}), \\ \left\langle H_3^2 \right\rangle_{\mathcal{A}} &= A(V_3 - 1)^2 + I_\alpha(U_\alpha - U_{0\alpha}), \end{aligned}$$

which finally yields the desired result

$$\begin{aligned} 2\left\langle \Psi_{\rm SVK}(\hat{\mathbf{E}}') \right\rangle_{\mathcal{A}} &\approx EA(V_3 - 1)^2 + GA(V_1^2 + V_2^2) \quad (27) \\ &+ EI_\alpha(U_\alpha - U_{0\alpha}) + GI_3(U_3 - U_{03}), \end{aligned}$$

corresponding exactly to the stored energy function (10) with effective stiffness parameters given by Eq. (2). The subsequent introduction of *shear correction factors* ($GA \to GA\varkappa_\alpha$) as well as the corresponding correction $GI_3 \to GJ_T = GI_3\varkappa_3$ of *torsional rigidity*[12] finally yields the stored energy function (10) with correspondingly modified effective stiffnesses as given by Eq. (7) (see also Sect. 4.1 for a more detailed discussion of this point).

3.6 Kinetic energy and energy balance for Cosserat rods

In general, the kinetic energy of a body is given by the volume integral $\int_{V_0} \mathrm{d}V \, \frac{1}{2}\rho_0 \boldsymbol{v}^2$, where $\rho_0(\boldsymbol{X})$ is the local mass density of the body in the reference volume, and $\boldsymbol{v}(\boldsymbol{X},t) = \partial_t \boldsymbol{x}(\boldsymbol{X},t)$ is the velocity of the respective material point. Using the kinematic ansatz (15) with the geometric approximation $J_0 \approx 1$, assuming a homogeneous mass density, and neglecting the contribution of cross section warping ($\boldsymbol{w} \equiv 0$), we obtain the integral expression $W_k = \int_0^L \mathrm{d}s \, \frac{1}{2}\rho_0[A(\partial_t\boldsymbol{\varphi})^2 + \left\langle \xi_\alpha^2 \right\rangle_{\mathcal{A}} (\partial_t \boldsymbol{a}^{(\alpha)})^2]$ for the kinetic energy of the rod as a quadratic functional of the time derivatives of its kinematic variables. The rotatory part may be reformulated in terms of the material components $\Omega_j = \boldsymbol{\omega} \cdot \boldsymbol{a}^{(j)}$ of the angular velocity vector $\boldsymbol{\omega} = \Omega_j \boldsymbol{a}^{(j)}$ of the rotating frame, which is implicitely defined by $\partial_t \boldsymbol{a}^{(k)} = \boldsymbol{\omega} \times \boldsymbol{a}^{(j)}$, by substituting $\left\langle \xi_\alpha^2 \right\rangle_{\mathcal{A}} (\partial_t \boldsymbol{a}^{(\alpha)})^2 = I_k \Omega_k^2$. This finally yields the familiar expression $W_k = \int_0^L \mathrm{d}s \, \frac{1}{2}\rho_0[A(\partial_t\boldsymbol{\varphi})^2 + I_k\Omega_k^2]$ for the kinetic energy of a Cosserat rod as given in Lang et al. (2011) with Ω_k expressed in quaternionic formulation. Altogether we obtain the approximation $\int_{V_0} \mathrm{d}V \, [\frac{1}{2}\rho_0 \boldsymbol{v}^2 + \Psi_e] \approx W_e + W_k =: W_m$ of the three–dimensional mechanical energy of a rod shaped body in terms of the corresponding sum of the kinetic and stored energy functions W_k and W_e of the Cosserat rod model as given above. In the absence of any dissipative effects, the mechanical energy must be conserved *exactly* in both the 3-D as well as the 1-D setting, such that the identities $\frac{\mathrm{d}}{\mathrm{d}t} \int_{V_0} \mathrm{d}V \, [\frac{1}{2}\rho_0 \boldsymbol{v}^2 + \Psi_e] = 0 = \frac{\mathrm{d}}{\mathrm{d}t} W_m$ hold identically as a consequence of the respective balance equations for both the 3-D volumetric body and the 1-D rod.

4 Kelvin–Voigt damping for Cosserat rods

Now we have collected all technical prerequisites and approximate results that enable us to derive the dissipation function (11) of a Cosserat rod from a three-dimensional Kelvin–Voigt model in analogy to the derivation of the stored energy function (10) in a consistent way.

In Landau and Lifshitz (1986) (see ch. V §34) the *dissipation function* $\int_V \mathrm{d}V \, \frac{1}{2}\eta_{ijkl}\dot{\varepsilon}_{ij}\dot{\varepsilon}_{kl}$ is considered as an appropriate model of dissipative effects within a solid body near thermodynamic equilibrium, with constant fourth order tensor components η_{ijkl} that are the viscous analogon of the components of the Hookean elasticity tensor. Transfering this ansatz to the formalism used in our paper, the dissipation function of Landau and Lifshitz (1986) becomes that of a *Kelvin–Voigt solid* as given in Lemaitre and Chaboche (1990)

$$D_{\rm KV} = \int_0^L ds \left\langle \Psi_{\rm KV}(\partial_t\hat{\mathbf{E}}) \right\rangle_{\mathcal{A}} = \int_0^L ds \frac{1}{2} \left\langle \partial_t\hat{\mathbf{E}} : \mathbb{V} : \partial_t\hat{\mathbf{E}} \right\rangle_{\mathcal{A}}, \quad (28)$$

which is a quadratic form in the material strain rate $\partial_t\hat{\mathbf{E}}$ defined as the time derivative of the Green–Lagrange strain tensor. The constant fourth order *viscosity tensor* $\mathbb{V}$ may be assumed to have the same symmetries as the Hookean tensor $\mathbb{H}$, with its components depending on *viscosity parameters* in the same way as the components of $\mathbb{H}$ depend on elastic

[12] The correction of torsional rigidity accounts for the contribution of out-of-plane cross section warping in terms of a corresponding torsional stress function $\Phi(\xi_1,\xi_2)$ and leads to an improved approximation of the strain and stress fields as well as the resulting elastic energy given by Eq. (10) compared to its 3-D volumetric counterpart. Similar arguments apply to an improved approximation of transverse shear strains and stresses as well as the associated part of the elastic energy density by accounting for additional contributions given by a corresponding pair of stress functions $\chi_\alpha(\xi_1,\xi_2)$. The classical results obtained by St.-Venant are given in ch. XIV of Love's treatise (Love, 1927) (see also ch. II §16 in Landau and Lifshitz, 1986). They are contained as a special (and simplified) case within Berdichevsky's more comprehensive and modern treatment in terms of his method of variational asymptotic analysis applied to rods (see Berdichevsky, 1981, 1983 and ch. 15 of Berdichevsky, 2009). Apart of Timoshenko's original treatment of shear correction factors, the article of Cowper (1966) is a classical reference on this subject, with correction factors obtained from pointwise (centroidal) and cross section averaged values of transverse shear stresses $\sigma_{\alpha 3}$ (see also the discussions in ch. II, section 11 of Villagio, 1997 and section 2.1 of Simo et al., 1984). More recently an alternative approach based on *energy balance* as utilized e.g. in (Gruttmann and Wagner, 2001) and likewise fits to our considerations, is considered as standard due to superior results. However, the issue of correction factors for transverse shear in Timoshenko-type rod models is still subject of discussion and research activities (see e.g. Dong et al., 2010).

moduli. The stress-strain relation of the Kelvin–Voigt model is given by $\hat{\mathbf{S}} = \mathbb{H} : \hat{\mathbf{E}} + \mathbb{V} : \partial_t \hat{\mathbf{E}}$, with the viscous stress[13] given by the term $\hat{\mathbf{S}}_v := \mathbb{V} : \partial_t \hat{\mathbf{E}} = \partial_{\partial_t \hat{\mathbf{E}}} \Psi_{\mathrm{KV}}(\partial_t \hat{\mathbf{E}})$.

The dissipation function for a Cosserat rod results by inserting the rate $\partial_t \hat{\mathbf{E}}'$ of the modified strain tensor (24) into the dissipation density function Ψ_{KV} of the Kelvin–Voigt model. We will compute this dissipation function explicitely in closed form for the special case of a *homogeneous and isotropic* material. In this special case, the viscosity tensor assumes the form

$$\mathbb{V}_{\mathrm{IKV}} = \zeta \hat{\mathbf{I}} \otimes \hat{\mathbf{I}} + 2\eta \mathbb{P} = (\zeta - \frac{2}{3}\eta) \hat{\mathbf{I}} \otimes \hat{\mathbf{I}} + 2\eta \mathbb{I}\,, \tag{29}$$

depending on two constant parameters: *bulk viscosity* ζ and *shear viscosity* η.

To compute $\partial_t \hat{\mathbf{E}}'$ we use the expression (24) for the modified Green–Lagrange strain tensor of a Cosserat rod including the small strain approximation (19), with the result

$$\partial_t \hat{\mathbf{E}}' \approx \frac{1}{2}\left[\partial_t \boldsymbol{H} \otimes \boldsymbol{a}_0^{(3)} + \boldsymbol{a}_0^{(3)} \otimes \partial_t \boldsymbol{H}\right] - \nu\, \partial_t H_3 \left[\hat{\mathbf{I}} - \boldsymbol{a}_0^{(3)} \otimes \boldsymbol{a}_0^{(3)}\right] \tag{30}$$

depending on the time derivative $\partial_t \boldsymbol{H}(\xi_1, \xi_2, s, t) = \partial_t H_k(\xi_1, \xi_2, s, t)\, \boldsymbol{a}_0^{(k)}(s)$ of the material strain vector with components

$$\begin{aligned} \partial_t H_1(\xi_2, s, t) &= \partial_t V_1(s,t) - \xi_2\, \partial_t U_3(s,t)\,, \\ \partial_t H_2(\xi_1, s, t) &= \partial_t V_2(s,t) + \xi_1\, \partial_t U_3(s,t)\,, \\ \partial_t H_3(\xi_1, \xi_2, s, t) &= \partial_t V_3(s,t) + \xi_2\, \partial_t U_1(s,t) - \xi_1\, \partial_t U_2(s,t)\,, \end{aligned} \tag{31}$$

i.e.: $\hat{\mathbf{R}}_0^T \cdot \partial_t \boldsymbol{H} = (\partial_t H_k)\, \boldsymbol{e}_k = \partial_t \boldsymbol{V} - \xi_\alpha \boldsymbol{e}_\alpha \times \partial_t \boldsymbol{U}$, written as a cartesian vector w.r.t. the global basis $\{\boldsymbol{e}_1, \boldsymbol{e}_2, \boldsymbol{e}_3\}$.

Inserting Eqs. (30) and (31) into the dissipation density function $\Psi_{\mathrm{IKV}}(\partial_t \hat{\mathbf{E}}') = \frac{1}{2}\, \partial_t \hat{\mathbf{E}}' : \mathbb{V}_{\mathrm{IKV}} : \partial_t \hat{\mathbf{E}}'$ of the isotropic Kelvin–Voigt model, analogous computational steps as those done for the derivation of the stored energy $\Psi_{\mathrm{SVK}}(\hat{\mathbf{E}}')$ in the previous subsection yield the expression

$$2\Psi_{\mathrm{IKV}}(\partial_t \hat{\mathbf{E}}) \approx \eta_E (\partial_t H_3)^2 + \eta \left[(\partial_t H_1)^2 + (\partial_t H_2)^2\right],$$

with the *extensional viscosity* parameter η_E as defined in Eq. (5) appearing as the prefactor[14] of $(\partial_t H_3)^2$. The computation of the cross section integrals of the squared time derivatives $(\partial_t H_k)^2$ yields the expressions

$$\begin{aligned} \left\langle (\partial_t H_3)^2 \right\rangle_{\mathcal{A}} &= A(\partial_t V_3)^2 + I_\alpha (\partial_t U_\alpha)^2\,, \\ \left\langle (\partial_t H_1)^2 + (\partial_t H_2)^2 \right\rangle_{\mathcal{A}} &= A\left[(\partial_t V_1)^2 + (\partial_t V_2)^2\right] + I_3 (\partial_t U_3)^2\,, \end{aligned}$$

from which we obtain the desired cross section integral of the dissipation density function:

$$\begin{aligned} 2\left\langle \Psi_{\mathrm{IKV}}(\partial_t \hat{\mathbf{E}}) \right\rangle_{\mathcal{A}} &\approx \eta_E A (\partial_t V_3)^2 + \eta_E I_\alpha (\partial_t U_\alpha)^2 \\ &+ \eta A\left[(\partial_t V_1)^2 + (\partial_t V_2)^2\right] + \eta I_3 (\partial_t U_3)^2\,. \end{aligned} \tag{32}$$

The dissipation function (11) of the Cosserat rod with diagonal damping coefficient matrices (3) and damping parameters (4) is then obtained as $D_v = D_{\mathrm{IKV}} := \int_0^L \mathrm{d}s \left\langle \Psi_{\mathrm{IKV}}(\partial_t \hat{\mathbf{E}}') \right\rangle_{\mathcal{A}}$.

4.1 Modification by shear correction factors

There is obviously a high degree of formal algebraic similarity in the derivations of the stored energy function (10) as presented in Sect. 3.5 and the dissipation function (11) as presented above: both functionals result by inserting the specific strain tensor (24) of a Cosserat rod or respectively its rate (30) into a volume integral over the 3-D body domain of a density function defined as a quadratic form given by constant isotropic fourth order material tensors $\mathbb{H}$ and $\mathbb{V}$, making use of the same geometric as well as "*small strain*" approximations implied by the specific kinematical ansatz (15) for the configurations of a Cosserat rod. The formal analogy in the derivation procedure leads to a dissipation density (32) that may be obtained from its elastic counterpart (27) by substituting viscosity parameters for corresponding elastic moduli ($G \to \eta$, $E \to \eta_E$) and strain rates for strain measures.

In the case of the stored energy function (10) the effective stiffness parameters (2) of the rod model are obtained from a derivation using a kinematical ansatz that completely neglects out-of-plane warping (i.e.: $w_3 = 0 = \partial_k w_3$) due to transverse shearing and twisting, but accounts for in-plane warping (i.e.: $w_\alpha \neq 0$) in a simplified way by assuming a uniform lateral contraction (ULC) of the cross section according to the linear elastic theory (see Sect. 3.4). Softening effects due to out-of-plane warping are then accounted for by introducing *shear correction factors* $0 < \varkappa_j \leq 1$, which in the case of a homogeneous and isotropic material enter the model as multipliers $A \to A_\alpha = A\varkappa_\alpha$ and $I_3 \to J_T = I_3 \varkappa_3$ of the area A

[13] Note that $\hat{\mathbf{S}}_v : \partial_t \hat{\mathbf{E}} = 2\Psi_{\mathrm{KV}}(\partial_t \hat{\mathbf{E}})$ corresponds to the *viscous stress power density*, such that the integral $P_v(t) := 2\int_V \mathrm{d}V\, \Psi_{\mathrm{KV}}(\partial_t \hat{\mathbf{E}})$ over the body volume yields the (time dependent) rate at which a Kelvin–Voigt solid dissipates mechanical energy under approximately isothermal conditions near thermodynamic equilibrium, (see ch. V §34 and §35) of Landau and Lifshitz, 1986). For a thorough discussion of the role of the dissipation function within the theory of small fluctuations near thermodynamic equilibrium from the viewpoint of statistical physics we refer to the the corresponding paragraphs in ch. XII in Landau and Lifshitz (1980) (in particular §121), as well as V. Berdichevsky's recent article 2003. In section VI of the latter, the author points out that a Kelvin–Voigt type constitutive relation holds also at *finite* strains, with the dissipative part governed by a fourth order viscosity tensor $\mathbb{V}[\hat{\mathbf{E}}, \partial_t \hat{\mathbf{E}}]$ depending on the local strain and its rate. While a dependence of $\mathbb{V}$ on the invariants of $\partial_t \hat{\mathbf{E}}$ in general prevents the existence of a dissipation function, the latter *does* indeed exist according to V.B.'s arguments if $\mathbb{V} = \mathbb{V}[\hat{\mathbf{E}}]$ is independent of the strain rate. This holds e.g. in the case of the Kelvin–Voigt limit of constitutive laws belonging to the class of *finite linear viscoelasticity* (Coleman and Noll, 1961) at sufficiently small strain rates (i.e. sufficiently slow deformations of a body).

[14] The term $K(1-2\nu)^2 + \frac{4}{3}G(1+\nu)^2 = E$ analogously appears as the prefactor of H_3^2 in the expression (26) of the stored energy function of a Cosserat rod for the St.-Venant-Kirchhoff material.

and polar moment I_3 of the cross section and – according to the linear theory – depend *solely* on the *cross section geometry*. The modified stiffness constants (7) are obtained in combination with the elastic moduli $G=\mu$ and E, the latter appearing instead of $\lambda+2\mu$ due to the enforcment of vanishing in-plane stresses by allowing for ULC according to Eq. (24).

Although the derivation of explicit formulas[15] for $\varkappa_j$ is carried out for *static* boundary value problems, the same $\varkappa_j$, as well as the kinematic ansatz accounting for ULC, may be used for *dynamic* problems, due to the negligible influence of dynamic effects on the warping behaviour of cross sections, provided that the rod geometry is sufficiently slender. Therefore the geometric modifications $A \to A_\alpha = A\varkappa_\alpha$ and $I_3 \to J_T = I_3\varkappa_3$, which have already been used to provide modified stiffness parameters (7) for an improved approximation of the 3-D (volumetric) *elastic energy* by the stored energy function (10) in the static as well as in the dynamic case, remain likewise valid to achieve a comparable improvement for the approximation of the 3-D integrated *viscous stress power* by the dissipation function (11), with modified damping parameters given by Eq. (8), leading to the modified expressions (9) for the effective viscosity matrices.

This completes our derivation of the Kelvin–Voigt type dissipation function of a Cosserat rod. Although the arguments given above would certainly benefit from a mathematical confirmation by rigorous (asymptotic) analysis, the latter is beyond the scope of this work.

4.2 An (erroneous) alternative derivation approach

The formulation of the Cosserat rod model given by Simo (1985) introduces spatial force and moment vectors $\boldsymbol{f}$ and $\boldsymbol{m}$, usually denoted as *stress resultants* and *stress couples*, as the cross section integrals

$$\begin{aligned} \boldsymbol{f}(s,t) &= \left\langle \hat{\mathbf{P}}(\xi_1,\xi_2,s,t)\cdot \boldsymbol{a}_0^{(3)}(s)\right\rangle_{\mathcal{A}} , \\ \boldsymbol{m}(s,t) &= \left\langle \boldsymbol{\xi}(s)\times\hat{\mathbf{P}}(\xi_1,\xi_2,s,t)\cdot \boldsymbol{a}_0^{(3)}(s)\right\rangle_{\mathcal{A}} \end{aligned}$$

of the traction forces of the 1st Piola–Kirchhoff stress tensor acting on the cross section area and the corresponding moments generated by the Piola–Kirchhoff tractions w.r.t. the cross section centroid, which are obtained by means of the "lever arm" vector $\boldsymbol{\xi}(s)=\xi_\alpha \boldsymbol{a}_0^{(\alpha)}(s)$. Both integrants may be expressed in terms of the 2nd Piola–Kirchhoff stress by means of the transformation $\hat{\mathbf{P}}=\hat{\mathbf{F}}\cdot\hat{\mathbf{S}}$ with the deformation gradient. In view of the small strain approximation $\hat{\mathbf{P}}\approx\hat{\mathbf{R}}_{\text{rel}}\cdot\hat{\mathbf{S}}$ with $\hat{\mathbf{S}}\approx\mathbb{H}:\hat{\mathbf{E}}$ discussed in Sect. 3.3 we obtain the relations

$$\begin{aligned} \hat{\mathbf{R}}_0(s)\cdot\boldsymbol{F}(s,t) &\approx \left\langle \hat{\mathbf{S}}(\xi_1,\xi_2,s,t)\cdot \boldsymbol{a}_0^{(3)}(s)\right\rangle_{\mathcal{A}} , \\ \hat{\mathbf{R}}_0(s)\cdot\boldsymbol{M}(s,t) &\approx \left\langle \boldsymbol{\xi}(s)\times\hat{\mathbf{S}}(\xi_1,\xi_2,s,t)\cdot \boldsymbol{a}_0^{(3)}(s)\right\rangle_{\mathcal{A}} \end{aligned}$$

connecting the spatial stress resultants $\boldsymbol{f}=\hat{\mathbf{R}}\cdot\boldsymbol{F}$ and stress couples $\boldsymbol{m}=\hat{\mathbf{R}}\cdot\boldsymbol{M}$ to their material counterparts rotated to the local reference frame $\hat{\mathbf{R}}_0(s)=\boldsymbol{a}_0^k(s)\otimes\boldsymbol{e}_k$.

Expanding the material force and moment vectors w.r.t. the local ONB given by the reference frame $\hat{\mathbf{R}}_0(s)$ as $\hat{\mathbf{R}}_0(s)\cdot\boldsymbol{F}(s,t)=F_k(s,t)\,\boldsymbol{a}_0^k(s)$ and $\hat{\mathbf{R}}_0(s)\cdot\boldsymbol{M}(s,t)=M_k(s,t)\,\boldsymbol{a}_0^k(s)$ yields their components in terms of the cross section integrals

$$F_j = \left\langle S_{j3}\right\rangle_{\mathcal{A}} , \quad M_1 = \langle \xi_2 S_{33}\rangle_{\mathcal{A}} , \; M_2 = \langle -\xi_1 S_{33}\rangle_{\mathcal{A}} , \quad M_3 = \langle \xi_1 S_{23}-\xi_2 S_{13}\rangle_{\mathcal{A}}$$

of the components of $\hat{\mathbf{S}}$ w.r.t. this basis. To compute these components of the material force and moment vectors in closed form for the special case $\hat{\mathbf{S}}'=\mathbb{H}_{\text{SVK}}:\hat{\mathbf{E}}'+\mathbb{V}_{\text{IKV}}:\partial_t\hat{\mathbf{E}}'=\hat{\mathbf{S}}'_{\text{SVK}}+\hat{\mathbf{S}}'_{\text{IKV}}$ with the approximate expressions (24) and (30) of the Green–Lagrange strain tensor and its rate and the constant isotropic material tensors $\mathbb{H}_{\text{SVK}}=K\hat{\mathbf{I}}\otimes\hat{\mathbf{I}}+2G\mathbb{P}$ and $\mathbb{V}_{\text{IKV}}=\zeta\hat{\mathbf{I}}\otimes\hat{\mathbf{I}}+2\eta\mathbb{P}$, we have to evaluate the cross section integrals with the stress components $S'_{\alpha 3}=GH_\alpha+\eta\partial_t H_\alpha$ and $S'_{33}=EH_3+\tilde{\eta}_E\partial_t H_3$, with $\tilde{\eta}_E:=(1-2\nu)\zeta+(1+\nu)\frac{4}{3}\eta$ multiplying the strain rate $\partial_t H_3\approx\partial_t E_{33}$.

Therefore $\tilde{\eta}_E$ has to be interpreted as extensional viscosity, but obviously differs from the expression η_E given in Eq. (5) and derived above by computing the dissipation function. Therefore the corresponding retardation time constant $\tilde{\tau}_E:=\tilde{\eta}_E/E=\frac{1}{3}(\tau_B+2\tau_S)$, which is independent of the value of Poisson's ratio ν, likewise differs from the expression of the extensional retardation time τ_E given in Eq. (6). Both expressions $\tilde{\eta}_E$ and η_E yield extensional viscosity as a combination of shear and bulk viscosity, but agree only in the special case $\nu=0$. The same assertion likewise holds for the corresponding retardation times, of course. However, only η_E yields the correct incompressible limit $\eta_E\to 3\eta$ for $\nu\to\frac{1}{2}$, while $\tilde{\eta}_E$ tends to the smaller (and incorrect) value of 2η in this case.

The resulting expressions for the material force components are given by

$$F_\alpha = GA\left[V_\alpha + \tau_S\,\partial_t V_\alpha\right] , \quad F_3 = EA\left[(V_3-1)+\tilde{\tau}_E\,\partial_t V_\alpha\right] ,$$

and the material moment components correspondingly by

$$\begin{aligned} M_\alpha &= EI_\alpha\left[(U_\alpha-U_{0\alpha})+\tilde{\tau}_E\,\partial_t U_\alpha\right] , \\ M_3 &= GI_3\left[(U_3-U_{03})+\tau_S\,\partial_t U_3\right] . \end{aligned}$$

A comparison with the stiffness and damping parameters (2) and (6) entering the constitutive equations (1) shows that the derivation approach sketched above correctly yields *all* of the stiffness parameters as well as the damping parameters associated to transverse and torsional shear deformations. However, the damping parameters governed by normal stresses and extensional viscosity do not agree due to the appearance of $\tilde{\tau}_E$ instead of the correct time constant τ_E.

The discrepancy between the results of both derivation approaches can be traced back to the fact that the integration

[15] We refer to footnote 12 for a discussion of this issue.

of the traction forces and their associated moments over the cross section fails to account for the non-vanishing contributions of the in-plane strain rates $\partial_t E'_{\alpha\alpha} = -\nu\partial_t H_3$ associated to uniform lateral contraction to the total energy dissipation of the rod. Paired with the corresponding viscous stress components $S'_{\alpha\alpha} = [(1-2\nu)\zeta-(1+\nu)\eta]\partial_t H_3$ these result in the (in general non-vanishing) contribution

$$\begin{aligned} S'_{\alpha\alpha}(\partial_t E'_{\alpha\alpha}) &= -2\nu[(1-2\nu)\zeta-(1+\nu)\eta](\partial_t H_3)^2 \\ &= (\eta_E - \tilde{\eta}_E)(\partial_t H_3)^2 \end{aligned}$$

to the dissipation function. As the cross section integrals given above involve only the stress components $S'_{\alpha 3}$ and S'_{33}, this additional source of damping is, by definition, not contained in the resulting formulas for the material force and moment components F_j and M_j obtained via this approach.

However, this deficiency affects only the *viscous* part of the constitutive equations. The elastic part does not show any discrepancy, as the modified strain tensor (24) by construction provides vanishing in-plane elastic stress components (see Sect. 3.4), such that the stored energy function does not contain any contributions from non-vanishing in-plane elastic stresses to the elastic energy, and the cross section integrals of the traction forces and their moments yield *all* stiffness parameters correctly.

In summary, the considerations above suggest that, also in the case of more general viscoelastic constitutive laws, our approach to derive effective constitutive equations for Cosserat rods by computing the stored energy and dissipation functions is superior to the alternative approach based on a direct computation of the forces and moments as resultant cross section integrals of the traction forces and associated moments, as the latter yields an effective extensional viscosity which is systematically too small for partially compressible and incompressible solids (i.e.: $0 < \nu \leq \frac{1}{2}$).

4.3 ANCF beams with Kelvin–Voigt damping

In the recent article of Abdel-Nasser and Shabana (2011), a damping model for geometrically nonlinear beams given in the ANCF (absolute nodal coordinates) formulation has been proposed. The authors obtained their model by inserting the 3-D isotropic Kelvin–Voigt model as described above into their ANCF element ansatz. They used the Lamé parameters λ and μ as elastic moduli, and introduced corresponding viscosity parameters λ_v and μ_v, which they related to the elastic moduli by *dissipation factors* γ_{v1} and γ_{v2}. From the context it seems clear that in our notation $\gamma_{v2} = \tau_S$, such that $\mu_v = G\tau_S = \eta$. Likewise we may identify $\gamma_{v1} = \tau_B$, such that $\lambda_v = K\tau_B - \frac{2}{3}G\tau_S = \zeta - \frac{2}{3}\eta$, and the viscosities are related by the same relation as the elastic moduli (i.e.: $\lambda = K - \frac{2}{3}G$). If the ANCF ansatz chosen in Abdel-Nasser and Shabana (2011) handles lateral contraction effects correctly, both models should behave similar and yield similar simulation results. However, the appearance of the unmodified elastic moduli $\lambda = 2\mu\nu/(1-2\nu)$ and $\lambda + 2\mu = 2\mu(1-\nu)/(1-2\nu)$ in the element stiffness matrix (see Eq. 25 of the paper) indicates that the formulation chosen in Abdel-Nasser and Shabana (2011) may have problems in the case of incompressible materials ($\nu \to \frac{1}{2}$). A clarifying investigation of this issue as well as a detailed comparison of both models remains to be done in future work.

4.4 Validity of the Kelvin–Voigt model

As remarked already in Landau and Lifshitz (1986), the modelling of viscous dissipation for solids by a dissipation function of Kelvin–Voigt type is valid only for relatively slow processes near thermodynamic equilibrium, which means that the temperature within the solid should be approximately constant, and the macroscopic velocities of the material particles of the solid should be sufficiently slow w.r.t. the time scale of all internal relaxation processes.

To illustrate and quantify this statement, we briefly discuss the one-dimensional example of a linear viscoelastic stress-strain relation $\sigma(t) = \int_0^\infty d\tau G(\tau)\dot{\varepsilon}(t-\tau)$ governed by the relaxation function $G(\tau) = G_\infty + \sum_{j=1}^N G_j \exp(-\tau/\tau_j)$ (i.e.: a *Prony series*) of a *generalized Maxwell model*. By Fourier transformation we obtain the relation $\hat{\sigma}(\omega) = \hat{G}(\omega)\hat{\varepsilon}(\omega)$ in the frequency domain, where the real and imaginary parts of the complex modulus function $\hat{G}(\omega) = G_\infty + \sum_{j=1}^N G_j \frac{i\tau_j\omega}{1+i\tau_j\omega}$ model the frequency dependent stiffness and damping properties of the material.

Using a 1-D Kelvin–Voigt model $\sigma_{KV}(t) = G\varepsilon(t) + \eta\dot{\varepsilon}(t)$ we obtain the simple expression $\hat{\sigma}_{KV}(\omega) = [G + i\eta\omega]\hat{\varepsilon}(\omega)$, which approximates the generalized Maxwell model at sufficiently low frequencies with $G = G_\infty$ and $\eta = \sum_{j=1}^N G_j\tau_j$. The deviation between the generalized Maxwell model and its Kelvin–Voigt approximation may be estimated as

$$|\sigma(t) - \sigma_{KV}(t)| \leq \frac{1}{\pi}\sum_{j=1}^N G_j \int_0^\infty \mathrm{d}\omega \frac{|\hat{\varepsilon}(\omega)|(\tau_j\omega)^2}{\sqrt{1+(\tau_j\omega)^2}}\,.$$

This deviation may indeed become small, provided that the modulus $|\hat{\varepsilon}(\omega)|$ of the strain spectrum, which appears as a weighting factor for the terms of the sum on the r.h.s., takes on non–vanishing values only at frequencies much smaller than those given by the discrete spectrum of the inverse relaxation times $\omega_j = 1/\tau_j$. The estimate given above also shows that in this case the Kelvin–Voigt model provides a *low frequency approximation* of second order accuracy.

5 Numerical examples

To illustrate the behaviour of our damping model, we show the results of numerical simulations of nonlinear vibrations of a cantilever beam in Fig. 3 obtained with the discrete Cosserat rod model presented in Lang et al. (2011).

The parameters of the beam are: length $L = 30\,\mathrm{cm}$, quadratic cross-section area $A = 1\times 1\,\mathrm{cm}^2$, mass density $\rho =$

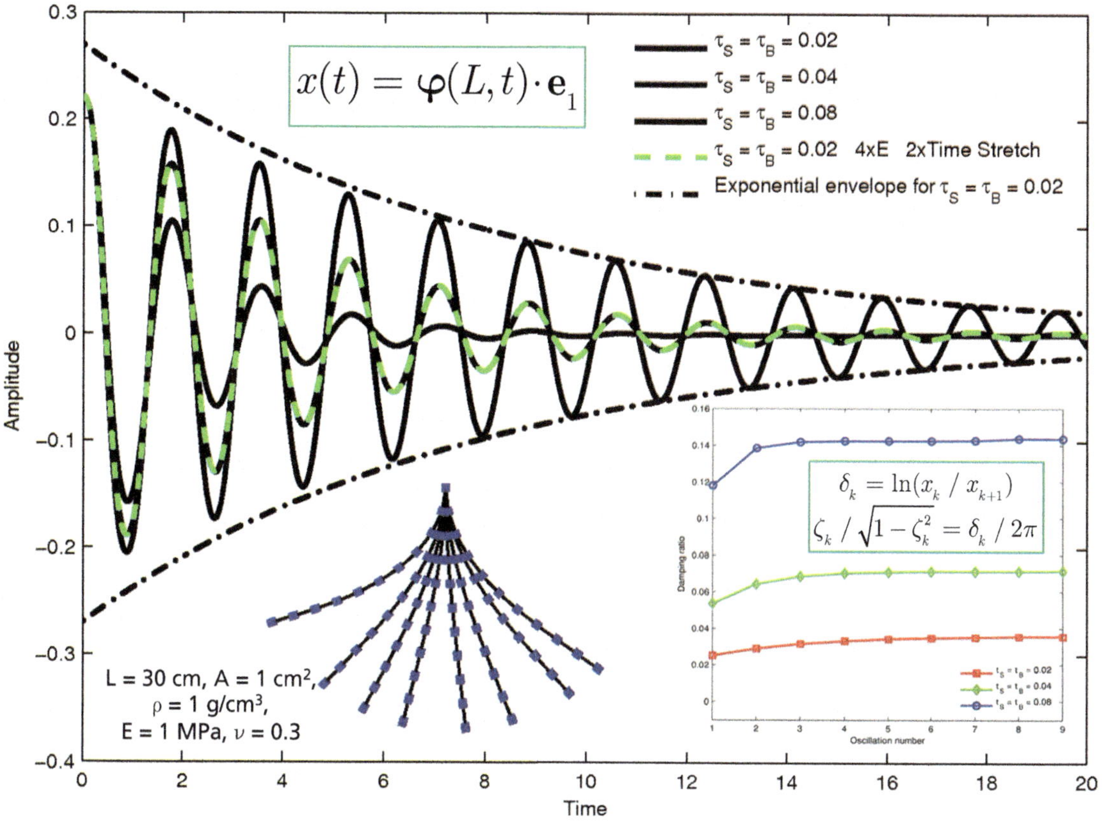

Figure 3. Damped non-linear bending vibrations of a clamped cantilever beam (see text for further details).

$1\,\mathrm{g\,cm^{-3}}$, Young's modulus $E = 1\,\mathrm{MPa}$, and Poisson's ratio $\nu = 0.3$. We assume that $\zeta/\eta = K/G$ holds for the viscosity parameters, such that according to our model (6) the values of all retardation time constants are equal ($\tau_B = \tau_S = \tau_E$). The tests were performed with three different values (0.02 s, 0.04 s, and 0.08 s) of $\tau_E = \tau_{B/S}$. No gravitation is present.

The beam is fully clamped at one end, the other end is initially pulled sideways by applying a force $\boldsymbol{f}_L = F\boldsymbol{e}_1$ of magnitude $F = 0.05\,\mathrm{N}$ to the other end. The resulting initial deformation state in static equilibrium[16] deviates far from the linear range of deformations governed by (infinitesimally) small displacements and rotations w.r.t. the reference configuration, while local strains are small in accordance with the constitutive assumptions. Starting from this initial equilibrium configuration, the beam is then released to vibrate transversally. The deformations of the beam shown in the inset of Fig. 3 are snapshots taken during the first half period of the oscillations which illustrate that in the initial phase of the oscillations substantial geometric nonlinearities are present. During the vibrations the beam remains in the plane of its initial deformation, such that all deformations are of plane bending type, and the extensional viscosity $\eta_E = E\tau_E$ becomes the main influence for damping.

As expected, the plots of the transverse oscillation amplitude $x(t) = \boldsymbol{e}_1 \cdot \boldsymbol{\varphi}(L,t)$ recorded at the free end of the beam show an exponential dying out in the range of small amplitudes (linear regime). The deviations from the exponential envelope adapted to the linear regime that are observed during the initial phase clearly show the influence of geometric nonlinearity. The plots also suggest that damping becomes weaker in the nonlinear range. However, linear behaviour seems to start already with the fifth oscillation period, where the amplitude still has a large value of $\approx L/3$.

This may be further analyzed by evaluating the *logarithmic decrements* $\delta_k = \ln(x(t_k)/x(t_{k+1}))$ recorded between succesive maxima $x(t_k)$ of the amplitude as well as the corresponding *damping ratios* ζ_k implicitely defined (see Craig and Kurdila, 2006, ch. 3.5, p. 75) by $\delta_k = 2\pi\zeta_k/\sqrt{1-\zeta_k^2}$. The plots for the values of ζ_k determined in this way are shown in the inset of Fig. 3. As expected, the ratios approach constant values in the linear regime, which scale as $1:2:4$ proportional to the values of the time constant τ_E used in the simulations. The simulations also show that the decrements become lower in the range of large amplitudes, which confirms

[16]A highly accurate approximation of this equilibrium configuration may be obtained as the curve $s \mapsto \boldsymbol{\varphi}_{\mathrm{el}}(s)$ and adapted frame $\hat{\mathbf{R}}_{\mathrm{el}} = (\boldsymbol{e}_2 \times \partial_s \boldsymbol{\varphi}_{\mathrm{el}}) \otimes \boldsymbol{e}_1 + \boldsymbol{e}_2 \otimes \boldsymbol{e}_2 + \partial_s \boldsymbol{\varphi}_{\mathrm{el}} \otimes \boldsymbol{e}_3$ of an *inextensible Euler elastica*, which may be computed analytically in closed form in terms of Jacobian elliptic functions and elliptic integrals (see Love, 1927, ch. XIX §260–263 or Landau and Lifshitz, 1986, ch. II §19).

the observation that the damping effect of our Kelvin–Voigt model is extenuated by the presence of geometrical nonlinearity. Nevertheless, ζ_k still scales approximately proportional to τ_E also in the nonlinear range.

To investigate the influence of a variation of the bending stiffness on the damping behaviour, an additional test with quadrupled Young's modulus $E = 4\,\text{MPa}$ was performed. In the corresponding amplitude plot shown in Fig. 3 the time axis of the plot with quadrupled E was streched twofold, such that the oscillations could be compared directly. After time stretching the $(E = 4\,\text{MPa}, \tau_E = 0.02\,\text{s})$ plot coincides with the $(E = 1\,\text{MPa}, \tau_E = 0.04\,\text{s})$ plot, surprisingly even throughout the whole nonlinear range. Since the oscillation period T of the four times stiffer $(E = 4\,\text{MPa})$ beam is twice smaller than that of the softer $(E = 1\,\text{MPa})$ beam, this suggests that the damping ratio varies proportional to the ratio τ_E/T. Again this would be the expected behaviour in the linear regime, but is observed here in the nonlinear range as well.

For small amplitudes, the oscillation period may be estimated as $T \approx (2\pi/3.561)\,L^2\,\sqrt{\rho A/EI}$ using the well known formula for the fundamental transverse vibration frequency of a cantilever beam obtained from *Euler–Bernoulli* theory (see Craig and Kurdila, 2006, ch. 13.2, Ex. 13.3, eq. 8). Inserting the parameters assumed above, we get $T \approx 1.81\,\text{s}$ as an estimate, which correponds well to the time intervals of approximately 1.8 s between successive maxima shown in Fig. 3 that are also observed throughout the range of geometrically nonlinear deformations. For linear vibrations, damping ratio values $\zeta \approx 1$ correspond to a *critical* damping of the vibrating system, while values $0 < \zeta \ll 1$ indicate a *weak* damping. According to that, the values ζ_k observed in our experiments are in the range of weak to moderate damping, and are well approximated by the empirical formula $\zeta \approx \frac{1}{\pi}\tau_E/T$. This provides a rough guideline for estimating the strenght of damping, or likewise an adjustment of the the retardation time τ_E relative to the fundamental period T, if the Kelvin–Voigt model is utilized to provide artificial viscous damping in the sense of Antman (2003). According to this, a critical damping of transverse bending vibrations would be observed at a value of $\tau_E \approx \pi T$.

Corresponding experiments for axial or torsional vibrations are limited to the range of small vibrations amplitudes, similar to the ones shown by Abdel-Nasser and Shabana (2011), as for large amplitudes one would inevitably induce buckling to bending deformations, such that all deformation modes would occur simultaneously, which greatly hampers a systematic investigation of different damping effects in the geometrically nonlinear range. Nevertheless, experiments at small amplitudes are helpful to determine the ranges of weak, moderate and critical damping for the respective deformation modes, quantifyable by explicit formulas similar to the one given above for the case of transverse vibrations. These could then be used e.g. to adjust damping of different deformation modes to experimental obervations.

6 Conclusions

In our paper we presented the derivation of a viscous Kelvin–Voigt type damping model for geometrically exact Cosserat rods. For homogeneous and isotropic materials, we obtained explicit formulas for the damping parameters given in terms of the stiffness parameters and retardation time constants, assuming moderate reference curvatures, small strains and sufficiently low strain rates. In numerical simulations of vibrations of a clamped cantilever beam we observed a slightly weakening influence of geometric nonlinearities on the damping of the oscillation amplitudes. We also found that the variation of retardation time and bending stiffness has a similar effect on the damping ratio as in the linear regime. In view of the limitations of the Kelvin–Voigt model w.r.t. higher frequencies it would be worthwile to develop more complex viscoelastic models (e.g. of generalized Maxwell type) for Cosserat rods. Our approach to derive Kelvin–Voigt damping for Cosserat rods may be helpful to obtain such models from three-dimensional continuum theory in an analogous way.

Appendix A

Measuring 3-D strains and stresses for rods

From a mathematical point of view, the tensor $\hat{\mathbf{C}}$ may be regarded as the fundamental quantity to decribe the *shape* of a body, as it corresponds to the *metric* which determines the shape up to rigid body motions, provided that certain integrability conditions (i.e.: the vanishing of the Riemann curvature tensor) are satisfied. Other strain measures may be obtained as invertible functions of $\hat{\mathbf{C}}$ via its *spectral decomposition*. As a supplement to the brief discussion given in Sect. 3.1, we mention a few alternatives to measure 3-D strains and stresses used elsewhere in connection with geometrically exact rod theory.

A1 The Biot strain and its approximation

In the case of small strains, the *Biot* strain tensor defined as $\hat{\mathbf{E}}_B := \hat{\mathbf{U}} - \hat{\mathbf{I}}$, with the *right stretch tensor* $\hat{\mathbf{U}}$ given *implicitely* either by the polar decompostion $\hat{\mathbf{F}} = \hat{\mathbf{R}}_{\text{pd}} \cdot \hat{\mathbf{U}}$ of the deformation gradient, or as $\hat{\mathbf{U}} = \hat{\mathbf{C}}^{1/2}$ in terms of the right Cauchy–Green tensor, is likewise an appropriate alternative choice of a frame-indifferent material strain measure. Due to the algebraic identity $\hat{\mathbf{E}} = \frac{1}{2}(\hat{\mathbf{U}}^2 - \hat{\mathbf{I}}) = \frac{1}{2}(\hat{\mathbf{I}} + \hat{\mathbf{U}}) \cdot \hat{\mathbf{E}}_B$ the Biot and Green–Lagrange strains agree up to leading order for small strains, i.e.: $\hat{\mathbf{E}} \approx \hat{\mathbf{E}}_B$ holds whenever $\hat{\mathbf{U}} \approx \hat{\mathbf{I}}$.

One might argue that for small strains it is preferable to use $\hat{\mathbf{E}}_B$ as a strain measure, as it is *linear* in $\hat{\mathbf{U}}$ and therefore a first order quantity in terms of in the principal stretches, different from $\hat{\mathbf{E}}$, which is quadratic in $\hat{\mathbf{U}}$. However, while (18) provides a *kinematically exact* expression for $\hat{\mathbf{I}} + 2\hat{\mathbf{E}} = \hat{\mathbf{C}} = \hat{\mathbf{U}}^2$, a comparably simple closed form expression for $\hat{\mathbf{U}}$ itself is not available. In general the tensor $\hat{\mathbf{U}}$ has to be constructed via the spectral decomposition of $\hat{\mathbf{C}}$, which in 3-D cannot be expressed easily[17] in closed form.

For special simplified problems, like the *plane* deformation of an extensible Kirchhoff rod as discussed by Irschik and Gerstmayr (2009) and Humer and Irschik (2011), it is possible to derive simple, kinematically exact closed form expressions[18] for $\hat{\mathbf{U}}$ and $\hat{\mathbf{R}}_{\rm pd}$ by inspection of the deformation gradient. Also in the more general case of $\hat{\mathbf{C}}$ given by Eq. (18) an analytical solution of the spectral problem is possible: by inspection $\boldsymbol{N}_3 := \boldsymbol{H}\times\boldsymbol{a}_0^{(3)}/(H_1^2+H_2^2)^{1/2}$ is found to be one of its eigenvectors, with eigenvalue $\lambda_3^2 = 1$. The remaining 2-D spectral problem may then be solved analytically by a *Jacobi rotation* which diagonalizes the matrix representing $\hat{\mathbf{C}}$ w.r.t. the ONB in the plane orthogonal to $\boldsymbol{H}\times\boldsymbol{a}_0^{(3)}$ given by $\boldsymbol{a}_0^{(3)}$ and the unit vector along the direction of the projection $\boldsymbol{a}_0^{(3)}\times(\boldsymbol{H}\times\boldsymbol{a}_0^{(3)}) = H_\alpha\,\boldsymbol{a}_0^{(\alpha)}$ of the material strain vector $\boldsymbol{H}$ onto the local reference cross section. The resulting analytical formulas[19] for the two eigenvalues $\lambda_\pm^2$ and orthonormal eigenvectors $\boldsymbol{N}_{1/2}$ of $\hat{\mathbf{C}}$, which we present below without providing further details of their derivation, are given by:

$$\begin{aligned}
\lambda_\pm^2 - 1 &= \tilde{H}_3 + \|\tilde{\boldsymbol{H}}\|^2/2 \pm \sqrt{(\tilde{H}_3 + \|\tilde{\boldsymbol{H}}\|^2/2)^2 + (\tilde{H}_1^2+\tilde{H}_2^2)}\,,\\
\boldsymbol{N}_1 &= \cos(\phi)\,H_\alpha\,\boldsymbol{a}_0^{(\alpha)}/(H_1^2+H_2^2)^{1/2} + \sin(\phi)\,\boldsymbol{a}_0^{(3)}\,,\\
\boldsymbol{N}_2 &= -\sin(\phi)\,H_\alpha\,\boldsymbol{a}_0^{(\alpha)}/(H_1^2+H_2^2)^{1/2} + \cos(\phi)\,\boldsymbol{a}_0^{(3)}\,,
\end{aligned}$$

with $\tilde{\boldsymbol{H}} := \boldsymbol{H}/J_0$, and the angle ϕ given implicitely by

$$\sqrt{\tilde{H}_1^2+\tilde{H}_2^2}\,\cos(2\phi) + (\tilde{H}_3 + \|\tilde{\boldsymbol{H}}\|^2/2)\sin(2\phi) = 0\,.$$

They provide the spectral decomposition $\hat{\mathbf{C}} = \sum_{k=1}^3 \lambda_k^2\,\boldsymbol{N}_k\otimes\boldsymbol{N}_k$ of the right CG tensor (see Gurtin, 1981, ch. I and II), and the closed form expression $\hat{\mathbf{E}}_B = \sum_{k=1}^3(\lambda_k - 1)\,\boldsymbol{N}_k\otimes\boldsymbol{N}_k$ of the Biot strain tensor, as $\hat{\mathbf{U}} = \hat{\mathbf{C}}^{1/2}$.

These considerations confirm that, although a kinematically exact closed form expression of $\hat{\mathbf{E}}_B$ for deformed configurations of a Cosserat rod ($\boldsymbol{H} \neq \mathbf{0}$) may be derived in this way, it consists of algebraically rather complicated expressions in terms of the vector $\boldsymbol{H}/J_0$ and its components, compared to the relatively simple formula (18) for the Green–Lagrange strain. Otherwise, it is straightforward to show that $\hat{\mathbf{U}} \approx \hat{\mathbf{I}} + \frac{1}{2J_0}\left[\boldsymbol{H}\otimes\boldsymbol{a}_0^{(3)} + \boldsymbol{a}_0^{(3)}\otimes\boldsymbol{H}\right]$ provides an *approximate* expression for the right stretch tensor of leading order in $\boldsymbol{H}/J_0$, as its square agrees with the exact expression for $\hat{\mathbf{C}}$ up to terms of order $O(\boldsymbol{H}^2/J_0^2)$. Therefore, we obtain $\hat{\mathbf{E}}_B \approx \frac{1}{2J_0}\left[\boldsymbol{H}\otimes\boldsymbol{a}_0^{(3)} + \boldsymbol{a}_0^{(3)}\otimes\boldsymbol{H}\right]$ as an *approximate* expression for the Biot strain, which reduces to Eq. (19) for $J_0 \approx 1$ and in this way provides an alternative interpretation of Eq. (19). Within the same order we may use $\hat{\mathbf{R}}_{\rm pd}(\xi_1,\xi_2,s,t) \approx \hat{\mathbf{R}}_{\rm rel}(s,t)$ to approximate the rotational part of the polar decomposition of $\hat{\mathbf{F}}$.

[17] Whereas analytical expressions for the *eigenvalues* of a 3-D symmetric matrix are provided by Cardano's formulas, we are not aware of any simple closed form expression for the *eigenvectors*.

[18] In this special case, or likewise for spatial deformations of extensible *Elastica* without twisting, $H_\alpha \equiv 0 \Rightarrow \boldsymbol{H} = H_3\,\boldsymbol{a}_0^{(3)}$ holds, such that the exact expressions $\hat{\mathbf{R}}_{\rm pd} = \hat{\mathbf{R}}_{\rm rel}$ and $\hat{\mathbf{U}} = \hat{\mathbf{I}} + \frac{H_3}{J_0}\boldsymbol{a}_0^{(3)}\otimes\boldsymbol{a}_0^{(3)}$ may be read off directly from Eq. (16) due to the uniqueness of the polar decomposition.

[19] The spectral problem for the modified tensor $\hat{\mathbf{C}}' = \hat{\mathbf{I}} + 2\hat{\mathbf{E}}'$ given by Eq. (24), which accounts for uniform lateral contraction and vanishing in-plane stresses, may also be solved analytically in the same way. The corresponding formulas, which we omit here, are very similar to the ones given above, with the term $(\tilde{H}_3 + \|\tilde{\boldsymbol{H}}\|^2/2)$ multiplied by factors $(1 \pm \nu)$.

A2 Relation of the material strain vector to the Biot strain

Following Kapania and Li (2003), Mata et al. (2007, 2008) use the spatial vector quantity

$$(\hat{\mathbf{F}} - \hat{\mathbf{R}}_{\rm rel})\cdot\boldsymbol{a}_0^{(3)} = \frac{1}{J_0}\hat{\mathbf{R}}_{\rm rel}\cdot\boldsymbol{H} = \frac{1}{J_0}H_k\,\boldsymbol{a}^{(k)}$$

with $\hat{\mathbf{F}}$ given by a kinematically exact expression for the deformation gradient of a Cosserat rod equivalent to Eq. (16) to measure the strain at the individual points of a cross section. Its material counterpart $J_0^{-1}\hat{\mathbf{R}}_0^T\cdot\boldsymbol{H} = J_0^{-1}H_k\,\boldsymbol{e}_k$ as well as objective rates of both vector quantities are then used by these authors to formulate inelastic constitutive laws for their rod model on the 3-D level, which are required for a subsequent numerical evaluation of the spatial stress resultants and couples of the rod in its deformed configurations by numerical integration over the cross section areas.

Following our discussion of the Biot strain and its approximation given above, one recognizes that the strain measure used by Mata et al. (2008) likewise may be interpreted in terms of an approximation of the Biot strain via

$$\hat{\mathbf{F}} - \hat{\mathbf{R}}_{\rm rel} \approx \hat{\mathbf{F}} - \hat{\mathbf{R}}_{\rm pd} = \hat{\mathbf{R}}_{\rm pd}\cdot\hat{\mathbf{E}}_B \approx \hat{\mathbf{R}}_{\rm rel}\cdot\hat{\mathbf{E}}_B\,.$$

Using $\hat{\mathbf{F}} - \hat{\mathbf{R}}_{\rm pd}$ as a strain measure is directly related to the geometric idea to quantify the strains caused by the deformation of a body by the deviation of a deformation mapping to a rigid body motion, as discussed by Chao et al. (2010). For a given deformation gradient $\hat{\mathbf{F}}$ with positive determinant, this deviation may be measured by the distance of $\hat{\mathbf{F}}$ to the group SO(3) of proper rotations defined as $\min_{\hat{\mathbf{R}}\in SO(3)}\|\hat{\mathbf{F}} - \hat{\mathbf{R}}\|_F$, where $\|\cdot\|_F$ denotes the Frobenius norm. It can be shown that the minimum is actually reached for the unique rotation $\hat{\mathbf{R}} = \hat{\mathbf{R}}_{\rm pd}$ provided implicitely by the polar decomposition of $\hat{\mathbf{F}}$, such that $\min_{\hat{\mathbf{R}}\in SO(3)}\|\hat{\mathbf{F}} - \hat{\mathbf{R}}\|_F = \|\hat{\mathbf{R}}_{\rm pd}\cdot(\hat{\mathbf{U}} - \hat{\mathbf{I}})\|_F = \|\hat{\mathbf{E}}_B\|_F$ holds due to the invariance of the norm under rotations. Altogether these considerations, combined with the approximation $\hat{\mathbf{R}}_{\rm pd} \approx \hat{\mathbf{R}}_{\rm rel}$, provide a geometric interpretation for the strain measure considered by Mata et al. (2008) and its relation to the Biot strain.

A3 The Biot stress and its approximation

In some works dealing with geometrically exact rods, e.g. in the articles of Irschik and Gerstmayr (2009) and Humer and

Irschik (2011), 3-D stress distributions within cross sections are analyzed in terms of the (unsymmetric) *Biot stress* tensor $\hat{\mathbf{T}}_B := \hat{\mathbf{R}}_{\text{pd}}^T \cdot \hat{\mathbf{P}} = \hat{\mathbf{U}} \cdot \hat{\mathbf{S}}$, which is related to the (*true*) Cauchy stress $\hat{\boldsymbol{\sigma}}$ via the co-rotational stress tensor $\hat{\mathbf{R}}_{\text{pd}}^T \cdot \hat{\boldsymbol{\sigma}} \cdot \hat{\mathbf{R}}_{\text{pd}} = J^{-1}\hat{\mathbf{T}}_B \cdot \hat{\mathbf{U}}$. The Biot stress tensor $\hat{\mathbf{T}}_B$ as well as its *symmetric* part $\hat{\mathbf{T}}_B^{(s)} := \frac{1}{2}(\hat{\mathbf{T}}_B + \hat{\mathbf{T}}_B^T)$ are both work–conjugate stresses related to the Biot strain $\hat{\mathbf{E}}_B$, as $\left[\hat{\mathbf{T}}_B - \hat{\mathbf{T}}_B^{(s)}\right] : \delta\hat{\mathbf{E}}_B = 0$ holds, such that both yield identical virtual work expressions.

Small strain approximations of these stress quantities are obtained by substituting $\hat{\mathbf{U}} \approx \hat{\mathbf{I}}$ (implying $\hat{\mathbf{F}} \approx \hat{\mathbf{R}}_{\text{pd}}$ and $J \approx 1$) into the various transformation identities for the stresses as given above. This yields the set of approximate relations $\hat{\mathbf{T}}_B \approx \hat{\mathbf{R}}_{\text{pd}}^T \cdot \hat{\boldsymbol{\sigma}} \cdot \hat{\mathbf{R}}_{\text{pd}} \approx \hat{\mathbf{T}}_B^{(s)} \approx \hat{\mathbf{S}}$, which are valid to leading order, analogous to the approximate relations $\hat{\mathbf{E}}_B \approx \hat{\mathbf{E}}$ for the corresponding strain quantities. The approximate stress relations (21) are obtained by the additional approximation $\hat{\mathbf{F}} \cdot \hat{\mathbf{U}}^{-1} = \hat{\mathbf{R}}_{\text{pd}} \approx \hat{\mathbf{R}}_{\text{rel}}$, likewise valid to the same order, which effectively amounts to applying the approximation $\hat{\mathbf{F}} \approx \hat{\mathbf{R}}_{\text{rel}}$ (implying $J \approx 1$) within all transformations of stress tensors.

In summary, due to the assumption of small strains, the Biot and 2nd Piola–Kichhoff stress tensors approximately coincide to leading order (i.e.: $\hat{\mathbf{T}}_B \approx \hat{\mathbf{S}}$), such that both stresses approximately correspond to the co-rotational stress tensor given by the components of the Cauchy stress (i.e.: $\hat{\mathbf{T}}_B \approx \hat{\mathbf{R}}_{\text{rel}}^T \cdot \hat{\boldsymbol{\sigma}} \cdot \hat{\mathbf{R}}_{\text{rel}} \approx \hat{\mathbf{S}}$) w.r.t. the approximate material basis (i.e.: $\boldsymbol{G}_k \approx \boldsymbol{a}_0^{(k)} \approx \boldsymbol{G}^k$) given by the reference frames $\hat{\mathbf{R}}_0(s)$.

Appendix B

Discretizations of the Kelvin–Voigt model

Our recent articles (Lang et al., 2011; Lang and Arnold, 2012) provide one concrete example of an implementation of a discrete version of our constitutive model (1), as an integral part of (and taylored to) our specific *continuum formulation* of the Cosserat rod model using unimodular quaternions, our specific *spatial* discretization approach – finite differences for the centerline, finite quotients for the quaternion field, both on a staggered grid – applied on the level of the stored energy (10), kinetic energy (see Sect. 3.6) and the dissipation function (11), the specific formulation of the resulting *semidiscrete* system as a first order DAE or ODE (depending on the kind of internal kinematical constraints and their treatment), and the class of *time integration methods* we choose to solve the semidiscrete equations for various initial-boundary value problems.

Our treatment differs substantially from other approaches as discussed e.g. in the textbooks (Géradin and Cardona, 2001 and Bauchau, 2011) or the article (Bauchau et al., 2008) already mentioned in the introduction, which mainly use *finite elements* (of first or higher order) for the spatial discretization, but again differ among each other in the treatment of *rotational variables* and the related *interpolation* strategy. In addition, other model variants for geometrically nonlinear rods or beams exist, like the already mentioned ANCF approach used by Abdel-Nasser and Shabana (2011), or the recent approach of *dynamic splines* investigated by Theetten et al. (2008) and Valentini and Pennestri (2011), where geometrically exact extensible Kirchhoff rods, which require only a single angle variable to account for twisting, are desribed using computer-aided geometrical design functions, very similar to the usage of cubic Hermite splines on the element level as employed by Weiss (2002b).

In view of the great variety of discretization approaches applied to different geometrically exact rod models, a corresponding discussion of *discrete* versions of our Kelvin–Voigt model (1) for each variant is clearly beyond the scope of this article. In general, any implementation may be obtained most easily by a *semidiscrete* approach in terms of *material* strain quantities as used in Eq. (1). In this way one circumvents the technically rather complicated issue of constructing (and implementing) objective strain rates, which for the discretized *material* strain measures $\boldsymbol{U}_h$ and $\boldsymbol{V}_h$ of a Cosserat rod are given by simple partial time derivatives $\partial_t\boldsymbol{U}_h$ and $\partial_t\boldsymbol{V}_h$. An adaption of Eq. (1) for the dynamic spline model mentioned above is obtained by setting the transverse shear strains and their rates to zero ($V_\alpha = 0 = \partial_t V_\alpha$), such that $V_3 = \|\partial_s\boldsymbol{\varphi}(s,t)\|$ remains as the measure for elongational strain.

Acknowledgements. This work was supported by German BMBF with the research project *NeuFlexMKS* (FKZ: 01|S10012B).

This article is an extended version of the paper presented by Linn et al. (2012) at the IMSD 2012 conference.

Edited by: A. Tasora

References

Abdel-Nasser, A. M. and Shabana, A. A.: A nonlinear visco-elastic constitutive model for large rotation finite element formulations, Multibody Syst. Dyn., 26, 57–79, 2011.

Antman, S. S.: Invariant dissipative mechanisms for the spatial motion of rods suggested by artificial viscosity, J. Elasticity, 70, 55–64, 2003.

Antman, S. S.: Nonlinear Problems of Elasticity, 2nd Edn., Springer, 2005.

Arne, W., Marheineke, N., Schnebele, J., and Wegener, R.: Fluid-fiber-interaction in Rotational Spinning Process of Glass Wool Production, Journal of Mathematics in Industry, 1:2, doi:10.1186/2190-5983-1-2, 2011.

Bauchau, O. A.: Flexible Multibody Dynamics, Springer, 2011.

Bauchau, O. A., Epple, A., and Heo, S.: Interpolation of Finite Rotations in Flexible Multibody Dynamics Simulations, P. I. Mech. Eng. K-J. Mul., 222, 353–366, 2008.

Berdichevsky, V. L.: On the energy of an elastic rod, J. Appl. Math. Mech. (PMM), 45, 518–529, 1981.

Berdichevsky, V. L.: Structure of equations of macrophysics, Phys. Rev. E, 68, 1–26, 2003.

Berdichevsky, V. L.: Variational Principles of Continuum Mechanics, Vol. I: Fundamentals, Vol. II: Applications, Springer, 2009.

Berdichevsky, V. L. and Staroselsky, L. A.: On the theory of curvilinear Timoshenko-type rods, J. Appl. Math. Mech. (PMM), 47, 809–816, 1983.

Bergou, M., Wardetzky, M., Robinson, S., Audoly, B. and Grinspun, E.: Discrete Elastic Rods, ACM Transaction on Graphics (SIGGRAPH), 27, 63:1–63:12, 2008.

Bergou, M., Audoly, B., Vouga, E., Wardetzky, M., and Grinspun, E.: Discrete Viscous Threads, ACM Transaction on Graphics (SIGGRAPH), 29, 116:1–116:10, 2010.

Chao, I., Pinkall, U., Sanan, P., and Schröder, P.: A Simple Geometric Model for Elastic Deformations. ACM Transaction on Graphics (SIGGRAPH), 29, 38:1–38:6, 2010.

Coleman, B. D. and Noll, W.: Foundations of Linear Viscoelasticity, Rev. Mod. Phys., 33, 239–249, 1961.

Cowper, G.R.: The shear coefficient in Timoshenko's beam theory, J. Appl. Mech., 33, 335–340, 1966.

Craig, R. R. and Kurdila, A. J.: Fundamentals of Structural Dynamics, 2nd Edn., John Wiley & Sons, 2006.

Dong, S. B., Alpdogan, C., and Taciroglu, E.: Much ado about shear correction factors in Timoshenko beam theory. Int. J. Solids Struct., 47, 1651–1665, 2010.

Géradin, M. and Cardona, A.: Flexible Multibody Dynamics: A Finite Element Approach, John Wiley & Sons, 2001.

Gruttmann, F. and Wagner, W.: Shear correction factors in Timoshenko's beam theory for arbitrary shaped cross sections, Comput. Mech., 27, 199–207, 2001.

Gurtin, M. E.: An Introduction to Continuum Mechanics, Academic Press, 1981.

Hodges, D. H.: Nonlinear Composite Beam Theory, AIAA, 2006.

Humer, A. and Irschik, H.: Large deformation and stability of an extensible elastica with an unknown length, Int. J. Solids Struct., 48, 1301–1310, 2011.

Irschik, H. and Gerstmayr, J.: A continuum mechanics based derivation of Reissner's large-displacement finite-strain beam theory: The case of plane deformations of originally straight Bernoulli–Euler beams, Acta Mech., 206, 1–21, 2009.

Jung, P., Leyendecker, S., Linn, J., and Ortiz, M.: A discrete mechanics approach to the Cosserat rod theory – Part 1: static equilibria, Int. J. Numer. Methods Eng., 85, 31–60, 2011.

Kapania, R. K. and Li, J.: On a geometrically exact curved/twisted beam theory under rigid cross-section assumption, Comput. Mech., 30, 428–443, 2003.

Klar, A., Marheineke, N., and Wegener, R.: Hierarchy of Mathematical Models for Production Processes of Technical Textiles, ZAMM, 89, 941–961, 2009.

Landau, L. D. and Lifshitz, E. M.: Statistical Physics – Part 1, Course of Theoretical Physics Vol. 5, 3rd Edn., Butterworth Heinemann, 1980.

Landau, L. D. and Lifshitz, E. M.: Theory of Elasticity, Course of Theoretical Physics Vol. 7, 3rd Edn., Butterworth Heinemann, 1986.

Lang, H. and Arnold, M.: Numerical aspects in the dynamic simulation of geometrically exact rods, Appl. Numer. Math., 62, 1411–1427, 2012.

Lang, H., Linn, J., and Arnold, M.: Multibody dynamics simulation of geometrically exact Cosserat Rods, Multibody System Dynamics, 25, 285–312, 2011.

Lemaitre, J. and Chaboche, J.-L.: Mechanics of Solid Materials, Cambridge Universtity Press, 1990.

Linn, J., Stephan, T., Carlsson, J., and Bohlin, R.: Fast Simulation of Quasistatic Rod Deformations for VR Applications, in: Progress in Industrial Mathematics at ECMI 2006, edited by: Bonilla, L. L., Moscoso, M., Platero, G., and Vega, J. M.: 247–253, ISBN 978-3-540-71992-2, Springer, 2008.

Linn, J., Lang, H., and Tuganov, A.: Geometrically exact Cosserat rods with Kelvin–Voigt type viscous damping, in: Proceedings of the 2nd Joint International Conference on Multibody System Dynamics (IMSD2012), Stuttgart, Germany, edited by: Eberhard, P. and Ziegler, P., ISBN 978-3-927618-32-9, 2012.

Lorenz, M., Marheineke, N., and Wegener, R.: On an asymptotic upper-convected Maxwell model for a viscoelastic jet, Proc. Appl. Math. Mech. (PAMM), 12, 601–602, 2012.

Love, A. E. H.: A Treatise on the Mathematical Theory of Elasticity, 4th Edn. (1927), reprinted by Dover, New York, 1963.

Mata, P., Oller, S., and Barbat, A. H.: Static analysis of beam structures under nonlinear geometric and constitutive behavior, Comput. Meth. Appl. Mech. Eng., 196, 45–48, 4458–4478, 2007.

Mata, P., Oller, S., and Barbat, A. H.: Dynamic analysis of beam structures considering geometric and constitutive nonlinearity, Comput. Meth. Appl. Mech. Eng., 197, 857–878, 2008.

Marheineke, N. and Wegener, R.: Asymptotic Model for the Dynamics of Curved Viscous Fibers with Surface Tension, J. Fluid Mech., 622, 345–369, 2009.

Panda, S., Marheineke, N., and Wegener, R.: Systematic Derivation of an Asymptotic Model for the Dynamics of Curved Viscous Fibers, Math. Meth. Appl. Sci., 31, 1153–1173, 2008.

Petrie, C.J.S: Extensional viscosity: A critical discussion. J. Non-Newtonian Fluid Mech., 137, 15–23, 2006.

Ribe, N. M.: Coiling of viscous jets, Proc. R. Soc. Lond. A, 460, 3223–3239, 2004.

Romero, I.: A comparison of finite elements for nonlinear beams: the absolute nodal coordinate and geometrically exact formulations, Multibody Syst. Dyn., 20, 51–68, 2008.

Schulze, M., Dietz, S., Tuganov, A., Lang, H., and Linn, J.: Integration of nonlinear models of flexible body deformation in Multibody System Dynamics, in: Proceedings of the 2nd Joint International Conference on Multibody System Dynamics (IMSD2012), Stuttgart, Germany, edited by: Eberhard, P. and Ziegler, P., ISBN 978-3-927618-32-9, 2012.

Simo, J. C.: A finite strain beam formulation: the three dimensional dynamic problem – Part I, Comput. Meth. Appl. Mech. Eng., 49, 55–70, 1985.

Simo, J. C., Hjelmstad, K. D., and Taylor, R. L.: Numerical formulations of elasto-viscoplastic response of beams accounting for the effect of shear, Comput. Meth. Appl. Mech. Eng., 42, 301–330, 1984.

Theetten, A., Grisoni, L., Andriot, C., and Barsky, B.: Geometrically exact dynamic splines, Computer Aided Design, 40, 35–48, 2008.

Trouton, F. T.: On the coefficient of viscous traction and its relation to that of viscosity, Proc. R. S. Lond. A, 77, 426–440, 1906.

Valentini, P. P. and Pennestri, E.: Modelling Elastic Beams Using Dynamic Splines, Multibody Syst. Dyn., 25, 271–284, 2011.

Villagio, P.: Mathematical Models for Elastic Structures, Cambridge University Press, 1997 (digital paperback reprint: 2005).

Weiss, H.: Dynamics of Geometrically Nonlinear Rods: I – Mechanical Models and Equations of Motion, Nonlinear Dyn., 30, 357–381, 2002A.

Weiss, H.: Dynamics of Geometrically Nonlinear Rods: II – Numerical Methods and Computational Examples, Nonlinear Dyn., 30, 383–415, 2002B.

Zupan, E., Saje, M., and Zupan, D.: The quaternion-based three-dimensional beam theory, Comput. Meth. Appl. Mech. Eng., 198, 3944–3956, 2009.

Zupan, E., Saje, M., and Zupan, D.: Quaternion-based dynamics of geometrically nonlinear spatial beams using the Runge–Kutta method, Finite Elem. Anal. Des., 54, 48–60, 2012A.

Zupan, E., Saje, M., and Zupan, D.: Dynamics of spatial beams in quaternion description based on the Newmark integration scheme, Comput. Mech., 51, 47–64, 2012B.

12

New empirical stiffness equations for corner-filleted flexure hinges

Q. Meng[1], **Y. Li**[1,2], **and J. Xu**[1]

[1]Department of Electromechanical Engineering, Faculty of Science and Technology, University of Macau, Av. Padre Tomas Pereira, Taipa, Macao SAR, China
[2]School of Mechanical Engineering, Tianjin University of Technology, Tianjin 300384, China

Correspondence to: Y. Li (ymli@umac.mo)

Abstract. This paper investigates the existing stiffness equations for corner-filleted flexure hinges. Three empirical stiffness equations for corner-filleted flexure hinges (each fillet radius, r, equals to $0.1\,l$; l, the length of a corner-filleted flexure hinge) are formulated based on finite element analysis results for the purpose of overcoming these investigated limitations. Three comparisons made with the existing compliance/stiffness equations and finite element analysis (FEA) results indicate that the proposed empirical stiffness equations enlarge the range of rate of thickness (t, the minimum thickness of a corner-filleted flexure hinge) to length (l), t/l ($0.02 \le t/l \le 1$) and ensure the accuracy for each empirical stiffness equation under large deformation. The errors are within 6 % when compared to FEA results.

1 Introduction

Flexure-based compliant mechanisms (FCMs) are becoming increasingly popular due to their remarkable advantages such as part-count reduction, reduced assembly time, and simplified manufacturing processes in terms of cost reduction, increased precision and reliability, reduced wear and weight, and increased performance (Palmieri et al., 2012; Lobontiu, 2002; Berselli, 2009). In addition, FCMs are the perfect substitutions of traditional rigid mechanisms, compared to the other types of compliant mechanisms. A FCM can not only transform a traditional rigid mechanism in terms of its functions and structure, but also implement high precision and high frequency while the traditional rigid mechanism cannot do. For instance, FCMs are increasingly used in the fields of micro-scale and nano-scale technologies, such as smaller and high precision positioning devices in automobiles, telecommunications, medical, biology, optics or computer industries (Ma et al., 2006; Ivanov and Corves, 2010; Yin and Ananthasuresh, 2003; Lobontiu and Garcia, 2003; Dong et al., 2008, 2005; Xu and King, 1996; Li and Xu, 2009).

A FCM relies on the elastic deformation of its connectors, i.e. the flexure hinges, to perform its functions of transmitting and/or transforming motion and force (Yin and Ananthasuresh, 2003). The flexure hinges which are utilized in connecting the rigid components are regarded as the traditional joints in a mechanism, therefore, play a key role in realizing the roles of the mechanism. In spite that flexure hinges own numerous attractive attributes over traditional rigid joints, however, FCMs are not used as widely as rigid-body mechanisms. The main limitation might be the lack of materials and processing techniques that enable structures deform considerably with adequate strength. A lot of researchers put efforts on changing the shape of flexure hinges and studying the stiffness and stress characteristics of these flexure hinges. So far, the flexure hinges can be classified into two categories: the primitive flexure hinges, such as circular flexure hinges, corner filleted flexure hinges, elliptical flexure hinges, parabolic flexure hinges, hyperbolic flexure hinges, "V" shape flexure hinges, right circular elliptical flexure hinges, right circular corner-filleted flexure hinges, two axes flexure hinges, multiple axes flexure hinges, and the complex flexure hinges such as cross axis flexure hinges, cartwheel flexure hinges. The shapes of these flexure hinges can be found in the literatures (Meng et al., 2012; Zettl et al., 2005; Chen et al., 2005; Lobontiu and Paine, 2002; Lobontiu et al., 2002a, 2001, 2002b), this paper will not present them in detail again. Among these flexure hinges, circular flexure

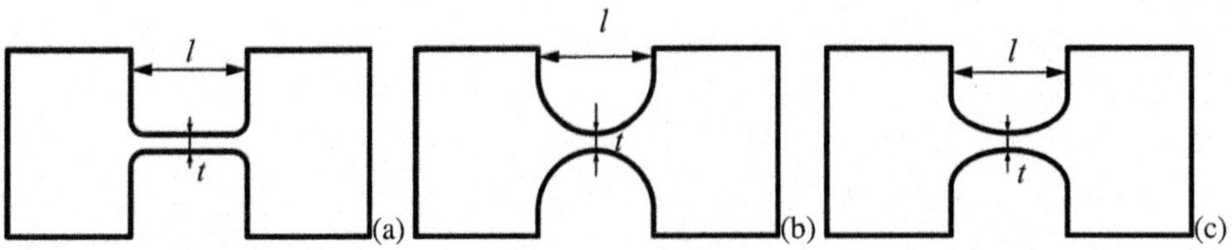

Figure 1. Corner-filleted flexure hinge (**a**), circular flexure hinge (**b**) and elliptical flexure hinge (**c**).

hinges, elliptical flexure hinges and corner-filleted flexure hinges as shown in Fig. 1 are commonly used in FCMs because they are simple and easy to control (Smith et al., 1997; Yong et al., 2008).

This paper addresses the corner-filleted flexure hinges. The stiffness characteristics of corner-filleted flexure hinges with a wide range of t/l are analyzed under different loads applied at the free-end. The influence introduced by shearing is taken into account during the whole analysis. Different ratios t/l are investigated in order to overcome the influence induced by shearing.

The remaining sections of this paper are organized as follows. Section 2 summarizes simply existing research results and some problems need to be resolved in future about corner-filleted flexure hinges. In Sect. 3, fifty finite element analysis models of corner-filleted flexure hinges are built up and their static plane stress analysis is simulated. In Sect. 4, three stiffness empirical equations are formulated by fitting the FEA results. A comparison of the new results together with the previous results is made to FEA in Sect. 5. In the end, the conclusions are drawn in Sect. 7.

2 Literature review

It can be dated back to 1965, Paros and Weisbord firstly presented the compliance-based approach to symmetric circular and right circular flexure hinges by giving the compliance equations and the approximate engineering formulas for these flexure hinges. The analytical approach of monolithic flexure hinges is the landmark in the research of flexure hinges. Particularly, the angles and linear deflections produced on all three axes are expressed in terms of the corresponding external loading.

Ragulskis et al. (1989) analyzed the filleted flexure hinges by applying the static finite element analysis method in order to calculate their compliances (Lobontiu et al., 2001).

In the early work of Howell and Midha (1994), they presented a computer-aided design method to pseudo-rigid-body model that included short length beam. A short length beam can be considered as a corner-filleted flexure hinges without filleted corners at the junction of flexible part and rigid part.

In 1996, Xu and King utilized the topology method to design a flexure-based amplifier for piezo-actuators. Right circular, corner-filleted, and elliptical flexure hinges were analyzed by static finite element analysis (FEA) method. By comparing both the FEA models to the traditional right-circular flexure hinges, three useful points were presented in their conclusions. They revealed that the corner-filleted flexure hinges are the most accurate in terms of motions relative to the elliptical flexure hinges, the elliptical flexure hinges have less stress for the same displacement and right circular flexure hinges are the stiffest. It is worth to note that the motion described here does not mean the precision of rotation. Even though the corner-filleted flexure hinges provide more accuracy on motion than the elliptical flexure hinges, they present less precision in terms of rotation. The deviation of the rotation center point of a corner-filleted flexure hinges is bigger than a circular flexure hinge.

Lobontiu et al. (2001) did a lot of works on the analysis of corner-filleted flexure hinges. They presented an analytical approach to corner-filleted flexure hinges in order to implement corner-filleted flexure hinges used in piezoelectric-driven amplification mechanisms. In this paper, closed-form in-plane compliance factor equations were formulated based on Castigliano's second theorem. A comparison was made with right circular flexure hinges which revealed that the corner-filleted flexure hinges are more bending-compliant and induce lower stresses but less precise in rotation. Lobontiu and Paine (2002) as well as Lobontiu et al. (2002a, b) also developed the other type of flexure hinges by means of the similar approach as corner-filleted flexure hinges, for instance, cross-section corner-filleted flexure hinges in three-dimensional compliant mechanisms applications, conic-section (circular, elliptical, parabolic and hyperbolic) flexure hinges. In 2004, Lobontiu et al. proposed the closed-form stiffness equations that can be used to characterize the static model and dynamic behavior of single-axis corner-filleted flexure hinges based on Castiliagno's first theorem. The new stiffness equations reflect sensitivity to direct- and cross-bending, axial loading, and torsion. Compared to their previous works, the resulting equations for the stiffness factors are more accurate and completely define the elastic response of corner-filleted flexure hinges. No matter how the previous stiffness equations or the refined stiffness equations were presented by the researchers, these equations can be valid only under some assumptions. Two main assumptions are that the deflection subject to shearing is negligible for beam-like structures and the deformations of a flexure hinge are small. Lobontiu and Garcia (2003) presented an analytical model for displacement and stiffness calculations of planar compliant mechanisms with single-axis flexure hinges relying on the strain energy and Castigliano's displacement theorem. Specifically, circular flexure hinges and corner-filleted flexure hinges as typical symmetric single-axis flexure hinges are contained in the amplifiers in order to verify the deduced stiffness equations.

Du et al. (2011) proposed that a new class of flexure hinges named elliptical-arc-fillet flexure hinges, which covers elliptical arc, circular-arc-fillet, elliptical-fillet, elliptical, circular, circular-fillet (in the other words, corner-filleted, the major axis equals to the minor axis of the elliptical arc), and right circular flexure hinges together under one set of equations. The closed-form equations for compliance and precision matrices of elliptical-arc-fillet flexure hinges were derived in their works. In consequence, the analytical results were within 10 % error compared to the FEA results and within 8 % error compared to the experimental results. It is worth to note that the closed-form equations were derived based on the small-deformation theory. Therefore, these equations can be valid only when the deformations of the flexure hinges are small enough, or the thickness t to length l ratios are small enough (Du et al., 2011).

With the development of FCMs, the refined design equations under small deflections for typical flexure hinges cannot meet the needs of designers. Large deformation started to be a key problem to develop the typical flexure hinges. Nevertheless, the stiffness characteristics of flexure hinges under large deformation are complex due to shearing deformation. Scholars seek to stay away from the problem in the research. For instance, Trease et al. (2005) proposed a new compliant translational joint in order to overcome the drawbacks of typical flexure hinges such as limited range of motion, axis drift and off-axis stiffness. Compared with the typical flexure hinges, however, the new designed joint is complex. Howell (2002) studied short slender beams under large deformation due to a force or a moment on its free-end. The length of the short beam should be much shorter than the length of the rigid part, while the flexible part should be more compliant than the rigid part. It signifies that the thickness of the short beam should be much less than its length ($t/l \leq 0.1$) (as shown in Fig. 1a).

Tian et al. presented a dimensionless design graph for circular, corner-filleted and cross flexure hinges, based on finite element analysis. The maximum stiffness properties from different hinges in identical situations were described by the FEA results. It revealed that a corner-filleted flexure hinge is preferred over a circular flexure hinge for stiffness demands in a single direction, while the medium stiffness is a cross flexure hinge (Qin et al., 2013; Tian et al., 2010b).

In real applications, however, the existed research results cannot meet some requirements such as stiffness and structure magnitude (the different ratio t/l) determination. Meng et al. (2012) studied the corner-filleted flexure hinges by synthesizing each kind of situation, i.e. the different ratio t/l and large deformation, under only a pure moment on the free-end.

From the previous works, it can be noted that there are no more accurate design formulas at the stage to estimate stiffness/compliances in the x and y directions for $t/l \geq 0.1$ and the rotational stiffness equation for a vertical force applied at the free end. Therefore, general empirical stiffness equations (named K_θ, K_x and K_y) are formulated in the Sect. 4 based on FEA results to evaluate the stiffness in x and y directions for a wide range of t/l ratios. Also, the rotational empirical stiffness equation will be presented in this section.

3 FEA modeling of corner-filleted flexure hinge

According to the descriptions aforementioned, there are three basic research approaches to the flexure hinges. The first one is based on displacement theorem, the second one is the pseudo rigid body model (PRBM), and the last one is the finite element analysis method. FEA method is used as a benchmark for calculating the rotational stiffness and the stiffness in the x and y directions for flexures. Also, FEA is an important approach to verify these formulas proposed based on the first two research methods. The accuracy of these FEA models was verified by Lobontiu and was with the maximum 8 % error compared to three experimental results (Lobontiu et al., 2004). Therefore, this paper studies the stiffness characteristics for corner-filleted flexure hinge with a wide range of t/l under large deformation by means of FEA.

COMSOL software was used to do with FEA of flexure hinges. Fifty corner-filleted flexure hinge models were generated by using two-dimensional, plane stress, parametric analysis, which were moved in the x and y directions. The ratio of t/l for these fifty models are $0.02, 0.04, 0.06, \ldots, 1$, respectively. Please note that the fillet radius, r, for each corner-filleted flexure model is specified to $0.1l$ in this paper because the fillet radius, r, is used in eliminating the stress concentration at each corner of a hinge. In the work of Meng et al. (2012), the rate of r/l was investigated and the results indicated that the minimum stress happened when the rate of r/l equals to 0.1 under identical material, t/l rate, and deformation conditions. The geometry of one of these analysis models is shown in Fig. 2. The modeled corner-filleted flexure hinges had a depth of 5 mm with a Young's modulus (E) of 1.135 GPa and a Poisson ratio (υ) of 0.33. Triangle element type is more suitable to model irregular shapes and is chosen in this paper to generate the model mesh. Four times refined meshing technique was used to produce automatically refined meshing at parts that high stress concentrations were most likely to occur in order to increase the analytical accuracy (see Fig. 3). The analysis of flexure hinges always is assumed to be cantilever beam. One end is fixed and the other one is free-end. It is easy to figure out that the length of a fixed rigid part cannot be too long because the accuracy of the FEA results can be significantly influenced by the

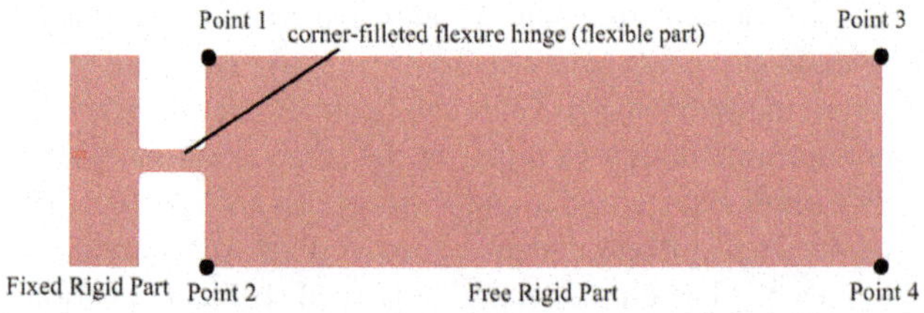

Figure 2. An analytical corner-filleted flexure hinge model.

Figure 3. FEA meshing.

deformation of this part. On the other hand, the free rigid part should keep it rigid in order to ensure the accuracy of the results. Therefore, the FEA models were designed with a short rigid part at the fixed end and a long rigid part of the fore-end as shown in Fig. 2. The research of corner-filleted flexure hinges under a pure moment applied at the free-end was presented in the previous work. This paper studies corner-filleted flexure hinges under a force in the x or y direction. It is well known that the accuracy of the FEA results can also be influenced by the way the point conditions are assigned to the model. For instance, when a point force in the y direction is applied at the end point of the whole model, an extra moment can be produced in terms of the rotational center point. As for the position where force is applied in the y direction, in this paper, a couple of forces are loaded at the point 1 and point 2, respectively, as shown in Fig. 2. While as for the position where force loaded in the x direction, a couple of forces are applied at the point 3 and point 4. Such loading method can decrease the FEA results error.

4 Empirical stiffness/compliance equations

From the previous section, FEA models with different t/l ratios, which were set from 0.02 to 1 with an increment of 0.02, were generated in COMSOL software. Forces in terms of F_x and F_y were applied at each model and the corresponding deformations, δx and δy were read. Figure 4a, b and c, which shows the relationship between the applied force, the deformation/deflection, and the rate of t/l, indicates that the stiffness is increasing with the increasing deformation/deflection and the increasing geometric parameter t/l. According to the compliance/stiffness equations calculated based on the Castiliagno's displacement theorem (Lobontiu, 2002), the Young's modulus, E, and the width of hinge, w, are proportional to the stiffness around z axis and in x, y directions. Therefore, the product of the two parameters can be divided. Such this, the stiffness can be transformed into a dimensionless quantity. The dimensionless design approach is very popular in the design of FCMs. In addition, the ratio of height to length, t/l, is an important parameter to the stiffness.

4.1 Empirical rotational stiffness equation

The rotational stiffness for corner-filleted here is under the condition of a perpendicular force applied at the end. According to the data in terms of δx and δy read from FEA results, the rotational angles were calculated by means of deforming geometric relationship. The data shown in Fig. 4a is rearranged for stiffness characteristic and the relationship is about rotational stiffness dimensionless design parameter (K_θ/Ew) and the rate t/l as shown in Fig. 5.

It is easy to figure out that the dimensionless design parameter is nearly a function of the rate t/l. While the deviation lies at the end of the relationship curve indicates that the stiffness is related to the deformation. The fourth polynomial, the fifth polynomial, and the sixth polynomial functions were fitted with fitting target of the norm of the residuals, 0.61475,0.61323,0.61321, respectively, by means of the data points to formulate empirical rotational stiffness equations. The fitting errors of these fitting functions are shown in Fig. 6. It can be observed from the figure that the minimum fitting error is produced by the sixth polynomial fitting function. The maximum fitting error produced by this fitting function is 2.3474 %. In consequence, the sixth polynomial fitting function is chosen as the rotational stiffness design equation as shown in Eq. (1) and its coefficients are shown in Table 1.

$$\frac{K_\theta}{Ew} = \sum_{i=0}^{6} a_i \left(\frac{t}{l}\right)^i \tag{1}$$

It is worth to mention that the maximum rotational angle of these FEA models during their deforming is 23°. It means that the influence about shearing deformation was taken into account in the empirical stiffness equation.

4.2 Empirical stiffness equation in x direction

By following the similar procedure as the rotational stiffness, the relationship between the dimensionless design parameter for stiffness in x direction, $K{-}x/Ew$, and the rate t/l is shown in Fig. 7.

According to the data δx read from FEA results, third, fourth, fifth, sixth, and seventh degrees of polynomial functions were fitted to the fitting target of the norm of the residuals, 0.04339, 0.015942, 0.010882, 0.010003, 0.0098163, respectively. Figure 8 shows the fitting errors for each fitting function. We can figure out from this figure that the minimum fitting errors were produced by the seventh degree

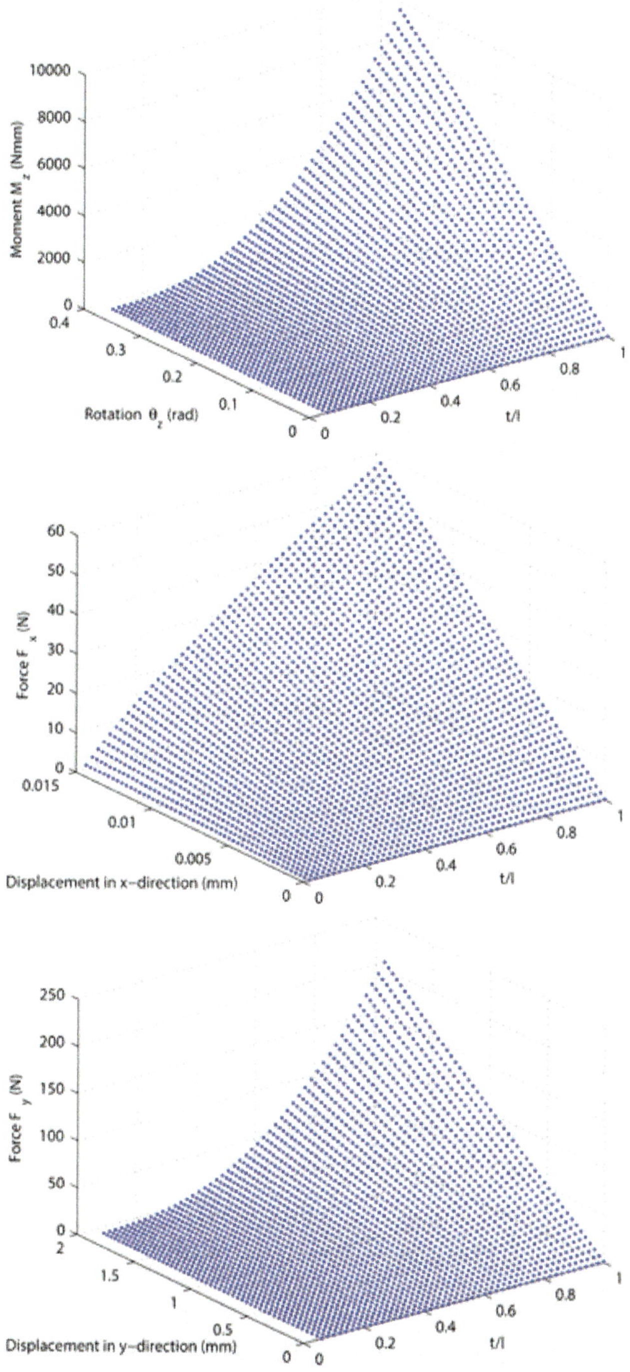

Figure 4. **(a)** The relationship between M_z, t/l, and rotation angle, θ_z; **(b)** The relationship between F_x, t/l, and displacement in x direction; **(c)** The relationship between F_y, t/l, and displacement in y direction.

polynomial function and the maximum fitting error produced by this function is 0.7213 %. Consequently, the seventh degree polynomial function is chosen as the dimensionless stiffness design equation as shown in Eq. (2) and its coefficients are listed in Table 1.

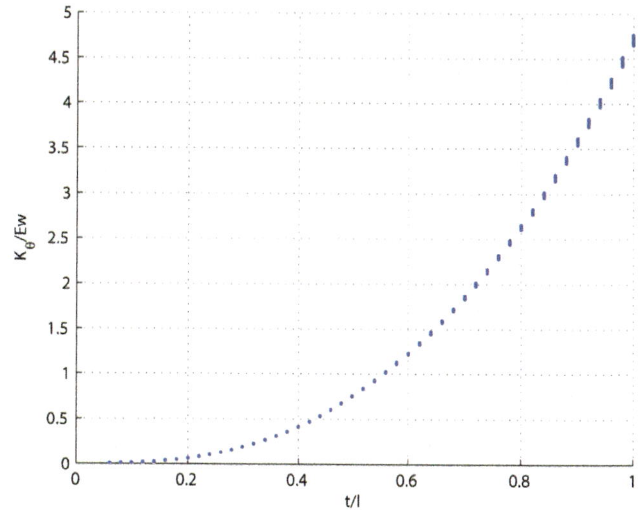

Figure 5. Rotational stiffness dimensionless design parameter.

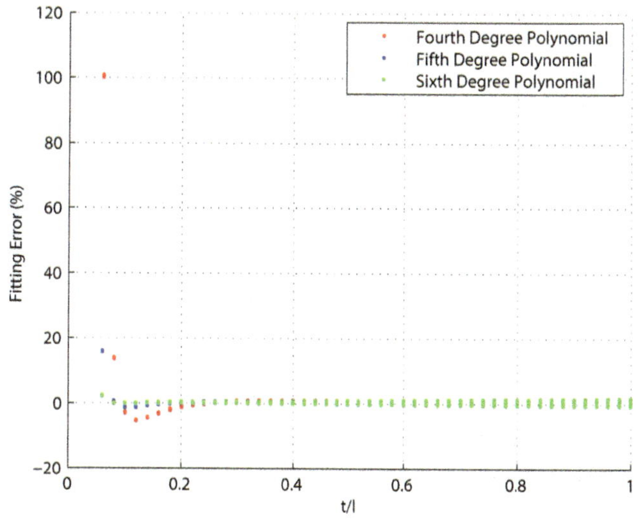

Figure 6. Fitting errors.

$$\frac{K_x}{Ew} = \sum_{i=0}^{7} b_i \left(\frac{t}{l}\right)^i \tag{2}$$

4.3 Empirical stiffness equation in y direction

Figure 4c is rearranged for stiffness equation in y direction as shown in Fig. 9. It can be found from this figure that the dimensionless design stiffness parameter K_y/Ew also can be considered as a function of the rate t/l.

However, the deviation which lies on the end part of the curve is larger than the rotational stiffness parameter. It indicates that the stiffness in y direction is related to the deformation of the hinge. In this paper, a simple fitting function

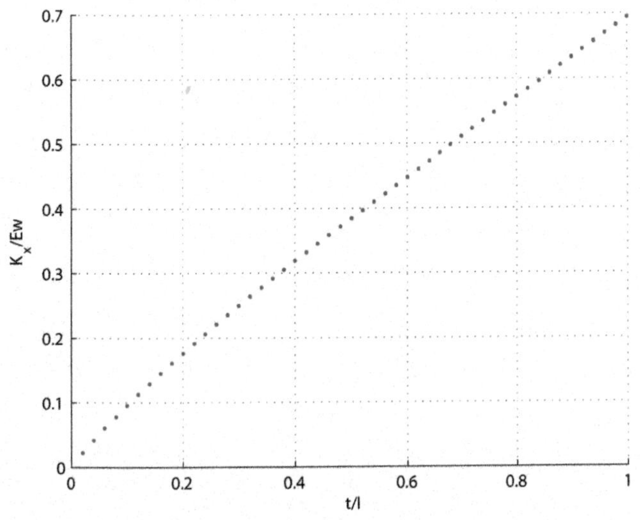

Figure 7. Stiffness in x direction dimensionless design parameter.

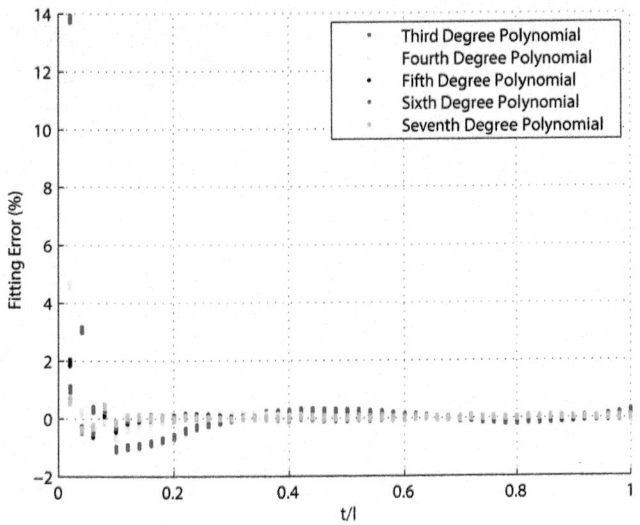

Figure 8. Fitting errors.

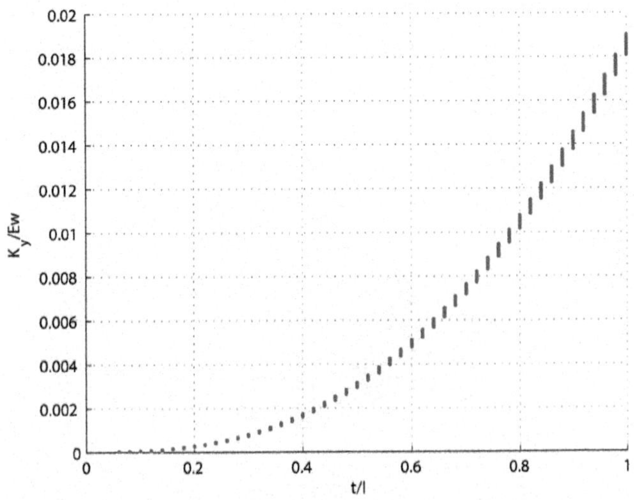

Figure 9. Stiffness in y direction dimensionless design parameter.

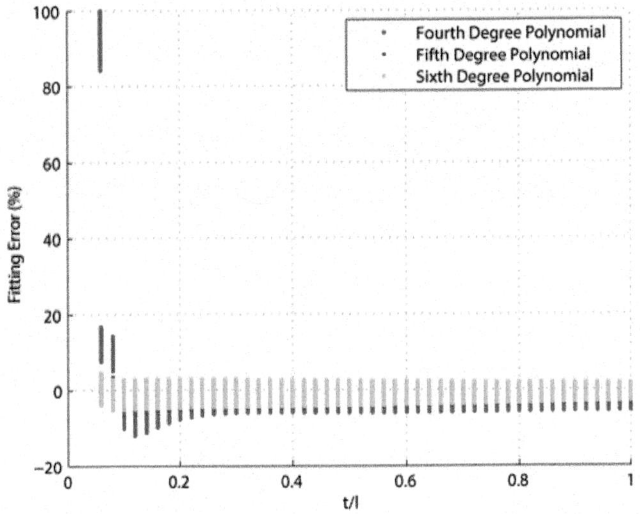

Figure 10. Fitting errors.

was chosen at first. The fourth, fifth, sixth degree polynomial function was fitted with the fitting target of the norm of the residuals, 0.0064273, 0.0064261, and 0.0064261, respectively. The fitting errors are shown in Fig. 10 that shows the minimum errors were produced by the sixth degree polynomial function and the maximum error is 5.2914 %. Therefore, as for the simple fitting function, the sixth degree polynomial function is chosen as the dimensionless stiffness in y direction design equation as shown in Eq. (3) and its coefficients are listed in Table 1.

$$\frac{K_y}{Ew} = \sum_{i=0}^{6} c_i \left(\frac{t}{l}\right)^i \tag{3}$$

In order to obtain more accurate fitting stiffness equations, the displacement δ_y is taken into account. By following the procedure mentioned above, the fitting equation is shown in Eq. (4) and its coefficients are listed in Table 1. Figure 11 shows the fitting error together with the fitting error produced by the simple fitting equation. It shows that the maximum error produced by the complex one is 1.8233 % when the rate t/l is larger than 0.08, while the error is much larger than the one produced by simple fitting equation when the rate t/l is smaller than 0.08. Therefore, the designer can choose the design equation according to the design rate t/l.

$$\frac{K_{cy}}{Ew} = \sum_{i,j=0}^{5} \mu_{ij} \left(\frac{t}{l}\right)^i (\Delta y)^j,\ j \leq 2 \tag{4}$$

It should be noticed that the influence about shearing deformation is taken into account in the empirical stiffness equation.

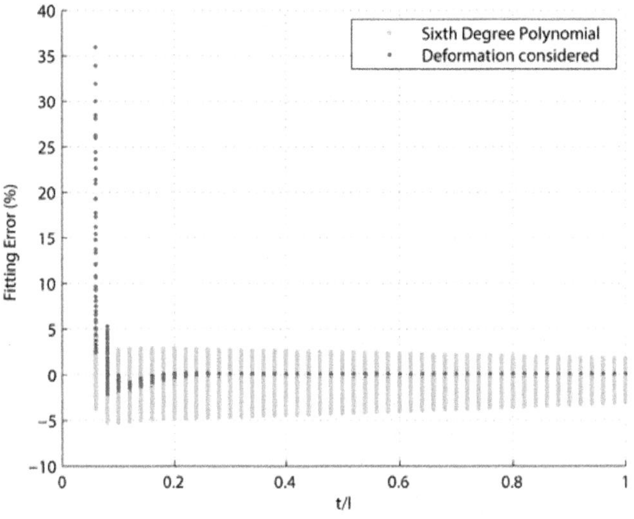

Figure 11. Fitting errors.

Table 1. Coefficients for Eqs. (1), (2), (3), and (4).

Eq. (1)		Eq. (2)		Eq. (3)	
a_0	6.10E-04	b_0	1.64E-03	c_0	−1.60E-03
a_1	−1.87E-02	b_1	1.02E+00	c_1	7.78E-03
a_2	2.25E-01	b_2	−1.11E+00	c_2	−1.85E-02
a_3	7.42E+00	b_3	2.28E+00	c_3	3.00E-02
a_4	−4.57E+00	b_4	−3.61E+00	c_4	8.28E-04
a_5	2.11E+00	b_5	3.75E+00	c_5	−6.27E-05
a_6	−4.59E-01	b_6	−2.17E+00	c_6	1.92E-06
		b_7	5.27E-01		
		Eq. (4)			
μ_{00}	1.10E-05	μ_{12}	−1.13E-02	μ_{31}	8.71E-03
μ_{01}	−1.28E-04	μ_{20}	1.94E-03	μ_{32}	−1.39E-02
μ_{02}	7.63E-04	μ_{21}	−6.77E-03	μ_{40}	−1.18E-02
μ_{10}	−2.35E-04	μ_{22}	5.02E-02	μ_{41}	−3.80E-03
μ_{11}	1.75E-03	μ_{30}	2.55E-02	μ_{50}	2.67E-03

5 Comparison of compliance/stiffness results with the previous work and FEA

From Sect. 2, there are a lot of researches in the study of compliance/stiffness characteristics for corner-filleted flexure hinges. This paper adopts the stiffness/compliance equations proposed by Howell (2002), Lobontiu (2002), Du et al. (2011) and Meng et al. (2012) in order to compare to the new empirical stiffness equation. Howell and Midha (1994) proposed stiffness equations for slender beam (it can be considered as a corner-filleted flexure hinge without filleted corners) by means of the pseudo rigid body model. The thickness-length ratio (t/l) is limited to less than 0.1. However, the deflection is not limited. Lobontiu proposed compliance equations by integrating the linear differential equation of a beam. The thickness-length ratio has not been limited. But the deformation must be small. Du et al. (2011) derived the compliance equations based on matrix methods. These closed-form compliance equations for the special configuration which is circular-fillet flexure hinges can be utilized when the thickness(t)-length(l) ratio is less than 0.1. Stiffness equations proposed by Meng et al. (2012) are fitted by means of FEA results based on Lobontiu's equations. Full stiffness equation and simple stiffness equation can be valid even though the deformation is large. However, it is noticed that the loading approach for equations of Howell (2002) and Meng et al. (2012) is a pure moment applied at the end. Therefore, these equations cannot describe the rotational stiffness accurately when a vertical force is applied at the free-end. These stiffness equations are shown in Appendix A, B, C and D.

5.1 Comparison of rotational stiffness equations, M_z/θ_z

Stiffness, M_z/θ_z (or its inverse, compliance, θ_z/M_z) of the corner-filleted flexure hinge as mentioned before was calculated using design equations of Lobontiu (2002), Howell and Midha (1994), Du et al. (2011), full and simple equation of Meng et al. (2012) and the new empirical equation. Their results were compared with the FEA results by the simulation method used in this paper. A corner-filleted flexure hinge, which was with thickness-length ratio t/l equaled to 0.5 and the maximum deforming rotational angle equaled to 23°, is chosen as an instance in order to show the errors induced by these equations. The comparison results are shown in Fig. 12. It shows that the results calculated by the equation of Howell (2002) and Lobontiu diverged greatly from the FEA results. The maximum errors induced by the equation of Howell (2002) and Lobontiu were 40.2252 % and 45.787 %, respectively, happened on the maximum rotation. The error induced by the compliance equation of Du et al. (2011) is less than Howell's and Lobontiu's, and it is 32.57 %. However, the errors induced by full and simple stiffness equations deduced by Meng et al. (2012) are small, and the maximum errors are 3.3292 % and 3.6826 %, respectively. The minimum error is produced by the new empirical rotational stiffness equation and it is only 0.7955 % at the identical situation.

Percentage errors of the comparison of various ratios of thickness-length were plotted in Fig. 13. The figure shows that the equations of Howell (2002) and Du et al. (2011) respectively keep accurate when the thickness-length ratio t/l is less than 0.1; the equation of Lobontiu only keeps accurate when the rotational angle is small enough; and the equations proposed by Meng et al. (2012) can keep accurate whatever high ratio of t/l or high rotation. However, the equations of Meng et al. (2012) are more complex than the new empirical rotational stiffness equations and are less accurate than the

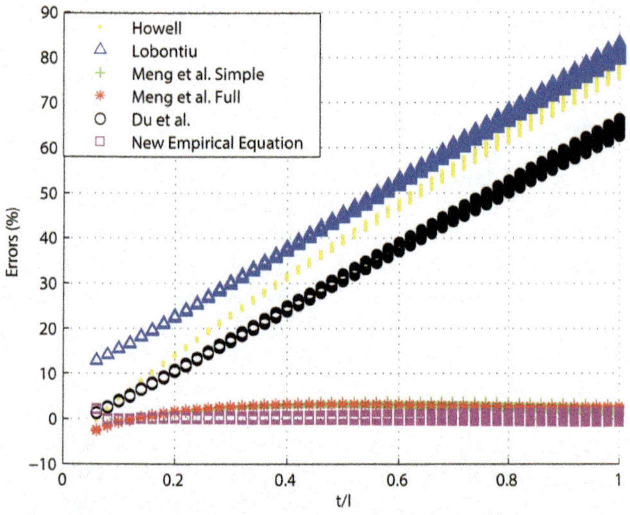

Figure 12. Comparison between FEA results and stiffness equations results $t/l = 0.5$.

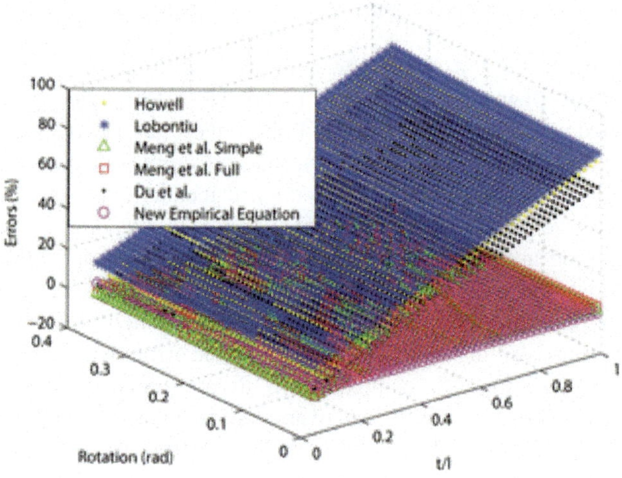

Figure 13. Errors of comparison results.

new one. The maximum errors induced by these equations of Howell (2002), Lobontiu (2002), Du et al. (2011), Meng et al. (2012) full and simple, and new empiric are 79.3530 %, 83.3920 %, 66.3578 %, 3.7248 %, 3.3292 %, and 2.2474 %, respectively.

5.2 Comparison of stiffness equations in x direction, F_x/δ_x

According to the previous comparison approach, the stiffness in x direction, F_x/δ_x of the corner-filleted flexure hinge was calculated using design equations of Lobontiu, Du et al. (2011) and new empirical stiffness equation. Their results were compared with the FEA results by the simulation method used in this paper. As what's described above, the corner-filleted flexure hinge, which was with thickness-length ratio t/l being 0.5 and the maximum displacement in x direction being 0.0147 mm, was chosen as an instance in order to show the errors induced by these equations. The comparison results are shown in Fig. 14. We can observe that the results calculated by the equations of Lonbontiu and Du et al. (2011) diverged greatly from the FEA results. The maximum error induced by the equation of Lobontiu (2002) and Du et al. (2011) were 32.3346 % and 32.1432, respectively, happened on the maximum displacement. However, the new empirical rotational stiffness equation produced only 0.0763 % at the identical situation.

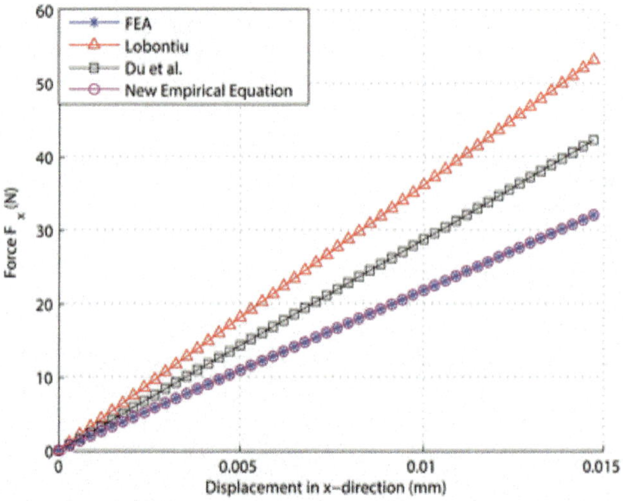

Figure 14. Comparison between FEA results and stiffness equations results $t/l = 0.5$.

Percentage errors of the comparison for various ratios of thickness-length are plotted in Fig. 15. The figure shows that the equations of Lobontiu and Du et al. (2011) only keeps accurate when the ratio of thickness to length is small enough. However, the results calculated by the new empirical stiffness equation in x direction have not diverged too much from the FEA results. The maximum errors induced by these equations of Lobontiu, Du et al. (2011), and new empirical are 45.2946 %, 45.0941 %, 0.7213 %, respectively.

5.3 Comparison of stiffness equations in y direction, F_y/δ_y

The stiffness in y direction, F_y/δ_y of the corner-filleted flexure hinge was calculated using design equations of Lobontiu (2002), Du et al. (2011) and new empirical stiffness equations (Eqs. 3 and 4) in y direction. Their results were compared with the FEA results by simulation method as described above. It is noticed that the corner-filleted flexure hinge, which was with thickness-length ratio t/l being 0.5 and the maximum displacement in y direction being 1.9226 mm, was chosen as an instance in order to show the errors induced by these equations. The comparison results are shown in

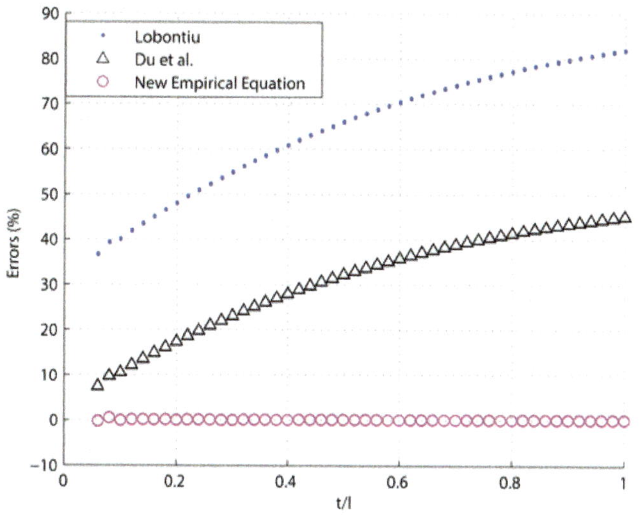

Figure 15. Errors of comparison results.

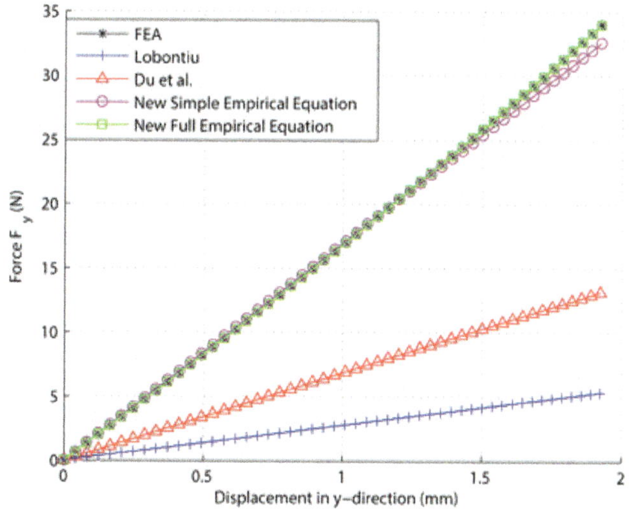

Figure 16. Comparison between FEA results and stiffness equations results $t/l = 0.5$.

Fig. 16. We can see that the results calculated by equation of Lobontiu (2002) and Du et al. (2011) deviate greatly from the FEA results. The maximum error induced by the equation of Lobontiu and Du et al. (2011) were 84.3457 % and 61.2894 %, respectively, happened on the maximum displacement. However, the new empirical rotational stiffness equations produced only 2.49 % and 0.0622 %, respectively, at the identical position.

Percentage errors of the comparison for various ratios of thickness-length were plotted in Fig. 17. The figure shows that the equation of Lobontiu and Du et al. (2011) were not accurate whenever the ratio of thickness to length is small or large. However, the results calculated by the new empirical stiffness equation in y direction have not deviated too much from the FEA results. The maximum errors induced

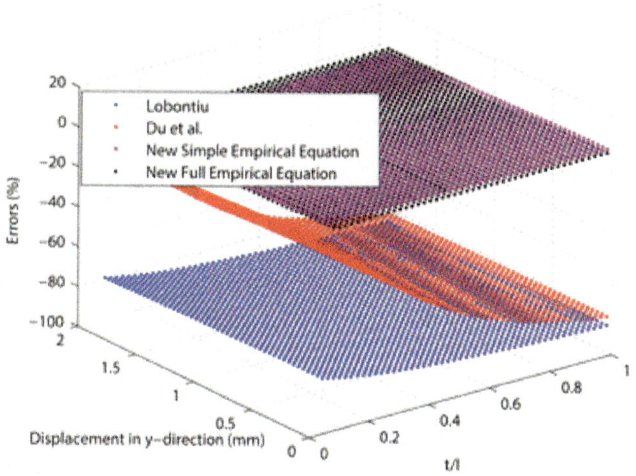

Figure 17. Errors of comparison results.

by these equations of Lobontiu, and new empirical equations are 87.1453 %, 82.8926 %, 5.2914 %, and 1.8233 %, respectively. The errors induced by equations of Lobontiu existed because that the shear compliances/stiffness were not considered in the compliance equations.

To sum up, the suggested stiffness/compliances equations to be used for any particular t/l range and the corresponding minimum, maximum percent errors, were summarized in Table 2.

6 Conclusions

This paper proposes three empirical stiffness equations for a wide range of t/l ratios ($0.02 \leq t/l \leq 1$) and large deformation for corner-filleted flexure hinges with fillet radius $0.1l$ based on FEA results. These three empirical stiffness equations are rotational stiffness equation when a vertical force was applied at the end; stiffness equation in x direction when a force in the x direction was applied at the end; and stiffness equation in y direction when a force in the y direction was applied at the end. These equations were compared to FEA results together with rotational stiffness equations of Howell, stiffness equations proposed by Lobontiu, compliance equations derived by Du et al. (2011) and rotational equations of Meng et al. (2012). Based on the results of comparisons, the proposed empirical stiffness equations are not only more simple than the existed stiffness equations, but also more accurate than others. The percentage errors of these empirical equations were found to be less than 6 % when compared to FEA results. The new proposed empirical stiffness design equations own simple shape and high design precision, which can be easily used in design of flexure-based compliant mechanisms.

Table 2. Suggested stiffness/compliance equations for a particular t/l range.

	t/l	K_θ				K_x			K_y		
		θ_z (rad)	Form	Err_{min} (%)	Err_{max} (%)	Form	Err_{min} (%)	Err_{max} (%)	Form	Err_{min} (%)	Err_{max} (%)
Howell (2002)	(0,0.1]	0.38	Simple	0.13	4.48	N/A	N/A	N/A	N/A	N/A	N/A
Lobontiu (2002a)	(0,0.3]	0.38	Complex	12.57	22.8	Complex	7.28	22.8	Complex	74.21	80.23
Chen et al. (2005)	(0,0.1]	0.38	Complex	1.13	4.13	Complex	7.28	10.64	Complex	9.84	18.87
Meng et al. (2012) (Full)	(0,1]	0.38	Complex	0.12	3.33	N/A	N/A	N/A	N/A	N/A	N/A
Meng et al. (2012) (Simple)	(0,1]	0.38	Complex	0.03	3.72	N/A	N/A	N/A	N/A	N/A	N/A
New Empirical	(0.02,1]	0.38	Simple	3.40E-04	2.25	Simple	5.70E-05	0.5	Simple	2.30E-04	5.29
Equation	(0.08,1]	0.38	N/A	N/A	N/A	N/A	N/A	N/A	Complex	2.50E-05	1.82

Appendix A

Du et al.

Compliance equation in x, y and α_z direction

$$\frac{\Delta x}{F_x} = \frac{a}{Ewb}N_1 + \frac{l}{Ewt}$$

$$\frac{\Delta y}{F_y} = \frac{ka}{Gwb}N_1 + \frac{6L^2a - 12a(cl + c^2 - a^2)}{Ewb^3}N_2 - \frac{12a^3}{Ewb^3}N_4 + \frac{24a^2l}{Ewb^3}N_6 + \frac{kl}{Gwt} + \frac{6L^2l - 2l(l^2 + 6c^2 + 6cl)}{Ewt^3}$$

$$\frac{\Delta\alpha_z}{M_z} = \frac{12a}{Ewb^3}N_2 + \frac{12l}{Ewt^3}$$

where,

$$N_1 = \frac{2(2s+1)}{\sqrt{4s+1}}\arctan\left(\sqrt{4s+1}\tan\frac{\phi_m}{2}\right) - \phi_m$$

$$N2 = \frac{12s^4(2s+1)}{(4s+1)^{5/2}}\arctan\left(\sqrt{4s+1}\tan\tfrac{\phi_m}{2}\right) + \frac{2s^3(2s+1)(6s^2+4s+1)\sin\phi_m}{(4s+1)^2(1+2s-2s\cos\phi_m)^2} - \frac{2s^4(12s^2+4s+1)\sin\phi_m\cos\phi_m}{(4s+1)^2(1+2s-2s\cos\phi_m)^2}$$

$$N_4 = \frac{48s^5 + 8s^4 + 20s^3 + 30s^2 + 10s + 1}{2(4s+1)^{5/2}} \times \arctan(\sqrt{4s+1}\tan\tfrac{\phi_m}{2}) - [\frac{s(2s+1)^2(12s^3 - 12s^2 - 3s)\cos\phi_m}{2(4s+1)^2(2s+1-2s\cos\phi_m)^2} + \frac{-12s^3 + 2s^2 + 6s + 1}{2(4s+1)^2(2s+1-2s\cos\phi_m)^2}] \times \sin\phi_m - \tfrac{\phi_m}{4}$$

$$N_6 = \frac{s^3(\cos\phi_m - 1)[(2s+1)\cos\phi_m - 2s - 1]}{2(2s+1-2s\cos\phi_m)^2}$$

Appendix B

Howell et al.

- Rotational stiffness equation
  $$K_H = \frac{Ewt^3}{12l}$$

Appendix C

Lobontiu

- Compliance equation in x direction
  $$C_{x,L} = \frac{1}{Ew}\left[\frac{l-2r}{t} + \frac{2(2r+t)}{\sqrt{t(4r+t)}}\arctan\sqrt{1+\frac{4r}{t}} - \frac{\pi}{2}\right]$$
- Rotational compliance equation
  $$C_{\theta_z,L} = \frac{12}{Ewt^3}\left\{l - 2r + \frac{2r}{(2r+t)(4r+t)^3}\left[t(4r+t)\left(6r^2+4rt+t^2\right) + 6r(2r+t)^2\sqrt{t(4r+t)}\arctan\sqrt{1+\frac{4r}{t}}\right]\right\}$$

- Compliance equation in y direction

$$
\begin{aligned}
C_{y,L} = \tfrac{3}{Ew}\{ & \frac{4(l-2r)\left(l^2-lr+r^2\right)}{3t^3} \\
& + \sqrt{t(4r+t)}\left[-80r^4+24r^3t+8(3+2\pi)r^2t^2+\right. \\
& \left. 4(1+2\pi)rt^3+\pi t^4\right]/4\sqrt{t^5(4r+t)^5} \\
& + \frac{(2r+t)^3\left(6r^2-4rt-t^2\right)\arctan\sqrt{1+\frac{4r}{t}}}{\sqrt{t^5(4r+t)^5}} \\
& + \left[-40r^4+8lr^2(2r-t)+12r^3t+4(3+2\pi)r^2t^2\right. \\
& \left. +2(1+2\pi)rt^3+\frac{\pi t^4}{2}\right]/2t^2(4r+t)^2 \\
& + \frac{4l^2r\left(6r^2+4rt+t^2\right)}{t^2(2r+t)(4r+t)^2} - \\
& \frac{(2r+t)\left[-24(l-r)^2r^2-8r^3t+14r^2t^2+8rt^3+t^4\right]}{\sqrt{t^5(4r+t)^5}} \\
& \arctan\sqrt{1+\frac{4r}{t}}\}
\end{aligned}
$$

Appendix D

Meng et al.

- Full rotational compliance equation

$$\Gamma_{\text{Full}} = \sum_{i,j=0}^{3} u_{ij}\theta_z^i\left(\frac{h}{L}\right)^j \quad \text{where } u_{ij}=0 \text{ if } i+j\geq 4$$

$K_{M,\text{Full}} = \Gamma_{\text{Full}} K_{\theta_z,L}$
Where the coefficients u_{ij} are shown in Table 3.

- Simple rotational compliance equation

$$\Gamma_{\text{Simple}} = \sum_{k=0}^{3} v_k\left(\frac{h}{L}\right)^k$$

$K_{M,\text{Simple}} = \Gamma_{\text{Simple}} K_{\theta_z,L}$
Where the coefficients v_{ij} are shown in Table 3.

Table A1. Coefficients for Full and Simple Equation.

u_{00}	1.016065	u_{03}	−0.043752	u_{20}	0.009576
u_{01}	−0.680692	u_{10}	0.002410	u_{21}	−0.095298
u_{02}	0.292380	u_{11}	−0.018696	u_{30}	−0.003720
u_{12}	0.018095				
v_0	1.018856	v_1	−0.713719	v_2	0.350531
v_3	−0.081827				

Acknowledgements. This work was supported in part by National Natural Science Foundation of China (Grant No. 61128008), Macao Science and Technology Development Fund (Grant No. 016/2008/A1), Research Committee of University of Macau (Grant no. MYRG203(Y1-L4)-FST11-LYM, MYRG183(Y1-L3)FST11-LYM).

Edited by: G. Hao

References

Berselli, G.: On designing compliant actuators based on dielectric elastomers, University of Bologna, Ph.D. thesis, 2009.

Chen, G. M., Jia, J. Y., and Li, Z. W.: Right-circular corner-filleted flexure hinges, in: Proceedings of the IEEE International Conference on Automation Science and Engineering, 249–253, 2005.

Dong, W., Du, Z., and Sun, L.: Stiffness influence atlases of a novel flexure hinge-based parallel mechanism with large workspace, in: IEEE/RSJ International Conference on Intelligent Robots and Systems, 856–861, 2005.

Dong, W., Sun, L., and Du, Z.: Stiffness research on a high-precision, large-workspace parallel mechanism with compliant joints, Precis. Eng., 32, 222–231, 2008.

Du, Y., Chen, G., and Liu, X.: Elliptical-arc-fillet flexure hinges: toward a generalized model for compmonly used flexure hinges, J. Mech. Design, 133, 1–9, 2011.

Howell, L. L.: Compliant Mechanisms, A Wiley-Interscience Publicationb, John Wiley & SONS., INC., 2002.

Howell, L. L. and Midha, A.: A method for the design of compliant mechanisms with small-length flexural pivots, J. Mech. Design, 116, 280–290, 1994.

Ivanov, I. and Corves, B.: Flexure hinge-based parallel manipulators enabling high-precision micro manipulations, Mechanisms and Machine Science, 49–60, 2012.

Li, Y. and Xu, Q.: Design and analysis of a totally decoupled flexure-based XY parallel micromanipulator, IEEE T. Robot., 25, 645–657, 2009.

Lobontiu, N.: Compliant Mechanisms: Design of Flexure Hinges, CRC Press, 2002.

Lobontiu, N. and Garcia, E.: Analytical model of displacement amplification and stiffness optimization for a class of flexure-based compliant mechanisms, Comput. Struct., 81, 2797–2810, 2003.

Lobontiu, N. and Paine, J. S. N.: Design of circular cross-section corner-filleted flexure hinges for three-dimensional compliant mechanisms, J. Mech. Design, 124, 479–484, 2002.

Lobontiu, N., Paine, J. S. N., Garcia, E., and Goldfarb, M.: Corner-filleted flexure hinges, J. Mech. Design, 123, 346–352, 2001.

Lobontiu, N., Paine, J. S. N., Garcia, E., and Goldfarb, M.: Design of symmetric conic-section flexure hinges based on closed-form compliance equations, Mech. Mach. Theory, 37, 477–498, 2002a.

Lobontiu, N., Paine, J. S. N., O' Malley, E., and Samuelson, M.: Parabolic and hyperbolic flexure hinges: flexibility, motion precision and stress characterization based on compliance closed-form equations, Journal of the International Societies for Precision Engineering and Nanotechnology, 26, 183–192, 2002b.

Lobontiu, N., Garcia, E., Hardau, M., and Bal, N.: Stiffness characterization of corner-filleted flexure hinges, Rev. Sci. Instrum., 75, 4896–4905, 2004.

Ma, H. W., Yao, S. M., Wang, L. Q., and Zhong, Z.: Analysis of the displacement amplification ratio of bridge-type flexure hinge, Sensor. Actuat. A-Phys., 132, 730–736, 2006.

Meng, Q., Berselli, G., Vertechy, R., and Castelli, P. V.: An improved method for design flexure-baesd nonlinear springs, in: ASME International Design Engineering Technical Conferences & Computers and Information in Engineering Conference, DETC2012-70367, 2012.

Palmieri, G., Palpacelli, M., and Callegari, M.: Study of a fully compliant u-joint designed for minirobotics applications, J. Mech. Design, 134, 1–9, 2012.

Paros, J. M. and Weisbord, L.: How to design flexure hinges, Mach. Des., Nov., 25, 151–156,1965.

Qin, Y., Shirinzadeh, B., Zhang, D., and Tian, Y.: Compliance modeling and analysis of statically indeterminate symmetric flexure structures, Precis. Eng., 37, 415–424, 2013.

Ouyang, P. R., Zhang, W. J., and Gupta, M. M.: Design of a new compliant mechanical amplifier, in: ASME International Design Engineering Technical Conferences & Computers and Information in Engineering Conference, DETC2005-84371, 2005.

Ragulskis, K. M., Arutunian, M. G., Kochikian, A. V., and Pogosian, M. Z.: A Study of Fillet Type Flexure Hinges and their Optimal Design, Vibr. Eng., 3, 447–452, 1989.

Smith, T. S., Badami, V. G., Dale, J. S., and Xu, Y.: Elliptical flexure hinges, Rev. Sci. Instrum., 68, 1474–1483, 1997.

Tian, Y., Shirinzadeh, B., Zhang, D., and Zhong, Y.: Three flexure hinges for compliant mechanism designs based on dimensionless graph analysis, Precis. Eng., 34, 92–100, 2010a.

Tian, Y., Shirinzadeh, B., and Zhang, D.: Closed-form equations of the filleted V-shaped flexure hinges for compliant mechanism designs, Precis. Eng., 34, 408–418, 2010b.

Trease, B. P., Moon, Y. M., and Kota, S.: Design of large-displacement compliant joints, J. Mech. Design, 127, 788–798, 2005.

Xu, W. and King, T.: Flexure hinges for piezoactuator displacement amplifiers: flexibility, accuracy, and stress considerations, Precis. Eng., 19, 4–10, 1996.

Yin, L. Z. and Ananthasuresh, G. K.: Design of distributed compliant mechanisms, Mech. Based Des. Struc., 31, 151–179, 2003.

Yong, Y. K., Lu, T. F., and Handley, D. C.: Review of circular flexure hinge design equations and derivation of empirical formulations, Precis. Eng., 32, 63–70, 2008.

Zettl, B., Szyszkowski, W., and Zhang, W. J.: On systematic errors of two-dimensional finite element modeling of right circular planar flexure hinges, J. Mech. Design, 127, 782–787, 2005.

13

A novel director-based Bernoulli–Euler beam finite element in absolute nodal coordinate formulation free of geometric singularities

P. G. Gruber[1], K. Nachbagauer[2], Y. Vetyukov[1], and J. Gerstmayr[1]

[1]Linz Center of Mechatronics GmbH, Altenberger Straße 69, 4040 Linz, Austria

[2]Institute of Technical Mechanics, Johannes Kepler Universität Linz, Altenbergerstraße 69, 4040 Linz, Austria

Correspondence to: P. G. Gruber (peter.gruber@lcm.at)

Abstract. A three-dimensional nonlinear finite element for thin beams is proposed within the absolute nodal coordinate formulation (ANCF). The deformation of the element is described by means of displacement vector, axial slope and axial rotation parameter per node. The element is based on the Bernoulli–Euler theory and can undergo coupled axial extension, bending and torsion in the large deformation case. Singularities – which are typically caused by such parameterizations – are overcome by a director per element node. Once the directors are properly defined, a cross sectional frame is defined at any point of the beam axis. Since the director is updated during computation, no singularities occur. The proposed element is a three-dimensional ANCF Bernoulli–Euler beam element free of singularities and without transverse slope vectors. Detailed convergence analysis by means of various numerical static and dynamic examples and comparison to analytical solutions shows the performance and accuracy of the element.

1 Introduction

In the present paper, a Bernoulli–Euler beam finite element based on the absolute nodal coordinate formulation (ANCF) is introduced. The ANCF has been developed by Shabana (1997, 2005) as an alternative to the classical large rotation vector formulation for the modeling of large deformation structural problems in two and three dimensions. Originally, thin (Bernoulli–Euler) ANCF elements led to a different solution as the classical Euler Elastica, however in the two-dimensional case, Gerstmayr and Irschik (2008) showed how to write the work of elastic forces, such that classical solutions can be retrieved by ANCF elements. In the three-dimensional case, however, there are no singularity-free Bernoulli–Euler ANCF elements so far (in contrast to fully parametrized shear deformable ANCF elements, see Schwab and Meijaard (2010), who present and validate a locking free 3-D beam element with the usage of an elastic lien approach and the Whu-Washizu variational principle). Among earlier works on Bernoulli–Euler ANCF elements we mention von Dombrowski (2003); Dmitrochenko (2005); Gerstmayr and Shabana (2006). The formulation of von Dombrowski (2003) contains a time-dependent mass matrix and due to the parameterization of rotations it suffers from singularities. A similar approach as von Dombrowski (2003) is chosen by Dmitrochenko (2005); Dmitrochenko and Pogorelov (2003). In the latter references, the usage of a Frenet basis leads to singularities in the inflection points, because the torsion angle is not uniquely defined. Gerstmayr and Shabana (2006) presented an efficient approach for thin structural problems, however, torsion has not been included in their formulation. Note that in von Dombrowski (2003); Dmitrochenko (2005); Gerstmayr and Shabana (2006), the material measure of curvature has not been utilized, therefore the axial extension and bending are not decoupled, which may lead to erroneous results under large axial forces, see Gerstmayr and Irschik (2008). The original three-dimensional approach by Yakoub and Shabana (2001) describes a shear-deformable beam element, which involves a constant mass matrix and does not suffer from singularities.

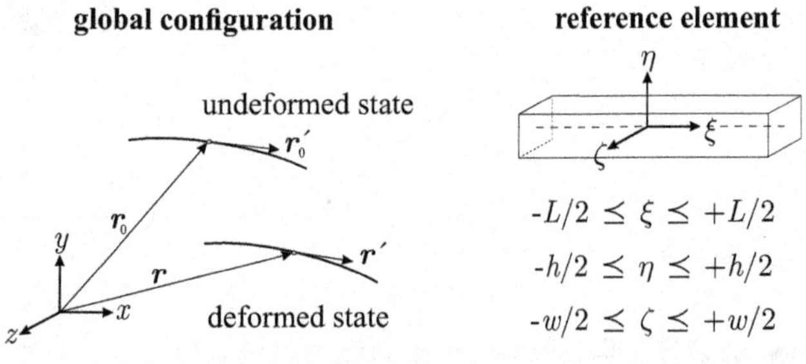

Figure 1. Finite element configurations.

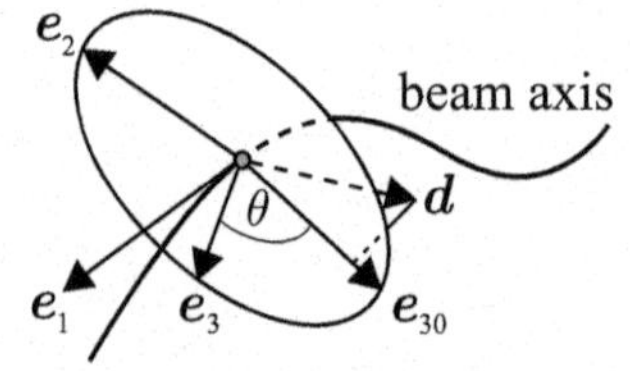

Figure 2. In order to have a uniquely defined orientation of the cross section about the beam axis direction $\boldsymbol{e}_1 = \frac{\boldsymbol{r}'}{|\boldsymbol{r}'|}$ a non-collinear director $\boldsymbol{d}$ is utilized. The local frame $(\boldsymbol{e}_1, \boldsymbol{e}_2, \boldsymbol{e}_3)$ is defined by the normalized projection $\boldsymbol{e}_{30}$ of $\boldsymbol{d}$ into the normal plane of $\boldsymbol{e}_1$, and a subsequent rotation around a torsional angle θ.

The present ANCF formulation features strain measures for axial extension, bending and torsion. For the consistent derivation of such strain measures see Eliseev (2003). The principle of virtual work includes the kinematic relations for the strain measures of the rod, evaluated at a material line. For similar formulations of the direct approach to a material line see e.g. the works of Reissner (1973) or Antman (1972).

The parameterization of the kinematics of a spatial curved rod without shear was presented by Krommer and Vetyukov (2009), however there, the calculation was performed using a global Ritz approach. The kinematic description and the virtual work for the proposed element is chosen in a way similar to Vetyukov and Eliseev (2010) or rather Vetyukov (2008). In the latter paper, only static problems are regarded, while in the present paper, also dynamic behavior is analyzed. In contrast to standard ANCF elements, in which the mass matrix is constant, here the exact mass matrix is time-dependent, since a rigid cross section is assumed. In addition, a constant mass matrix, similar to Dmitrochenko (2005), can be applied for linear problems or small deformation problems, e.g. small oscillations, while for nonlinear problems a non-constant mass matrix is utilized. The geometry of the beam is described by a curve, representing the beam's axis, as well as by an orientation of the cross section at each point of this axis, see Fig. 1. The orientation of the cross section is defined by an angle of axial rotation relative to a certain reference direction, which is introduced in order to perform the rotation in a correct manner and to avoid singularities originating from the usage of the rotation parameter. There exist different interpolation procedures in literature, in which displacements and rotations (see Simo and Vu-Quoc, 1986), displacements and slopes (see Shabana, 1997) or strains (see Gams et al., 2007; Zupan and Saje, 2003) are used as basic interpolated variables. In contrast to the above-mentioned formulations, the present approach is based on the interpolation of displacements, slopes and a rotation around the beam axis. For the interpolation of displacements and slopes, cubic shape functions are chosen, while the angle of axial rotation is interpolated linearly with respect to the beam's axis.

In the following, the geometric description of the finite element, the choice of degrees of freedom, as well as the definition of the strain energy for the ANCF beam element are presented. In contrast to previous Bernoulli Euler ANCF elements, the proposed element is investigated using a large number of numerical examples, many of them suggested by Schwab and Meijaard (2009). In this work, the examples are restricted to small and large deformation static problems, buckling and linearized dynamic (eigenvalue) problems. In addition to the content of Nachbagauer et al. (2011), the exact dynamic terms of the beam element and a nonlinear dynamic example are provided in this work.

2 Geometric description

The geometry of the beam is described by a curve, representing the beam's axis, and a cross section at each point of this axis, see Fig. 1. In the present paper, a Bernoulli–Euler beam is considered, which means, that the cross section of the beam is thin, undeformed, and orthogonal to the beam axis (since shear deformation can be neglected) at any time and for any point of the beam axis. Moreover we assume, that the beam's axis intersects each cross section exactly at the cross section's centroid. The position $\boldsymbol{p}$ of an arbitrary point (or particle) in the cross section may thus be computed by

$$\boldsymbol{p}(\xi,\eta,\zeta) = \boldsymbol{r}(\xi) + \mathbf{A}(\xi)\begin{bmatrix} 0 \\ \eta \\ \zeta \end{bmatrix}. \tag{1}$$

Here, $\boldsymbol{r}(\xi)$ denotes the axial position (i.e., where the cross section belonging to $\boldsymbol{p}$ meets the beam's axis). The rotation tensor $\mathbf{A}(\xi)$ is defined by

$$\mathbf{A}(\xi) = \begin{bmatrix} \boldsymbol{e}_1(\xi) & \boldsymbol{e}_2(\xi) & \boldsymbol{e}_3(\xi) \end{bmatrix}, \tag{2}$$

in which, $\mathbf{e}_i(\xi)$ denotes the i-th base vector of the local axis frame at ξ (see Fig. 2). These vectors can be defined in terms of the slope vector $\boldsymbol{r}' = \frac{\partial \boldsymbol{r}}{\partial x}$, the reference direction (director) $\boldsymbol{d}$ and an angle θ as follows (for easier reading, the dependencies on ξ are omitted):

$$\boldsymbol{e}_1 = \frac{\boldsymbol{r}'}{|\boldsymbol{r}'|}, \tag{3}$$

$$\boldsymbol{e}_2 = \boldsymbol{e}_{20}\cos(\theta) + \boldsymbol{e}_{30}\sin(\theta), \tag{4}$$

$$\boldsymbol{e}_3 = \boldsymbol{e}_{30}\cos(\theta) - \boldsymbol{e}_{20}\sin(\theta), \tag{5}$$

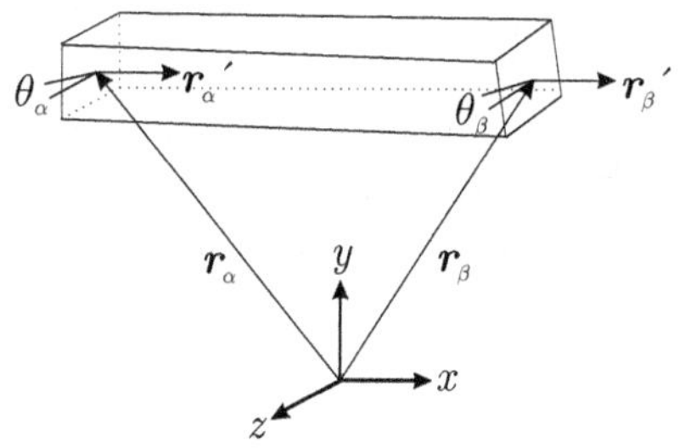

Figure 3. Degrees of freedom for one beam finite element in the nodes α and β: axial position vector $\boldsymbol{r}$, slope vector $\boldsymbol{r}'$, and cross section orientation angle θ.

in which $\boldsymbol{e}_{30}$ denotes the normalized projection of the director $\boldsymbol{d}$ into the normal plane of the slope $\boldsymbol{r}'$, i. e.,

$$\boldsymbol{e}_{30} = \frac{\hat{\boldsymbol{e}}_{30}}{|\hat{\boldsymbol{e}}_{30}|}, \qquad \hat{\boldsymbol{e}}_{30} = \boldsymbol{d} - (\boldsymbol{d}^T \boldsymbol{e}_1)\boldsymbol{e}_1, \tag{6}$$

and $\boldsymbol{e}_{20}$ is defined by the cross product

$$\boldsymbol{e}_{20} = \boldsymbol{e}_{30} \times \boldsymbol{e}_1. \tag{7}$$

Note that a similar parameterization of the cross section orientation has already been used by von Dombrowski (2003), however, not regarding a director. Summarizing, the geometry of a Bernoulli–Euler beam can be fully described by the axial position $\boldsymbol{r}$, slope $\boldsymbol{r}'$, and cross section orientation angle θ, for a fixed choice of a director $\boldsymbol{d}$. Equation (6) says, that geometric singularities occur, if (and only if) the slope $\boldsymbol{r}'$ is collinear with the director $\boldsymbol{d}$. A safe strategy for preventing singularities throughout the deformation process using an updating procedure is addressed in Sect. 4. Let us first turn to the finite element discretization.

3 Finite element discretization

The proposed beam finite element is defined by a pair of axial position vectors, axial slope vectors, and axial rotation angles, i.e.,

$$\boldsymbol{r}_\alpha = \boldsymbol{r}\left(\xi = -\frac{L}{2}\right), \qquad \boldsymbol{r}_\beta = \boldsymbol{r}\left(\xi = \frac{L}{2}\right), \tag{8}$$

$$\boldsymbol{r}'_\alpha = \boldsymbol{r}'\left(\xi = -\frac{L}{2}\right), \qquad \boldsymbol{r}'_\beta = \boldsymbol{r}'\left(\xi = \frac{L}{2}\right), \tag{9}$$

$$\theta_\alpha = \theta\left(\xi = -\frac{L}{2}\right), \qquad \theta_\beta = \theta\left(\xi = \frac{L}{2}\right). \tag{10}$$

All together those are 14 degrees of freedom, which we collect in the vector of generalized coordinates

$$\boldsymbol{q} = \Big[\underbrace{\boldsymbol{r}_\alpha^T\ \boldsymbol{r}_\alpha'^T\ \boldsymbol{r}_\beta^T\ \boldsymbol{r}_\beta'^T}_{\boldsymbol{q}_r}\ \underbrace{\theta_\alpha\ \theta_\beta}_{\boldsymbol{q}_\theta}\Big]^T, \tag{11}$$

see Fig. 3 for a sketch of the element. The element coordinates are chosen similar to von Dombrowski (2003) and Dmitrochenko (2005), however due to the use of the director, the meaning of the axial rotation angle is different in the present approach.

For $\xi \in [-L/2, L/2]$ the axial position vector $\boldsymbol{r}(\xi)$ is interpolated by cubic polynomials,

$$\boldsymbol{r}(\xi) = S_1(\xi)\boldsymbol{r}_\alpha + S_2(\xi)\boldsymbol{r}'_\alpha + S_3(\xi)\boldsymbol{r}_\beta + S_4(\xi)\boldsymbol{r}'_\beta, \tag{12}$$

with

$$\begin{aligned}
S_1(\xi) &= \frac{1}{4}\left(2 - \frac{6\xi}{L} + \frac{8\xi^3}{L^3}\right),\\
S_2(\xi) &= \frac{1}{2L}\left(1 - \frac{2\xi}{L} - \frac{4\xi^2}{L^2} + \frac{8\xi^3}{L^3}\right),\\
S_3(\xi) &= \frac{1}{4}\left(2 + \frac{6\xi}{L} - \frac{8\xi^3}{L^3}\right),\\
S_4(\xi) &= \frac{1}{2L}\left(-1 - \frac{2\xi}{L} + \frac{4\xi^2}{L^2} + \frac{8\xi^3}{L^3}\right),
\end{aligned} \tag{13}$$

denoting the positional shape functions, and the rotation angle $\theta(\xi)$ is interpolated by linear polynomials,

$$\theta(\xi) = S_5(\xi)\theta_\alpha + S_6(\xi)\theta_\beta, \tag{14}$$

with

$$S_5(\xi) = \frac{1}{2} - \frac{\xi}{L}, \qquad S_6(\xi) = \frac{1}{2} + \frac{\xi}{L}, \tag{15}$$

denoting the rotational shape functions. The position and orientation of an arbitrary point is computed by

$$\begin{bmatrix} r_x & r_y & r_z & \theta \end{bmatrix}^T = \mathbf{S}\boldsymbol{q}, \tag{16}$$

with the shape function matrix

$$\mathbf{S} = \begin{bmatrix} \mathbf{S}_r & 0 \\ 0 & \mathbf{S}_\theta \end{bmatrix}. \tag{17}$$

The sub matrices $\mathbf{S}_r$ and $\mathbf{S}_\theta$ are defined by

$$\mathbf{S}_r = \begin{bmatrix} S_1\mathbf{I} & S_2\mathbf{I} & S_3\mathbf{I} & S_4\mathbf{I} \end{bmatrix}, \qquad \mathbf{S}_\theta = \begin{bmatrix} S_5 & S_6 \end{bmatrix}, \tag{18}$$

with $\mathbf{I}$ denoting the identity in $\mathbb{R}^3$. The director $\boldsymbol{d}$ is also defined by the pairs,

$$\boldsymbol{d}_\alpha = \boldsymbol{d}\left(\xi = -\frac{L}{2}\right), \qquad \boldsymbol{d}_\beta = \boldsymbol{d}\left(\xi = \frac{L}{2}\right), \tag{19}$$

and, accordingly to the axial rotation angle θ, interpolated linearly for ξ in the interval $[-L/2,\ L/2]$ by

$$\boldsymbol{d}(\xi) = \begin{bmatrix} S_5(\xi)\mathbf{I} & S_6(\xi)\mathbf{I} \end{bmatrix} \begin{bmatrix} \boldsymbol{d}_\alpha \\ \boldsymbol{d}_\beta \end{bmatrix}. \tag{20}$$

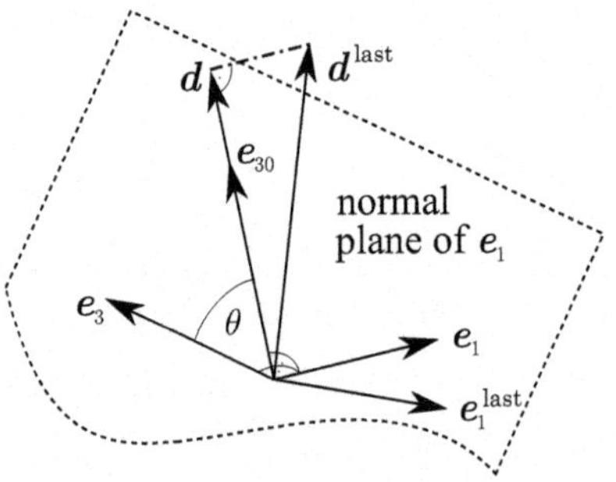

Figure 4. In large deformation problems it is crucial to prevent the director $\boldsymbol{d}$ from becoming collinear with the beam axis direction $\boldsymbol{e}_1$. Therefore a director update in form of a projection into the normal plane of $\boldsymbol{e}_1$ is suggested at every time or load step of the simulation. Note, that an update of this kind does not change the configuration of the local frame, see also Eq. (6) and Fig. 2.

4 Director update

Due to Eq. (6) the local frame is well defined if and only if the director $\boldsymbol{d}$ is not (numerically) collinear with the axial slope $\boldsymbol{r}'$ at any point on the beam axis. A safe strategy to avoid this situation is to successively update the director at the FE-nodes at every load/time step. As experienced in the scope of this work, a simple projection of the directors at the FE-nodes into the cross section of the beam seems to be sufficient, see Fig. 4. In case the axial direction $\boldsymbol{e}_1$ becomes collinear during a load or time step of the simulation, it helps to repeat this same iteration step with a reduced load or time increment, and to update the director as explained at the intermediate step. Moreover, a subsequent rotation of the projected director around the beam axis by the torsional angle θ, and subsequently resetting θ to zero, provides an easy way to decrease the complexity of the terms appearing in the equations of motion and thereby speed up the static/dynamic calculations. As it seems, a simple linear interpolation of the director between the FE-nodes is sufficient for even complex scenarios and large deformations, given that a reliable initial configuration was chosen. For instance, two neighbouring nodal directors in opposite directions would lead to a vanishing director at the center of the finite element. Let be finally mentioned, that the proposed Bernoulli–Euler beam finite element provides C^1-continuity along element borders only for the beam axis, whereas the torsion of the cross section, i.e. angle θ, is just C^0-continuous. Hence, a 4th-order convergence will be prohibited particularly in problems with significant torsional effects. A fully C^1 continuous setting, requiring the rate of the torsional angle θ' at the FE-nodes to serve as a generalized coordinate, together with a conforming interpolation of the director along the beam axis is left for further investigation.

5 Equations of motion

The equations of motion are written in the form of the Lagrange equations of second kind:

$$\frac{\mathrm{d}}{\mathrm{d}t}\left(\frac{\partial T}{\partial \dot{q}_i}\right) - \frac{\partial T}{\partial q_i} + \frac{\partial \Pi}{\partial q_i} = Q_i, \tag{21}$$

where T denotes the kinetic energy, Π the strain energy, Q_i are generalized external forces, q_i are the generalized coordinates, as defined in Eq. (11), and $\dot{q}_i$ denotes the partial derivative of q_i with respect to the time variable t. In the case of static problems, the kinetic energy T vanishes, such that the equations of the static equilibrium are

$$\frac{\partial \Pi}{\partial q_i} = Q_i. \tag{22}$$

5.1 Strain energy

With the principle of virtual work applied to the one-dimensional Cosserat continuum (see, e.g., Antman, 1972; Eliseev, 2003; Reissner, 1973) one can prove that the strain energy per unit length is a function of two strain measures, one responsible for shear and axial extension, and the second one responsible for bending and torsion. In the considered model, the shear is kinematically constrained: the rotation of particles is in correspondence with the variation of the tangential direction $\boldsymbol{r}'$. Conforming to Simo (1985, Eq. 4.10), the axial extension is described by the axial strain

$$\varepsilon = \frac{\mathrm{d}s}{\mathrm{d}S} - 1 = \frac{|\boldsymbol{r}'| - |\boldsymbol{r}_0'|}{|\boldsymbol{r}_0'|}, \tag{23}$$

in which $\mathrm{d}s = |\boldsymbol{r}'(\xi)|\,\mathrm{d}\xi$ and $\mathrm{d}S = |\boldsymbol{r}_0'(\xi)|\,\mathrm{d}\xi$ denote the arc length in the deformed and undeformed state, respectively (see Fig. 1).

Utilizing Einstein's summation convention, the vector of bending and torsional strain (see Eliseev, 2003) reads:

$$\boldsymbol{\kappa} = \kappa_i \boldsymbol{e}_i = (k_i - k_{0i})\boldsymbol{e}_i, \tag{24}$$

in which the torsional strain κ_1 and bending strains κ_2, κ_3 are expressed as the difference of the components of the vector of twist and curvature in deformed and undeformed state,

$$\boldsymbol{k} = \boldsymbol{e}_i \times \boldsymbol{e}_i'/2 = k_i \boldsymbol{e}_i, \quad \boldsymbol{k}_0 = \boldsymbol{e}_{0i} \times \boldsymbol{e}_{0i}'/2 = k_{0i}\boldsymbol{e}_{0i}, \tag{25}$$

As an important geometrical property the derivative of the basis with respect to the beam's axis may be determined by $\boldsymbol{e}_i' = \boldsymbol{k} \times \boldsymbol{e}_i$ and $\boldsymbol{e}_{0i}' = \boldsymbol{k}_0 \times \boldsymbol{e}_{0i}$, respectively. Notice, the components κ_i are considered in the local basis $\boldsymbol{e}_i$, i.e., there holds

$$(\kappa_i)_{i=1}^3 = \mathbf{A}^T \boldsymbol{\kappa}. \tag{26}$$

Interpolated values for the axial strain are computed by $\boldsymbol{r}'(\xi) = \mathbf{S}_r'(\xi)\boldsymbol{q}_r$, whereas curvature components are functions of the local frame $(\boldsymbol{e}_i)_{i=1}^3$, and thus of $\mathbf{S}(\xi)\boldsymbol{q}$, see Eqs. (3)–(7) and Eq. (16).

If the local basis vectors $\boldsymbol{e}_2, \boldsymbol{e}_3$ are chosen in the directions of the principal axis of the cross section, then the following quadratic approximation of the strain energy can be adopted:

$$\Pi = \frac{1}{2}\int_{-L/2}^{L/2}\left(EA\varepsilon^2 + GJ\kappa_1^2 + EI_y\kappa_2^2 + EI_z\kappa_3^2\right)d\xi \tag{27}$$

with EA, GJ, EI_y and EI_z being correspondingly the stiffnesses for axial extension, torsion and bending in the two principal directions. The form of the strain energy in Eq. (27) corresponds to the numerical formulation of Simo and Vu-Quoc (1986) if neglecting (cross sectional) transverse shear. The strain energy in Eq. (27) is also similar to that suggested by von Dombrowski (2003) and Dmitrochenko (2005), however the curvature components are different in the present approach. Additional theoretical argumentation for the strain energy in the form in Eq. (27) and the kinematical definitions in Eqs. (23, 24) are provided by the asymptotic analysis of the three-dimensional problem of the theory of elasticity for a naturally curved and twisted rod, presented by Yeliseyev and Orlov (1999).

The variation of the strain energy in Eq. (27) follows as

$$\delta\Pi = \int_{-L/2}^{L/2}\left(EA\varepsilon\,\delta\varepsilon + GJ\kappa_1\,\delta\kappa_1 + EI_y\kappa_2\,\delta\kappa_2 + EI_z\kappa_3\,\delta\kappa_3\right)d\xi. \tag{28}$$

The variation of ε in Eq. (23) is easily computed as

$$\delta\varepsilon = \frac{1}{|\boldsymbol{r}_0'||\boldsymbol{r}'|}\boldsymbol{r}'^T\mathbf{S}'\delta\boldsymbol{q}. \tag{29}$$

The variation of κ_i in Eq. (26) has to be computed column-wise by

$$\delta\kappa_i = \frac{\partial\kappa_i}{\partial\boldsymbol{v}}\frac{\partial\boldsymbol{v}}{\partial\boldsymbol{q}}\delta\boldsymbol{q}, \tag{30}$$

in which $\boldsymbol{v} = [\boldsymbol{r}', \boldsymbol{r}'', \theta, \theta']^T$. Here, $\frac{\partial\boldsymbol{v}}{\partial\boldsymbol{q}}$ can be easily computed by help of the shape functions in Eqs. (13, 15).

5.2 Kinetic energy in dynamic problems

In case of dynamic problems, the full Lagrange Equations in Eq. (21) have to be solved. The kinetic energy T is defined as

$$T = \frac{1}{2}\int_V \rho\,\dot{\boldsymbol{p}}^T\dot{\boldsymbol{p}}\,dV, \tag{31}$$

in which ρ denotes the material density, and $\boldsymbol{p}$, as defined in Eq. (1), points to an arbitrary point of the volume V of the element in undeformed state, see Fig. 1. Since $\dot{\boldsymbol{p}}$ depends linearly on $\dot{\boldsymbol{q}}$, i. e.,

$$\dot{\boldsymbol{p}}(\boldsymbol{q},\dot{\boldsymbol{q}}) = \left(\frac{\partial\boldsymbol{r}}{\partial\boldsymbol{q}}(\boldsymbol{q}) + \frac{\partial\boldsymbol{e}_2}{\partial\boldsymbol{q}}(\boldsymbol{q})\eta + \frac{\partial\boldsymbol{e}_3}{\partial\boldsymbol{q}}(\boldsymbol{q})\zeta\right)\dot{\boldsymbol{q}}, \tag{32}$$

Equation (31) turns into

$$T = \dot{\boldsymbol{q}}^T\mathbf{M}(\boldsymbol{q})\dot{\boldsymbol{q}}, \tag{33}$$

where the mass matrix $\mathbf{M}$ is defined by the integral

$$\mathbf{M} = \frac{1}{2}\int_{-L/2}^{L/2}\mathbf{L}_0^T(\boldsymbol{q})\mathbf{D}\mathbf{L}_0(\boldsymbol{q})|\boldsymbol{r}_0'|d\xi, \tag{34}$$

and the matrices $\mathbf{D}$ and $\mathbf{L}_0$ are defined

$$\mathbf{D} = \begin{bmatrix}\rho A & 0 & 0\\ 0 & \rho I_z & 0\\ 0 & 0 & \rho I_y\end{bmatrix}, \quad \mathbf{L}_0^T = \left[\mathbf{S}_r \;\; \frac{\partial\boldsymbol{e}_2}{\partial\boldsymbol{q}} \;\; \frac{\partial\boldsymbol{e}_3}{\partial\boldsymbol{q}}\right], \tag{35}$$

in which the moments of inertia read $I_z = \frac{h^3 w}{12}$ and $I_y = \frac{w^3 h}{12}$. The kinetic energy T appears in the following two terms in the Lagrange's equations Eq. (21):

$$-\frac{\partial T}{\partial\boldsymbol{q}} \quad \text{and} \quad \frac{\mathrm{d}}{\mathrm{d}t}\left(\frac{\partial T}{\partial\dot{\boldsymbol{q}}}\right), \tag{36}$$

which means, that the following two terms have to be implemented regarding the kinetic energy:

$$-\frac{\partial T}{\partial q_k} = -\sum_{i,j}\frac{1}{2}\dot{q}_i^T\frac{\partial M_{ij}}{\partial q_k}\dot{q}_j, \tag{37}$$

$$\frac{\mathrm{d}}{\mathrm{d}t}\left(\frac{\partial T}{\partial\dot{\boldsymbol{q}}}\right) = \sum_k\dot{\boldsymbol{q}}^T\frac{\partial\mathbf{M}}{\partial q_k}\frac{\partial q_k}{\partial t} + \ddot{\boldsymbol{q}}^T\mathbf{M}. \tag{38}$$

6 Numerical examples

The proposed ANCF element has been implemented in the framework of the multibody and finite element research code HOTINT[1]. In order to show the performance and accuracy of the proposed ANCF element, several numerical examples are considered. Many of the following examples are based on the beam benchmark problems proposed by Schwab and Meijaard (2009).

As a first example, a cantilever beam under a tip load combined with a bending and torsional moment leading to small deformations is investigated. Further, we discuss large bending and torsion problems, Euler and lateral buckling, as well as the linear dynamic problem of an eigenfrequency analysis in the case of a simply supported beam with pre-stress as well as a nonlinear dynamic pendulum.

[1]http://tmech.mechatronik.uni-linz.ac.at/staff/gerstmayr/hotint.html

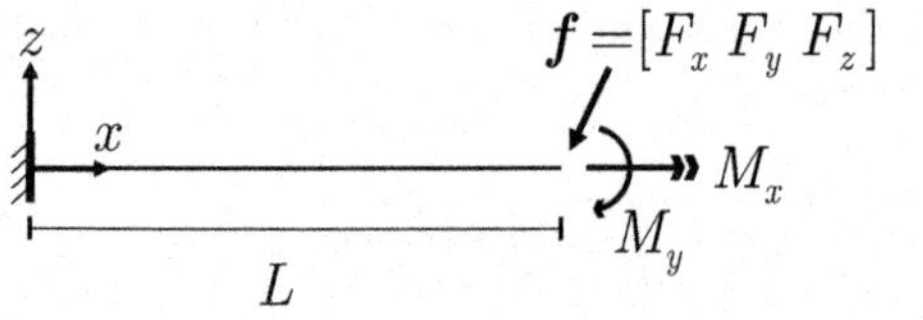

Figure 5. Cantilever beam with tip force and moment resultants.

If not mentioned differently in the description of the examples, we assume a cantilever beam, see Fig. 5 with length L in x direction, width w in y direction, and height h in z direction. By default all components of the tip resultant force $\boldsymbol{f} = \left[F_x, F_y, F_z\right]^T$ and the tip resultant moment $\mathbf{M} = \left[M_x, M_y, M_z\right]^T$ are zero, Young's modulus is assumed to be $E = 2.1 \times 10^{11}\,\mathrm{N\,m^{-2}}$, the Poisson ratio is $\nu = 0.3$, and the director points into z direction, $\boldsymbol{d} = [0, 0, 1]^T$, on the whole axis of the cantilever.

6.1 Small bending and torsion

As a first static example, a cantilever beam loaded at the tip by a vertical force, by a bending and torsional moment is investigated according to Schwab and Meijaard (2009, Sect. 2.1). The chosen load case leads to small displacements and small rotations. The cantilever beam has length $L = 1$ m, width $w = 0.01$ m and height $h = 0.02$ m. The vertical tip load is chosen as $F_z = 1 \times 10^{-4}$ N, and the bending moment M_y and the torsional moment M_x are of the same magnitude $M_x = M_y = 1 \times 10^{-4}$ Nm. The torsional stiffness of the rectangular cross section is set to $GJ = Ghw^3/3$, where G denotes the standard shear modulus $G = E/(2 + 2\nu)$. The theoretical value of the tip displacement in z direction u_z, as well as the axial rotation θ and the rotation around the y axis, φ_y, can be computed by the following formulas:

$$u_z = \frac{F_z L^3}{3 E I_y} - \frac{M_y L^2}{2 E I_y}, \tag{39}$$

$$\theta = \frac{M_x L}{GJ}, \tag{40}$$

$$\varphi_y = \frac{F_z L^2}{2 E I_y} - \frac{M_y L}{E I_y}. \tag{41}$$

In our tests, the difference between theoretical and numerical solution was less than 10^{-6} m for one element, and less than 10^{-12} for two and more elements.

6.2 Large bending

A cantilever beam with length $L = 2$ m, width and height $w = h = 0.1$ m under a transversal tip load $F_y = 12 E I_y / L^2$ is investigated. The obtained tip displacement is compared to a solution computed with arbitrary precision given in Gerstmayr and Irschik (2008), see Table 1. The convergence rate is of order 4, as can be seen in Fig. 6

Table 1. Convergence analysis (order 4) of the axial and transverse displacement at the tip of the cantilever beam in Sect. 6.2 compared to the exact solution stated in Gerstmayr and Irschik (2008).

Elements	$\mathbf{u}_x$	$\mathbf{u}_y$
1	−0.3411725115810615	0.9654494547661615
2	−0.4879599317854074	1.1714616527622450
4	−0.5075492277225774	1.2053708868794728
8	−0.5085245204356039	1.2071998231055112
16	−0.5085375347924183	1.2072390085269251
32	−0.5085373396910754	1.2072398289564636
64	−0.5085373073884966	1.2072398533822040
128	−0.5085373045949709	1.2072398544836476
256	−0.5085373043521027	1.2072398545459371
exact	−0.5085373043258772	1.207239854549824

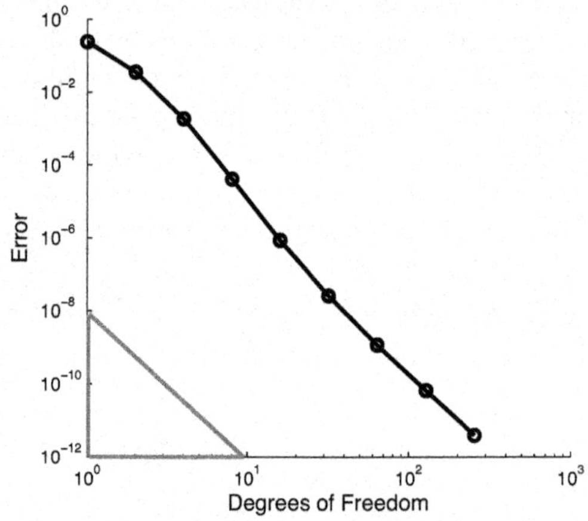

Figure 6. The error $|\mathbf{u}_k - \mathbf{u}|$ between computed and exact solution is plotted versus the degrees of freedom for the large bending example in Sect. 6.2. Actually, fourth order convergence may be observed only in plane problems.

6.3 Large bending and torsion

A cantilever beam with length $L = 1$ m, width $w = 0.005$ m, and height $h = 0.02$ m is investigated. At the tip, a bending moment $M_y = 50$ Nm, and a torsional moment $M_x = 12.5$ Nm is applied. The torsional stiffness of the cross section is supposed to be $GJ = Ghw^3/3$ with G again denoting the shear modulus. The obtained tip displacement is compared in Table 2 to the numerical solution based on an ANSYS implementation with 40 beam elements. Notice, this example is almost identical with a benchmark problem in Schwab and Meijaard (2009, Sect. 2.3). Since it was not possible for the software package ANSYS to calculate a numerical solution for the full amount of loading, only 50 percent were prescribed. Thereafter, the solution of ANSYS was compared to the solution by using the proposed beam finite element. Note, that the solution of the proposed element coincides with the solution of ANSYS up to 5 digits, which significantly

Table 2. Displacement of the tip in case of large bending and torsion in Sect. 6.3 compared to a numerical solution with 40 beam elements in the software package ANSYS.

Elements	u_x ($\times 10^{-3}$)	u_y ($\times 10^{-2}$)	u_z ($\times 10^{-2}$)
1	−1.70988	−3.21322	−3.83156
2	−1.70974	−2.82968	−3.82723
4	−1.72399	−2.84476	−3.82940
8	−1.72968	−2.85337	−3.83113
16	−1.73116	−2.85572	−3.83160
32	−1.73153	−2.85631	−3.83172
64	−1.73162	−2.85646	−3.83175
128	−1.73165	−2.85649	−3.83176
ANSYS	−1.7316	−2.8564	−3.8318

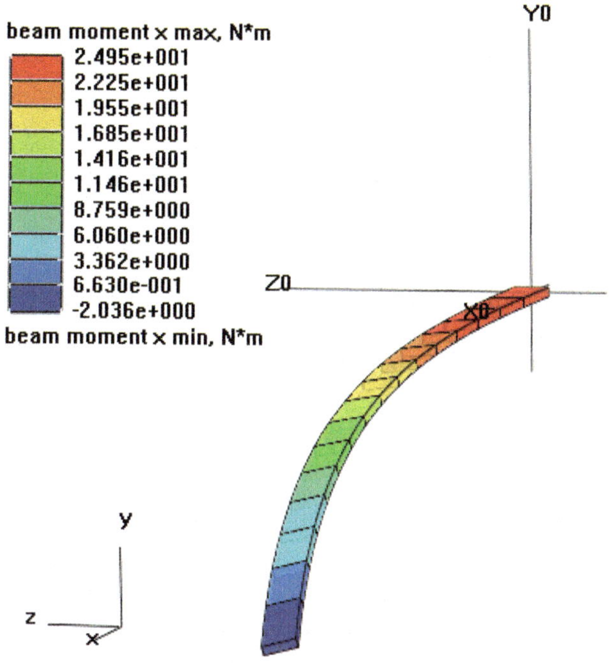

Figure 7. Deformation of the beam in Sect. 6.3 with color-plot of the torsional moment.

behaves better than an approach by von Dombrowski (2003, Fig. 3).

6.4 Full circle bending problem

A cantilever beam with length $L = 2$ m, width and height $w = h = 0.1$ m under a tip bending moment $M_z = 2\pi E I_z / L$ is analyzed. Table 3 reports on the convergence behavior.

6.5 Euler buckling

As a first buckling problem, a Euler buckling of a cantilever beam under a compressive normal force F_x is analyzed. The dimensions of the beam are chosen as length $L = 1$ m, width $w = 0.01$ m and height $h = 0.02$ m, according to the benchmark problem in Schwab and Meijaard (2009, Sect. 2.5). The theoretical buckling load can be calculated by $F_{\text{th}} = (\pi/4)(E I_{\min}/L^2)$, in which $I_{\min}$ denotes the minimum of the bending stiffnesses $I_{\min} = \min(I_y, I_z)$. Notice, that differently to Schwab and Meijaard (2009) the eigenvalues are numerically calculated in the deformed state which corresponds to the numerical compressive normal force F_{num}. In other words, the buckling loads were determined by observing the lowest bending mode of the system, which corresponds to the deformed state of the beam (since the simulation sofware we used does not provide linearized equations directly for the eigenmode analysis, but the system matrix is assembled automatically for the deformed state of the beam).

Thus, in order to allow a comparison to the benchmark results, another test was done, in which the axial stiffness of the beam is penalized (multiplied by the factor 100), such that buckling occurs at a less deformed state of the beam. Both cases (original and penalized axial stiffnesses) are compared to a benchmark example by Schwab and Meijaard (2009) in Table 4.

Table 3. Displacements u_x and u_y for the full circle bending in Sect. 6.4 converge at order 4.

Elements	u_x	u_y
1	−1.5105395426085659	1.3841149699588149
2	−2.2167244240620558	0.1408464658361039
4	−2.0302356356499369	−0.0005393788083298
8	−2.0026639789559821	−0.0000566264761093
16	−2.0001814751535658	−0.0000009572511065
32	−2.0000117896528211	−0.0000000147330922
64	−2.0000007741038281	−0.0000000002256560
128	−2.0000000527750790	−0.0000000000034961
256	−2.0000000038609360	0.0000000000036471
exact	−2	0

Table 4. The ratio of the numerical to the theoretical buckling load $F_{\text{num}}/F_{\text{th}}$ in case of Euler buckling, Sect. 6.5, compared to the results of Schwab and Meijaard (2009). Note, that the ratio $F_{\text{num}}/F_{\text{th}}$ goes up after using 8 elements when using the original value for EA, which is due to the fact, that the numerical buckling loads could only be determined by solving an eigenvalue problem for the deformed system. A second test for a stiffer value of EA showed order 4 convergence.

Number of	Schwab	Proposed Beam	
elements	Meijaard	penalized EA	original EA
1	1.00752232	1.00752255	1.00754320
2	1.00051214	1.00051330	1.00053272
4	1.00003276	1.00003298	1.00005325
8	1.00000206	1.00000226	1.00000050
16	1.00000012	1.00000038	1.00002053

Table 5. The ratio of the numerical to theoretical buckling load $F_{\text{num}}/F_{\text{th}}$ in case of lateral buckling, Sect. 6.6, compared to the results of Schwab and Meijaard (2009). Alike in Sect. 6.5, the numerical buckling loads could only be determined by solving an eigenvalue problem for the deformed system. In order to compare the results, a thinner beam has been utilized, such that buckling occured already at a less deformed state.

Number of elements	Schwab Meijaard	Proposed Beam $w = 0.0002$ m	Proposed Beam $w = 0.002$ m
1	1.495290	1.493133	–
2	1.069138	1.068821	1.073653
4	1.015367	1.015119	1.031665
8	1.003862	1.003774	1.019491
16	1.000969	1.000810	1.016410

6.6 Lateral buckling

As a further buckling problem, the lateral buckling of a cantilever beam under a lateral force F_z is investigated. The dimensions of the beam are chosen as length $L = 1$ m, width $w = 0.002$ m and height $h = 0.02$ m, as in the benchmark example by Schwab and Meijaard (2009, Sect. 2.6). The theoretical buckling load can be calculated by $F_{\text{th}} = 4.012599344\sqrt{EI_{\min}GJ}/L^2$, in which $EI_{\min}$ is the smaller flexural stiffness with $I_{\min} = \min(I_y, I_z)$ and $GJ = Ghw^3/3$ the torsional stiffness of the rectangular cross section, in which G denotes the shear modulus. The numerical solution is compared to the convergence rate of the nonlinear beam finite element proposed by Schwab and Meijaard (2009), see Table 5. Differently to Schwab and Meijaard (2009), the eigenvalues are calculated in the deformed state, which corresponds to the applied numerical buckling load F_{num}. Since buckling occurs already at a less deformed state as the ratio $\min\{w,h\}/\max\{w,h\}$ becomes smaller, the width w of the beam was chosen to be $w = 0.0002$ m (which is 100 times smaller than in the benchmark example by Schwab and Meijaard, 2009). The ratio $F_{\text{num}}/F_{\text{th}}$ for both the cases $w = 0.0002$ m and $w = 0.002$ m are outlined in Table 5.

6.7 Eigenfrequencies of a simply supported beam

According to Schwab and Meijaard (2009, Sect. 2.9), the eigenfrequencies of a simply supported beam with dimensions length $L = 1$ m, width $w = 0.02$ m and height $h = 0.02$ m with pre-stress F_x are computed. For a sketch of the problem setup, see Fig. 8. The exact values for the zero-load frequency w_0, the non-dimensional pre-stress α and the first non-dimensionalized frequency w_B can be calculated by the formulas

$$\begin{aligned} w_0 &= \pi^2\sqrt{EI_y/(\rho AL^4)}, \\ \alpha &= FL^2/(\pi^2 EI_y), \qquad (42) \\ w_B &= w_0\sqrt{1+\alpha}. \end{aligned}$$

Table 6. First non-dimensionalized eigenfrequency analysis for a simply supported beam with pre-stress in Sect. 6.7 versus number (#) of finite elements.

#	alpha 0	0.01	0.1	1
1	0.88134885	0.88057103	0.87418005	0.84110859
2	1.00841241	1.00834836	1.00780544	1.00410601
4	1.00067751	1.00067135	1.00061937	1.00026875
8	1.00004464	1.00004121	1.00001170	0.99978926
16	1.00000050	0.99999951	0.99997012	0.99975442
32	1.00000006	0.99999678	0.99996850	0.99975346

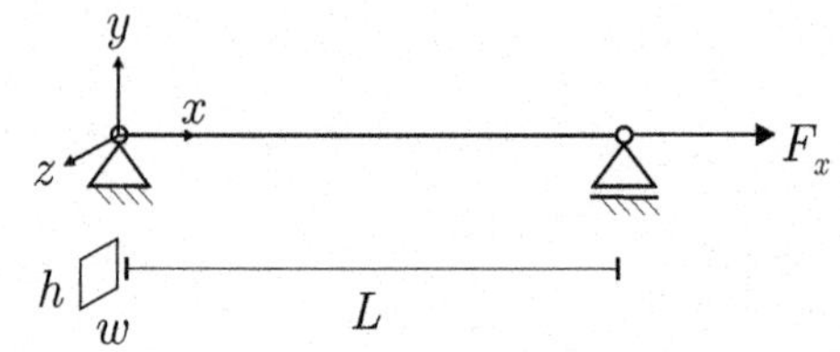

Figure 8. Simply supported beam under axial pre-stress in Sect. 6.7 investigated for eigenfrequency analysis.

The convergence analysis in dependency on the non-dimensional pre-stress α is outlined in Table 6. Here, the numerical eigenvalue w_{num} is divided by the theoretical eigenvalue w_B. Differently to Schwab and Meijaard (2009), the numerical eigenvalues are calculated in deformed state (by using the stiffness matrix of the deformed geometry), which corresponds to the load $F = \alpha\pi^2 EI_y L^{-2}$. Thus, the numerical solution is different from the theoretical solution w_B, but converging as α tends to zero.

6.8 Dynamic example of a double pendulum

In this example the nonlinear dynamic behavior of the proposed elements is tested by a rigid flexible pendulum as in Sugiyama et al. (2003). Note, that the original example is used for verifying thick beam finite elements including shear terms in the variational formulation according to Timoshenko's beam theory. In contrast to that, in this work the example serves for the experimental verification of the dynamic solution of the proposed Bernoulli Euler beam finite element, by comparing it to another solution obtained by already verified thick ANCF-beam finite elements based on Timoshenko's theory, see Nachbagauer and Gerstmayr (2012, 2013). The theory says, that both solutions (according to Bernoulli–Euler's and Timoshenko's theory, respectively) converge to each other, as the spatial (thickness to length) ratio of the beams goes to zero. Two tests are done, one for the original experiment setup, where the beams' spatial ratio is 1 : 5, and another test for the ratio of 1 : 50 and an appropriate adjustment of the material parameters, which is necessary to

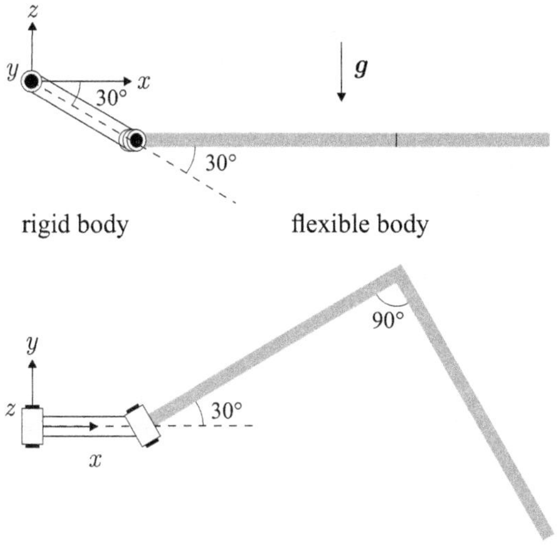

Figure 9. Geometry of a double pendulum consisting of a rigid and a flexible body, as proposed in Sugiyama et al. (2003).

obtain a comparable solution. For both tests a sufficiently refined spatial and temporal discretization is chosen, such that the discretization error has no significant influence.

6.8.1 Test A

The considered multibody system consists of a rigid body connected to a very flexible right angle frame, see Fig. 9.

The involved components are a rigid and a flexible body, and two revolute joints, which connect the rigid body both to the ground and to the right angle frame. The rigid body has a length of $L^r = 0.2\,\mathrm{m}$, a cross section area of $A^r = 2.5 \times 10^{-3}\,\mathrm{m}^2$, and a material density of $\rho^r = 7200\,\mathrm{kg\,m}^{-3}$. The flexible right angle frame is modeled by two beams connected by a rigid joint. Each beam in the original example (see Sugiyama et al., 2003; Nachbagauer and Gerstmayr, 2013) has a length of $L^f = 0.5\,\mathrm{m}$, a cross section area of $A^f = 1\times10^{-2}\,\mathrm{m}^2$, a material density of $\rho^f = 7200\,\mathrm{kg\,m}^{-3}$, a Young's modulus of $E^f = 2.0\times10^6\,\mathrm{Pa}$, a Poisson ratio of $\nu^f = 0.3$, and a second moment of area $I^f_A = 8.33 \times 10^{-6}\,\mathrm{m}^4$. Apart from gravity, there is no external force acting on the multibody system. The rigid body is rotated 30 degrees around the y-axis. Since the revolute joint which connects the rigid body to the ground allows rotation around the global y-axis, the motion of the rigid body is executed in the global (x, z)-plane. The flexible beam is rotated 30 degrees around the z-axis, and coherently the axis of the second revolute joint is chosen 30 degrees off the global y-axis. Therefore, the flexible body performs an in- and out-of-plane motion, which causes torsion. Two Finite Element simulations are performed for this Test: one simulation with an FE-approximation by 32 of the proposed Bernoulli–Euler beam finite elements, and another simulation with an FE-approximation by 32 already verified Timoshenko beam finite elements (see Nachbagauer and Gerstmayr, 2013). In Plot (a) of Fig. 10 the tip position (x, y, and z components) are plotted versus time for this Test A. The blue curves correspond to the discretization by Bernoulli–Euler beam finite elements, whereas the black curves correspond to Timoshenko beam finite elements.

6.8.2 Test B

The thickness of the pendulum is divided by the factor 10 compared to Test A, such that the cross section area of the rigid part measures $A^r = 2.5 \times 10^{-5}\,\mathrm{m}^2$ and the cross section area of the flexible part $A^f = 1 \times 10^{-4}\,\mathrm{m}^2$. In order to obtain a similar dynamic solution as in Test A, Young's modulus is chosen $E^f = 2.0 \times 10^{10}\,\mathrm{Pa}$, whereas the density of both the rigid and the flexible part is multiplied by the factor 10^2, such that $\rho^r = \rho^f = 720000\,\mathrm{kg\,m}^{-3}$. Again, as in Test A, two Finite Element simulations are performed. The dynamic behavior of the double pendulum is outlined in the Plot (b) of Fig. 10. In the first simulation (blue curves) Bernoulli–Euler beam finite elements are used, whereas the second simulation utilizes Timoshenko beam finite elements (black curves).

6.9 Large bending and torsion: director update

Table 7. Convergence Table Sect. 6.9. In the first column the number of Finite Elements is displayed. The next two columns show the error of the Finite Element iterates compared the converged solution $|\mathbf{u}^{\mathrm{FE}}_{\mathrm{A/B}} - \mathbf{u}^*_{\mathrm{A/B}}|$ in Test A and Test B, whereas the third column reports on the difference of the iterates $|\mathbf{u}^{\mathrm{FE}}_{\mathrm{A}} - \mathbf{u}^{\mathrm{FE}}_{\mathrm{B}}|$. As converged solution, the FE-solution for 512 Elements was used. The converged solutions in Test A and B differ by $|\mathbf{u}^*_{\mathrm{A}} - \mathbf{u}^*_{\mathrm{B}}| = 2.4050 \times 10^{-7}$.

Elements	Test A	Test B	difference
1	2.6272×10^{-1}	2.4675×10^{-1}	1.6558×10^{-2}
2	1.2809×10^{-1}	1.1399×10^{-1}	1.4723×10^{-2}
4	3.5333×10^{-2}	3.1031×10^{-2}	4.4026×10^{-3}
8	7.0811×10^{-3}	6.0522×10^{-3}	1.0373×10^{-3}
16	1.5113×10^{-3}	1.2660×10^{-3}	2.4719×10^{-4}
32	3.6922×10^{-4}	3.0828×10^{-4}	6.1547×10^{-5}
64	9.1062×10^{-5}	7.6005×10^{-5}	1.5388×10^{-5}
128	2.1677×10^{-5}	1.8092×10^{-5}	3.8473×10^{-6}
256	4.3353×10^{-6}	3.6182×10^{-6}	9.6185×10^{-7}

In this final example the impact of the director update, as suggested in Sect. 4, is studied. The geometry of the beam and its material properties are chosen as in Sect. 6.3 (large bending and torsion). In difference to Sect. 6.3 a tip load is applied with components $F_y = 25\,\mathrm{N}$ and $F_z = 250\,\mathrm{N}$ instead of moments. In the simulation exactly two load steps are performed, the first step with the half, and the second step with the full loading. For the convergence analysis at each FE-node the director is chosen to point in z direction. Two tests are performed: In Test A the directors are kept constant, in Test B the directors are updated at each node (see Sect. 4) by

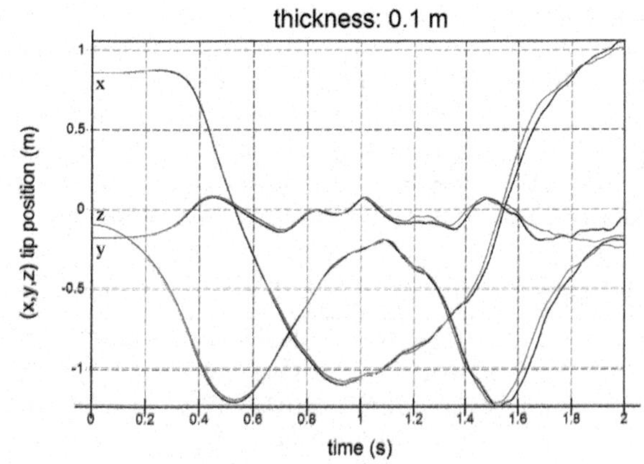

a)

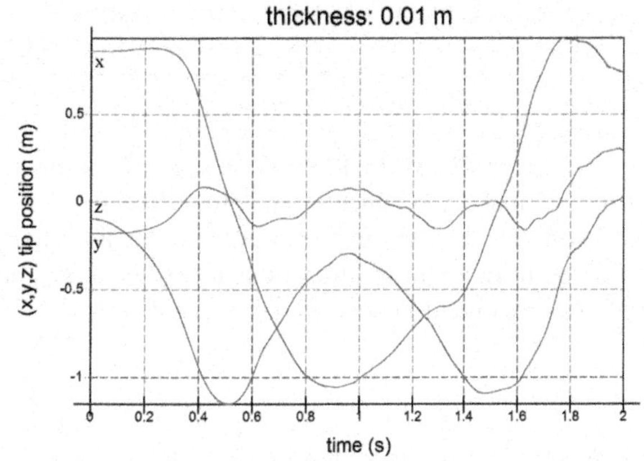

b)

Figure 10. Trajectories of the tip of the double pendulum in Test A **(a)** and in Test B **(b)** in Sect. 6.8. For each of the two tests either two simulations are performed: one by using an FE-discretization with 32 Bernoulli–Euler Finite Elements (3 blue curves for x, y, and z position components), and another by using 32 Timoshenko beam finite elements (3 black curves). The three black curves cannot be seen in Plot **(b)**, since the blue curves are on top of them. In other words, the FE-solutions converge as the ratio of thickness to length goes to zero.

their projection into the cross section. Table 7 shows clearly, that the update of the directors does not add any additional inaccuracy to the discretized solutions.

A remark on the use of constraints

The considered example includes a right angle joint of two beam segments. The demand on identity of rotations on the end points of both segments is trivial in the framework of Timoshenko theory, in which rotations appear as independent degrees of freedom. An efficient implementation of this condition for Bernoulli–Euler elements under consideration has been realized via Lagrange multipliers for six scalar conditions of equivalence of translations and rotations in the adjacent cross-sections of the two rod segments.

7 Conclusions

In the present paper, a spatial thin beam element with axial, bending, and torsional deformation is presented. The proposed element underlies the absolute nodal coordinate formulation, which is designed for large displacements, large rotations and large deformations and multibody dynamics problems. The element kinematics and deformation energy are chosen according to Vetyukov and Eliseev (2010), including an additional rotation director, which is updated during computation process. The investigation of numerical examples, both in the static and dynamic case, shows the expected accuracy and a fast performance of the proposed element. Moreover, in contrast to already existing 3-D ANCF elements, it does not suffer from geometrical singularities.

Acknowledgements. P. Gruber and K. Nachbagauer acknowledge support from the Austrian Science Funds FWF via the project I337-N18, Y. Vetyukov and J. Gerstmayr from the K2-Comet Austrian Center of Competence in Mechatronics ACCM.

Edited by: O. Brüls

References

Antman, S.: The theory of rods, in: Handbuch der Physik, edited by: Flügge, S. and Truesdell, C., Vol. VIa/2, 641–703, Berlin-Heidelberg-New York, Springer, 1972.

Dmitrochenko, O.: A new finite element of thin spatial beam in absolute nodal coordinate formulation, in: Proceedings of Multibody Dynamics, ECCOMAS Thematic Conference Madrid, Spain, 2005.

Dmitrochenko, O. and Pogorelov, D.: Generalization of Plate Finite Elements for Absolute Nodal Coordinate Formulation, Multibody Syst. Dyn., 10, 17–43, doi:10.1023/A:1024553708730, 2003.

Eliseev, V.: Mechanics of elastic bodies, St. Petersburg State Polytechnical University Publishing House, St. Petersburg, 2003 (in Russian).

Gams, M., Planinc, I., and Saje, M.: The strain-based beam finite elements in multibody dynamics, J. Sound Vib., 305, 194–210, 2007.

Gerstmayr, J. and Irschik, H.: On the correct representation of bending and axial deformation in the absolute nodal coordinate formulation with an elastic line approach, J. Sound Vib., 318, 461–487, 2008.

Gerstmayr, J. and Shabana, A.: Analysis of thin beams and cables using the absolute nodal coordinate formulation, Nonlinear Dynam., 45, 109–130, 2006.

Krommer, M. and Vetyukov, Y.: Adaptive sensing of kinematic entities in the vicinity of a time-dependent geometrically nonlinear pre-deformed state, Int. J. Solids Struct., 46, 3313–3320, 2009.

Nachbagauer, K. and Gerstmayr, J.: Nonlinear Dynamic Analysis of Three-Dimensional Shear Deformable ANCF Beam Finite Elements, Proceedings of the IMSD2012 – The 2nd Joint International Conference on Multibody System Dynamics, 29 May–1 June 2012, Stuttgart, Germany, 2012.

Nachbagauer, K. and Gerstmayr, J.: Structural and Continuum Mechanics Approaches for a 3D Shear Deformable ANCF Beam Finite Element: Application to Buckling and Nonlinear Dynamic Examples, Special Issue of the Journal of Computational and Nonlinear Dynamics on Flexible Multibody Dynamics, accepted , 2013.

Nachbagauer, K., Gruber, P., Vetyukov, Y., and Gerstmayr, J.: A spatial thin beam finite element based on the absolute nodal coordinate formulation without singularities, in: Proceedings of the ASME 2011 International Design Engineering Technical Conferences, Computers and Information in Engineering Conference IDETC/CIE 2011, Paper No. DETC2011/MSNDC-47732, Washington, DC, USA, 2011.

Reissner, E.: On One-Dimensional Large-Displacement Finite-Strain Beam Theory, Stud. Appl. Math., LII, 87–95, 1973.

Schwab, A. and Meijaard, J.: Beam benchmark problems for validation of flexible multibody dynamics, in: Multibody Dynamics, ECCOMAS Thematic Conference, Warsaw, Poland 29 June–2 July 2009, edited by: Wojtyra, M., Arczewski, K. and Fraczek, J., 2009.

Schwab, A. and Meijaard, J.: Comparison of three-dimensional flexible beam elements for dynamic analysis: classical finite element formulation and absolute nodal coordinate formulation, J. Comput. Nonlin. Dyn., 5, 011010, doi:10.1115/1.4000320, 2010.

Shabana, A.: Definition of the slopes and the finite element absolute nodal coordinate formulation, Multibody Syst. Dyn., 1, 339–348, 1997.

Shabana, A.: Dynamics of Multibody Systems (3rd Edn.), Cambridge University Press, New York, 2005.

Simo, J. and Vu-Quoc, L.: A Three-Dimensional Finite-Strain Rod Model. Part II: Computational Aspects, Comput. Method. Appl. M., 58, 79–116, 1986.

Simo, J. C.: A Finite Strain Beam Formulation. The Three-Dimensional Dynamic Problem. Part I, Comput. Method. Appl. M., 49, 55–70, 1985.

Sugiyama, H., Escalona, J., and Shabana, A.: Formulation of Three-Dimensional Joint Constraints Using the Absolute Nodal Coordinates, Nonlinear Dynam., 31, 167–195, 2003.

Vetyukov, Y.: Direct approach to elastic deformations and stability of thin-walled rods of open profile, Acta Mech., 200, 167–176, 2008.

Vetyukov, Y. and Eliseev, V.: Modeling of building frames as spatial rod structures with geometric and physical nonlinearities (in Russian), Computational Continuum Mechanics, 3, 32–45, http://www.icmm.ru/journal/download/CCMv3n3a3.pdf, 2010.

von Dombrowski, S.: Analysis of large flexible body deformation in multibody systems using absolute coordinates, Multibody Syst. Dyn., 8, 409–432, 2003.

Yakoub, R. Y. and Shabana, A. A.: Three dimensional absolute nodal coordinate formulation for beam elements, ASME J. Mech. Design, 123, 606–621, 2001.

Yeliseyev, V. and Orlov, S.: Asymptotic splitting in the three-dimensional problem of linear elasticity for elongated bodies with a structure, J. Appl. Math. Mech., 63, 85–92, 1999.

Zupan, D. and Saje, M.: Finite-element formulation of geometrically exact three-dimensional beam theories based on interpolation of strain measures, Comput. Method. Appl. M., 192, 5209–5248, 2003.

14

Segmental contributions to the ground reaction force in the single support phase of gait

D. S. Mohan Varma and S. Sujatha

Department of Mechanical Engineering, Indian Institute of Technology Madras, Chennai, India

Correspondence to: S. Sujatha (sujsree@iitm.ac.in)

Abstract. An inverse dynamics model for the single support (SS) phase of gait is developed to study segmental contributions to the ground reaction force (GRF). With segmental orientations as the generalized degrees of freedom (DOF), the acceleration of the body's center-of-mass is expressed analytically as the summation of the weighted kinematics of individual segments. The weighting functions are constants that are functions of the segment masses and center-of-mass distances. Using kinematic and anthropometric data from literature as inputs, and using the roll-over-shape (ROS) to model the foot-ground interaction, GRF obtained from the inverse model are compared with measured GRF data from literature. The choice of the generalized coordinates and mathematical form of the model provides a means to weigh individual segment contributions, simplify models and choose more kinetically accurate inverse dynamics models. For the kinematic data used, an anthropomorphic model that includes the frontal plane rotation of the pelvis in addition to the sagittal DOF of the thigh and shank most accurately captures the vertical component of the GRF in the SS phase of walking. Of the two ROS used, the ankle-foot roll-over shape provides a better approximation of the kinetics in the SS phase. The method presented here can be used with additional experimental studies to confirm these results.

1 Introduction

In inverse dynamics models of gait, measured kinematics, estimated body segment parameters (BSP) and measured ground reaction force (GRF) are usually used as inputs to compute the internal joint forces and moments (Winter, 2005). GRF are external forces acting on the body, and introducing them as inputs in the inverse model overdetermines the problem (Hatze, 1981). This overdeterminacy results in a residual error in the joint kinetics computations (Riemer et al., 2008). In this work, an inverse dynamics model that uses only the kinematics and BSP as inputs is developed. The measured GRF is used to validate the model.

The motivation for this work is a mathematical model that can be extended to study asymmetric gait with only kinematic measurements as inputs. This model for normal walking can serve as a baseline model for asymmetric gait, such as the gait of a prosthesis user, since errors in the anthropomorphic model are accounted for.

Mathematical models of gait vary from simple planar link-segment models (Onyshko and Winter, 1980; Winter, 2005; Mochon and McMahon, 1980) to complex musculoskeletal models (Anderson and Pandy, 2001). A simplified model for gait cannot capture all the characteristics of human walking, while a complex model becomes analytically intractable. Ideally, a mathematical model for gait should include all the degrees of freedom (DOF) that are major contributors to the activity. Six quantities, namely, the rotation of pelvis in the frontal and transverse planes, knee flexion in the stance phase, rotations of the foot and knee and the lateral movement of the pelvis were identified as the six determinants of gait (Saunders et al., 1953). Although, the influence of these DOF on the vertical center-of-mass (COM) motion and energy cost has been debated (Kuo, 2007), it is accepted that these six quantities are characteristic features of gait. The individual contributions of these and possibly other DOF to the overall kinetics needs to be quantified. By knowing the DOF that are active contributors, one can determine the kinematics that must be most accurately measured and reduce or add DOF to the mathematical model to make it simpler or more accurate.

Various researchers have used forward or inverse models to predict the GRF. Pandy and Berme (1989b) used a three-dimensional forward dynamics model that used five of the six determinants of gait for the simulation of normal gait to quantify the influence of individual gait determinants on the GRF and studied (Pandy and Berme, 1989a) the compensatory actions required when each of these DOF was systematically removed from the model. The model used was not an anthropomorphic model, and the joint moments were heuristically selected. Mochon and McMahon (1980) used a ballistic model for the swing phase, which assumed zero moments at the joints, to study the influence of the determinants of gait on the GRF. Thornton-Trump and Daher (1975) obtained GRF and joint kinetics from an inverse model and used these data as inputs for the analysis of a polycentric prosthetic knee. Zarrugh (1981) obtained GRF from an inverse model to study joint powers but did not include a pelvis link. Winiarski and Rutkowska-Kucharska (2009) used inverse dynamics to study the suitability of estimating GRF from the kinematics of the COM in normal and pathological gait. Pillet et al. (2010) obtained GRF from an inverse model for a force-plate-less estimation of center of pressure (COP) trajectory during gait. Oh et al. (2013) used an artificial neural networks-based analysis for force-plate-less estimation of GRF.

While forward dynamic models require joint moments (which cannot be measured directly) to be specified; inverse dynamics models need kinematic data as inputs. The general trends of joint/segment angles in normal gait have been studied extensively and are readily available in the literature (Saunders et al., 1953; Inman et al., 1981; Perry, 1992). Therefore, an inverse dynamics model that incorporates all the sagittal plane DOF of the lower limb segments, the frontal and transverse plane rotation of the pelvis, lateral movement of the legs and a single head-arms-and-trunk (HAT) segment is developed in this work. Newton's equation is rewritten to express GRF as a weighted summation of segmental kinematics. This enables the study of contributions of each of the DOF/segment kinematics to the GRF in a more direct way. Expressing the model in terms of contributions of individual segments will enable extension of the model to study asymmetric gait, such as the gait of a prosthesis user, where the properties and kinematics of the prosthetic limb are likely to differ from those of the unaffected side.

Many inverse dynamics models suffer from errors in kinematics (noise in measurements, errors due to data filtering, curve fitting technique used and derivative computation) and BSP (mass, inertia and COM location) estimates. While some studies (Pàmies-Vilà et al., 2012; Reinbolt et al., 2007) suggest that errors in kinematic data influence joint moment estimates more than the errors in BSP estimates, others (Rao et al., 2006; Pearsall and Costigan, 1999) show that the influence of BSP errors cannot be neglected. In all the above cases, the sensitivity of the joint reactions to the perturbations in the kinematics and BSP are studied using the overdetermined inverse model. Since joint reactions cannot be measured experimentally, the validity of these studies is dependent on the accuracy of the model. In this work, comparison of the GRF from the model to the measured values provides a means to determine the accuracy of the model, and the use of segmental orientations as the generalized DOF enables us to obtain analytical expressions for the contributions of each segment's kinematics to the GRF results. By systematically eliminating the DOF whose contribution is not significant, an optimal number of measurements needed for a reasonable estimation of GRF can be obtained. This can reduce the number of kinematic measurements and BSP estimates and thereby reduce the corresponding errors.

Apart from kinematics and BSP, the COP and foot kinematics (Mccaw and Devitat, 1995; Silva and Ambrosio, 2004) have also been found to influence the joint moment computations. Simulation studies on gait have used models such as the triple rocker (Perry, 1992), polynomial fits (Ju and Mansour, 1988; Ren et al., 2007) and COP-based fits (Koopman et al., 1995; Srinivasan et al., 2008). McGeer (1990) used circular arcs for the plantar surface of the foot in passive dynamic walkers, and Vanderpool et al. (2008) found that rolling feet are energetically efficient and can account for the loss of ankle motion. In their experimental studies, Hansen et al. (2004) define foot roll-over shape (FROS) and ankle-foot roll-over shape (AFROS) based on COP data. Their experimental studies found that the AFROS remains the same irrespective of changes in walking speed (Hansen et al., 2004), heel height (Hansen and Childress, 2004) and load carriage (Hansen and Childress, 2005). They also found that a person controls the ankle kinematics to maintain a consistent ROS (Wang and Hansen, 2010) and that in the case of a trans-tibial prosthesis user, an alignment for better walking performance for a given type of prosthetic foot can be determined using this ROS (Hansen et al., 2000). Srinivasan et al. (2009) used Hansen's ROS in a forward dynamics model for the gait of trans-tibial prosthesis users. Since the FROS and AFROS shapes are backed by extensive experimental studies, we use these models to model the foot-ground interface in our work.

Optimization is used in this work to ensure that the input data satisfies the kinematic constraint of the swing foot clearing the ground during the SS phase. Unlike other optimization-based inverse models (Koopman et al., 1995; Ren et al., 2007) in which optimization is used in parallel with the inverse model to predict the kinematics, kinetics or both, here optimization is used only to correct any errors in data that cause violation of the kinematic constraints. The effects of using the FROS and AFROS (Hansen et al., 2004) models on the GRF computations are studied.

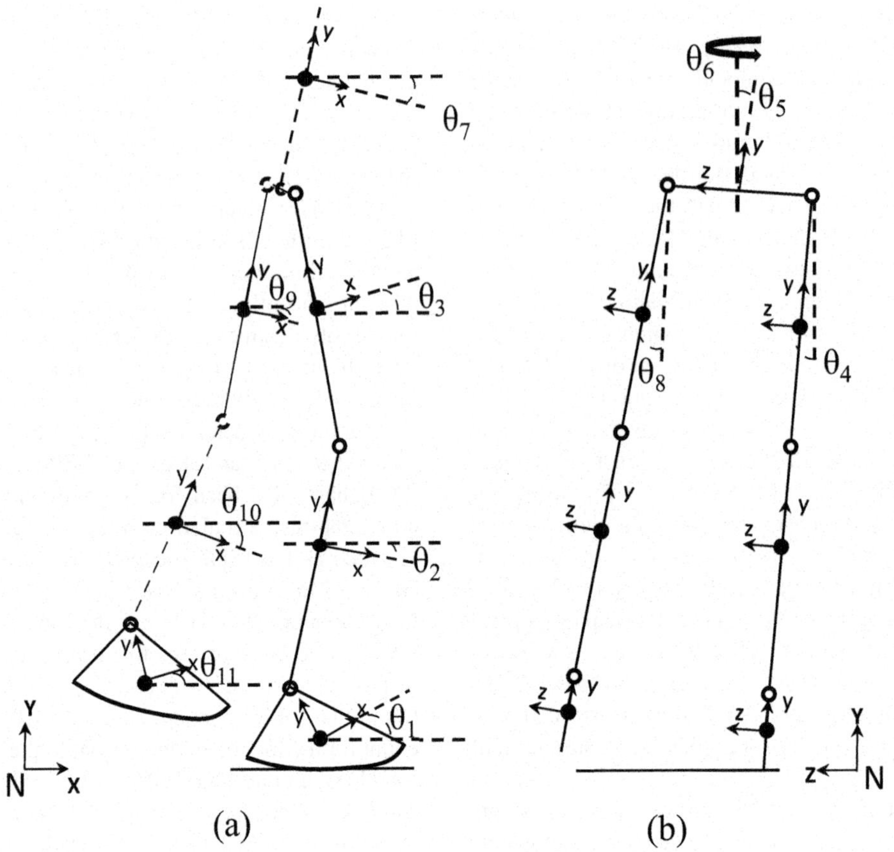

Figure 1. Stick figures showing the angles in the **(a)** sagittal and **(b)** frontal planes. All angles are measured with respect to N. (HAT segment is not shown in the frontal plane **b**.)

2 Methods

2.1 Mathematical model

The anthropomorphic model of gait consists of eight segments and eleven DOF (Fig. 1). The eight segments are: the foot, shank and thigh of each leg, the pelvis and the HAT. The segments of the leg – foot, shank and thigh – are assumed to remain in the same plane and are connected by revolute joints at the hip, knee and ankle. Transverse plane rotation of the leg is assumed to be zero. Thus, the stance leg has 4 DOF: sagittal plane rotations of the foot (θ_1), shank (θ_2) and thigh (θ_3) and the frontal plane rotation of the leg as a whole (θ_4). The swing leg also has 4 DOF – sagittal plane rotations of the foot (θ_{11}), shank (θ_{10}) and thigh (θ_9) and the frontal plane rotation of the leg as a whole (θ_8). The pelvis has two DOF – rotation in the frontal(θ_5) and transverse planes (θ_6). The HAT segment is assumed to be rotating only in the sagittal plane (θ_7). The angles in the frontal, transverse and sagittal planes are measured as the orientation of the body-fixed x, y or z axis with respect to the ground-fixed X, Y and Z axes respectively. The body-fixed coordinate systems of the segments are as shown in Fig. 1. The pelvis is modeled as a massless link, and the HAT segment is assumed to be connected to the mid-point of the pelvis. The foot-ground interaction is modeled using two different foot models:

1. A rolling contact foot model with Hansen's Foot ROS (Fig. 2).
2. A rolling contact foot model using Hansen's Ankle-Foot ROS (Fig. 3).

The FROS is obtained by transforming the COP data from N (the coordinate system fixed to the ground) to a coordinate system fixed to the foot at the ankle, and the AFROS is obtained by transforming the COP data from N to a coordinate system fixed to the shank at the ankle. In the latter case, the AFROS takes into account ankle flexion and models the net motion as rolling motion of the shank and foot complex with respect to the ground so that the ankles of the swing and stance legs can be assumed to be rigid in the mathematical model. The angles of the foot segment θ_1 and θ_{11} are then determined using $\theta_1 = \theta_2 - \theta_{ank}$ and $\theta_{11} = \theta_{10} - \theta_{ank}$. The value θ_{ank} is the angle between the shank and the foot segments that remains constant since the AFROS accounts for the ankle flexion. Therefore, while the model using the FROS has eleven DOF, the model using AFROS has only nine DOF.

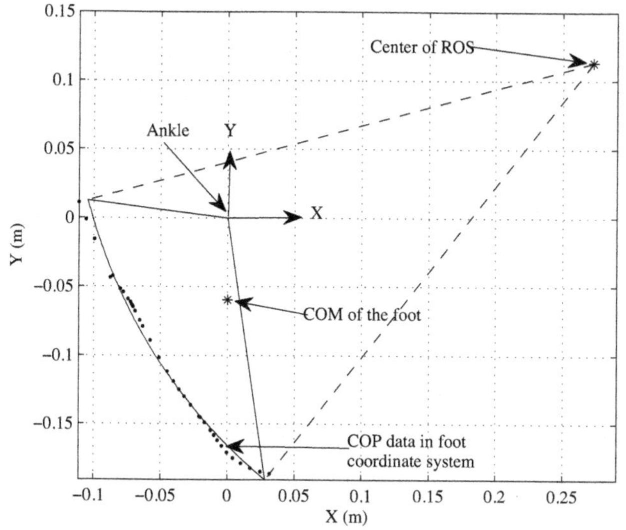

Figure 2. FROS with y axis (of the coordinate system fixed to the foot) lying along the line joining the ankle and the COM of the foot.

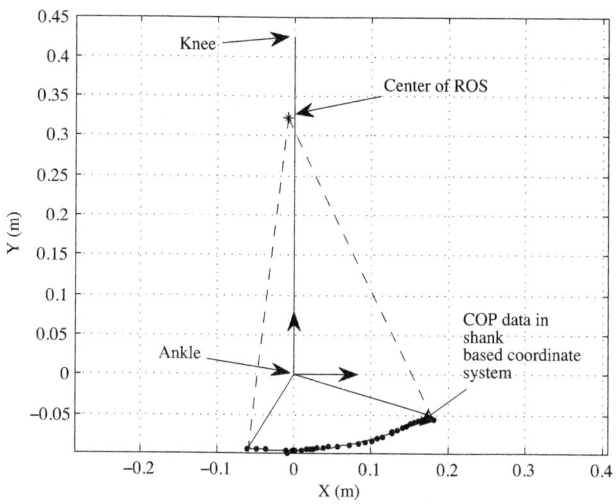

Figure 3. AFROS with y axis (of the coordinate system fixed to the shank) along the shank length.

2.2 Kinematics

The position, velocity and acceleration (in N) of a point BP on a rigid body B moving in space with respect to another point BQ on the same body (expressed in N) are given by

$$\begin{aligned} &\boldsymbol{P}_{\mathrm{BP/BQ}}=\boldsymbol{P}_{\mathrm{BP}}-\boldsymbol{P}_{\mathrm{BQ}}=\mathbf{T}_{\mathrm{B}}\boldsymbol{r}^{\mathrm{B}},\\ &\boldsymbol{V}_{\mathrm{BP/BQ}}=\boldsymbol{V}_{\mathrm{BP}}-\boldsymbol{V}_{\mathrm{BQ}}=\boldsymbol{\omega}_{\mathrm{B}}\times\mathbf{T}_{\mathrm{B}}\boldsymbol{r}^{\mathrm{B}} \text{ and} \\ &\boldsymbol{A}_{\mathrm{BP/BQ}}=\boldsymbol{A}_{\mathrm{BP}}-\boldsymbol{A}_{\mathrm{BQ}}=\boldsymbol{\alpha}_{\mathrm{B}}\times\mathbf{T}_{\mathrm{B}}\boldsymbol{r}^{\mathrm{B}}+\boldsymbol{\omega}_{\mathrm{B}}\times\left(\boldsymbol{\omega}_{\mathrm{B}}\times\mathbf{T}_{\mathrm{B}}\boldsymbol{r}^{\mathrm{B}}\right)\end{aligned} \tag{1}$$

respectively, where $\boldsymbol{\alpha}_{\mathrm{B}}$ is the angular acceleration, $\boldsymbol{\omega}_{\mathrm{B}}$ is the angular velocity of body B in N (expressed in N), $\mathbf{T}_{\mathrm{B}}$ is the 3×3 transformation matrix obtained using space-fixed Z-Y-X transformations (Kane et al., 1983) and $\boldsymbol{r}^{\mathrm{B}}$ is the position vector from point BQ to BP expressed in B.

The above equations can be rewritten as

$$\begin{aligned} &\boldsymbol{P}_{\mathrm{BP/BQ}}=\mathbf{T}_{\mathrm{B}}\boldsymbol{r}^{\mathrm{B}},\\ &\boldsymbol{V}_{\mathrm{BP/BQ}}=\mathbf{C}_{\mathrm{B}}\boldsymbol{r}^{\mathrm{B}} \text{ and}\\ &\boldsymbol{A}_{\mathrm{BP/BQ}}=\mathbf{E}_{\mathrm{B}}\boldsymbol{r}^{\mathrm{B}}\end{aligned} \tag{2}$$

where $\boldsymbol{C}_{\mathrm{B}}=\tilde{\omega}_{\mathrm{B}}\mathbf{T}_{\mathrm{B}}$, $\mathbf{E}_{\mathrm{B}}=\tilde{\alpha}_{\mathrm{B}}\mathbf{T}_{\mathrm{B}}+\tilde{\omega}_{\mathrm{B}}\mathbf{C}_{\mathrm{B}}$ and the skew-symmetric matrix $\tilde{\mathbf{q}}$ of any general vector $\boldsymbol{q}=(q_x, q_y, q_z)'$ is given as (Shabana, 2010)

$$\tilde{\mathbf{q}}=\begin{bmatrix} 0 & -q_z & q_y\\ q_z & 0 & -q_x\\ -q_y & q_x & 0\end{bmatrix}. \tag{3}$$

The product $\mathbf{E}_{\mathrm{B}}\boldsymbol{r}^{\mathrm{B}}$ of Eq. (2) can also be written in column vector notation as

$$\mathbf{E}_{\mathrm{B}}\boldsymbol{r}^{\mathrm{B}}=r_x^{\mathrm{B}}\begin{bmatrix} | \\ \mathbf{E}_{\mathrm{B}} \\ | \end{bmatrix}_{c1}+r_y^{\mathrm{B}}\begin{bmatrix} | \\ \mathbf{E}_{\mathrm{B}} \\ | \end{bmatrix}_{c2}+r_z^{\mathrm{B}}\begin{bmatrix} | \\ \mathbf{E}_{\mathrm{B}} \\ | \end{bmatrix}_{c3} \tag{4}$$

where the subscripts $c1$, $c2$ and $c3$ indicate the first, second and third columns of the matrix $\mathbf{E}_{\mathrm{B}}$, respectively. Using the above notation, the acceleration of the COM can be expressed as (Appendix A)

$$\boldsymbol{A}_{\mathrm{com}}={}^{\mathrm{st}}\boldsymbol{A}_{\mathrm{ank}}+\sum k\begin{bmatrix} | \\ \mathbf{E} \\ | \end{bmatrix} \tag{5}$$

where ${}^{\mathrm{st}}\boldsymbol{A}_{\mathrm{ank}}$ is the acceleration of the ankle of the stance leg, ks are functions of mass fractions and COM distances and $\mathbf{E}$s are dependent on the kinematics of the segments.

2.3 Segmental contributions to GRF

Newton's equation of motion for the whole body in the SS phase is given by

$$\boldsymbol{GRF}=M\boldsymbol{A}_{\mathrm{com}}-M\boldsymbol{g}=M\left(\boldsymbol{A}_{\mathrm{com}}-\sum m_i\boldsymbol{g}\right)=\sum \boldsymbol{R}_i \tag{6}$$

where m_i is the mass fraction of the ith segment, M is the total mass and $\boldsymbol{g}$ is the gravity vector. Expressing all the accelerations in terms of $\mathbf{E}$s, GRF can be rewritten as the summation of contributions from each segment's kinematics as (Appendix A)

$$\begin{aligned}\boldsymbol{GRF}={}&{}^{\mathrm{st}}\boldsymbol{R}_{\mathrm{ft}}+{}^{\mathrm{st}}\boldsymbol{R}_{\mathrm{sk}}+{}^{\mathrm{st}}\boldsymbol{R}_{\mathrm{th}}+\boldsymbol{R}_{\mathrm{pel}}\\ &+\boldsymbol{R}_{\mathrm{hat}}+{}^{\mathrm{sw}}\boldsymbol{R}_{\mathrm{th}}+{}^{\mathrm{sw}}\boldsymbol{R}_{\mathrm{sk}}+{}^{\mathrm{sw}}\boldsymbol{R}_{\mathrm{ft}}\end{aligned} \tag{7}$$

where $\boldsymbol{R}_i$s are the segmental contributions and the superscripts "sw" and "st" indicate swing leg and stance leg, respectively, and the subscripts "ank", "ft", "sk", "th", "hat" and "pel" indicate the ankle, foot, shank, thigh, HAT and pelvis segments, respectively. The contributions of the segmental kinematics towards the GRF are given by Eqs. (A8) to (A15). The segmental contribution $\boldsymbol{R}_i$ has units of force.

Each of the $\boldsymbol{R}_i$s is expressed as a summation of a variable acceleration term ($M k_i \mathbf{E}_i$) and a constant gravity term ($m_i M \boldsymbol{g}$). The k_is in the acceleration terms are functions of the mass fractions of multiple segments and the distance of the COM from the joints (Eq. A5), and each $\mathbf{E}_i$ is a function of the kinematics of segment i. For a given set of anthropometric data (estimated segment masses and COM distances), each $\boldsymbol{R}_i$ is a function of only the kinematics of a particular segment. The foot segment's contribution ${}^{\text{st}}\boldsymbol{R}_{\text{ft}}$ is also dependent on the ankle acceleration (${}^{\text{st}}\boldsymbol{A}_{\text{ank}}$), which is determined by the ROS used (Appendix B). Using the anthropometric data from Winter (2005) in Eq. (A5), we get (in m) $k_1 = -0.0009$, $k_2 = 0.41$, $k_3 = 0.28$, $k_4 = 0.144$, $k_5 = 0.194$, $k_6 = -0.033$, $k_7 = -0.0147$ and $k_8 = -0.0009$.

The segmental contributions to the vertical component of GRF (GRFy) are given by (y components in Eqs. A8 to A15)

$$
\begin{aligned}
{}^{\text{st}}R_{\text{fty}} &= M\,{}^{\text{st}}A_{\text{anky}} + Mk_1\left(-S_1\alpha_1 - \left(\omega_4^2+\omega_1^2\right)C_1C_4 \right.\\
&\quad \left. -\alpha_4 S_4 C_1\right) + {}^{\text{st}}m_{\text{ft}}Mg,\\
{}^{\text{st}}R_{\text{sky}} &= Mk_2\left(-S_2\alpha_2 - \left(\omega_4^2+\omega_2^2\right)C_2C_4 - \alpha_4 S_4 C_2\right)\\
&\quad + {}^{\text{st}}m_{\text{sk}}Mg,\\
{}^{\text{st}}R_{\text{thy}} &= Mk_3\left(-S_3\alpha_3 - \left(\omega_4^2+\omega_3^2\right)C_3C_4 - \alpha_4 S_4 C_3\right)\\
&\quad + {}^{\text{st}}m_{\text{th}}Mg,\\
R_{\text{pely}} &= Mk_4\left(\omega_5\omega_6 S_6 + \omega_5^2 S_5 C_6 - \alpha_5 C_5 C_6\right),\\
R_{\text{haty}} &= Mk_5\left(-S_7\alpha_7 - \omega_7^2 C_7\right) + m_{\text{hat}}Mg,\\
{}^{\text{sw}}R_{\text{thy}} &= Mk_6\left(-S_9\alpha_9 - \left(\omega_8^2+\omega_9^2\right)C_9C_8 - \alpha_8 S_8 C_9\right)\\
&\quad + {}^{\text{sw}}m_{\text{th}}Mg,\\
{}^{\text{sw}}R_{\text{sky}} &= Mk_7\left(-S_{10}\alpha_{10} - \left(\omega_8^2+\omega_{10}^2\right)C_{10}C_8 - \alpha_8 S_8 C_{10}\right)\\
&\quad + {}^{\text{sw}}m_{\text{sk}}Mg \text{ and}\\
{}^{\text{sw}}R_{\text{fty}} &= Mk_8\left(-S_{11}\alpha_{11} - \left(\omega_8^2+\omega_{11}^2\right)C_{11}C_8 - \alpha_8 S_8 C_{11}\right)\\
&\quad + {}^{\text{sw}}m_{\text{ft}}Mg
\end{aligned}
\tag{8}
$$

where, $S_i = \sin\theta_i$, $C_i = \cos\theta_i$, $\omega_i = \dot{\theta}_i$, $\alpha_i = \ddot{\theta}_i$, M is the total mass and ${}^{\text{st}}A_{\text{anky}}$ is the vertical component of the acceleration of the ankle of the stance leg given by Eq. (B1). The segmental contributions to the anterior-posterior component of GRF (GRFx) and medio-lateral component of GRF (GRFz) can also be obtained from Eqs. (A8) to (A15).

Each segmental contribution is a function of one or two DOF. To study the contribution of a DOF of a particular segment, the other DOF is set equal to zero. For instance, the contributions of pelvis to GRFx and GRFy are given by,

$$
\begin{aligned}
R_{\text{pelx}} &= Mk_4\left(-\omega_6^2 S_6 - \omega_5\omega_6 S_5 C_6 + \alpha_6 C_5 C_6\right)\\
R_{\text{pely}} &= Mk_4\left(\omega_5\omega_6 S_6 + \omega_5^2 S_5 C_6 - \alpha_5 C_5 C_6\right).
\end{aligned}
\tag{9}
$$

When θ_5, ω_5 and α_5 are set equal to zero, the contributions of transverse plane rotation of the pelvis (θ_6) are given by,

$$
\begin{aligned}
R_{\text{pelx}} &= Mk_4\left(-\omega_6^2 S_6 + \alpha_6 C_6\right)\\
R_{\text{pely}} &= 0
\end{aligned}
\tag{10}
$$

and when θ_6, ω_6 and α_6 are set equal to zero, the contributions of frontal plane rotation of the pelvis (θ_5) are given by,

$$
\begin{aligned}
R_{\text{pelx}} &= 0\\
R_{\text{pely}} &= Mk_4\left(\omega_5^2 S_5 - \alpha_5 C_5\right).
\end{aligned}
\tag{11}
$$

The above equations show that the transverse plane rotation of the pelvis has no effect on the GRFy and the frontal plane rotation has no effect on the GRFx.

2.4 Unknown kinematics

The kinematic data from Winter (2005) are used to study the segmental contributions to the GRF. The sagittal plane rotations of the foot, shank, thigh and HAT are obtained from Winter (2005). The frontal plane rotation of the leg (θ_4 and θ_8) is assumed based on general gait trends in normal gait (Inman et al., 1981; Perry, 1992). In normal gait, the trajectory of any point on the pelvis, including the hip joints, in the transverse plane is a sine curve (Inman et al., 1981). Therefore, the trajectory in the transverse plane of any point on the pelvis can be given by $Z = C \sin(\omega t + \phi)$, where $\omega = 2\pi/T$ and T is the period of the gait cycle. The maximum lateral translation of the pelvis is approximately 0.02 m and is reached slightly after mid-stance (Inman et al., 1981). Using these two assumptions, the values of C and ϕ were determined. The height above the ground of the hip joint is known from the hip trajectory data (Winter, 2005), while the equation for Z gives the amount of lateral displacement. Using this information, the frontal plane rotation of the leg is determined for the entire stance phase. A Fourier series curve for the full gait cycle is fit using this data to determine the leg angle in the frontal plane during swing.

Using the available data and the fact that gait is symmetric for a normal person, the frontal plane rotation (θ_5) and the transverse plane rotation (θ_6) of the pelvis are obtained as follows: from the hip trajectory data (Winter, 2005), it is seen that the height above the ground (y coordinate) of the hip joint at ipsilateral and contralateral heel contact (HC) is approximately the same. This equality implies that in symmetric gait (as is the case here) at HC the height above the ground of the right and left hips is the same, indicating that the pelvis is level with (i.e. parallel to) the ground in the frontal plane at HC. Also, Inman et al. (1981) reported gait characteristics of six adult males walking at moderate walking speed and found that, in general, the pelvis is level with the ground in the frontal plane at HC. Therefore, the pelvis rotation in the frontal plane (θ_5) at HC is assumed to be zero.

From the hip trajectory data, the hip displacement in the direction of progression (x direction) from ipsilateral HC to contralateral HC (say d_1) and the hip displacement in the direction of progression from contralateral HC to ipsilateral HC (say d_2) are known. Again, due to the symmetry of normal gait, the distance along the x direction between the right and left hips at right HC $(x_1 - x_2)$ must be the same as that at left HC $(x_2 + d_2) - (x_1 + d_1)$. Since d_1 and d_2 are known quantities, this equality can be solved to obtain $\delta x = x_1 - x_2$. This value δx, in turn can be used to obtain the transverse plane angular position (θ_6) of the pelvis at HC. For the kinematic data from Winter (2005), the value of θ_6 at HC is 10.3°.

Since the relative position of the hips at HC is known, a unit vector along the line joining the hip joints, $\boldsymbol{h}$, at HC is known. Using the sagittal plane hip trajectory data from Winter (2005) and Z, the 3-D trajectories of the right and left hips, and therefore the vector $\boldsymbol{h}$, are known for the full gait cycle. For a Z-Y-X transformation with no rotation about the Z axis, the DOF θ_5 and θ_6 can be computed using

$$\theta_5 = \text{atan}\left(-h_y/h_z\right) \text{ and}$$
$$\theta_6 = \text{atan}\left(-h_x S_5/h_y\right). \tag{12}$$

Fourier series curves are fit to all the kinematic data (θ_1 through θ_{11}). In order to determine the number of Fourier coefficients n required for each fit, the least squares errors are determined as n is increased. For each of the fits, the n (as it is increased) that results in a change in the least squares error that is less than one degree is chosen. In order to ensure symmetry of the gait data, the time period used for the Fourier series fit for the HAT segment is half that of the others, and only odd harmonics of the Fourier series are used for the pelvis DOF.

2.5 Optimization

In the single support phase of walking the stance leg is in contact with the ground while the swing leg clears the ground as it moves forward. In experimentally obtained kinematic data, due to errors, the data may indicate that the swing leg contacts or digs into the ground. This is a violation of the kinematic constraint that the swing leg remains above the ground in the SS phase. To address this issue, in this work, optimization is used to modify the segment angles such that the swing leg remains above the ground during SS phase. No kinetic constraints or energy criteria are employed, and the optimization merely serves to render the kinematics realistic. With the input kinematic data and FROS, the kinematic constraints are not violated. However, when AFROS is used, the constraints are violated in the swing phase just before HC. The optimization problem statement for this case is given as

$$\begin{aligned} &\text{Minimize } |\boldsymbol{x} - \boldsymbol{x}_0| \\ &\text{such that, } y_{\min} \geq \Delta \\ &\boldsymbol{x} - \delta\boldsymbol{x} \leq \boldsymbol{x} \leq \boldsymbol{x} + \delta\boldsymbol{x} \end{aligned} \tag{13}$$

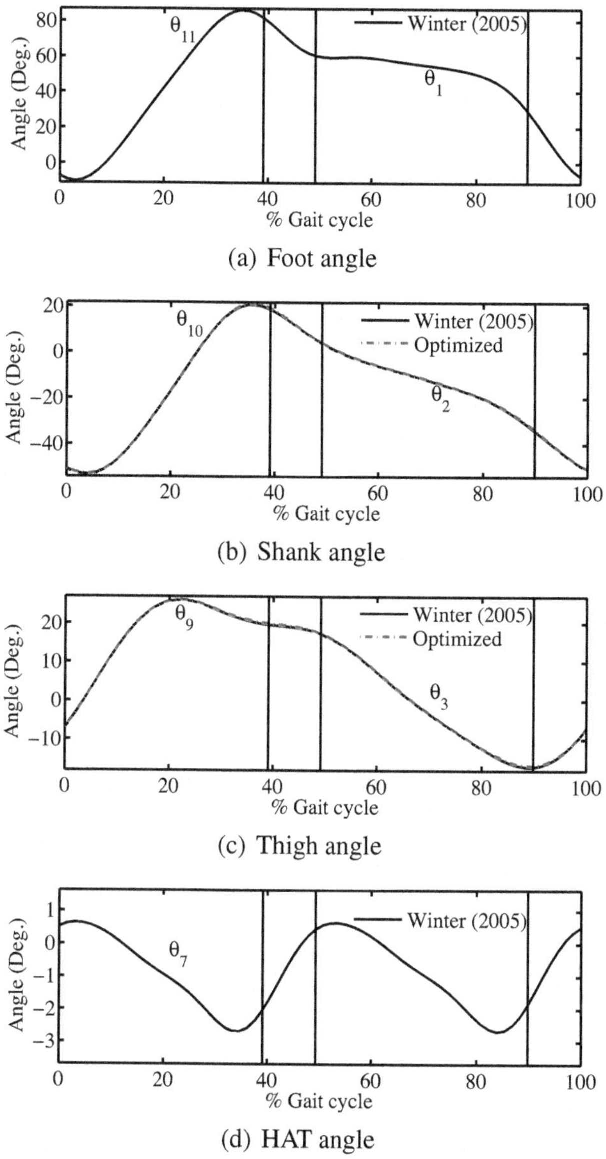

Figure 4. Sagittal plane kinematic data for full gait cycle (toe-off to toe-off) (Winter, 2005). **(a)** Foot, **(b)** shank, **(c)** thigh and **(d)** HAT angles. For **(a)**–**(c)** the data from 0 to 39 % gait cycle indicate data for the swing leg and angles from 50 to 89 % gait cycle indicate data for the stance leg in the SS phase.

where $\boldsymbol{x}$ is the vector $[\theta_2, \theta_3, \theta_4, \theta_5, \theta_6, \theta_8, \theta_9, \theta_{10}]$ and $y_{\min}$ is the y coordinate of the lowest point on the foot. To ensure minimal deviation from known trends, small values of 1° and 1 mm are chosen for $\delta\boldsymbol{x}$ and Δ, respectively. Every instance of the SS phase is checked for constraint violation, and optimization is performed whenever there is a violation. Fourier series curves are fit again to the optimized kinematic data. The optimization and curve fitting are thus performed iteratively until a set of kinematics that satisfies the constraints are obtained. It can be seen from Figs. 4 and 5 that after optimization, the kinematic data show little or no variation from

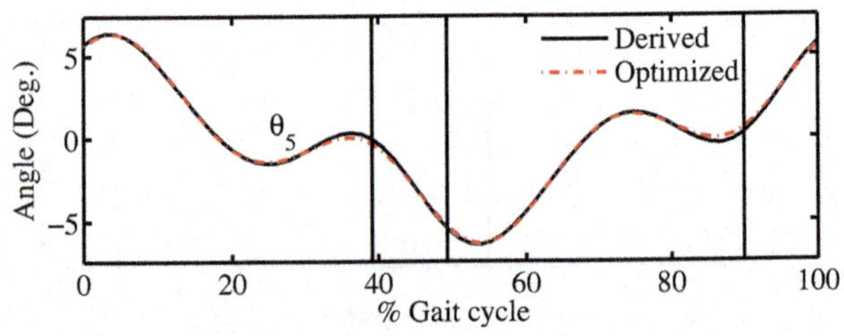

(a) Pelvis angle in frontal plane

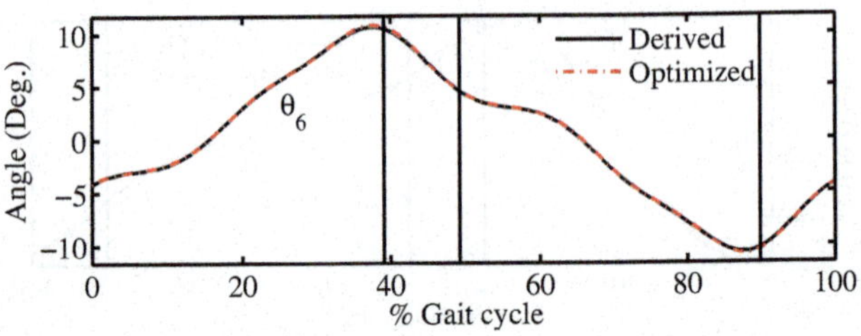

(b) Pelvis angle in transverse plane

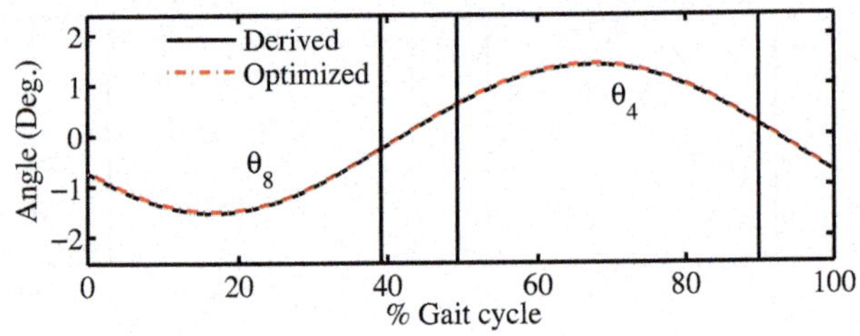

(c) Leg angle in the frontal plane

Figure 5. Derived frontal plane and transverse plane kinematic data for full gait cycle (toe-off to toe-off). Pelvis angles **(a, b)** in the frontal and transverse planes and leg angle **(c)** in the frontal plane. For **(c)** data from 0 to 39 % gait cycle indicate data for the swing leg and data from 50 to 89 % gait cycle indicate data for the stance leg in the SS phase.

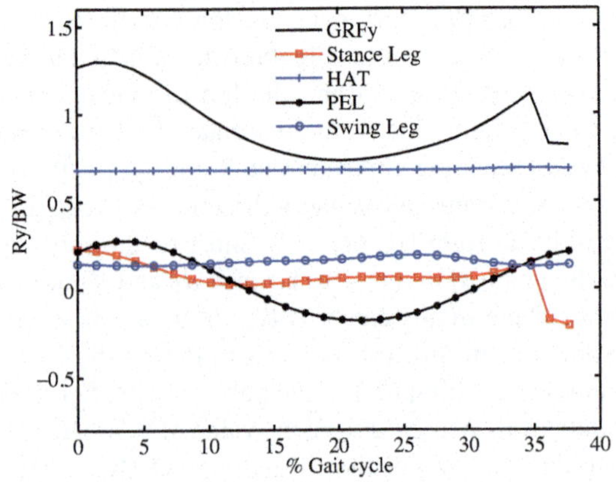

(a)

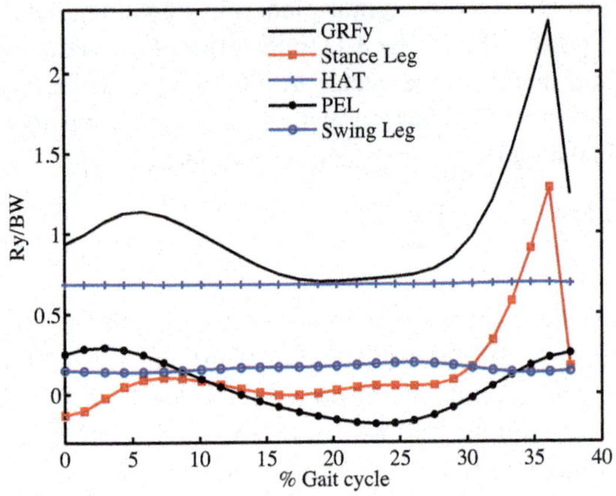

(b)

Figure 6. GRFy and stance leg, HAT, pelvis and swing leg contributions during SS phase as a fraction of body weight with **(a)** AFROS and **(b)** FROS. (Gait cycle starts at toe-off of the other leg, that is, at the beginning of SS.)

the initial data except for pelvis rotation in the frontal plane (Fig. 5a) around HC of right leg (39 % of gait cycle). Once a valid set of kinematics are obtained, these data are used to study the contribution of kinematics to the GRF.

3 Results

The segmental contributions of the stance leg (foot, shank and thigh), pelvis, swing leg and HAT (expressed as a fraction of body weight – BW) to GRFy and GRFx from the models using AFROS and FROS are shown in Figs. 6 and 7, respectively. Between the models using AFROS and FROS, the contributions of the stance leg segments showed variation (Fig. 8) while the contributions from the segments of the swing leg remained similar as expected. The contributions of the segments of the swing leg to GRFy for the model that uses AFROS are shown in Fig. 9. The variations in the swing leg contributions to GRFy are much smaller, and the contributions are almost constant compared to the contributions of the stance leg and pelvis (Fig. 6). The contributions of the HAT to GRFy are also relatively constant (Fig. 6). The variable acceleration terms ($k_i \mathbf{E}_i$s) of $\boldsymbol{R}_i$ of the swing leg segments (Eqs. A13 to A15) are rendered negligible by the small values of the corresponding k_is. In the case of the HAT segment, the $k_i \mathbf{E}_i$ of the contribution (Eq. A12) is small due to minimal change in the HAT angle (Fig. 4d, HAT angle has a range of 4°). Hence, only the segment weights (that is, the gravity terms in Eqs. A12 to A15) of the HAT and swing leg segments contribute to GRFy. The difference between RMS values (Table 1) with all the DOF and without the DOF of the swing leg and HAT is approximately 0.01 BW for GRFy. This difference is negligible given that the GRFy varies around 1 BW in the SS phase of normal walking. Therefore, the DOFs of the swing leg and HAT, θ_7 to θ_{11}, can be eliminated from the equation for GRFy.

The contribution of the transverse plane rotation of the pelvis to GRFy (Eq. 10, Fig. 10) is zero, and the contribution of the frontal plane rotation of the pelvis to GRFy (Eq. 11, Fig. 10) is the same as when both the DOF are included. In

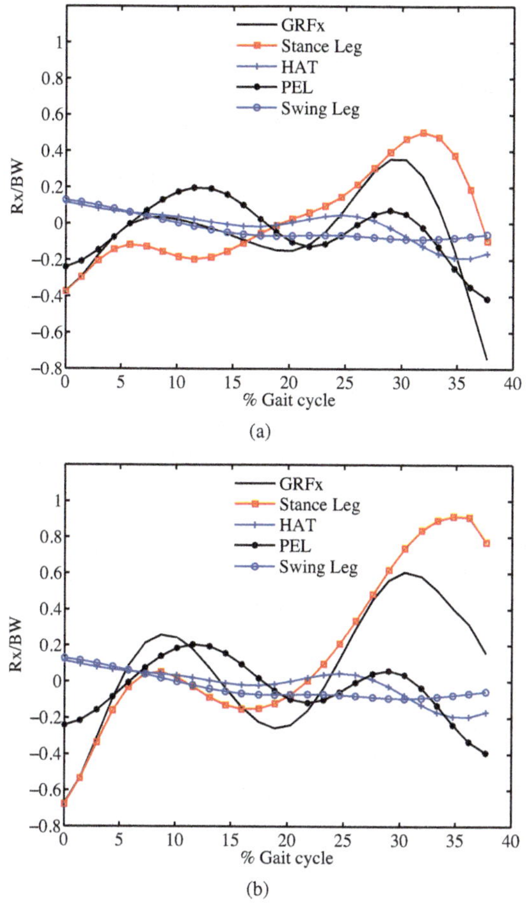

Figure 7. GRFx and stance leg, HAT, pelvis and swing leg contributions to GRFx during SS phase as a fraction of body weight with **(a)** AFROS and **(b)** FROS. (Gait cycle starts at toe-off of the other leg, that is, at the beginning of SS.)

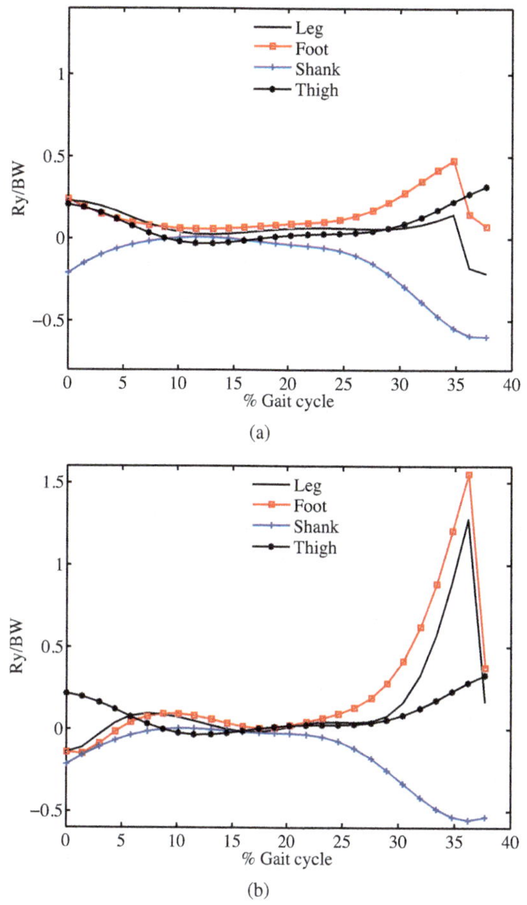

Figure 8. Contributions from foot, shank and thigh of the stance leg and sum of all three (Leg) to GRFy in SS phase as a fraction of body weight with **(a)** AFROS and **(b)** FROS models for the foot. (Gait cycle starts at toe-off of the other leg, that is, at the beginning of SS.)

other words, the entire contribution from the pelvis segment to GRFy comes from the frontal plane rotation θ_5. The RMS values also show a minimal change in GRFy (Table 1) when θ_6 is ignored, while the difference is considerable when θ_5 is ignored. Also, the contribution of the frontal plane rotation of the stance leg to GRFy (Fig. 11) is almost zero (no difference in the RMS value with all DOF and when $\theta_4 = 0$ in Table 1), confirming the dominance of the sagittal plane DOF of the leg. Therefore, the measurement of θ_4 and θ_6 is not required for the computation of GRFy.

In the case of GRFx, the pelvis and the stance leg are the most dominant contributors (Fig. 7). The contributions of the swing leg and HAT, however, are not constant as in the case of GRFy. Moreover, the GRFx obtained from the models gave an RMS error of 0.25 and 0.28 BW for AFROS and FROS, respectively, and the results showed a poor match with the experimental data (Fig. 13).

4 Discussion

4.1 Reduced model for GRFy

Most mathematical models include only sagittal plane kinematics (Selles et al., 2001; Ren et al., 2007; Martin and Schmiedeler, 2014) and avoid the pelvis link. Mochon and McMahon (1980) use a ballistic swing phase model but do not include the frontal plane rotation of the pelvis or ankle plantarflexion in their model. Their results for GRFy do not show the characteristic double hump, and they concluded that frontal plane rotation of the pelvis is probably necessary to capture that. Our result confirms that conclusion and shows that the frontal plane rotation of the pelvis is a major contributor to the characteristic shape of GRFy and must be included in a mathematical model for gait. One contradictory result is from Pandy and Berme (1989b) who, in their forward dynamics simulation, found that the pelvic list (frontal plane rotation of the pelvis) is not a dominant dynamical determinant of GRFy. However, their model was not anthropomorphic

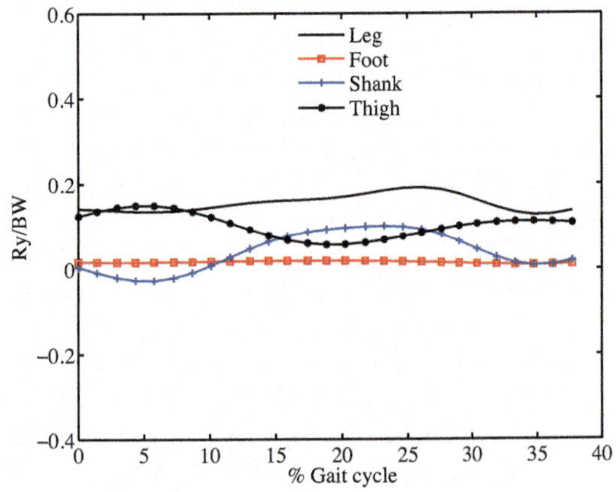

Figure 9. Contributions from foot, shank and thigh of the swing leg and sum of all three (Leg) to GRFy is SS phase as a fraction of body weight with AFROS.The results from the model with FROS are similar. (Gait cycle starts at toe-off of the other leg, that is, at the beginning of SS.)

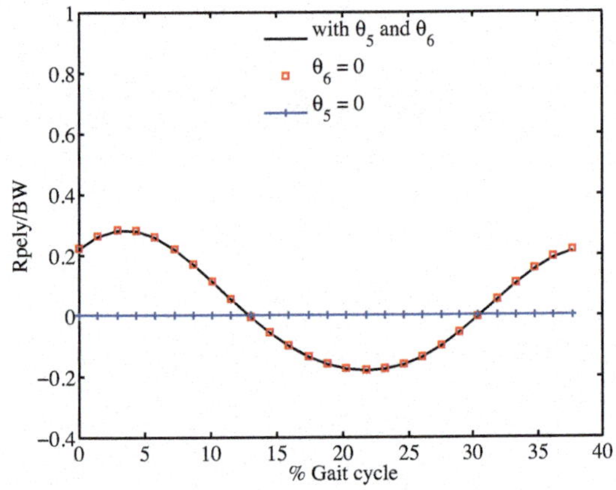

Figure 10. Contributions of pelvis DOF – θ_5 and θ_6 to GRFy in SS phase expressed as a fraction of body weight with AFROS. The results from the model with FROS are similar. (Gait cycle starts at toe-off of the other leg, that is, at the beginning of SS.)

because they neglected the effect of the foot and attached a massless damped spring between the hip and ankle of the stance leg.

The results of the segmental contributions showed that the sagittal plane DOFs of the stance leg and the frontal plane rotation of the pelvis link are the most dominant DOFs for the prediction of GRFy. Based on these observations, the analytical expression for GRFy in SS phase of gait can be reduced to

$$\begin{aligned}
\text{GRFy} &\approx Mg + Ms_x\left(\alpha_1 C_1 - \omega_1^2 S_1\right) \\
&+ M\left(k_1 + s_y\right)\left(-\alpha_1 S_1 - \omega_1^2 C_1\right) \\
&+ Mk_2\left(-S_2\alpha_2 - \omega_2^2 C_2\right) + Mk_3\left(-S_3\alpha_3 - \omega_3^2 C_3\right) \\
&+ Mk_4\left(\omega_5^2 S_5 - \alpha_5 C_5\right).
\end{aligned} \tag{14}$$

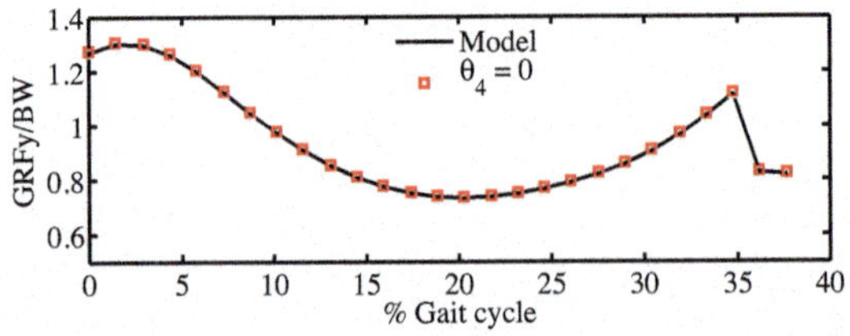

Figure 11. GRFy results for SS phase with all DOF and without the frontal plane rotation of the leg ($\theta_4 = 0$) with AFROS. The results from the model using FROS are similar. (Gait cycle starts at toe-off of the other leg, that is, at the beginning of SS.)

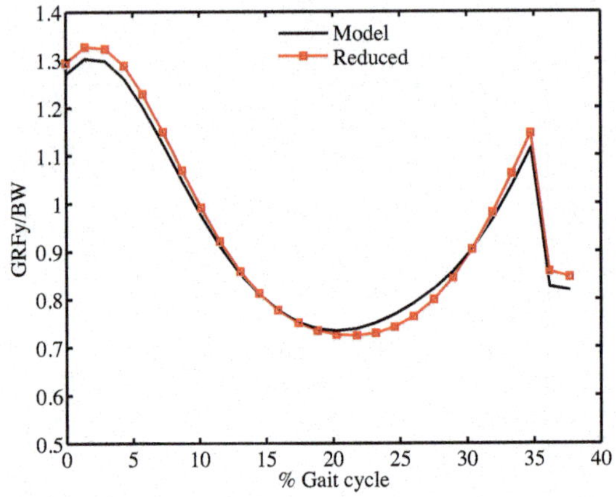

Figure 12. GRFy results when all the DOFs are used and with minimal DOFs in the model with AFROS. The results for the model using FROS are similar. (Gait cycle starts at toe-off of the other leg, that is, at the beginning of SS.)

When FROS is used for the foot-ground interface, the GRFy estimation requires measurement of four DOF (θ_1, θ_2, θ_3 and θ_5). When AFROS is used, the number of required DOF is reduced to three since $\theta_1 = \theta_2 - \theta_{\text{ank}}$, $\omega_1 = \omega_2$ and $\alpha_1 = \alpha_2$. The GRFy estimated using minimal kinematics match closely with the GRFy from the model with all the DOFs (Fig. 12). The RMS error between GRFy from the model with all DOF and with minimal kinematics are 0.0202 BW for FROS and 0.0198 BW for AFROS. Therefore, measurement of sagittal plane kinematics of the stance leg and the frontal plane rotation of the pelvis is sufficient to obtain a good approximation of GRFy.

4.2 GRFx for inverse dynamics analysis

The GRFx obtained from the model cannot be used in an inverse dynamics model as this would give large errors in the horizontal joint forces and joint moments. Research (Herr and Popovic, 2008) shows that the angular momentum regulation in gait influences the horizontal component of GRF. Koopman et al. (1995) ensured that the kinematics are such that the GRF vector passes though the COM and obtained GRFx and GRFz that matched well with the experimental

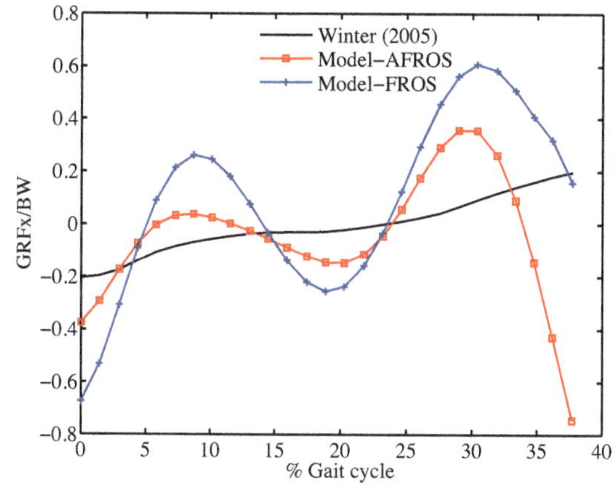

Figure 13. GRFx results for SS phase – experimental data and results using AFROS and FROS foot models. (Gait cycle starts at toe-off of the other leg, that is, at the beginning of SS.)

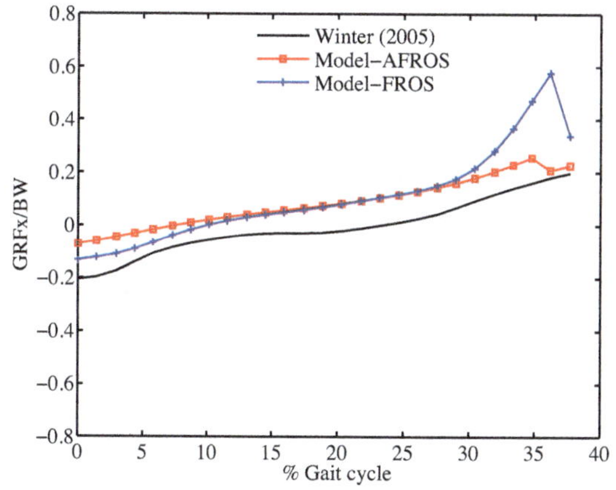

Figure 14. GRFx results for SS phase – experimental data and results for AFROS and FROS foot models using GRFy and the assumption that the GRF vector passes through the COM. (Gait cycle starts at toe-off of the other leg, that is, at the beginning of SS.)

data. Making this assumption, the GRFx at every instance is computed as

$$\text{GRFx} = (u_x/u_y)\,\text{GRFy} \tag{15}$$

where $\boldsymbol{u}$ is the vector from the contact point to the COM. The GRFx obtained using this assumption (Fig. 14) show a better match with the experimental data than the GRFx obtained from the model (Fig. 13). The GRFx obtained using this assumption gave an RMS error of 0.09BW and 0.13BW for AFROS and FROS models, respectively.

4.3 Foot roll over shapes

The input kinematic data for both the models are similar, and hence, the segmental contributions for all the segments showed similar patterns (Figs. 6 and 7) except for the contributions from the stance foot (Fig. 8). This discrepancy can

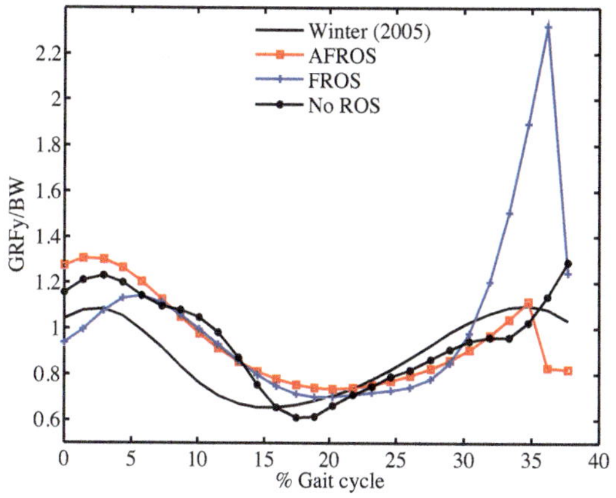

Figure 15. GRFy results for SS phase – experimental data (Winter, 2005) and results using rolling foot models and without using any ROS model. (Gait cycle starts at toe-off of the other leg, that is, at the beginning of SS.)

be attributed to the ROS models used. From Eq. (A6), we observe that the ankle acceleration has a direct correlation to GRF while the rest of the kinematics (**E**s) are scaled by a factor less than one. Therefore, the accuracy with which the foot model predicts the acceleration of the ankle joint plays a significant role in accurate GRF computation. The GRFy obtained using the FROS model reaches a very high value towards the end of the SS phase when compared to GRFy when the AFROS model was used (Fig. 15). This difference could be due to the fact that the ankle is assumed to be rigid for AFROS model but not for the FROS model. The circular ROS takes into account the plantarflexion of the ankle joint, and since the FROS model has both rolling motion and ankle plantarflexion, the combined effect has likely resulted in overprediction of GRFy towards the end of the SS phase (Fig. 15).

In order to understand the effect of using a ROS, the GRF is computed using the ankle acceleration directly from Winter (2005). In other words, the ${}^{st}\boldsymbol{A}_{ank}$ in Eq. (A6) is obtained directly from experimental data as opposed to the models with ROS in which case the ${}^{st}\boldsymbol{A}_{ank}$ is computed using Eq. (B1). In all three cases, there is a consistent overprediction of GRFy in the first half of the SS phase (Fig. 15). When FROS is used, the first peak of the GRFy curve is shifted forward, while the model with AFROS gives a peak at approximately the same % gait cycle as the experimental values. Even in the case where no ROS is used, the GRFy curve does not match the experimental values well (RMS error = 0.1399 BW), possibly due to measurement errors. The RMS error for GRFy is lower in the case where no ROS is used (0.1399 BW) when compared to RMS error when the FROS and AFROS models are used for the foot-ground interaction (Table 1). This result shows that direct measurement of the ankle kinematics is better than using a model for the foot. However, using a ROS to model the foot-ground

Table 1. RMS values of predicted GRFy (normalized by BW) with respect to the experimental values.

GRFy	All DOF	θ_7 to $\theta_{11} = 0$	$\theta_6 = 0$	$\theta_5 = 0$	$\theta_4 = 0$	Reduced model θ_4, θ_6 and θ_7 to $\theta_{11} = 0$
AFROS	0.1571	0.1664	0.1570	0.1777	0.1567	0.1660
FROS	0.3195	0.3308	0.3204	0.2840	0.3198	0.3322

interaction would enable comparison of a prosthetic foot to a normal foot or comparisons involving different prosthetic feet and varying alignments without collecting gait data each time. Since the purpose of this study is to extend the model to the analysis of the gait of a prosthesis user, the two ROS models are compared.

Both the FROS and AFROS are derived using the same COPx data. Although the ROS captures the geometric shape of the rolling foot, the dynamics associated with the FROS and AFROS are different, resulting in changes in the corresponding GRFy. In other words, the ROS models the displacement of the foot with respect to the ground more accurately than it models the time derivatives of the displacement. The discrepancy is less pronounced in the case of the AFROS model. The RMS error for GRFy from the two models (0.1571 BW with AFROS foot model and 0.3195 BW with FROS model in Table 1) show that the AFROS gives a better approximation than the FROS.

While the model's ability to accurately predict the timing of the first peak is sensitive to the foot model used (which in turn is dependent on the COP trajectory), multiple factors could be influencing the overprediction in the first half of the SS phase. The contribution from the HAT segment to GRFy is essentially constant and is dependent on its estimated mass. Any change in the HAT segment's estimated mass would increase or decrease the magnitude of GRFy without changing the shape of the peak. Also, the head, arms and trunk are modeled as a single segment. A model that includes more segments (that is, more DOFs) in the upper body could improve the results. All the segments are assumed to be rigid, and the overprediction could be due to the lack of compliance in the model. The use of a damped spring attached to the segments of the stance leg as in Pandy and Berme (1988) to account for the damping in the leg could potentially improve the GRF results.

5 Conclusions

The choice of segmental orientations with respect to a ground reference frame as the generalized DOF enables the expression of the GRF vector as a linear combination of the columns of segmental kinematics (**E** matrices). This mathematical form enables the study of the influence of the kinematics of a chosen segment or DOF, resulting in a reduced model for GRFy.

Results from this work show that the pelvic rotation in the frontal plane, the rolling foot model and the sagittal plane rotations of the segments of the stance leg are the major contributors to GRFy in the SS phase of gait. Of the two ROS, the AFROS provides a better approximation of GRFy. A minimal 3-DOF (rotations of the thigh and shank in the sagittal plane, and pelvic rotation in the frontal plane) anthropomorphic model can be used to approximate GRFy. Using the reduced model can enable gait studies in settings other than a gait lab by using minimal kinematic measurements and predetermined ROS characteristics (radius, arc length and center of ROS location) of the foot (Hansen et al., 2000).

A systematic reduction of the mathematical model to obtain a simplistic yet effective model is possible if a complete set of 3-D kinematic data and BSP are available. In this work, due to the lack of a full set of kinematic data, the frontal plane rotation of the legs was assumed, and the pelvis DOF were derived using the hip trajectory in the sagittal plane and the fact that gait is symmetric. However, this is not a limitation of the modeling procedure itself. While the lack of full data limited the validation of the model to the one set of gait data (Winter, 2005), the typical patterns of the joint angles and the GRF for normal walking are well-established (Inman et al., 1981; Perry, 1992; Srinivasan et al., 2008), which lends confidence to the model's validity.

A limitation of the model is the inaccuracy in predicting the GRFx directly from the model. Balance considerations, however, are known to play a role in influencing GRFx (Herr and Popovic, 2008; Firmani and Park, 2013). Using the GRFy from the reduced model with the assumption of angular momentum control for balance better approximates the GRFx, indicating that this condition is necessary for the model to be useful for inverse dynamic analysis.

Even though this work does not perfectly model normal gait, its usefulness is as a baseline for asymmetric gait. The form of the model enables easy extension to model asymmetric gait where the BSP and kinematics of the segments of the right and left sides vary, as in the case of prosthesis users. The use of ROS enables the analysis of the gait of prosthesis users with different prosthetic feet since experimental data for ROS of different prosthetic feet have been reported (Hansen et al., 2000, 2004; Hansen and Childress, 2004). The radius of the ROS and the center-of-ROS location are representative of the compliance and the alignment of the prosthetic limb, respectively (Hansen et al., 2000; Srinivasan et al., 2009). The effect of different prosthetic feet (modeled by different ROS), alignments and mass distributions of the prosthetic limb, etc., can be studied. The results can be compared using the developed normal gait model as the baseline model. This comparison would be more appropriate than comparison of a model for asymmetric gait to experimental normal gait data since the baseline model accounts for errors due to modeling assumptions.

The methodology described can be used with more sets of experimental data to determine statistically the minimum DOFs and ideal ROS for the model. The developed model can be extended to the double support (DS) phase by incorporating additional kinematic constraints in the optimization since some of the DOF will be dependent as a result of the closed loop kinematic chain formed by the limb segments in DS. Modeling DS would also require assumptions on the load sharing between the two legs (Ren et al., 2007; Koopman et al., 1995). Future work will include the study of the double support phase and sensitivity of joint moments and forces to perturbations in the kinematic data such as in the case of the gait of prosthesis users.

Appendix A: Derivation of analytical expressions for segmental contributions

The position vectors $\boldsymbol{r}_2$ through $\boldsymbol{r}_{14}$ (Fig. A1) expressed in the respective body-fixed coordinate systems are given by-

$$\boldsymbol{r}_2=\left[0 r_{2y} 0\right]', \ \boldsymbol{r}_3=\left[0 r_{3y} 0\right]', \ \boldsymbol{r}_4=\left[0 r_{4y} 0\right]',$$
$$\boldsymbol{r}_5=\left[0 r_{5y} 0\right]', \ \boldsymbol{r}_6=\left[0 r_{6y} 0\right]', \ \boldsymbol{r}_7=\left[0 r_{7z} 0\right]',$$
$$\boldsymbol{r}_8=\left[0 r_{8y} 0\right]', \ \boldsymbol{r}_9=\left[0 r_{9z} 0\right]', \ \boldsymbol{r}_{10}=\left[0 r_{10y} 0\right]',$$
$$\boldsymbol{r}_{11}=\left[0 r_{11y} 0\right]', \ \boldsymbol{r}_{12}=\left[0 r_{12y} 0\right]', \ \boldsymbol{r}_{13}=\left[0 r_{13y} 0\right]' \text{ and}$$
$$\boldsymbol{r}_{14}=\left[0 r_{14y} 0\right]'. \tag{A1}$$

The acceleration of the COM $\boldsymbol{A}_{\text{com}}$ can be written as

$$\begin{aligned}\boldsymbol{A}_{\text{com}} =& {}^{\text{st}}m_{\text{ft}}\,{}^{\text{st}}\boldsymbol{A}_{\text{ft}} + {}^{\text{st}}m_{\text{sk}}\,{}^{\text{st}}\boldsymbol{A}_{\text{sk}} + {}^{\text{st}}m_{\text{th}}\,{}^{\text{st}}\boldsymbol{A}_{\text{th}} \\ &+ m_{\text{pel}}\boldsymbol{A}_{\text{pel}} + m_{\text{hat}}\boldsymbol{A}_{\text{hat}} \\ &+ {}^{\text{sw}}m_{\text{th}}\,{}^{\text{sw}}\boldsymbol{A}_{\text{th}} + {}^{\text{sw}}m_{\text{sk}}\,{}^{\text{sw}}\boldsymbol{A}_{\text{sk}} + {}^{\text{sw}}m_{\text{ft}}\,{}^{\text{sw}}\boldsymbol{A}_{\text{ft}}\end{aligned} \tag{A2}$$

where m denotes the mass fraction and the subscripts "ft", "sk", "th", "pel" and "hat" indicate the foot, shank, thigh, pelvis and HAT segments, respectively, and the superscripts "sw" and "st" indicate swing and stance legs, respectively. Using Eq. (2) the accelerations of the COM of each segment can be expressed as

$$\begin{aligned}{}^{\text{st}}\boldsymbol{A}_{\text{ft}} =& {}^{\text{st}}\boldsymbol{A}_{\text{ank}} + {}^{\text{st}}\mathbf{E}_{\text{ft}}(-\boldsymbol{r}_2),\\ {}^{\text{st}}\boldsymbol{A}_{\text{sk}} =& {}^{\text{st}}\boldsymbol{A}_{\text{ank}} + {}^{\text{st}}\mathbf{E}_{\text{sk}}(-\boldsymbol{r}_3),\\ {}^{\text{st}}\boldsymbol{A}_{\text{th}} =& {}^{\text{st}}\boldsymbol{A}_{\text{ank}} + {}^{\text{st}}\mathbf{E}_{\text{sk}}(-\boldsymbol{r}_3+\boldsymbol{r}_4) + {}^{\text{st}}\mathbf{E}_{\text{th}}(-\boldsymbol{r}_5),\\ \boldsymbol{A}_{\text{hat}} =& {}^{\text{st}}\boldsymbol{A}_{\text{ank}} + {}^{\text{st}}\mathbf{E}_{\text{sk}}(-\boldsymbol{r}_3+\boldsymbol{r}_4) + {}^{\text{st}}\mathbf{E}_{\text{th}}(-\boldsymbol{r}_5+\boldsymbol{r}_6)\\ &+ \mathbf{E}_{\text{pel}}(-\boldsymbol{r}_7) + \mathbf{E}_{\text{hat}}\boldsymbol{r}_8,\\ {}^{\text{sw}}\boldsymbol{A}_{\text{th}} =& {}^{\text{st}}\boldsymbol{A}_{\text{ank}} + {}^{\text{st}}\mathbf{E}_{\text{sk}}(-\boldsymbol{r}_3+\boldsymbol{r}_4) + {}^{\text{st}}\mathbf{E}_{\text{th}}(-\boldsymbol{r}_5+\boldsymbol{r}_6)\\ &+ \mathbf{E}_{\text{pel}}(-\boldsymbol{r}_7+\boldsymbol{r}_9) + {}^{\text{sw}}\mathbf{E}_{\text{th}}(-\boldsymbol{r}_{10}),\\ {}^{\text{sw}}\boldsymbol{A}_{\text{sk}} =& {}^{\text{st}}\boldsymbol{A}_{\text{ank}} + {}^{\text{st}}\mathbf{E}_{\text{sk}}(-\boldsymbol{r}_3+\boldsymbol{r}_4) + {}^{\text{st}}\mathbf{E}_{\text{th}}(-\boldsymbol{r}_5+\boldsymbol{r}_6)\\ &+ \mathbf{E}_{\text{pel}}(-\boldsymbol{r}_7+\boldsymbol{r}) + {}^{\text{sw}}\mathbf{E}_{\text{th}}(-\boldsymbol{r}_{10}+\boldsymbol{r}_{11})\\ &+ {}^{\text{sw}}\mathbf{E}_{\text{sk}}(-\boldsymbol{r}_{12}) \text{ and}\\ {}^{\text{sw}}\boldsymbol{A}_{\text{ft}} =& {}^{\text{st}}\boldsymbol{A}_{\text{ank}} + {}^{\text{st}}\mathbf{E}_{\text{sk}}(-\boldsymbol{r}_3+\boldsymbol{r}_4) + {}^{\text{st}}\mathbf{E}_{\text{th}}(-\boldsymbol{r}_5+\boldsymbol{r}_6)\\ &+ \mathbf{E}_{\text{pel}}(-\boldsymbol{r}_7+\boldsymbol{r}_9) + {}^{\text{sw}}\mathbf{E}_{\text{th}}(-\boldsymbol{r}_{10}+\boldsymbol{r}_{11})\\ &+ {}^{\text{sw}}\mathbf{E}_{\text{sk}}(-\boldsymbol{r}_{12}+\boldsymbol{r}_{13}) + {}^{\text{sw}}\mathbf{E}_{\text{ft}}(-\boldsymbol{r}_{14}).\end{aligned} \tag{A3}$$

Expressing all $\mathbf{E}\boldsymbol{r}$s in Eq. (A3) in column vector notation (Eq. 4) and using in Eq. (A2) gives

$$\begin{aligned}\boldsymbol{A}_{\text{com}} =& {}^{\text{st}}\boldsymbol{A}_{\text{ank}} + k_1\begin{bmatrix} | \\ {}^{\text{st}}\mathbf{E}_{\text{ft}} \\ | \end{bmatrix}_{c2} + k_2\begin{bmatrix} | \\ {}^{\text{st}}\mathbf{E}_{\text{sk}} \\ | \end{bmatrix}_{c2}\\ &+ k_3\begin{bmatrix} | \\ {}^{\text{st}}\mathbf{E}_{\text{th}} \\ | \end{bmatrix}_{c2} + k_4\begin{bmatrix} | \\ \mathbf{E}_{\text{pel}} \\ | \end{bmatrix}_{c3} + k_5\begin{bmatrix} | \\ \mathbf{E}_{\text{hat}} \\ | \end{bmatrix}_{c2}\\ &+ k_6\begin{bmatrix} | \\ {}^{\text{sw}}\mathbf{E}_{\text{th}} \\ | \end{bmatrix}_{c2} + k_7\begin{bmatrix} | \\ {}^{\text{sw}}\mathbf{E}_{\text{sk}} \\ | \end{bmatrix}_{c2} + k_8\begin{bmatrix} | \\ {}^{\text{sw}}\mathbf{E}_{\text{ft}} \\ | \end{bmatrix}_{c2}\end{aligned} \tag{A4}$$

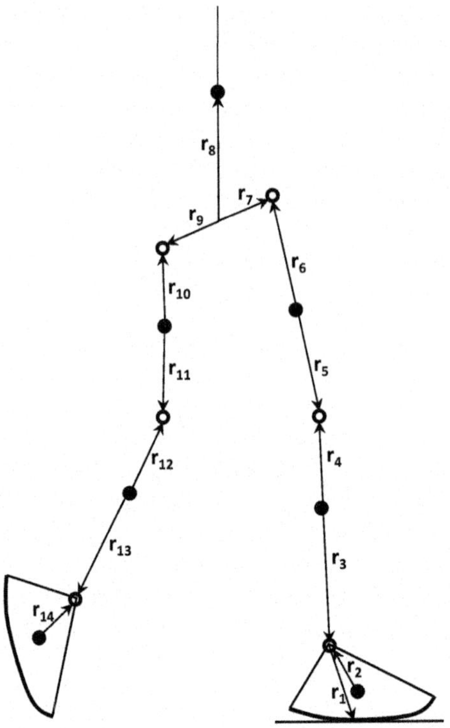

Figure A1. Link segment model showing the position vectors in each segment.

where

$$\begin{aligned}k_1 &= -{}^{\text{st}}m_{\text{ft}}r_{2y},\\ k_2 &= -\left(1-{}^{\text{st}}m_{\text{ft}}\right)r_{3y} + \left(1-{}^{\text{st}}m_{\text{ft}}-{}^{\text{st}}m_{\text{sk}}\right)r_{4y},\\ k_3 &= -\left(1-{}^{\text{st}}m_{\text{ft}}-{}^{\text{st}}m_{\text{sk}}\right)r_{5y} + \left(1-{}^{\text{st}}m_{\text{ft}}-{}^{\text{st}}m_{\text{sk}}-{}^{\text{st}}m_{\text{th}}\right)r_{6y},\\ k_4 &= -\left(1-{}^{\text{st}}m_{\text{ft}}-{}^{\text{st}}m_{\text{sk}}-{}^{\text{st}}m_{\text{th}}\right)r_{7z} + \left({}^{\text{sw}}m_{\text{ft}}+{}^{\text{sw}}m_{\text{sk}}+{}^{\text{sw}}m_{\text{th}}\right)r_{9z},\\ k_5 &= m_{\text{hat}}r_{8y},\\ k_6 &= -\left({}^{\text{sw}}m_{\text{ft}}+{}^{\text{sw}}m_{\text{sk}}+{}^{\text{sw}}m_{\text{th}}\right)r_{10y} + \left({}^{\text{sw}}m_{\text{ft}}+{}^{\text{sw}}m_{\text{sk}}\right)r_{11y},\\ k_7 &= -\left({}^{\text{sw}}m_{\text{ft}}+{}^{\text{sw}}m_{\text{sk}}\right)r_{12y} + {}^{\text{sw}}m_{\text{ft}}r_{13y} \text{ and}\\ k_8 &= -{}^{\text{sw}}m_{\text{ft}}r_{14y}.\end{aligned} \tag{A5}$$

Each $\mathbf{E}$ is a function of the kinematics of the corresponding segment. The terms ${}^{\text{st}}\boldsymbol{A}_{\text{ank}}$ and ${}^{\text{st}}\mathbf{E}_{\text{ft}}$ are functions of foot kinematics when the FROS model is used. However, when AFROS is used (since $\theta_1=\theta_2-\theta_{\text{ank}}$), ${}^{\text{st}}\boldsymbol{A}_{\text{ank}}$ and ${}^{\text{st}}\mathbf{E}_{\text{ft}}$ are functions of shank kinematics. Using Eq. (A2) in Newton's equation (Eq. 6) we get

$$\begin{aligned}\boldsymbol{GRF} &= M\boldsymbol{A}_{\text{com}} - M\boldsymbol{g}\\ &= M\,{}^{\text{st}}\boldsymbol{A}_{\text{ank}} + M\sum k\begin{bmatrix} | \\ \mathbf{E} \\ | \end{bmatrix} - M\boldsymbol{g}.\end{aligned} \tag{A6}$$

Therefore, GRF is given by

$$\begin{aligned}\boldsymbol{GRF} =& {}^{\text{st}}\boldsymbol{R}_{\text{ft}} + {}^{\text{st}}\boldsymbol{R}_{\text{sk}} + {}^{\text{st}}\boldsymbol{R}_{\text{th}} + \boldsymbol{R}_{\text{pel}}\\ &+ \boldsymbol{R}_{\text{hat}} + {}^{\text{sw}}\boldsymbol{R}_{\text{th}} + {}^{\text{sw}}\boldsymbol{R}_{\text{sk}} + {}^{\text{sw}}\boldsymbol{R}_{\text{ft}}\end{aligned} \tag{A7}$$

where

$$ {}^{st}\boldsymbol{R}_{ft} = M\,{}^{st}\boldsymbol{A}_{ank} + Mk_1 \begin{bmatrix} | \\ {}^{st}\mathbf{E}_{ft} \\ | \end{bmatrix}_{c2} - {}^{st}m_{ft}M\boldsymbol{g}, \tag{A8} $$

$$ {}^{st}\boldsymbol{R}_{sk} = Mk_2 \begin{bmatrix} | \\ {}^{st}\mathbf{E}_{sk} \\ | \end{bmatrix}_{c2} - {}^{st}m_{sk}M\boldsymbol{g}, \tag{A9} $$

$$ {}^{st}\boldsymbol{R}_{th} = Mk_3 \begin{bmatrix} | \\ {}^{st}\mathbf{E}_{th} \\ | \end{bmatrix}_{c2} - {}^{st}m_{th}M\boldsymbol{g}, \tag{A10} $$

$$ \boldsymbol{R}_{pel} = Mk_4 \begin{bmatrix} | \\ \mathbf{E}_{pel} \\ | \end{bmatrix}_{c3}, \tag{A11} $$

$$ \boldsymbol{R}_{hat} = Mk_5 \begin{bmatrix} | \\ \mathbf{E}_{hat} \\ | \end{bmatrix}_{c2} - m_{hat}M\boldsymbol{g}, \tag{A12} $$

$$ {}^{sw}\boldsymbol{R}_{th} = Mk_6 \begin{bmatrix} | \\ {}^{sw}\mathbf{E}_{th} \\ | \end{bmatrix}_{c2} - {}^{sw}m_{th}M\boldsymbol{g}, \tag{A13} $$

$$ {}^{sw}\boldsymbol{R}_{sk} = Mk_7 \begin{bmatrix} | \\ {}^{sw}\mathbf{E}_{sk} \\ | \end{bmatrix}_{c2} - {}^{sw}m_{sk}M\boldsymbol{g} \text{ and} \tag{A14} $$

$$ {}^{sw}\boldsymbol{R}_{ft} = Mk_8 \begin{bmatrix} | \\ {}^{sw}\mathbf{E}_{ft} \\ | \end{bmatrix}_{c2} - {}^{sw}m_{ft}M\boldsymbol{g}. \tag{A15} $$

The general form of the column vectors in the above equations are given below. The second column vector of the **E** matrix for the foot, shank and thigh segments is given by

$$ \begin{bmatrix} | \\ {}^{q}\mathbf{E}_p \\ | \end{bmatrix}_{c2} = \begin{bmatrix} \omega_z^2 S_z - \alpha_z C_x C_z + \omega_x \omega_z S_z C_z \\ -S_z \alpha_z - \left(\omega_x^2 + \omega_z^2\right) C_x C_z - \alpha_x S_x C_z \\ -\omega_x \omega_z S_z + \alpha_x C_x C_z - \omega_x^2 S_x C_z \end{bmatrix} \tag{A16} $$

where

$x=4$ and $z=1$	for	$q=$ st and $p=$ ft,
$x=4$ and $z=2$	for	$q=$ st and $p=$ sk,
$x=4$ and $z=3$	for	$q=$ st and $p=$ th,
$x=8$ and $z=9$	for	$q=$ sw and $p=$ th,
$x=8$ and $z=10$	for	$q=$ sw and $p=$ sk,
$x=8$ and $z=11$	for	$q=$ sw and $p=$ ft,
$\theta_x=0$ and $z=7$	for	$p=$ hat,

$S_i = \sin\theta_i$, $C_i = \cos\theta_i$, $\omega_i = \dot{\theta}_i$ and $\alpha_i = \ddot{\theta}_i$. The first column vector of the ${}^{st}\mathbf{E}_{ft}$ matrix for the foot segment is given by

$$ \begin{bmatrix} | \\ {}^{st}\mathbf{E}_{ft} \\ | \end{bmatrix}_{c1} = \begin{bmatrix} \omega_1^2 C_1 - \alpha_1 C_4 S_1 + \omega_4 \omega_1 S_1 S_4 \\ C_1 \alpha_1 - \left(\omega_1^2 + \omega_4^2\right) C_4 S_1 - \alpha_4 S_4 S_1 \\ -\omega_4 \omega_1 C_1 + \alpha_4 C_4 S_1 - \omega_4^2 S_4 S_1 \end{bmatrix} \tag{A17} $$

and third column vector of $\mathbf{E}_{pel}$ is given by

$$ \begin{bmatrix} | \\ \mathbf{E}_{pel} \\ | \end{bmatrix}_{c3} = \begin{bmatrix} -\omega_6^2 S_6 - \omega_5 \omega_6 S_5 C_6 + \alpha_6 C_5 C_6 \\ \omega_5 \omega_6 S_6 + \omega_5^2 S_5 C_6 - \alpha_5 C_5 C_6 \\ -\alpha_6 S_6 - \alpha_5 S_5 C_6 - \left(\omega_5^2 + \omega_6^2\right) C_5 C_6 \end{bmatrix}. \tag{A18} $$

Appendix B: Rolling foot kinematics

In the case of the rolling foot (analogous to the rolling coin problem (Greenwood, 1988)), the acceleration of the contact point, center of ROS and the ankle of the rolling foot are determined as

$$ \begin{aligned} \boldsymbol{A}_{cp} &= (\omega_z C_4)^2 R_{roll}\hat{u}, \\ \boldsymbol{A}_{o} &= \boldsymbol{A}_{cp} + \left({}^{st}\tilde{\alpha}_p + {}^{st}\tilde{\omega}_p\,{}^{st}\tilde{\omega}_p\right) R_{roll}\hat{u} \text{ and} \\ \boldsymbol{A}_{ank} &= \boldsymbol{A}_{o} + {}^{st}\mathbf{E}_p \boldsymbol{s} \end{aligned} \tag{B1} $$

where $\hat{u} = (0,\ C_4,\ S_4)'$ and R_{roll} is the radius of the roll-over shape, $p=$ ft and $z=1$ for FROS, $p=$ sk and $z=2$ for AFROS and $\boldsymbol{s}$ is the vector from the center of ROS to the ankle.

Edited by: J. Schmiedeler

References

Anderson, F. C. and Pandy, M. G.: Dynamic Optimization of Human Walking, J. Biomech. Eng., 123, 381–390, 2001.

Firmani, F. and Park, E. J.: Theoretical Analysis of the State of Balance in Bipedal Walking, J. Biomech. Eng., 135, 041003, doi:10.1115/1.4023698, 2013.

Greenwood, D. T.: Principles of Dynamics, Prentice Hall, New Jersey, 1988.

Hansen, A. H. and Childress, D. S.: Effects of shoe heel height on biologic roll-over characteristics during walking, J. Rehab. Res. Develop., 41, 547–554, 2004.

Hansen, A. H. and Childress, D. S.: Effects of adding weight to the torso on roll-over characteristics of walking, J. Rehab. Res. Develop., 42, 381–390, 2005.

Hansen, A. H., Childress, D. S., and Knox, E. H.: Prosthetic foot roll-over shapes with implications for alignment of trans-tibial prostheses, Prosthet. Orthot. Int., 24, 205–215, 2000.

Hansen, A. H., Childress, D. S., and Knox, E. H.: Roll-over shapes of human locomotor systems: effects of walking speed, Clin. Biomech., 19, 407–414, 2004.

Hatze, H.: A comprehensive model for human motion simulation and its application to the take-off phase of the long jump, J. Biomech., 14, 135–142, 1981.

Herr, H. and Popovic, M.: Angular momentum in human walking, J. Exp. Biol., 211, 467–81, 2008.

Inman, V. T., Ralston, H. J., and Todd, F.: Human Walking, Williams & Wilkins, Baltimore, London, 1981.

Ju, M. S. and Mansour, J. M.: Simulation of the double limb support phase of human gait, J. Biomech. Eng., 110, 223–229, 1988.

Kane, T. R., Likins, P. W., and Levinson, D. A.: Spacecraft Dynamics, McGraw Hill, Ithaca, New York, 1983.

Koopman, B., Grootenboer, H. J., and deJongh, H. J.: An inverse dynamics model for the analysis, reconstruction and prediction of bipedal walking, J. Biomech., 28, 1369–1376, 1995.

Kuo, A. D.: The six determinants of gait and the inverted pendulum analogy: A dynamic walking perspective, Human Move. Sci., 26, 617–656, 2007.

Martin, A. E. and Schmiedeler, J. P.: Predicting human walking gaits with a simple planar model, J. Biomech., 47, 1416–1421, 2014.

Mccaw, S. T. and Devitat, P.: Errors in alignment of center of pressure and foot coordinates affect predicted lower extremity torques, J. Biomech., 28, 985–988, 1995.

McGeer, T.: Passive Dynamic Walking, Int. J. Robot. Res., 9, 62–82, 1990.

Mochon, S. and McMahon, T. A.: Ballistic walking: an improved model, Math. Biosci., 52, 241–260, 1980.

Oh, S. E., Choi, A., and Hwan, J.: Prediction of ground reaction forces during gait based on kinematics and a neural network model, J. Biomech., 46, 2372–2380, 2013.

Onyshko, S. and Winter, D. A.: A mathematical model for the dynamics of human locomotion, J. Biomech., 13, 361–368, 1980.

Pàmies-Vilà, R., Font-Llagunes, J. M., Cuadrado, J., and Alonso, F. J.: Analysis of different uncertainties in the inverse dynamic analysis of human gait, Mech. Mach. Theory, 58, 153–164, 2012.

Pandy, M. G. and Berme, N.: Synthesis of human walking: A planar model for single support, J. Biomech., 21, 1053–1060, 1988.

Pandy, M. G. and Berme, N.: Quantitative assessment of gait determinants during single stance via a three-dimensional model – Part 2. Pathological gait, J. Biomech., 22, 725–733, 1989a.

Pandy, M. G. and Berme, N.: Quantitative assessment of gait determinants during single stance via a three-dimensional model – Part 1. Normal gait, J. Biomech., 22, 717–724, 1989b.

Pearsall, D. J. and Costigan, P. A.: The effect of segment parameter error on gait analysis results, Gait Posture, 9, 173–83, 1999.

Perry, J.: Gait Analysis Normal and Pathological Function, SLACK Inc., Thorofare, New Jersey, 1992.

Pillet, H., Bonnet, X., Lavaste, F., and Skalli, W.: Evaluation of force plate-less estimation of the trajectory of the centre of pressure during gait. Comparison of two anthropometric models, Gait Posture, 31, 147–152, 2010.

Rao, G., Amarantini, D., Berton, E., and Favier, D.: Influence of body segments' parameters estimation models on inverse dynamics solutions during gait, J. Biomech. 39, 1531–1536, 2006.

Reinbolt, J. A., Haftka, R. T., Chmielewski, T. L., and Fregly, B. J.: Are patient-specific joint and inertial parameters necessary for accurate inverse dynamics analyses of gait?, IEEE T. Biomed. Eng., 54, 782–793, 2007.

Ren, L., Jones, R. K., and Howard, D.: Predictive modelling of human walking over a complete gait cycle, J. Biomech., 40, 1567–1574, 2007.

Riemer, R., Hsiao-Wecksler, E. T., and Zhang, X.: Uncertainties in inverse dynamics solutions: a comprehensive analysis and an application to gait, Gait Posture, 27, 578–588, 2008.

Saunders, J. B., Inman, V. T., and Eberhart, H. D.: The major determinants in normal and pathological gait, J. Bone Joint Surge., 35A, 543–558, 1953.

Selles, R. W., Bussmann, J. B., Wagenaar, R. C., and Stam, H. J.: Comparing predictive validity of four ballistic swing phase models of human walking, J. Biomech., 34, 1171–1177, 2001.

Shabana, A. A.: Computational Dynamics, Wiley, New York, 2010.

Silva, M. P. and Ambrosio, J. A. C.: Sensitivity of the results produced by the inverse dynamic analysis of a human stride to perturbed input data, Gait Posture, 19, 35–49, 2004.

Srinivasan, S., Raptis, I. A., and Westervelt, E. R.: Low-dimensional sagittal plane model of normal human walking, J. Biomech. Eng. – T. ASME, 130, 051017, doi:10.1115/1.2970058, 2008.

Srinivasan, S., Westervelt, E. R., and Hansen, A. H.: A Low-Dimensional Sagittal-Plane Forward-Dynamic Model for Asymmetric Gait and Its Application to Study the Gait of Transtibial Prosthesis Users, J. Biomech. Eng – T. ASME, 131, 031003, doi:10.1115/1.3002757, 2009.

Thornton-Trump, A. and Daher, R.: The prediction of reaction forces from gait data, J. Biomech., 8, 173–178, 1975.

Vanderpool, M. T., Collins, S. H., and Kuo, A. D.: Ankle fixation need not increase the energetic cost of human walking, Gait Posture, 28, 427–433, 2008.

Wang, C. C. and Hansen, A. H.: Response of able-bodied persons to changes in shoe rocker radius during walking: Changes in ankle kinematics to maintain a consistent roll-over shape, J. Biomech., 43, 2288–2293, 2010.

Winiarski, S. and Rutkowska-Kucharska, A.: Estimated ground reaction force in normal and pathological gait, Acta Bioeng. Biomech., 11, 53–60, 2009.

Winter, D. A.: The biomechanics and motor control of human gait: normal, elderly and pathological, University of Waterloo Press, Waterloo, 2005.

Zarrugh, M.: Kinematic prediction of intersegment loads and power at the joints of the leg in walking, J. Biomech., 14, 713–725, 1981.

15

Sub-modeling approach for obtaining structural stress histories during dynamic analysis

T. T. Rantalainen, A. M. Mikkola, and T. J. Björk

Lappeenranta University of Technology, Department of mechanical engineering, Lappeenranta, Finland

Correspondence to: T. T. Rantalainen (tuomas.rantalainen@lut.fi)

Abstract. Modern machine structures are often fabricated by welding. From a fatigue point of view, the structural details and especially, the welded details are the most prone to fatigue damage and failure. Design against fatigue requires information on the fatigue resistance of a structure's critical details and the stress loads that act on each detail. Even though, dynamic simulation of flexible bodies is already current method for analyzing structures, obtaining the stress history of a structural detail during dynamic simulation is a challenging task; especially when the detail has a complex geometry. In particular, analyzing the stress history of every structural detail within a single finite element model can be overwhelming since the amount of nodal degrees of freedom needed in the model may require an impractical amount of computational effort. The purpose of computer simulation is to reduce amount of prototypes and speed up the product development process. Also, to take operator influence into account, real time models, i.e. simplified and computationally efficient models are required. This in turn, requires stress computation to be efficient if it will be performed during dynamic simulation. The research looks back at the theoretical background of multibody simulation and finite element method to find suitable parts to form a new approach for efficient stress calculation. This study proposes that, the problem of stress calculation during dynamic simulation can be greatly simplified by using a combination of Floating Frame of Reference Formulation with modal superposition and a sub-modeling approach. In practice, the proposed approach can be used to efficiently generate the relevant fatigue assessment stress history for a structural detail during or after dynamic simulation. Proposed approach is demonstrated in practice using one numerical example. Even though, examples are simplified the results show that approach is applicable and can be used as proposed.

1 Introduction

Multibody dynamic simulation represents a remarkable improvement in predicting machine performance compared to previous methods, which are often based on very simplified analytical models in combination with large safety margins for model uncertainties or empirical testing. With the development of more computationally powerful computers over recent decades, dynamic simulation has increasingly become a standard tool for comprehensive machine design. Furthermore, this continuously increasing computational power, combined with the availability of increasingly advanced codes, offers more possibilities for the dynamic analysis of complex structures.

Using dynamic simulation to determine stresses for flexible bodies also provides an opportunity to predict the fatigue life of a structure in practical applications. Currently in engineering applications, the prediction of fatigue life for a structure is thought to be a separate stage of design due its complexity and computational burden. If fatigue life prediction is implemented efficiently in multibody codes, it could be used throughout all stages of design without dedicating discrete designing stages to it. These two scenarios of the use of the fatigue analysis are depicted in Fig. 1.

Instead of being an explicit design step, fatigue analysis can be integrated into the design process (Fig. 1 right). This makes fatigue analysis an integrated part of the design process, and from a design point of view, fatigue analysis is

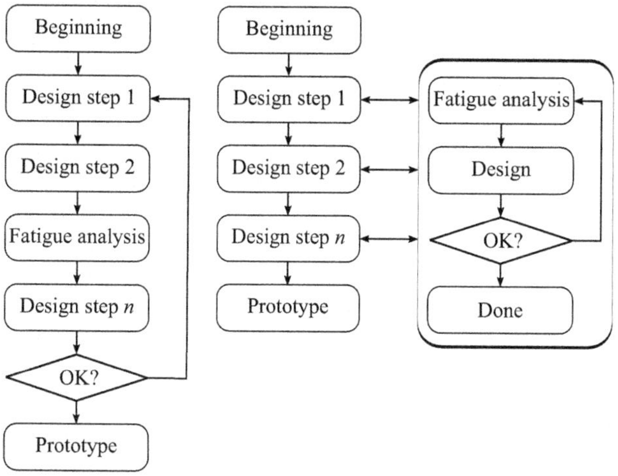

Figure 1. Workflow of a design process with integrated fatigue analysis.

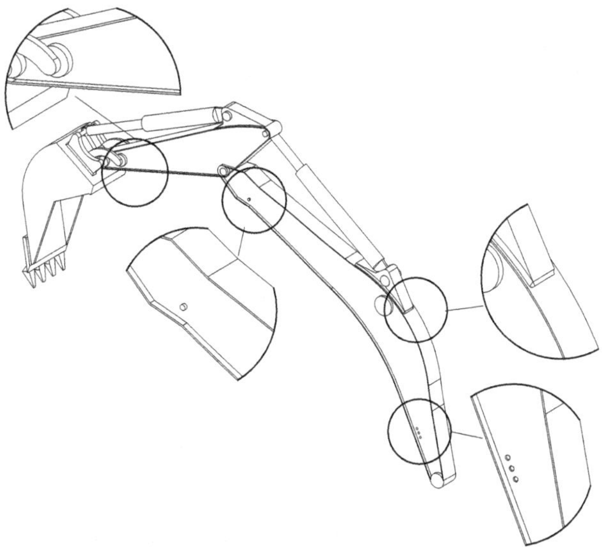

Figure 2. A crane with numerous discontinuities, which can be prone to fatigue.

taken into account automatically. To perform mutual integration, efficient methods for performing fatigue analysis are required.

Dynamically loaded structures such as booms are typically manufactured by welding. By definition, fluctuating loads result in fatigue damage to a structure. Without post-weld treatment, welds are prone to fatigue (Maddox, 1991; Haagensen and Maddox, 2011). In dynamic analysis, structural details that do not affect the dynamic behavior of a structure are usually neglected. Typically, this means that stress raisers are not analyzed in dynamic simulation even though they might be a possible location for fatigue. If treated separately, more work is required in fatigue analysis (Fig. 1 left). In Fig. 2, an excavator crane is depicted to illustrate the possible locations that might be vulnerable to fatigue and should be taken into consideration when predicting the fatigue life of a crane.

For stress history prediction, a welded structure requires particular attention. In addition, as computational power increases, there is increasing interest in predicting fatigue life for dynamically loaded structures. Even though computational capabilities have been greatly increased, the need remains for using coordinate reduction methods, especially in the case of large and complicated structures. Multibody dynamic simulation can be used to analyze the dynamics of complex mechanical systems. It can also be used to determine dynamic loads or even stresses for further fatigue analysis. Obtaining stress data for fatigue analysis from multibody system simulation is a main component of this work.

A new strategy developed in the proposed method combines three commonly used engineering approaches: the finite element method, the floating frame of reference formulation, and the sub-modeling approach. This strategy can be used when carrying out dynamic simulation to obtain, efficiently, the stress history of an arbitrary notch. In literature it is shown that computer simulations can be used for fault diagnosis (Korkealaakso et al., 2006). Moreover, this stress history can further be used as initial data for the fatigue analysis. In addition to efficient stress history calculation, the proposed combined methods strategy offers other beneficial features, such as the possibility of attaching structural details to a simulation model without modifying the simulation model. Furthermore, since structural detail does not affect the overall behavior of a simulation, the number of details can be changed arbitrary and separately.

To approximate, efficiently, the fatigue life of a structure or structural detail, some simplifying assumptions must be made. These assumptions decrease the quality of fatigue life estimation, especially in high frequency loading. For this work, a linear strain-stress relationship is assumed. In addition, presumptions have been made related to cumulative damage counting. For example, since fatigue damage can be linearly accumulated, structural failure is predicted when the entire available fatigue life is consumed. The focus here is on linear deformation. The strategy considers plastic deformation only local to the tip of a crack. Material nonlinearities fall outside the scope of this work. However, fatigue life estimation gives useful information when comparing alternative structures.

2 The floating frame of reference formulation

The floating frame of reference formulation is typically applicable to systems with large displacements and rotations and small deformations, even though the method can be used for large deformation problems (Wallrapp and Wiedemann, 2003). The method is based on describing the deformations of a flexible body with respect to a frame of reference. With

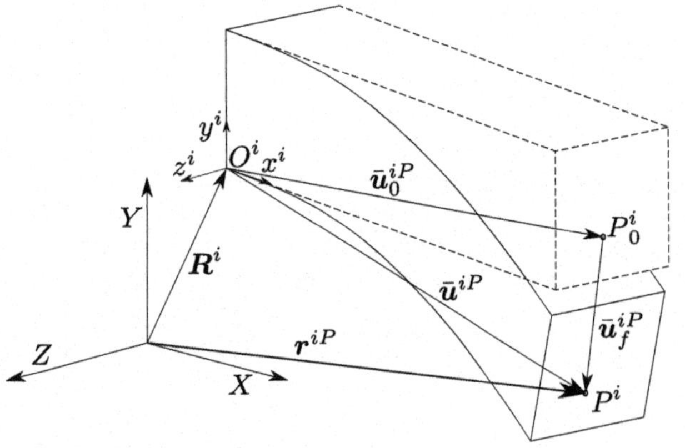

Figure 3. Position of particle P^i in a flexible body.

the frame of reference, large displacements and rotations can be described. The deformations of a flexible body in relation to its frame of reference can be described with a number of methods, but in the present study, deformation can be depicted with orthonormalized Craig-Bampton modes (Craig Jr. and Bampton, 1968). In it, eigenmodes are used together with static modes to describe structural deformation. The modes can be obtained using a finite element method.

The formulation separates the deformation of the body from the reference motion. The dynamics of the body can be generated using reference motion that is superposed by the deformation of the body. The interaction between the reference motion and deformation is accounted for with a mass matrix and quadratic velocity vector. This permits even mass distribution and inertia modeling (Shabana, 1998).

Figure 3 illustrates the position of particle P^i within a flexible body i. In the undeformed state, the position of the particle in the local reference frame of the body can be determined by vector $\bar{\boldsymbol{u}}_0^{iP}$.

As body i is deformed (Fig. 3), the position of particle P^i changes according to the vector $\bar{\boldsymbol{u}}_f^{iP}$. The global reference frame is represented (Fig. 3) using Cartesian coordinates X, Y, and Z. Respectively, the local reference frame of body i consists of coordinates x^i, y^i, and z^i. Therefore, the location of the particle in a global reference frame can be defined with the vector $\boldsymbol{r}^{iP}$ as follows:

$$\mathbf{r}^{iP} = \boldsymbol{R}^i + \mathbf{A}^i\left(\bar{\boldsymbol{u}}_0^{iP} + \bar{\boldsymbol{u}}_f^{iP}\right), \tag{1}$$

where $\boldsymbol{R}^i$ is translation of the local reference coordinate system of body i in the global coordinate system, and matrix $\mathbf{A}^i$ is the rotation matrix, which is expressed here in terms of four Euler parameters.

In Eq. (1), $\bar{\boldsymbol{u}}_0^{iP}$ is the position vector of particle P^i in the local reference coordinate system for the undeformed configuration, and $\bar{\boldsymbol{u}}_f^{iP}$ is the position vector in the local reference coordinate system for the deformed configuration. The behavior of the vector $\bar{\boldsymbol{u}}_f^{iP}$ can be described with a series of parallel differential equations. By separating the variables, if possible, the equation results in an infinite series that describes the deformations. For computational reasons, the infinite series cannot be applied to the analysis of flexible bodies. In practical application, the vector $\bar{\boldsymbol{u}}_f^{iP}$ is described using the finite element method.

The rotation matrix $\mathbf{A}^i$ using Euler parameters can be formulated as follows.

$$\mathbf{A}^i = \begin{bmatrix} \frac{1}{2}-\left(\theta_2^i\right)^2-\left(\theta_3^i\right)^2 & \theta_1^i\theta_2^i-\theta_0^i\theta_3^i & \theta_1^i\theta_3^i+\theta_0^i\theta_2^i \\ \theta_1^i\theta_2^i+\theta_0^i\theta_3^i & \frac{1}{2}-\left(\theta_1^i\right)^2-\left(\theta_3^i\right)^2 & \theta_2^i\theta_3^i-\theta_0^i\theta_1^i \\ \theta_1^i\theta_3^i-\theta_0^i\theta_2^i & \theta_2^i\theta_3^i+\theta_0^i\theta_1^i & \frac{1}{2}-\left(\theta_1^i\right)^2-\left(\theta_2^i\right)^2, \end{bmatrix} \tag{2}$$

where θ_0^i, θ_1^i, θ_2^i, and θ_3^i are Euler parameters. In this study, Euler parameters are used to avoid singular conditions, which can occur when Euler or Bryant angles are used (Nikravesh and Chung, 1982). The following mathematical constraint must be taken into consideration when Euler parameters are applied.

$$\left(\theta_0^i\right)^2+\left(\theta_1^i\right)^2+\left(\theta_2^i\right)^2+\left(\theta_3^i\right)^2=1 \tag{3}$$

The first time-derivative of the Euler parameters $\dot{\boldsymbol{\theta}}$ and the angular velocity vector $\bar{\omega}^i$ has the following linear connection.

$$\bar{\omega}^i = \bar{\mathbf{G}}^{iT}\dot{\boldsymbol{\theta}}^i \tag{4}$$

Matrices $\mathbf{A}^i$, and $\bar{\mathbf{G}}^i$ depend on the selected generalized coordinates. Using Euler parameters, the matrix $\bar{\mathbf{G}}^i$ can be expressed as.

$$\bar{\mathbf{G}}^i = \begin{bmatrix} -\theta_1^i & \theta_0^i & \theta_3^i & -\theta_2^i \\ -\theta_2^i & -\theta_3^i & \theta_0^i & \theta_1^i \\ -\theta_3^i & \theta_2^i & -\theta_1^i & \theta_0^i \end{bmatrix}, \tag{5}$$

where $\bar{\mathbf{G}}^i$ is the transformation matrix that relates the angular velocity $\bar{\omega}^i$ of a body and the first time derivative of the Euler parameter. Using the model reduction method, the position of an arbitrary particle P^i in the global coordinate system can be expressed as.

$$\boldsymbol{r}^{iP} = \boldsymbol{R}^i + \mathbf{A}^i\left(\bar{\boldsymbol{u}}_0^{iP} + \boldsymbol{\Phi}_R^{iP}\boldsymbol{p}^i\right) \tag{6}$$

Equation (6) is determined using a collection of modes. The vector $\bar{\boldsymbol{u}}_0^{iP}$ and the modal matrix $\boldsymbol{\Phi}_R^{iP}$ are constant with time. Consequently, they only need to be calculated once, at the beginning of the simulation.

The finite element model often consists of a large number of nodal degrees of freedom, and the use of large finite element models to describe flexibility may be computationally inefficient. For this reason, the floating frame of reference formulation is often used together with a modal reduction method in which the deformation is described with structural modes. Modes may be the presumed forms of deformation, but most often, they are eigenmodes of structural vibrations. The eigenmodes can be obtained from a finite element model of the structure. By employing a modal reduction method, the deformation vector $\bar{\boldsymbol{u}}_f^{iP}$ can be expressed in modal coordinates with a shape matrix.

$$\bar{\boldsymbol{u}}_f^{iP} = \boldsymbol{\Phi}_R^{iP}\, \boldsymbol{p}^i \tag{7}$$

$\boldsymbol{\Phi}_R^{iP}$ is the modal matrix whose columns describe the translation of particle P^i within the assumed deformation modes of the flexible body i (Shabana, 2005), and $\boldsymbol{p}^i$ is a vector of elastic coordinates. In general, the complete modal matrix $\boldsymbol{\Phi}^{iP}$ for body i obtained from the finite element method contains the location translation and orientation of particle P^i. In multibody dynamics, the modal matrix should separate translation and orientation descriptions into their own components.

The orthogonal shape matrix can be formulated from the eigenmodes of the body. Typically, the shape superposition technique yields acceptably accurate results even though only a few differential equations are applied. By approximating Eq. (7) with an n_p number of modal coordinates, the deformation vector $\bar{\boldsymbol{u}}_f^{iP}$ for a particle P^i can be written as follows.

$$\bar{\boldsymbol{u}}_f^{iP} \approx \sum_{j=1}^{n_p} \varphi_{R,j}^i p_j^i = \boldsymbol{\Phi}_R^{iP} \boldsymbol{p}^i \tag{8}$$

p_j^i is one modal coordinate in the modal coordinate vector that corresponds to the modal shape j. Rotations due to body deformation do not have any direct use in the floating frame of reference formulation, and therefore they are usually ignored. However, rotation modes can be used in the description of constraint equations applied to rotational degrees of freedom (Korkealaakso et al., 2009). Rotational modes are used here to connect sub-models to large-scale models. With the rotational modal matrix $\boldsymbol{\Phi}_\theta^{iP}$, the rotation change $\boldsymbol{\varepsilon}_f^{iP}$ resulting from deformation can be approximated as follows.

$$\boldsymbol{\varepsilon}_f^{iP} \approx \sum_{j=1}^{n_p} \varphi_{\theta,j}^i p_j^i \tag{9}$$

A rotation matrix $\mathbf{A}_f^{iP}$ that describes orientation due to deformation at the location of particle P^i with respect to the reference frame can be composed like this.

$$\mathbf{A}_f^{iP} = \mathbf{I} + \tilde{\boldsymbol{\varepsilon}}_f^{iP} \tag{10}$$

$\mathbf{I}$ is (3×3) identity matrix. The ~ symbol above a variable indicates the skew-symmetric form. The orientation at the location of particle P^i within the frame of reference can be expressed as follows.

$$\bar{\boldsymbol{v}}_f^{iP} = \mathbf{A}_f^{iP} \bar{\boldsymbol{v}}_0^{iP} \tag{11}$$

$\bar{\boldsymbol{v}}_0^{iP}$ is the orientation of the location of particle P^i in the undeformed state. The description of the rotation of the node has no direct use in formulating equations of motion, but the rotation may be needed to describe the constraint equations that are applied to rotational degrees of freedom. Taking into account the relation between first time derivative of Euler angels and angular velocity of the body i, Eq. (4), the generalized velocity vector of the flexible body i can be written as follows.

$$\dot{\boldsymbol{q}}^i = \begin{bmatrix} \dot{\boldsymbol{R}}^{i\mathrm{T}} & \bar{\boldsymbol{\omega}}^{i\mathrm{T}} & \dot{\boldsymbol{p}}^{i\mathrm{T}} \end{bmatrix} \tag{12}$$

The velocity of particle P^i can be determined by differentiating Eq. (1) with respect to time as follows.

$$\dot{\boldsymbol{r}}^{iP} = \begin{bmatrix} \mathbf{I} & -\mathbf{A}^i \bar{\mathbf{G}}^i \tilde{\bar{\boldsymbol{u}}}_f^{iP} & \mathbf{A}^i \boldsymbol{\Phi}_R^{iP} \end{bmatrix} \begin{bmatrix} \dot{\boldsymbol{R}}^i \\ \dot{\boldsymbol{\theta}}^i \\ \dot{\boldsymbol{p}}^i \end{bmatrix} \tag{13}$$

Note the vector, in the right hand of Eq. (13), describes the velocity of the generalized coordinates of a flexible body i. Differentiating the velocity of a particle Eq. (13) with respect to time, the acceleration of a particle can be written in this manner.

$$\begin{aligned} \ddot{\boldsymbol{r}}^{iP} = & \begin{bmatrix} \mathbf{I} & -\mathbf{A}^i \tilde{\bar{\boldsymbol{u}}}^{iP} \bar{\mathbf{G}}^i \mathbf{A}^i & \boldsymbol{\Phi}_R^{iP} \boldsymbol{R}^{iP} \end{bmatrix} \begin{bmatrix} \ddot{\boldsymbol{R}}^i \\ \ddot{\boldsymbol{\theta}}^i \\ \ddot{\boldsymbol{p}}^i \end{bmatrix} \\ & + \begin{bmatrix} \mathbf{0} & -\mathbf{A}^i \tilde{\bar{\boldsymbol{\omega}}}^i \tilde{\bar{\boldsymbol{u}}}^{iP} \bar{\mathbf{G}}^i & 2\mathbf{A}^i \tilde{\bar{\boldsymbol{\omega}}}^i \boldsymbol{\Phi}_R^{iP} \end{bmatrix} \begin{bmatrix} \dot{\boldsymbol{R}}^i \\ \dot{\boldsymbol{\theta}}^i \\ \dot{\boldsymbol{p}}^i \end{bmatrix}, \end{aligned} \tag{14}$$

where $\ddot{\boldsymbol{R}}^i$, $\ddot{\boldsymbol{\theta}}^{iE}$, and $\ddot{\boldsymbol{p}}^i$ are accelerations of translational coordinates, Euler parameters, and modal coordinates of body i.

According to the D'Alembert principle, inertial forces can be treated as external forces, thus forces of the body i can be written as follows.

$$\boldsymbol{F}^i = \int_{V^i} \rho^i \ddot{\boldsymbol{r}}^{iP} \mathrm{d}V^i, \tag{15}$$

where ρ^i is density and V^i is the volume of a body i, respectively. The virtual work done by the inertial forces can be represented as.

$$\delta W^i = \int_{V^i} \rho^i \delta \boldsymbol{r}^{iP\mathrm{T}} \ddot{\boldsymbol{r}}^{iP} \mathrm{d}V^i \tag{16}$$

The virtual displacement of the position vector $\delta \boldsymbol{r}^{iP}$ can be expressed as.

$$\delta \boldsymbol{r}^{iP} = \frac{\delta \boldsymbol{r}^{iP}}{\delta \boldsymbol{q}^i} \delta \boldsymbol{q}^i = \begin{bmatrix} \mathrm{I} & -\mathbf{A}^i \tilde{\bar{\boldsymbol{u}}}^{iP} \bar{\mathbf{G}}^i \mathbf{A}^i & \boldsymbol{\Phi}_R^{iP} \end{bmatrix} \delta \boldsymbol{q}^i, \tag{17}$$

By substituting the virtual displacement Eq. (17) into the equation for virtual work (Eq. 16) and separating terms, the following equation can be obtained.

$$\delta W^i = \delta \boldsymbol{q}^i \left[\mathbf{M}^i \ddot{\boldsymbol{q}}^i + \boldsymbol{Q}^{iv} \right], \tag{18}$$

where $\delta \boldsymbol{q}^i$ is virtual change of the generalized coordinates, $\boldsymbol{Q}^{iv}$ is quadratic velocity vector of body i, and $\mathbf{M}^i$ is the mass matrix.

The virtual work of externally applied forces can be defined as.

$$\delta W^{ie} = \int_{V^i} \delta \boldsymbol{r}^{iP^{\mathrm{T}}} \boldsymbol{F}^{iP} \mathrm{d}V^i = \delta \boldsymbol{q}^{i^{\mathrm{T}}} \boldsymbol{Q}^{ie}, \tag{19}$$

where $\boldsymbol{F}^{iP}$ is externally applied force per unit volume, and $\boldsymbol{Q}^{ie}$ is the vector of generalized forces, which can be expressed as follows.

$$\boldsymbol{Q}^{ie} = \begin{bmatrix} \boldsymbol{Q}_R^{ie} \\ \boldsymbol{Q}_\theta^{ie} \\ \boldsymbol{Q}_p^{ie} \end{bmatrix}, \tag{20}$$

where $\boldsymbol{Q}_R^{ie}$ is translational components, $\boldsymbol{Q}_\theta^{ie}$ is rotational components, and $\boldsymbol{Q}_p^{ie}$ is elastic components of the generalized force vector, respectively.

The elastic forces can be described using modal coordinates and the stiffness matrix in modal coordinates $\mathbf{K}^i$. The stiffness matrix in modal coordinates can be obtained using component mode synthesis. The virtual work of the elastic forces can be expressed as.

$$\delta W^{if} = \delta \boldsymbol{p}^{\mathrm{T}} \mathbf{K}^i \boldsymbol{p}^i \tag{21}$$

The vector of elastic forces can be represented as follows.

$$\boldsymbol{Q}^{if} = \begin{bmatrix} \mathbf{0} \\ \mathbf{0} \\ \mathbf{K}^i \boldsymbol{p}^i \end{bmatrix} \tag{22}$$

In multibody dynamics, different types of joints between bodies are accounted for with kinematic constraints applied on generalized coordinates. Algebraic equations are used for the description of constraints between bodies. By examining only holonomic constraints, constraint equations can be expressed as follows.

$$\boldsymbol{C}(\boldsymbol{q}) = \boldsymbol{0} \tag{23}$$

where, $\boldsymbol{C}$ is the constraint vector for the system. Equations of motion may be formulated using the widely known Lagrange method, in which kinematic constraints are accounted for as supplementary algebraic equations with the help of Lagrange multipliers. The method is called global formulation since it does not differentiate between open and closed kinematic chains, as topological methods do. After employing the concept of virtual work to externally applied forces and then introducing constraints with help of Lagrange multipliers, the equation of motion can be written in the form of a differential algebraic equation (DAE).

$$\mathbf{M}\ddot{\boldsymbol{q}} + \mathbf{K}\boldsymbol{q} + \mathbf{C}_q^{\mathrm{T}}\lambda = \boldsymbol{Q}^e + \boldsymbol{Q}^v - \boldsymbol{Q}^f \tag{24}$$

Equations (23) and (24) form a set of differential algebraic equations, which can be converted to ordinary differential equations (ODE) to solve for the dynamic response of the multibody system in the time domain. To be able to apply traditional ODE solvers to the system of equations, the constraint equations must be differentiated twice with respect to time.

$$\ddot{\boldsymbol{C}}(\boldsymbol{q}, \dot{\boldsymbol{q}}, \ddot{\boldsymbol{q}}) = \mathbf{C}_q \ddot{\boldsymbol{q}} + \left(\mathbf{C}_q \dot{\boldsymbol{q}}\right)_{\boldsymbol{q}} \dot{\boldsymbol{q}} = \boldsymbol{0}, \tag{25}$$

where $\boldsymbol{Q}^c = -\left(\mathbf{C}_q \dot{\boldsymbol{q}}\right)_{\boldsymbol{q}} \dot{\boldsymbol{q}}$ is the constraint force vector for the system. As a result, the final matrix form of equations of motion describing the system dynamics looks as following.

$$\begin{bmatrix} \mathbf{M} & \mathbf{C}_q^{\mathrm{T}} \\ \mathbf{C}_q & \mathbf{0} \end{bmatrix} \begin{bmatrix} \ddot{\boldsymbol{q}} \\ \lambda \end{bmatrix} = \begin{bmatrix} \boldsymbol{Q}^e + \boldsymbol{Q}^v - \boldsymbol{Q}^f \\ \boldsymbol{Q}^c \end{bmatrix} \tag{26}$$

2.1 Fatigue

Fatigue is a failure that occurs after cyclic loading, and it is a common cause of structural fracture. Fatigue damage is one of the most common faults in dynamically loaded structures. In principle, the entire development of fatigue damage can be described as follows: one or more cracks form in the material, and the cracks grow until fatigue failure takes place.

A fundamental design objective for any dynamically loaded structure exposed to cyclic loading or vibration is to avoid fatigue failure throughout its service life. Welding is one of the most efficient methods used to manufacture structures. Cranes, vehicle frames, and machines are just some examples of welded structures that are dynamically loaded. As a structural detail, a weld is initially very prone to fatigue due to the notch effect, high tensile residual stresses, and welding flaws. High-strength steels, which are seeing increasing use, are even more sensitive to this phenomenon. In general, high strength steels are chosen to achieve larger payloads with more slender structural elements. As a trade off, the slender structures are subject to increased nominal stresses and welds become more prone to fatigue due to the higher stresses. Typically, structural welds are at or near areas of structural discontinuity. The weld itself is a local discontinuity. Furthermore, welding processes typically introduce flaws in the weld or weld toe such as undercuts, the inclusion of impurities, and cold laps. These flaws are sources of incipient cracking.

Empirical testing with actual parts or complete systems is time consuming and laborious. Traditionally, the approach to avoiding fatigue failures in a new system is to fatigue test specific structural details before integrating them into the system design. In fact, many kinds of typical welded details can be found commonly in the literature including fillet welds, corner joints, and butt joints. This approach has several weaknesses. For example, all pertinent structural details to be used should be tested under various loading conditions if the fatigue evaluation is to be comprehensive, and the approach ignores other parameters that may relate to a particular joint, such as the number of weld beads or other geometrical and technical details. Obviously, for a complex system with an arbitrary number of structural details subjected to various multi-axial loading scenarios, it is practically impossible to use this traditional approach.

More recently, numerical methods have been developed to estimate fatigue life making empirical testing unnecessary and allowing the designer to more effectively consider the effects of fatigue (Haagensen and Maddox, 2011). Today, the finite element method has become a standard approach for estimating the fatigue life of a structure. Nonetheless, even though computational capacity is increasing all the time, applying the finite element approach to a complex structure subject to dynamic multi-axial loading presents an overwhelming computational burden.

Fatigue design approaches can be differentiated according to how cracks initiate. In some applications, such as rotating axles, fatigue life is equivalent to the duration of the crack initialization stage. Because of the notch effect, the crack in these applications quickly results in failure. For larger structures, such as machine frames or many welded structures, cracks are present from the beginning, so fatigue life is the determined by the length of time it takes the initial cracks to propagate.

The fatigue life of a structure under dynamic load can be estimated by assuming it to have some initial amount of fatigue endurance and then assuming that one load cycle will result in fatigue damage of some amount. This is commonly known as Palmgren-Miner's rule (Miner, 1945). It was suggested that fatigue damage could be accumulated linearly for a certain amplitude value. Finally, when all fatigue endurance has been depleted, failure is expected. A large amount of fatigue test data can be found in the literature, and stress histories can be obtained through experimentation of by simulation.

3 Sub-modeling

A common cause of structural damage is local stress concentration due to structural geometry. Practically all structural damage occurs where one or more stress raisers are present. Problematic details are often combinations of several geometries that concentrate hazardous stresses. Stress raisers result from local discontinuities in real structural features; such as welds, attachments grooves or holes. Typically, small discontinuities are neglected in multibody simulation, since their effect on overall behavior is relatively small compared to their contribution to computational burden. Traditionally, the problem of stress raisers is solved by calculating nominal stress levels and then taking into account the effect of stress raisers by applying predefined stress concentration factors. The stress concentration factor concept cannot be a general approach, since it is obvious that all possible geometrical shapes and their combinations combined with complex loading cases cannot be predefined, especially if a specific but arbitrary level of accuracy is needed. Obviously, the general approach covering all kinds of geometrical combinations in the finite element method is to model them as they appear in the structure. Since stress values change drastically in the neighborhood of a structural discontinuity, a refined element mesh is required, which will lead to a large number of degrees of freedom. This approach is impractical due to the computational burden, especially in the case of multibody simulation.

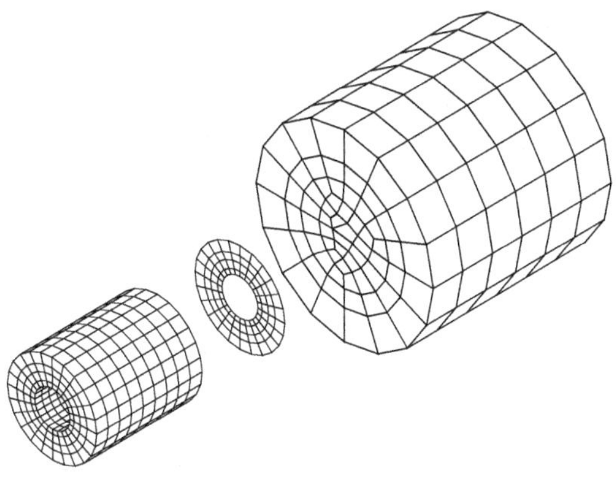

Figure 4. The sub-modeling approach for attaching dissimilar meshes.

The sub-modeling approach is commonly used, and it can overcome the previously mentioned problems. With the approach some new problems arise, but they will be discussed later. In principal, a sub-model is a model inside of or on top of a large scale model that describes a certain portion of the large scale model. It can be used to attach a locally refined element mesh to the larger scale model, which does not need to be changed. In sub-modeling, the simplified structural model is complemented by a more refined sub-model of structural details. The sub-models do not influence the operation of the system but get their boundary conditions and loading data from the larger simplified model. The sub-modeling approach is also referred to in the literature as the global/local approach (Knight et al., 1991). The sub-modeling approach can be used to connect dissimilar meshes together as shown schematically in Fig. 4.

Sub-modeling is an approach that is used together with the finite element method to combine two different finite element meshes. There are several reasons the sub-modeling approach is powerful. It can be used to connect finite models into a larger assembly. The approach does not require meshes to be similar and even element types can differ. Currently available methods do not require any coincident nodes. These beneficial features can be utilized to combine separately constructed models or refine the element mesh in a certain area without taking care of refining the mesh smoothly. In addition, a sub-model can be changed easily without modifying other parts of the model. Problems arise if the level of refinement differs significantly between two different models. A coarse mesh tends to be too stiff and displacements are underestimated, and if those underestimated displacements are used as boundary conditions for the refined model, calculated stress levels will be non-conservative.

Coupling between the sub-model and the large-scale model is assumed to be one directional, i.e., it is assumed that the behavior of the reduced model in dynamic simulation is not affected by the sub-model. That means, the large scale model is complemented with a sub-model of the desired detail and it does not affect the system's overall stiffness. This crucial simplification makes dynamic simulation and stress calculation independent from each other. Therefore, the computation can be straight forwardly parallelized. Displacement boundary conditions of the sub-model, however, are acquired from the large-scale model. In the proposed approach, during dynamic analysis the general behavior of the structure is calculated with a simplified model and details are examined as a separate problem. For assessing fatigue loads on a structure, this assumption is sufficient since any significant change in structural flexibility due to crack growth occurs only very late in the total life of a structure.

A sub-model can make the overall structure stiffer, or it can only carry boundary conditions without affecting the stiffness of a large-scale structure. Sub-models describe structural details, and only those that are interesting from a design point of view can be attached to a flexible multibody model. Connections between a multibody model and its sub-models are one-directional, guaranteeing that the multibody simulation is not affected by the sub-model. For instance, if a flexible multibody model is a part of a real-time simulation, it will be a real-time simulation even if sub-models are active. In this paper, one way of attaching sub-models to the larger multibody model is introduced, but the concept is general.

4 Stress in multibody dynamic simulation

Dimensioning components require information about loading and more accurately, stresses. Even though, stresses can somehow be obtained from rigid body dynamics by using simulated forces as force boundary conditions in finite element method. In general, to obtain structural stresses, structural flexibility should be taken into account. Concept of multibody dynamics gives attractive approach to simulate real operating conditions and thus obtain realistic loading conditions.

Stress recovery methods for a flexible multibody system can be divided into two main categories. One is the stress-mode-based method, which determines the body's stress state using a linear combination of stress modes and elastic coordinates. The other is the finite-element post-processing method, in which stresses are calculated by a finite-element code using forces or displacements obtained from multibody simulation. Both approaches have benefits, but in general, the method based on finite-element post processing is more accurate (Arczewski and Frączek, 2005).

For lightweight structures loaded by dynamic forces, accurate dynamic simulation is necessary to guarantee long-term reliability (fatigue life), the accuracy of control, and system usability. Deformation, even small deformations, must be taken into account to achieve the needed simulation precision. Arriving at an optimized design and understanding precisely how internal stresses vary over time leads to structures with improved fatigue life and whole systems that are safer. In the Floating Frame of Reference Formulation, bodies are loaded by numerous unique loads and moments; external forces, constraint forces, and inertial forces, for example. Forces produce deformation, and deformations set up internal stresses. The prediction of local stresses using dynamic simulation reveals structural weaknesses in the early design stages. In addition, dynamic simulation can analyze stress peaks in extreme cases, such as random overloading or component failure.

A stress history from a multibody dynamics simulation can be used as initial data for component dimensioning. Furthermore, it can provide loading data for the analyst that is otherwise difficult to obtain. Finally, the stress history can be used as input for the fatigue analysis of the component. In such cases, one should make sure that simulated operations describe the operating conditions of the machine with sufficient accuracy. With simulation, it is difficult to describe the impact of statistical issues, such as component wear and operator usage habits, on component loading. On the other hand, simulation helps to understand the causes and effects related to loading. This allows the use of optimization routines in component dimensioning. Simulation and measurement on a real-life machine can thus be considered to support each other, and using them together can help to reach an optimal solution.

The literature provides a number of alternative approaches to determining stress histories from multibody simulation. The first to combine the multibody dynamic approach and stress calculation was Melzer (1996). Yim and Lee (1996) obtained dynamic stress histories by using constraint forces solved in a multibody system. Dietz et al. (1997) described an approach for using multibody simulation to obtain all forces for finite element analysis. They also selected the most severe time steps for which stresses of the complete structure were later analyzed in finite element code. Later, Dietz et al. (1998) combined multibody simulation and fatigue life prediction. They obtained the load history from multibody simulation and calculated stress histories for selected locations using a stress load matrix. Stress histories were analyzed in a post-processing stage to predict fatigue life. Dietz (1999) presented a systematic way of combining multibody dynamic simulation and fatigue analysis using stress component modes. Claus (2001) generalizes the deformation-based stress recovery approach to multipurpose finite element codes. More recently, Jun et al. (2008) used the modal stress recovery approach to obtain stress for fatigue analysis. They also discussed the reliability of fatigue life calculation. Lee et al. (2009) studied the fatigue life for various parts of a guideway vehicle by coupling multibody dynamics and fatigue analysis. They determined stresses using the modal stress method or quasi-static force method depending on the loading conditions of the part. Braccesi and Cianetti (2005) used a modal approach to recover stresses. Arczewski and Frączek (2005) compared and discussed differences between force-based and deformation-based stress recovery methods in MBS. More recently, Tobias and Eberhard (2011) obtained stresses using a reduced MBS model and stress modes. They concluded the stress state in any particular point of a flexible body could be expressed as a linear combination of global shape functions for stresses and nodal coordinates.In experiments, fatigue life prediction is mainly related to uniaxial cyclic loading. This leads to discussion about damage hypothesis, and the question arises about which damage hypothesis should be used. This work focuses on welded structures, which can be studied under the assumption of cumulative damage counting. Welded structures without any post-weld treatments have large tensile residual stresses, even nominally as large as the yield strength of the material. In this case, initial compressive loading closes an incipient crack and residual stresses open it again. This leads to a situation in which even a fully compressive loading cycle will result in full fatigue damage. For the proposed method, linear hot spot extrapolation (Poutiainen et al., 2004) is selected, for both simplicity and robustness in use.

5 Empirical example

Fatigue damage typically originates from the points of discontinuity of the structure, especially if there is residual stress affecting the discontinuity. In order for the machine system to be simulated in real time, it must often be simplified, and details irrelevant in terms of structural stress neglected. The modeling of small details, such as welded clamp for a hydraulic pipe of a boom, increases the need for computational efficiency, and such details have only a localized importance. To illustrate the method for obtaining stress history of a structural detail, a practical example of a simple crane is studied. The crane and the structural detail are depicted in Fig. 5. In this example the simplified dynamic simulation and relative accurate model of notch are combined together using sub-modeling techniques.

The crane in Fig. 5 is hydraulically driven crane with one degree of freedom. Hydraulic system is modeled using the theory of lumped fluid (Watton, 1989), in which hydraulic system is divided into separate volumes in where pressure is equally distributed.

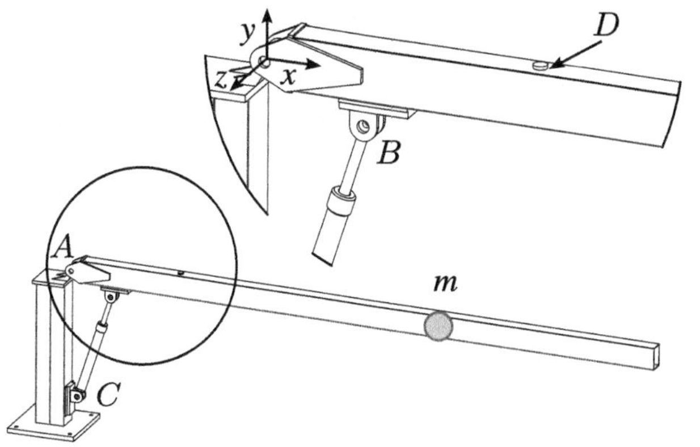

Figure 5. Crane parameters and the placement of strain gauges (D) – Letters A, B, and C refer to the joint locations and m is the added mass of 110 kg. Strain gauges are attached close to the welded notch so the first is $0.4t$ and the second is $1.0t$ from the notch. Plate thickness t is 4 mm.

A fundamental motivation of the introduced approach is to keep the dynamic simulation as numerically efficient as possible. Therefore, the hosting structural model is simplified and is then reduced with the Craig-Bampton method (Craig Jr. and Bampton, 1968). Even though the proposed approach is general, in this example boom-type structures that can be efficiently described with beam elements are studied. With beam models, obtaining boundary conditions for sub-models is straightforward. In the proposed approach, multibody simulation is used for producing displacement data for the sub-model, which is then analyzed and fatigue data is obtained.

Problem about computational burden is tried to overcome by combining sub-modeling and multibody dynamic approach in order to join computationally efficiency of multibody dynamic approach and accuracy of finite element method in observing damaging loads. Sub-model of the crane is shown in Fig. 6. Black circle shows the area where hot spot stress is obtained. Nodes used in hot spot extrapolation are show in Fig. 7.

The beam model represents the center line of a structural component. The sub-model is attached to the interpolated locations of the reduced model via rigid and massless beams. Due to the use of rigid beam webs, the cross-section is assumed to remain planar at the boundary condition points. The effect of this assumption, with respect to stresses in notch, is negligible due to Saint-Venant's principle (von Mises, 1945). In dynamic simulation, translational and rotational displacements are solved as boundary conditions for the sub-model. In general, sub-models are attached at arbitrary locations of the structural component, thus nodal displacement interpolation should be used.

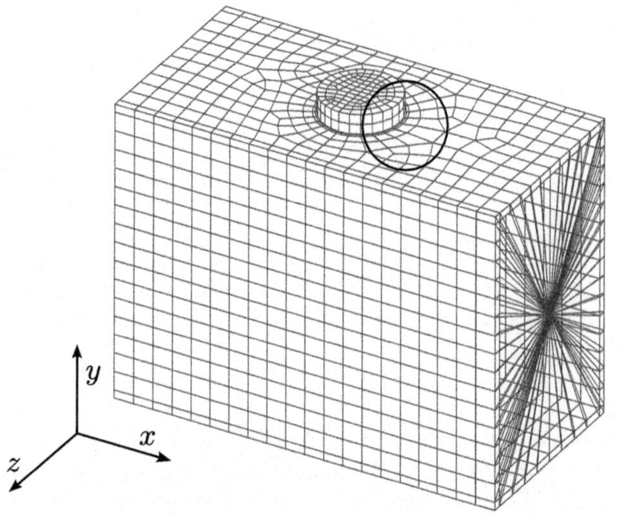

Figure 6. FE-model of the sub-model – the black circle indicates the location of the hot-spot nodes.

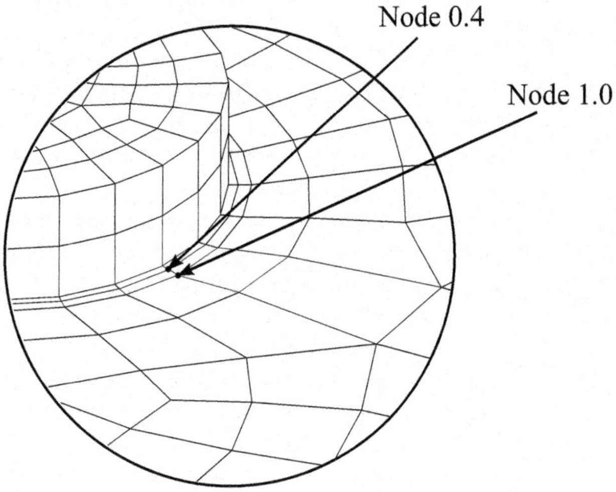

Figure 7. Hot spot extrapolation nodes on a sub-model.

5.1 Crane model composition and work cycle

The crane model contains four bodies, which are the crane, support and hydraulic cylinder and shaft for a cylinder. The lenght L of the crane is 4.5 m. depicted in Fig. 5. In addition to those parts model has added mass $m = 100$ kg and a hydraulic system. Hydraulic circuit can be neglected while comparing results, since simulation model was driven by predefined movement of the cylinder that is shown in Fig. 8. In Table 1, relevant dimensions and distances from revolute joint A along coordinates x, y, z, are presented.

The crane model consists of a beam model with 20 nodes. Craig-Bampton method is used to reduce coordinates of the beam model. Two connection nodes were selected as a masted nodes for Craig-Bampton. Structural flexibility was described using 10 lowest deformation modes. Geometrical properties of cross section of the crane are presented in Table 2.

Table 1. Dimensions of the crane.

Item	x coordinate [m]	y coordinate [m]
Revolute joint A	0	0
Revolute joint B	0.32	−0.125
Revolute joint C	0	−0.925
Mass m	2.5	0
Welded detail D	0.65	0.075
Gauge 0.4	0.672	0.075
Gauge 1.0	0.674	0.075

Table 2. Geometrical properties of cross section of the crane and material properties of the crane.

Property	Value	Unit
Profile height	0.15	m
Profile width	0.10	m
Profile area	1.9×10^{-3}	m^2
Plate thickness	4	mm
Area moment of inertia (yy)	6.17×10^{-6}	m^4
Area moment of inertia (zz)	3.29×10^{-6}	m^4
Elastic modulus	210	GPa
Poisson's ratio	0.3	
Density	7850	$kg\,m^{-3}$

5.2 Stress history

This displacement data is then used as a boundary condition for the sub-model. The sub-model contains the stress concentrations where fatigue damage can possibly occur. Nodal displacement history is applied as boundary conditions on the sub-model as a sequential set of static boundary conditions. The finite element mesh of the sub-model has 1800 linear, brick elements and 250 rigid, massless beams.

The gray lines seen on right side, in Fig. 6, represent rigid and massless beam webs and connect the cross section of the sub-model to the dynamical model via attachment nodes. Two attachment nodes are located on the middle line of the cross section. The use of rigid beam webs keeps the boundary cross section of the sub-model planar. This clearly simplifying assumption is made because in the beam model the cross section is assumed to remain planar. Generally, the problems with beam elements are the higher order deformations of cross section, such as warping and distortion, which are not included in low order beam elements.

Since the sub-model and the dynamic model have overlapping nodes, boundary conditions for the sub-model can be fixed based on nodal deformation from the dynamic simulation. In the case of non-overlapping nodes, interpolation of nodal translational deformation and rotation deformation between nodes is required. In this case, interpolation could be

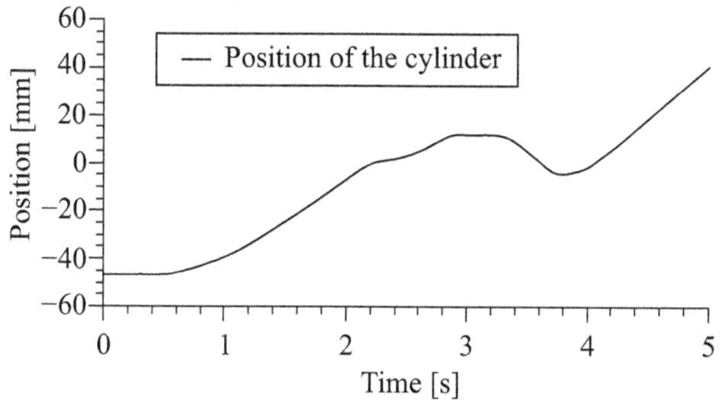

Figure 8. Measured position of the shaft of the hydraulic sylinder.

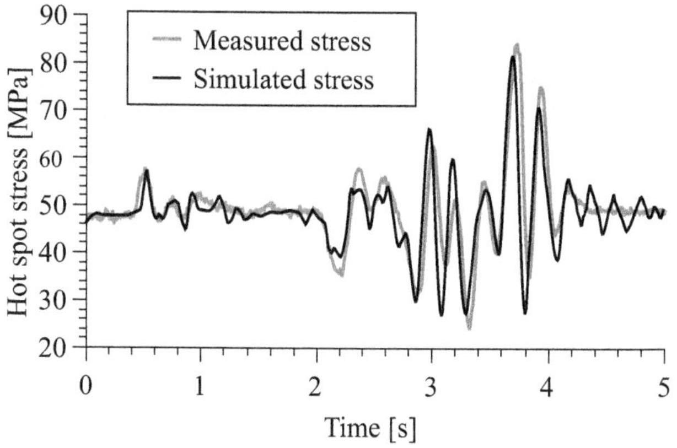

Figure 9. Stress history of a notch at the crane.

made linearly between the nodes. Since the sub-model only uses displacements obtained from the dynamic simulation as boundary conditions, it does not interfere with the overall behavior of the model.

For fatigue assessment the hot spot or structural stress is often used (Niemi et al., 2006). The crane was loaded using measured position of a hydraulic cylinder, shown in Fig. 8 A linear surface extrapolation for hot spot stress (Poutiainen et al., 2004) has been performed for obtaining stress histories shown in Fig. 9. The hot spot structural stress at the edge of the welded discontinuity, is based on a linear extrapolation of surface stresses at nodes 4 mm and 10 mm from the edge of the discontinuity. Axial direction (x-direction) is selected for extrapolating hot spot stress on the edge of the notch. The selection of the direction of hot spot extrapolation is made based on the assumption that the majority of stresses are acting in the axial direction.

The stress history obtained for the welded detail can later be processed using the rainflow counting algorithm, fatigue assessment or for any other post processing action.

The results of this numerical experiment show that the developed method can be used to determine the stresses of a structural detail using a real-time simulation model. This method enables a wide variety of uses from determining stresses from positions that cannot be measured from the real machine to determining the best practices for machine operation. Measuring bearing housing stresses, for example, in a real machine during an operation cycle is next to impossible. This method enables the determination of stresses during the entire operation cycle instead of just a suspected peak value. The method could be used to improve estimations on the machine durability as well as improving the machine durability already in the machine design phase. A practical example would be to use virtual prototyping in the machine product development phase. The model could then be used to run a series of reference operation cycles while recording displacement data from a structure. The recorded data could then be used to run analysis on several crucial parts of the structure in order to determine the life expectancy under operating conditions as well as different operators.

6 Conclusions

In this paper one approach of making fatigue analysis more usable among multibody simulation based product development is presented. In order to combine dynamic simulation and fatigue design this study introduces a novel approach for efficiently obtaining stress history from dynamic simulation.

This paper presents an approach in which the stress history for fatigue life estimation of an arbitrary structural discontinuity in a large-scale structure can be efficiently obtained in multibody simulation. In the proposed approach the structure is modeled with structured elements (i.e. planes or beams) in order to get rid small structural details to minimize nodal degrees of freedom. After that model is further reduced using component mode synthesis, in numerical example, Craig-Bampton method was used, this model is called as reduced model. Reduced model is used to represent flexible body in multibody simulation. Small structural details are modeled separately and are attached to reduced model using suitable methods. In this paper, in numerical examples sub-models were attached to reduced model using rigid beam webs. Sub-models were analyzed quasi-statically within finite element codes using displacements, obtained from dynamic simulation, as boundary conditions. This analysis can be made during the dynamic simulation or in post-processing phase. Computations involving sub-modeling allow the fatigue assessment calculation to be separated from the dynamic simulation and structural details can be analyzed independently.

In future work the integration of fatigue analysis and produced stress history could be improved. The way how stress data is analyzed and compared to real fatigue test results differs from the stress results that can be straightforwardly obtained from dynamic simulation. In numerical examples, this aspect is taken into account using linear surface extrapolation to estimate hot-spot stress. The problem using that approach is the difficulty of knowing the most probable crack growth direction. In reality, crack may change the direction of growing depending on geometry and/or loading conditions. Also, in future work methods of attaching sub-model into simulation model should be studied carefully. Possibility to use coordinate reduction for sub-models and what kinds of limitations it provides to attachment for sub-models.

Acknowledgements. The research is supported by the Academy of Finland, project 133154.

Edited by: A. Müller

References

Arczewski, K. and Frączek, J.: Friction Models and Stress Recovery Methods in Vehicle Dynamics Modelling, Multibody Syst. Dyn., 14, 205–224, 2005.

Braccesi, C. and Cianetti, F.: A procedure for the virtual evaluation of the stress state of mechanical systems and components for the automotive industry: development and experimental validation, Proceedings of the Institution of Mechanical Engineers, Part D: Journal of Automobile Engineering, 219, 633–643, 2005.

Claus, H.: A Deformation Approach to Stress Distribution in Flexible Multibody Systems, Multibody Syst. Dyn., 6, 143–161, 2001.

Craig Jr., R. R. and Bampton, M. C. C.: Coupling of Substructures for Dynamic Analyses, AIAA J., 6, 1313–1319, 1968.

Dietz, S.: Vibration and fatigue analysis of vehicle systems using component modes, Doctoral thesis, Technische Universität Berlin, 1999.

Dietz, S., Netter, H., and Sachau, D.: Fatigue life predictions by coupling finite element and multibody systems calculations, in: Proceedings of Design Engineering Technical Conferences, 1997.

Dietz, S., Netter, H., and Sachau, D.: Fatigue Life Prediction of a Railway Bogie under Dynamic Loads through Simulation, Vehicle Syst. Dyn., 29, 385–402, 1998.

Haagensen, P. J. and Maddox, S. J.: IIW Recommendations on Post Weld Fatigue Life Improvement of Steel and Aluminium Structures, International Institute of Welding, Paris, 2011.

Jun, K., Park, T., Lee, S., Jung, S., and Yoon, J.: Prediction of fatigue life and estimation of its reliability on the parts of an air suspension system, Int. J. Automot. Techn., 9, 741–747, 2008.

Knight, N. F., Ransom, J. B., Griffin, O. H., and Thompson, D. M.: Global/local methods research using a common structural analysis framework, Finite Elem. Anal. Des., 9, 91–112, 1991.

Korkealaakso, P., Mikkola, A., and Rouvinen, A.: Multi-body simulation approach for fault diagnosis of a reel, Proceedings of the Institution of Mechanical Engineers, Part K: Journal of Multi-body Dynamics, 220, 9–19, 2006.

Korkealaakso, P., Mikkola, A., Rantalainen, T., and Rouvinen, A.: Description of joint constraints in the floating frame of reference formulation, Proceedings of the Institution of Mechanical Engineers, Part K: Journal of Multi-body Dynamics, 223, 133–145, 2009.

Lee, S.-H., Park, T.-W., Park, J.-K., Yoon, J.-W., Jun, K.-J., and Jung, S.-P.: Fatigue life analysis of wheels on guideway vehicle using multibody dynamics, Int. J. Precis. Eng. Man., 10, 79–84, 2009.

Maddox, S. J.: Fatigue strength of welded structures, Abington Publishing, Cambridge, 2nd Edn., 1991.

Melzer, F.: Symbolic computations in flexible multibody systems, Nonlinear Dynam., 9, 147–163, 1996.

Miner, M. A.: Cumulative damage in fatigue, J. Appl. Mech., 12, 159–164, 1945.

Niemi, E., Fricke, W., and Maddox, S. J.: Fatigue analysis of welded components: Designer's guide to the structural hot-spot stress approach, Woodhead Publishing, Cambridge, 1st Edn., 2006.

Nikravesh, P. E. and Chung, I. S.: Application of Euler Parameters to the Dynamic Analysis of Three-Dimensional Constrained Mechanical Systems, J. Mech. Design, 104, 785–791, 1982.

Poutiainen, I., Tanskanen, P., and Marquis, G.: Finite element methods for structural hot spot stress determination - a comparison of procedures, I. J. Fatigue, 26, 1147–1157, 2004.

Shabana, A. A.: Dynamics of Multibody Systems, Cambridge University Press, Cambridge, 2nd Edn., 1998.

Shabana, A. A.: Dynamics of Multibody Systems, Cambridge University Press, Cambridge, 3rd Edn., 2005.

Tobias, C. and Eberhard, P.: Stress recovery with Krylov-subspaces in reduced elastic multibody systems, Multibody Syst. Dyn., 25, 377–393, 2011.

von Mises, R.: On Saint Venant's principle, B. Am. Math. Soc., 51, 555–562, 1945.

Wallrapp, O. and Wiedemann, S.: Comparison of Results in Flexible Multibody Dynamics Using Various Approaches, Nonlinear Dynam., 34, 189–206, 2003.

Watton, J.: Fluid power systems: Modeling, simulation, analog and microcomputer control, Prentice Hall, New York, 1st Edn., 1989.

Yim, H. J. and Lee, S. B.: An integrated CAE system for dynamic stress and fatigue life prediction of mechanical systems, J. Mech. Sci. Technol., 10, 158–168, 1996.

Permissions

All chapters in this book were first published in MS, by Copernicus Publications; hereby published with permission under the Creative Commons Attribution License or equivalent. Every chapter published in this book has been scrutinized by our experts. Their significance has been extensively debated. The topics covered herein carry significant findings which will fuel the growth of the discipline. They may even be implemented as practical applications or may be referred to as a beginning point for another development.

The contributors of this book come from diverse backgrounds, making this book a truly international effort. This book will bring forth new frontiers with its revolutionizing research information and detailed analysis of the nascent developments around the world.

We would like to thank all the contributing authors for lending their expertise to make the book truly unique. They have played a crucial role in the development of this book. Without their invaluable contributions this book wouldn't have been possible. They have made vital efforts to compile up to date information on the varied aspects of this subject to make this book a valuable addition to the collection of many professionals and students.

This book was conceptualized with the vision of imparting up-to-date information and advanced data in this field. To ensure the same, a matchless editorial board was set up. Every individual on the board went through rigorous rounds of assessment to prove their worth. After which they invested a large part of their time researching and compiling the most relevant data for our readers.

The editorial board has been involved in producing this book since its inception. They have spent rigorous hours researching and exploring the diverse topics which have resulted in the successful publishing of this book. They have passed on their knowledge of decades through this book. To expedite this challenging task, the publisher supported the team at every step. A small team of assistant editors was also appointed to further simplify the editing procedure and attain best results for the readers.

Apart from the editorial board, the designing team has also invested a significant amount of their time in understanding the subject and creating the most relevant covers. They scrutinized every image to scout for the most suitable representation of the subject and create an appropriate cover for the book.

The publishing team has been an ardent support to the editorial, designing and production team. Their endless efforts to recruit the best for this project, has resulted in the accomplishment of this book. They are a veteran in the field of academics and their pool of knowledge is as vast as their experience in printing. Their expertise and guidance has proved useful at every step. Their uncompromising quality standards have made this book an exceptional effort. Their encouragement from time to time has been an inspiration for everyone.

The publisher and the editorial board hope that this book will prove to be a valuable piece of knowledge for researchers, students, practitioners and scholars across the globe.

List of Contributors

S. K. Saha
Department of Mechanical Engineering, Indian Institute of Technology Delhi, Hauz Khas, New Delhi 110 016, India

S. V. Shah
Department of Mechanical Engineering, McGill University, Montreal, Canada

P. V. Nandihal
Department of Mechanical Engineering, Indian Institute of Technology Delhi, Hauz Khas, New Delhi 110 016, India

O. A. Bauchau
University of Michigan - Shanghai Jiao Tong University Joint Institute, Shanghai, 200240, China

S. Han
University of Michigan - Shanghai Jiao Tong University Joint Institute, Shanghai, 200240, China

E. V. Zahariev
Institute of Mechanics, Bulgarian Academy of Sciences, Sofia, Bulgaria

H. Mazhar
Simulation Based Engineering Lab, Department of Mechanical Engineering, University of Wisconsin, Madison, WI, 53706, USA

T. Heyn
Simulation Based Engineering Lab, Department of Mechanical Engineering, University of Wisconsin, Madison, WI, 53706, USA

A. Pazouki
Simulation Based Engineering Lab, Department of Mechanical Engineering, University of Wisconsin, Madison, WI, 53706, USA

D. Melanz
Simulation Based Engineering Lab, Department of Mechanical Engineering, University of Wisconsin, Madison, WI, 53706, USA

A. Seidl
Simulation Based Engineering Lab, Department of Mechanical Engineering, University of Wisconsin, Madison, WI, 53706, USA

A. Bartholomew
Simulation Based Engineering Lab, Department of Mechanical Engineering, University of Wisconsin, Madison, WI, 53706, USA

A. Tasora
Department of Industrial Engineering, University of Parma, V.G.Usberti 181/A, 43100, Parma, Italy

D. Negrut
Simulation Based Engineering Lab, Department of Mechanical Engineering, University of Wisconsin, Madison, WI, 53706, USA

S. S. Parsa
Department of Mechanical Engineering, University of New Brunswick, Fredericton, NB, Canada

J. A. Carretero
Department of Mechanical Engineering, University of New Brunswick, Fredericton, NB, Canada

R. Boudreau
Département de génie mécanique, Université de Moncton, Moncton, NB, Canada

L. Carbonari
Università Politecnica delle Marche, Via Brecce Bianche, 60131, Ancona, Italy

M. Battistelli
Università Politecnica delle Marche, Via Brecce Bianche, 60131, Ancona, Italy

M. Callegari
Università Politecnica delle Marche, Via Brecce Bianche, 60131, Ancona, Italy

M.-C. Palpacelli
Università Politecnica delle Marche, Via Brecce Bianche, 60131, Ancona, Italy

G. Kouroussis
Université de Mons - UMONS, Faculty of Engineering, Department of Theoretical Mechanics, Dynamics and Vibrations, Place du Parc 20, 7000 Mons, Belgium

O. Verlinden
Université de Mons - UMONS, Faculty of Engineering, Department of Theoretical Mechanics, Dynamics and Vibrations, Place du Parc 20, 7000 Mons, Belgium

A. Ratajczak
Institute of Computer Engineering, Control and Robotics, Wrocław University of Technology, ul. Janiszewskiego 11/17, 50-372 Wrocław, Poland

K. Tchón
Institute of Computer Engineering, Control and Robotics, Wrocław University of Technology, ul. Janiszewskiego 11/17, 50-372 Wrocław, Poland

J. F. Schorsch
Department of BioMechanical Engineering, Technical University Delft, Delft, the Netherlands

A. Q. L. Keemink
Laboratory of Biomechanical Engineering, University of Twente, Enschede, the Netherlands

A. H. A. Stienen
Laboratory of Biomechanical Engineering, University of Twente, Enschede, the Netherlands

F. C. T. van der Helm
Department of BioMechanical Engineering, Technical University Delft, Delft, the Netherlands

D. A. Abbink
Department of BioMechanical Engineering, Technical University Delft, Delft, the Netherlands

M. Arnold
Martin Luther University Halle-Wittenberg, NWF II - Institute of Mathematics, 06099 Halle (Saale), Germany

J. Linn
Fraunhofer Institute for Industrial Mathematics, Fraunhofer Platz 1, 67633 Kaiserslautern, Germany

H. Lang
Chair of Applied Dynamics, Univ. Erlangen-Nürnberg, Konrad-Zuse-Str. 3-5, 91052 Erlangen, Germany

A. Tuganov
Fraunhofer Institute for Industrial Mathematics, Fraunhofer Platz 1, 67633 Kaiserslautern, Germany

Q. Meng
Department of Electromechanical Engineering, Faculty of Science and Technology, University of Macau, Av. Padre Tomas Pereira, Taipa, Macao SAR, China

Y. Li
Department of Electromechanical Engineering, Faculty of Science and Technology, University of Macau, Av. Padre Tomas Pereira, Taipa, Macao SAR, China
School of Mechanical Engineering, Tianjin University of Technology, Tianjin 300384, China

J. Xu
Department of Electromechanical Engineering, Faculty of Science and Technology, University of Macau, Av. Padre Tomas Pereira, Taipa, Macao SAR, China

P. G. Gruber
Linz Center of Mechatronics GmbH, Altenberger Straße 69, 4040 Linz, Austria

K. Nachbagauer
Institute of Technical Mechanics, Johannes Kepler Universität Linz, Altenbergerstraße 69, 4040 Linz, Austria

Y. Vetyukov
Linz Center of Mechatronics GmbH, Altenberger Straße 69, 4040 Linz, Austria

J. Gerstmayr
Linz Center of Mechatronics GmbH, Altenberger Straße 69, 4040 Linz, Austria

D. S. Mohan Varma
Department of Mechanical Engineering, Indian Institute of Technology Madras, Chennai, India

S. Sujatha
Department of Mechanical Engineering, Indian Institute of Technology Madras, Chennai, India

T. T. Rantalainen
Lappeenranta University of Technology, Department of mechanical engineering, Lappeenranta, Finland

A. M. Mikkola
Lappeenranta University of Technology, Department of mechanical engineering, Lappeenranta, Finland

T. J. Björk
Lappeenranta University of Technology, Department of mechanical engineering, Lappeenranta, Finland

CPSIA information can be obtained
at www.ICGtesting.com
Printed in the USA
BVOW10*1539040716
454236BV00001BA/2/P